RESOURCE MANAGEMENT IN WIRELESS NETWORKING

Network Theory and Applications

Volume 16

RESOURCE MANAGEMENT IN WIRELESS NETWORKING

Edited by

MIHAELA CARDEI
Florida Atlantic University, Boca-Raton, FL

IONUT CARDEI
Florida Atlantic University, Boca-Raton, FL

DING-ZHU DU
University of Minnesota, Minneapolis, MN

Library of Congress Cataloging-in-Publication Data

A C.I.P. Catalogue record for this book is available
from the Library of Congress.

ISBN 0-387-23807-7 e-ISBN 0-387-23808-5 Printed on acid-free paper.

Printed in the United States of America.

9 8 7 6 5 4 3 2 1 SPIN 11161790

springeronline.com

Preface

Following the pattern of the Internet growth in popularity, started in the early 1990s, the current unprecedented expansion of wireless technology promises to have an even greater effect on how people communicate and interact, with considerable socio-economic impact all over the world. The driving force behind this growth is the remarkable progress in component miniaturization, integration, and also developments in waveforms, coding, and communication protocols.

Besides established infrastructure-based wireless networks (cellular, WLAN, satellite) ad-hoc wireless networks emerge as a new platform for distributed applications and for personal communication in scenarios where deploying infrastructure is not feasible. In ad-hoc wireless networks, each node is capable of forwarding packets on behalf of other nodes, so that multi-hop paths provide end-to-end connectivity. The increased flexibility and mobility of ad-hoc wireless networks are favored for applications in law enforcement, homeland defense and military.

In a world where wireless networks become increasingly interoperable with each other and with the high-speed wired Internet, personal communication systems will transform into universal terminals with instant access to variate content and able of handle demanding tasks, such as multimedia and real-time video. With users roaming between networks, and with wide variation in wireless link quality even in a single domain, the communications terminal must continue to provide a level of Quality of Service that is acceptable to the user and conforms to a contracted Service Level Agreement.

Taking these into considerations, the network must provide mechanisms for controlling connection admission, service differentiation, end-to-end communication delay and connection data rate. These functions are different aspects of network resource management that contribute to provisioning of Quality of Service.

For some applications, a critical element is application lifetime - the time the application remains operational before energy reserves at network nodes are depleted and normal operation is hindered. To extend the application lifetime, it is essential to have judicious power management at each node and to employ energy-efficient communication protocols that address the energy efficiency issue at the network scale. Extensive research has been performed over the last few years in the area of effective network resource management, QoS and energy-efficient protocols for wireless networks. This book presents a snapshot of representative work in the field and targets an audience that includes researchers, faculty members, students and other professionals interested in this field.

The chapters in this book cover key topics from the area of resource management and QoS in wireless networks, starting from channel scheduling, QoS-aware medium access control, to QoS routing, resource discovery, energy-efficient multicast and architectures for end-to-end QoS. The book content is structured as follows:

Section 1 chapters survey resource management architectures for multimedia transmission, for IP-based ad-hoc wireless networks and for UMTS cellular networks.
Section 2 presents research on channel allocation, packet scheduling, bandwidth management and a framework for end-to-end statistical delay guarantees in ad-hoc wireless networks.
Section 3 addresses issues in resource management and QoS for Medium Access Control protocols. The IEEE 802.11e QoS extensions and resource management for Bluetooth networks are analyzed in detail. Novel approaches for Spatial TDMA are introduced and energy-efficient wireless MAC protocols are surveyed.
Section 4 provides a comprehensive overview of routing and QoS in mobile and ad-hoc wireless networks. Chapters in this section also present research in resource localization and discovery for ad-hoc wireless networks.
Section 5 surveys the state of the art in energy-efficient broadcast and multicast protocols for ad-hoc wireless networks, increasingly important in the context of multimedia and content delivery in emerging 4G wireless networks.
Section 6 chapters describe techniques for improving the quality of wireless connections, at the link layer and at the transport layer, by improving the performance of TCP over wireless networks.

This book is the collective contribution of top world researchers in the field of wireless communications and networking. We would like to take the opportunity to thank the authors, the anonymous referees and the publisher who guided us through the process and made this book possible.

Mihaela Cardei, Ionut Cardei and Ding-Zhu Du

Florida Atlantic University and University of Minnesota

Contents

SECTION III Medium Access Control

SECTION V Broadcast and Multicast

SECTION VI Radio Link and Transport

QoS for Multimedia Services in Wireless Networks

Hua Zhu
Department of Electrical Engineering
University of Texas, Dallas, TX 75083
E-mail: `zhuhua@utdallas.edu`

Imrich Chlamtac
School of Engineering
University of Texas, Dallas, TX 75083
E-mail: `chlamtac@utdallas.edu`

Contents

1 Introduction

In general, Quality of Service (QoS) is not related to any dedicated network layer, rather it requires coordinated efforts from all layers. As such, QoS is a broad concept that covers virtually every layer and every component in the network and supporting services. The basic definition of QoS was given for the first time in [10] as '*the collective effect of service performance which determine the degree of satisfaction of a user of the service.*' In [1], Vogel et al. defined Quality of Service as follows: '*Quality of Service represents the set of those quantitative and qualitative characteristics of a distributed multimedia system necessary to achieve the required functionality of an application*', while in [2], the definition of QoS was '*a set of qualities related to the collective behaviour of one or more object*'. QoS is understood by IETF as '*a set of service requirements to be met by the network while transporting a flow.*' And in [3] Malamos et al. defined QoS as '*the basic measure of how well the system operates in order to meet the user's requirements.*' In [9] QoS is defined as '*an umbrella term for a number of techniques that intelligently match the needs of specific applications to the network resources available*'.

The dramatic advances in wireless communications and multimedia techniques have led to the development of next generation multimedia services over wireless networks giving QoS an added importance. Two main features characterize the next generation multimedia services: a wide range of customized multimedia applications and the coexistence of different access technologies, such as IEEE 802.11, Bluetooth, CDMA, satellite and infrared. As a result, research on QoS has been shifted to the wireless, mobile or heterogeneous environment, brining with it additional challenges.

This chapter shows the impetus behind the impressive market demands of multimedia services and applications with Quality of Service (QoS) guarantees over next generation wireless networks, and presents a representative collection of challenges and technological solutions of Quality of Service (QoS) in its different aspects. In the following section, we review the evolution and latest research activities of QoS. In Section 3, in order to have an overall picture of QoS, we define and discuss the various components involved in QoS provisioning. A summary of existing QoS architectures is presented in Section 4. The paper concludes with an outlook on challenges and opportunities for research on QoS in both near and long term future.

2 Evolution of QoS

In recent years, QoS has become ubiquitous, and omnipotent. The provisioning of QoS experiences the evolvement from QoS-aware to QoS enable, from local operating systems to networks, from wireline to wireless and mobile, from centralized to distributed or hybrid, from homogeneous networks to heterogeneous networks, and from separate support at different layers to integrated end-to-end QoS guarantee.

2.1 From Local Operating Systems to Networks

Along with the early and extremely limited multimedia techniques such as MEPG1 for digital storage media, the initial QoS was mainly concerned with providing a general operating system that equally supports digital audio, video, image, text, etc. Less effort had been made on issues of real-time programming and multimedia synchronization in a distributed network environment. After the rapid development of Internet and MEPG2 that supports both storage and transmission of multiple media streams, new contents, such as IntServ and DiffServ, were fed into the scope of QoS, to support traffic delivery over the best-effort Internet.

2.2 From Wireline to Wireless and Mobile

QoS mechanisms and protocols used in wireline environments usually cannot be directly applied to wireless or hybrid wireline/wireless environments due to the following characteristics of wireless communications:

- Scarcity of wireless channel bandwidth and the multihop link interference make it much harder to support high quality multimedia services. For example, for a 4-hops route in 802.11 ad hoc networks, the end-to-end capacity falls less than one quarter of the link saturation throughput, which is furthermore below the channel bandwidth

[66]. Therefore, the end-to-end capacity is very limited even for IEEE 802.11a/g 54 Mbps broadband services.

- Link instability due to wireless interference and mobility places challenges on QoS adaptation, error-resilient coding, etc.
- Longer delay and larger jitter make it much harder to support real-time services.
- Diversity and complexity of wireless access technologies lead to multitude solutions on different domains with different perspectives, which makes interoperability hard to be achieved.

2.3 From Centralized to Distributed or Hybrid

Centralized QoS management used to be the main stream solution. The desirability of a centralized QoS policy system from policy management point of view is clear, since by collecting global information, it considerably simplifies the job of the consistency check of the various policies stored and maintained at a central location. However, major disadvantages of the centralized QoS approach are:

- The smooth functioning of the entire policy framework is largely dependent on the central policy server. In the event the policy server becomes non-functioning, the entire QoS system may break down.
- The central policy server may become the bottleneck of the system if there are a large number of end users in the system or if the complexity of services is very high.
- With the rapid deployment of mobile networks, centralized approach exposes location-dependent errors. In forth-coming services in multi-hop ad hoc network, a mobile station may be several hops away from the central control station, making the overhead unacceptable.

Therefore, distributed QoS management is considered as a competitive solution to accommodate the highly complicated QoS task of current and next generation multimedia services, especially for the fully distributed mobile ad hoc network. In the distributed approach, end users are collaborating with each other and make local decisions. The incompleteness of local information is partially compensated by the distributive collaboration. From a statistical point of view, the stability of quality of service is guaranteed globally. Of course, the centralized and distributed approaches can be combined together to further improve the quality of service. An interesting discussion on non-cooperative networks where end users and applications are selfish has been studied in [7].

2.4 From Homogeneous to Heterogeneous

With the rapid development of various wireless access technologies, users may require multimedia services over a largely diversely heterogeneous environment, with the capability of roaming transparently from one network to the other. Thus, we expect a unified QoS framework that builds upon and reconciles the existing notion of QoS at different system levels and among different network architectures [8]. The generalized architecture must be configurable, predictable, maintainable and interoperable over all network architectures, while it must also provide a mechanism to specify and enforce the unique characteristics of each architecture. The significant difference among those access technologies brings tremendous challenges on how to couple different QoS specifications, resource limitations, signaling protocols, and finally QoS management and control.

2.5 From Separate Layer Support to End-to-End Guarantees

Although end-to-end QoS guarantee is the ultimate goal, reaching a common understanding about QoS mechanisms for inter-domain use, multi-vendor interoperability and expected service behaviors in a network is still challenging. We note that the cross layer interaction is a more efficient approach to guarantee and optimize the end-to-end quality of service.

3 Components of QoS

To further understand different QoS architectures and mechanisms being used or proposed, one needs to have an overall picture of the basic QoS components, including QoS specification, QoS mapping, QoS policing, QoS mechanism, QoS monitoring and adaptation, and other areas such as QoS routing, etc.

3.1 QoS Specifications

QoS specifications are declarative in nature. Applications specify what is required rather than how this is to be achieved by underlying QoS mechanisms [8]. The quality performance of the service or application is a composition of the performance of certain QoS parameters specified by the application, the infrastructure resources (both end system and network) and the user requirements. QoS parameters can be identified at different levels of abstraction. From the end user perspective the QoS parameters are of different nature (more qualitative oriented and subjective) from those at the system level (more quantitative oriented and objective) [3]. QoS may be divided into multiple levels, as shown in Table 1.

<table>
<tr><th>QoS Level</th><th colspan="2">QoS Parameter</th><th colspan="2">Characteristics</th><th colspan="2">Example</th></tr>
<tr><td>Commerce</td><td colspan="2">Cost and Feedback</td><td colspan="2">Service price, responses from provider to submitted complaints</td><td colspan="2">Fee charged for transmit 1Mb data; instant technical supports</td></tr>
<tr><td>Enterprise</td><td colspan="2">Subjective</td><td colspan="2">End user oriented QoS requirements (subjective & not formal)</td><td colspan="2">"Audio must be along video"; "DVD video quality"</td></tr>
<tr><td>Information</td><td colspan="2">Objective</td><td colspan="2">Precise statement of QoS requirements derived from subjective QoS specs. Often application independent</td><td colspan="2">25fps (PAL), 30fps (NTSC), Aspect ratio of 4:3 (PAL TV Format), Telephone quality 8Khz</td></tr>
<tr><td>Computational</td><td colspan="2">Application</td><td colspan="2">Describe QoS requirements for applications. Often specified in terms of media quality & media relations</td><td colspan="2">Lip synchronization between audio and video (skew ±80ms)</td></tr>
<tr><td>Engineering</td><td>System</td><td>Network</td><td>Describe QoS requirements from operating system</td><td>Describe QoS requirements from network</td><td>Buffer size, operations/s, Memory (Mb)</td><td>Throughput (Mbps), Delay and Jitter (ms)</td></tr>
<tr><td>Technology</td><td colspan="2">Infrastructure Properties</td><td colspan="2">Describe QoS properties of hardware, OS & network technologies</td><td colspan="2">Video device (PAL Format), Audio device (μ-law), BER (10^{-5})</td></tr>
</table>

Table 1: QoS Specifications

- *The Commerce QoS* concerns issues of service pricing, human customer support, etc. In general, it is non-technical, and business oriented, mainly influenced by factors of cost and profit.
- *The Enterprise QoS* is customer-orientated and subjective in nature. It deals with end users' expression of QoS requirements that are different for each end user and depend on the users' perception of the received or required services.
- *The Information QoS* describes the quality of information provided by the service, which is objective and quantitative. Information QoS is often application independent and media-oriented, which means different specifications for different types of information media. For example, specification for video data may include requirements such as gray or color, video size, resolution, and frame rate; specification for audio data may include requirements such as mono or stereo, audio equalization of bass or treble.
- *The Computational QoS* is application oriented, describing the QoS requirements for the specific application. Computational QoS is often in terms of media quality and media relations, which includes

source/sink characteristics such as playback rate, video PSNR (Peak Signal to Noise Ratio), and lip synchronization between audio and video, and transmission characteristics such as end-to-end delay and jitter.
- *The Engineering QoS* includes both system and network viewpoints of the service quality. System QoS viewpoint includes performance criteria of end systems such as memory size, hard disk buffer size, and MPOS (MegaOperations Per Second); network QoS viewpoint includes issues of throughput, goodput, transmission and queuing delay at a specific layer.
- *The Technology QoS* describes the QoS characteristics of devices, operating system and networks access technologies. It deals with issues of medium bit error rate, synchronous and asynchronous transmission, etc.

The classification of QoS specifications in Table 1 can be simplified into three classes, i.e. *assessed*, *perceived*, and *intrinsic* [11]. The *assessed* QoS class, corresponding to the Commerce level in Table 1, is mainly related to the customer's opinion on whether to choose the service or not. The customer's opinion depends on the perceived quality, the expectation based on the service price, and the responses of the provider to reported complaints and problems. Even a customer representative's attitude to a client may be an important factor in rating the assessed QoS. The assessed QoS class is beyond the scope of this chapter. The *perceived* QoS class, corresponding to the Enterprise level in Table 1, mainly reflects the customer's experience of using a particular service. Given a certain service price, the customer will compare the observed service performance with his/her expectation. However, this comparison is subjective and will be affected by the customer's previous experience of similar services from the same or different service providers. Therefore, various customers may perceive differently for the exact same service. The *intrinsic* QoS class, including Information, Computational, Engineering, and Technology levels, summarizes all technical aspects that influence the quality of service. It is quantitative and objective, and independent of user perception.

3.2 QoS Mapping

QoS mapping is critical because the ultimate goal of a QoS-enabled network should be to provide users with a set of end-to-end guarantees in the level of user perception, be they qualitative or (preferably) quantitative. In order to achieve this goal, QoS must be provided at all the layers. The quality of service at a given layer is the characterization (in absolute or relative term) of the expected quality to be achieved in the delivery of data units to the corresponding layer across the network. More precisely, it includes the delivery 'from the moment layer L data unit crosses the boundary from level L to L-1 at the source end-point to the moment it crosses the boundary from level L-1 to L at its destination end-point'.

The meaning of QoS mapping is twofold: first, QoS specification provided in the upper level needs to be translated into corresponding parameters in the lower level, vice versa. The mapping of parameters between levels may not necessarily be static. With network conditions changing at the lower level, the mapping relationships may change as well. Second, according to the service specifications, QoS control and managements of different levels may be varied as well. QoS specifications among different levels are related yet distinct. Thus, the required actions at different levels are not the same. For example, a QoS specification at the application level permits the user the selection of video quality. When mapped to the network level, it permits admission control and resource reservation. If further mapped to the infrastructure (or physical) level of next generation wireless networks, it may permit adaptive control on the PN (Pseudo Noise) code to maintain certain BER over the constantly changing wireless channel.

Several issues will impact the process of QoS mapping, among which [13]:

- *Segmentation, fragmentation and reassembly* – packets are often segmented into smaller lower-layer data units. For example, in IEEE 802.11, since the station in transmission cannot detect errors, if error occurs, channel will be taken and wasted until the end of the transmission of the corresponding error frame. In order to improve the channel utilization, large packets from the upper layer will be fragmented into shorter frames in MAC layer. In this condition, QoS mapping between different layers is necessary. Furthermore, even in the same layer, a packet (or data unit) may traverse networks with different maximum transfer unit (MTU), being segmented and re-assembled multiple times.
- *Hard and soft QoS constraints* – due to wireless interference and mobility, it is hard to offer hard (or quantitative) QoS constraints in presence of wireless and mobile environments. In this case, soft (or qualita-

tive) QoS constraints may be more appropriate. For instance, IEEE 802.11e EDCF provides service differentiation by prioritizing traffic flows. Higher priority will assure some users (or traffic flows) of receiving preferential treatment in channel contention over others. However, EDCF cannot even guarantee the minimum QoS level (throughput and end-to-end delay) for any flow.

- *Flow aggregation and multiplexing* – for scalability purposes, some QoS architectures aggregate data streams with similar performance requirements into a single flow, and then provide QoS guarantees to the aggregated flow. The effect of aggregation on the individual streams must be considered in QoS mapping.

3.3 QoS Policies

Essentially, there are two types of QoS policies, i.e., the global (or centralized) management and local (distributed) management. In centralized management, the global QoS agent collects all the necessary information of end systems and networks, and makes decisions on traffic admission, scheduling, and resource allocation. Based of the complete global knowledge, deterministic quality of service requirements can be guaranteed for applications. However, centralized management lacks scalability, and it incurs serious performance degradation with the increase of system size. Due to the distributed characteristic of the Internet and the enormous number of end systems and applications, centralized policy may not be appropriate for QoS management over Internet or heterogeneous networks. Instead, distributed QoS management with the collaborations among end systems has attracted much more attentions.

Collaboration is the information interchange, decision-making and interaction between end systems at physically disparate locations, working in dynamic heterogeneous environments with the purpose of accomplishing mutually beneficial activity. The essential requirement for collaboration is providing each end system with the ability to have direct and immediate access to all information defined by the client's needs, interests, resources and capabilities. However, the correctness and accuracy of the QoS-related decisions depend on how much information the end system can obtain. The complete information sharing may not be available in fully distributed environments, and thus, inaccurate or incorrect decisions may be made. As a result, deterministic or hard QoS guarantees may not be appropriate for distributed QoS policy. Instead, one can expect a reasonable degree of satisfaction by allowing tradeoffs with certain QoS requirements. For example, audio streaming applications are highly sensitive to jitter and hence can compromise the response time by buffering the data at the receiver before starting playback to smooth of the delay variations. In this case, the startup latency increases with the payoff of playback continuity. Moreover, with distributed QoS policy, the QoS requirements and performance are often measured statistically by tolerating certain temporary violations. For example, the guaranteed throughput (and delay) can be referred to the statistical mean value instead of the minimum (and maximum) value. The statistical QoS guarantee is also referred to *soft QoS* guarantee.

3.4 QoS Mechanisms

In general, Table 2 shows QoS building blocks and supporting mechanisms. Various types of QoS build blocks can be identified: multimedia content, operating system, communication subsystem and the underlying network. The network covers medium and access technologies, e.g., the MAC layer of WLAN or air interfaces in 3G networks. The communication subsystem includes all communication related functions to provide quality of service request by applications. For this purpose, communication subsystems need operating systems support, such as resource allocation and scheduling. The multimedia content deals with the fundamental coding technologies, which may determines what kind of QoS mechanisms can be chosen by other building blocks. For example, rate control and bandwidth allocation mechanisms may be used in communication subsystem and networks for video encoded in layers with bandwidth scalability, while these mechanisms cannot be used for traditional video data without bandwidth scalability. In order to provide a QoS architecture or platform for emerging applications, a tight integration of QoS mechanisms between building blocks is highly desirable [14]. The major reason for such integration can be seen in the QoS guarantees usually associated with performance-oriented parameters, such as throughput, delay, jitter, reliability and availability. Depending on the interests of the applications, different mechanisms may be chosen to guarantee the performance of some parameter or all of them.

3.4.1 Multimedia Content

Multimedia applications provide services and management of multiple types of data, such as audio, video, animation, image and text. All types of data

Building Blocks	Resources	Performance Parameters				
		Throughput	Delay	Jitter	Reliability	Availability
Multimedia content	Audio, video, image, text	Compression, bandwidth scalability, rate control, layered video, FGS (Fine Grained Scalability)			Error resilient coding	
Operating systems	Processor	Scheduling				
	Memory/hard disk	Storage management				
Communication subsystems	Protocol functions	Admission control, resource reservation, flow control, rate control, congestion control	Synchronization protocol		Error control	Data replication, caching & prefetching, pushing
	System functions	Protocol timer management, protocol memory management				
Networks	Bandwidth	Traffic shaping, policing, pricing, queueing and scheduling, congestion control, QoS routing			Error control	Disjointed multipath

Table 2: QoS Building Blocks and Supporting Mechanisms

except text may require compression for storage and transmission due to the large data size. However, traditional compression techniques may not be able to support the delivery and smooth playback in real-time multimedia applications with strict time constraints because of the bandwidth limitation and variations of wireless channel. Wireless channel bandwidth can vary significantly, depending on the signal strength and interference level that a user receives. As a result, when a user travels through different parts of the cell, different bandwidths may be dynamically assigned to the user. In addition, depending on the quality of service capability of the wireless network, multi-user sharing of the wireless channel with heterogeneous data types can also lead to significant user channel bandwidth variation. This unpredictability of available wireless channel bandwidth also introduces high delay jitter for multimedia applications and may lead to unacceptable quality of playback services. To overcome this bandwidth variation, bandwidth scalability is a very important feature for compressed multimedia data, particularly video data, usually used by rate control or synchronization protocols of communication subsystems and networks. The functionality of bandwidth scalable coding is twofold: (1) to avoid network congestion or saturation by decreasing source bit-rate, and to increase the utilization by increase the source bit-rate; and (2) to adjust the transmission time by change source bit-rate in order to comprise multimedia time constraints when necessary. Scalability is desired for progressive coding of images and video sent over heterogeneous networks, as well as for applications where the receiver is not capable of displaying the full resolution or full quality images or video sequences. This could for instance happen when processing power or display resolution is limited. Usually, scalability is provided by the ability to decode a part of a bitstream and reconstruct images or image sequences with: (1) reduced decoder complexity and thus reduced quality; (2) reduced spatial resolution; (3) reduced temporal resolution; and (4) equal temporal and spatial resolution but with reduced quality.

Moreover, due to the error prone characteristic of wireless communications, highly compressed multimedia data is highly vulnerable to propagation errors in wireless communications. Thus, in addition to the bandwidth scalability, error resilience, including issues of data separation and resynchronization, error concealment and error recovery, is another important feature for data compression technique to support QoS.

(i) MPEG-4

MPEG-4 [4] supports the coding of images and video objects, both with conventional rectangular as well as with arbitrary shape, with spatial scalability, temporal scalability, quality scalability and complexity scalability, which are explained as follows:

- *Spatial scalability* – is achieved by FGS (Fine Granularity Scalability) coding scheme, which allows decoders to decode a subset of the total bitstream generated by the encoder to reconstruct and display textures, images and video objects at reduced spatial resolution. A maximum of 11 levels of spatial scalability are supported in FGS, for video as well as textures and still images. In addition, object-based spatial scalability is also introduced in MPEG-4. It extends the 'conventional' types of scalability towards arbitrary shape objects, enabling flexible content-based scaling of video information.
- *Temporal scalability* – allows decoders to decode a subset of the total bitstream generated by the encoder to reconstruct and display video at reduced temporal resolution. A maximum of three levels are supported.
- *Quality scalability* – allows a bitstream to be parsed into a number of bitstream layers of different bit-rate such that the combination of a subset of the layers can still be decoded into a meaningful signal. The bitstream parsing can occur either during transmission or in the decoder. The reconstructed quality, in general, is related to the number of layers used for decoding and reconstruction.
- *Complexity scalability* – allows encoders/decoders of different complexity to encode/decode a given texture, image or video.

MPEG-4 provides error robustness and resilience to allow accessing image or video information over wireless and mobile networks. The error resilience tools can be divided into three major areas: resynchronization, data recovery, and error concealment.

- *Resynchronization*: Resynchronization is an effective mechanism to limit the error propagation in compressed video. Due to the compression, decoder will not be able to understand the bitstream after a residual error or errors. By using a pre-defined start code as the resynchronization marker, the decoder is able to restart the decoding process once it finds the nearest start code after the corrupted data. Generally, the data between the synchronization point prior to the error and the first point where synchronization is re-established, is discarded.
  Traditional *spatial resynchronization* approach adopted by the ITU-T standards of H.261 and H.263 inserts resynchronization markers at the beginning of GOBs (Group of Blocks). A GOB is defined as one or more rows of macroblocks. A potential problem with this approach is that since the encoding process is variable rate, these resynchronization markers will most likely be unevenly spaced throughout the bitstream. Therefore, certain portions of the scene, such as high motion areas, will be more susceptible to errors, which will also be more difficult to conceal. To solve this problem, MPEG-4 adopts a *video packet based resynchronization* scheme, which is based on providing periodic resynchronization markers throughout the bitstream. In other words, the length of the video packets are not based on the number of macroblocks, but instead on the number of bits contained in that packet. If the number of bits contained in the current video packet exceeds a predetermined threshold, then a new video packet is created at the start of the next macroblock.
- *Data Recovery*: MPEG-4 adopts Reversible Variable Length Codes (RVLC) to recover the corrupt data that in general would be lost between resynchronization markers. In this approach, the variable length codewords are designed in an error resilient manner such that they can be read both in the forward as well as the reverse direction. An example illustrating the use of a RVLC is given in Fig 1. Generally, in a situation such as this, where a burst of errors has corrupted a portion of the data, all data between the two synchronization points would be lost. However, as shown in Fig 1, an RVLC enables some of that data to be recovered.

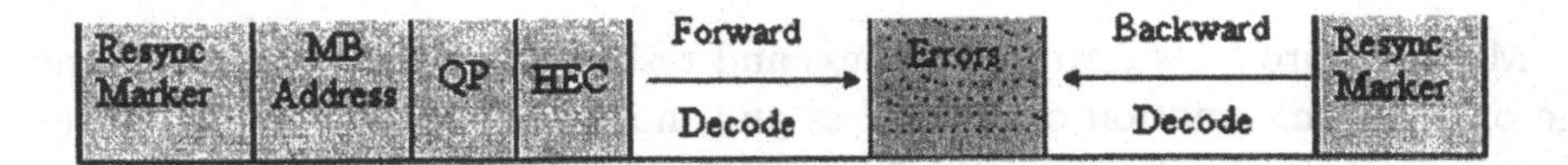

Figure 1: Example of Reversible Variable Length Code

- *Error Concealment*: In addition to the simple concealment strategy of copying blocks from the previous frame, MPEG-4 utilizes data partitioning to enhance the concealment capability. Specifically, a second resynchronization marker is inserted between motion and texture information to separate the motion and the texture. If the texture information is lost, the motion information can still be used to compensate the previous decoded frame and conceal these errors more accurately.

(ii) H.264 or MPEG-4 AVC

Another newly emerged standard for video coding standard with features of bandwidth scalability and error resilience is the MPEG-4 AVC (Advanced Video Codec) or ITU H.264 [16]. This standard is a joint effort between two standards bodies, i.e., ISO/IEC JTC1 and ITU-T. The main goal of this standard has been to improve the coding efficiency, at possibly an increased cost of computational complexity and memory requirements. For embedded systems, this cost could be significant. Current estimates are that MPEG-4 AVC requires about 2-2.5x factor in complexity over H.263+ Streaming Wireless Profile and MPEG-4 Simple Visual Profile, and 2.5-3x factor in complexity over H.263 baseline [15]. Furthermore, if a full implementation of the encoder is implemented (i.e., with multiple reference frames for motion estimation), the complexity can increase to as much as 10 times that of H.263 baseline. In addition, because of the increase in the number of reference frames that must be supported in MPEG-4 AVG (minimum of 3), additional frame buffers must be allocated in the decoder [17]. Considering the energy-saving issue for wireless end systems, the impact of MPEG-4 AVC needs to be carefully evaluated.

(iii) Conditional Replenishment Based Video Coding

Conditional replenishment has been proposed as a compression technique for taking advantage of the similarity between successive frames in video-telephony or video conferencing where video cameras typically are stationary and scenes usually change slowly. Compared with motion-estimation based video coding, conditional replenishment based scheme shows following advantages:

- Bandwidth scalability can be provided by dynamically changing the encoding parameter-the threshold.
- Error propagation in video decoding process can be limited by removing the motion estimation and inserting macro-block based resynchronization.
- To achieve an unobjectionable video, retransmission and error concealment are used to improve video spatial and temporal resolution, respectively.
- Lower computational complexity, and higher robustness in wireless channels can be achieved by eliminating inter-coding.

Given the above advantages, conditional replenishment based coding may be an alternative solution for video transmission over wireless channels and for mobile terminals with limited computing and energy resources [60].

(iv) SMIL

In addition to bandwidth scalability and error resilience, multimedia content possesses certain temporal properties, e.g., the playback time, the skew limit for lip synchronization, etc. A temporal specification are required to model the temporal properties of multimedia presentations. A multimedia presentation may consist of multiple multimedia objects such as audio, video, image and text, etc. The temporal specification must be able to describe the temporal relationship both within a single multimedia object, e.g., a video object, and between multimedia objects, e.g., the lip synchronization between audio and video. SMIL (Synchronized Multimedia Integration Language) [21], a widely accepted standard, enables simple authoring of interactive audiovisual presentations, and is typically used for "rich media"/multimedia presentations which integrate streaming audio and video with images, text or any other media type. By modeling complex temporal behaviors of multimedia presentations, especially if they are interactive ones, SMIL provides a feasible solution to quantitatively measure the performance of multimedia synchronization.

3.4.2 Operating System

CPU scheduling mechanisms in operating systems have been studied for many years, hence we will not elaborate them in this chapter. The main objective of scheduling is to support real-time applications by minimizing the process latency. In a mobile environment with scarce resources, optimizing resource management is also very important. The *memory and disk storage* used to be an important QoS issue of operating system. But with the fast development of computer hardware, it is not so important any more. The interested reader is referred to [67] for more details of operating system issues for real-time applications.

3.4.3 Communication Subsystem

QoS of multimedia services or applications is realized by CPU bandwidth scheduling and memory and disk storage management at underlying operating systems. At communication subsystem level, QoS guarantees may be specified in terms of transmission bandwidth and other link resources on each active connection; additional requirements regarding packet loss, inorder delivery, end-to-end delay, jitter, and service availability can also be specified. To ensure that each connection meets aforementioned QoS requirements of the supported applications, possible QoS mechanisms at communication subsystems includes admission control, resource reservation, flow control, rate control, error control, synchronization protocol, and data replication.

The general guide for designing QoS-enabled end communication subsystems is: (i) to provide perflow or perserviceclass guarantees, (ii) to maximize the aggregate utility of the communication service across all end systems, (iii) to gracefully adapt to transient overload, and (iv) to avoid, if possible, starving lower priority service classes during the period of sustained overload. Moreover, according to the specific QoS requirements of the application, only the minimum possible set of QoS mechanisms or protocol functions should be selected. All other unnecessary mechanisms or protocol functions should be turned off in order to improve the achievable performance of communication subsystems [14]. We will discuss several mechanisms in the following parts.

(i) Admission control and resource reservation/allocation

The objective of connection/call admission control (CAC) is to guarantee the newly admitted traffic not result in network overload or service degradation of existing traffic. An efficient CAC scheme should be relatively simple and robust to implement. In most cases, distributed schemes are preferred because they may result in smaller control overhead than centralized schemes. Resource reservation/allocation may have different functions in various networks. In wireless communications, resource reservation/allocation mainly focuses on the most important resource, i.e., the channel bandwidth. For cellular networks, it is designed to decrease the connection dropping probability, which may be caused by handoff. However, resource reservation also can be used as a method to differentiate services, by applying different policies of resource reservation to different services. Resource reservation can be either fixed or dynamic. In fixed schemes, resource may be wasted or insufficient when the network conditions change, which are very common in wireless networks. The signaling required to set up reservations for application flows can be provided by receiver initiated reservation protocols, such as RSVP, or sender initiated reservation protocols, such as STII [22]. Extensions and modifications of these signaling protocols can also be found in literature.

Admission control is very important for QoS guarantees in wireless networks with limited channel bandwidth, especially for IEEE 802.11, because the MAC protocol of 802.11 is load sensitive and only work efficiently under the media or low traffic load. Various schemes and algorithms for admission control, resource reservation, and corresponding and support signaling protocols have been proposed for cellular wireless networks and IEEE 802.11 WLANs [20, 5]. Comparing with those schemes, end-to-end admission control and resource reservation in multihop ad hoc networks is a relatively new area. The end-to-end capacity of the multihop ad hoc network is roughly $O(\frac{1}{\sqrt{n}})$, where n is the total number of nodes in the network [65]. With generally less capacity than single hop networks, admission control becomes more critical in multihop ad hoc networks. Unfortunately, in a multihop ad hoc network, because of wireless interference, a newly admitted service may affect other existing services even if they have disjoint pathes (i.e., no shared node). In addition, this influence may be time varying with topology changes due to power control and mobility. End-to-end admission control and resource reservation is proved to be a NP-complete problem, for both TDMA and 802.11-based multihop ad hoc networks [64]. All of the above issues make admission control for multihop ad hoc networks a highly complex open problem. In addition, end-to-end admission control and signaling for wired-cum-wireless networks, especially the integration with existing IntServ architecture of Internet, is also very important [6].

(ii) Flow control, rate control and congestion control

In general, flow control and rate control and congestion control are dealing with similar issues, which are keeping the load of the network under its capacity so that it can operate at an acceptable performance level. To some researchers, flow control is more general and covers not only link congestion control but also the control of other resources like memory, while rate control and congestion control are more specific for link congestion control. Since normally the memory is not the bottleneck of the system nowadays, we mainly focus on the link congestion control in this section. In order to provide service differentiation and guarantee the QoS of high priority traffic, ideally the source of traffic reduction comes from a user whose admission control priority is not critical. This may permit higher-priority traffic to continue to receive normal service. Therefore, congestion control may cooperate with admission control mechanisms.

There are many ways to classify congestion control schemes, such as window-based or rate-based; open-loop or feedback; source-based or router-based. An example of window-based scheme is TCP protocol, which multiplicatively decreases the size of the window when congestion occurs and cautiously increases the window when congestion subsides. In rate-based schemes, the destination node specifies the maximum rate (number of packets over a given time) at which the sources can send packets. Window-based schemes can be either end-to-end or hop-by-hop mechanism, while rate-based control schemes should be used as hop-by-hop mechanisms since all intermediate nodes should be made aware of the rate and enforce it. It has been argued that congestion control schemes with feedback may incur large delay, which led to the development of open-loop schemes. Similarly, the large delay in source-based schemes also led to the design of router-based schemes in which intermediate nodes, such as routers or other devices, may initiate control actions.

In IP-based networks, most of work focuses on congestion control in TCP, in which network congestion is identified as session packet loss or acknowledgement-timer expiration in wired networks. However, this causes significant performance degradation of TCP in wireless networks, where channels may experience errors from wireless channel interference and handoff. Therefore, necessary modifications of TCP are required to distinguish congestion errors with interference errors. A large number of wireless TCP protocols were proposed, and they were classified into three broad categories: end-to-end schemes, where the sender is aware of the wireless link; link layer schemes in providing reliability; and split-connection schemes, that break the end-to-end connection at the boundary of wired and wireless networks [30]. Major technologies (but not complete) used in above schemes are:

- *Snoop protocol* [31] introduces a snooping agent at the base station to observe and cache TCP packets going out to the mobile host as well as acknowledgements coming back. By comparing the cached packets and acknowledgements, the agent is able to determine what packets are lost on the wireless link and schedule a local link layer retransmission. In the same time, duplicate acknowledgements corresponding to wireless losses are suppressed to avoid triggering an end-to-end retransmission at the source. Snoop protocol can exactly find the cause of packet losses and take action to prevent the TCP sender from making unnecessary window reductions.

- *Selective acknowledgement (SACK)* (RFC1072, 2018) replaces the cumulative acknowledgements of TCP with selective acknowledgements. TCP congestion control action is still performed when losses occur, however, the sender will recover quickly from multiple packet losses within a single transmission window by the sufficient information provided by SACK.
- *Partial acknowledgment* [32] uses two types of acknowledgements to distinguish losses in the wired and the wireless links. The sender handles the two types of acknowledgements differently.
- *Explicit loss notification (ELN)* [33] uses a bit in TCP header to communicate the cause of packet losses to the sender, without caching any packet.

(iii) Error control

In addition to error resilience coding techniques of multimedia content, several error control schemes, e.g., forward error control (FEC) and retransmission (ARQ) based schemes are often employed in the link layer in providing reliability to wireless communications. The main advantage of link layer error control is that it fits naturally into the functionalities of layered network structure, and operates independently of higher layer protocols without maintenance of any per-connection state. The selection of error control mechanisms mainly depends on the reliability requirements of the applications and services. FEC-based schemes should be used if the service carries strict requirements of transmission delay, while ARQ-based schemes can be used for other services. For the best effort service, error control schemes may not even be necessary. Moreover, it is possible to combine FEC and ARQ into hybrid schemes, e.g., AIRMAIL protocol [34], for efficient loss recovery. Current digital cellular systems in the U.S., including both CDMA and TDMA, are primarily using ARQ techniques. Again, due to the changes of wireless channel conditions, adaptive mechanisms are highly desirable to achieve better quality of service and higher channel goodput.

(iv) Synchronization protocol

Synchronization protocol aims to achieve the temporal specification (i.e. the intra- and inter-stream synchronization) during the playback of distributed multimedia objects. Because of the increasing size of global networks and the integration of wireless services, end-to-end delays are gradually increasing. For streaming media services commonly used by the video-on-demand applications in these networks, the end-to-end delay may range from 5 to 10 seconds [23]. This may prevent the synchronization protocol from responding properly if asynchrony occurs. Thus, choosing a synchronization protocol to effectively achieve high quality playback performance under wireless environments with large end-to-end delays becomes a critical issue to wireless multimedia services.

Several synchronization protocols have been proposed in the literature. Ramanathan and Rangan [24] introduced the feedback techniques in the multimedia information retrieval. Hać et al. [26] proposed a feedback controller at the client buffer. Biersack et al. [27] presented a feedback control scheme including both distributed media servers and the client. Boukerche et al. [25] designed a MoSync system for cellular wireless and mobile networks.

These synchronization protocols use the idea of feedback control schemes to maintain intra- and inter-stream synchronization. Except for those by Hać et. al., all schemes rely on feedback loops between the client and media servers. Information of the asynchrony is collected at the client, and then sent to the server as feedback messages. Based on the feedback, media servers take appropriate actions to resynchronize the playback at the client. End-to-end delays from the client to media servers will cause feedback delays, called the deadtime of control schemes. Hać et. al. excluded media servers from the feedback loop to avoid a large system deadtime. However, this approach cannot provide a satisfactory solution due to the uncontrollability of the server side.

The problem of the gradually increasing end-to-end network delays presents new challenges for the synchronization schemes. Given a fixed finite buffer size, if the end-to-end delay in the system is large, there is no enough grace period for feedback control schemes to work properly. Being aware of the problem of large end-to-end delay in wireless and heterogeneous environments, Zhu et al. proposed a synchronization scheme, termed Synchronization protocol for Multimedia Applications for the wiReless InTernet (SMART) [28, 29] which provided a simple playback startup scheme and buffer determination based on the statistical knowledge of network condition, and required no feedback.

The client compensation buffer is pre-configured based on the knowledge of channel conditions and fixed since then in SMART. However, the fixed buffer management may not be efficient in the last mile wireless networks, where the traffic and channel conditions are constantly changing, e.g. the jitter may increase significantly from off-peak to peak time. In order to further save the valuable buffer resource at the wireless/mobile terminal, a new class of dynamic buffer management schemes has been proposed in [12]. The client compensation buffer is adjusted among a number of configurations based on the measurements of incorrect playbacks. The mapping relationship between incorrect playbacks and buffer configurations in SMART has been derived. By adopting the dynamic buffer management, SMART is able to achieve less incorrect playbacks comparing with fixed buffer management with equivalent buffer space.

(v) Data replication, caching, pre-fetching, server-pushing

Replication was commonly used in distributed systems to increase availability and fault tolerance. By allowing clients to perform data retrieval from their closest mirror site, replication helps to improve end-to-end service responsiveness (either in geographic term or network-wise), while at the same time balancing the load among the various servers. The large volume of requests arriving at popular sites can result in saturating even the most powerful web server, especially during peak hours. In order to tackle the problem, hot sites run multiple servers, potentially spanning the globe. One of the objectives of replication is to make al these servers appear to the user as a single highly capable web server. Replication began by manually mirroring web sites. The current target is to dynamically create, migrate and delete replicas among web hosts in response to changes in the access patterns. Main issues in implementing a replicated service are:

- *Consistency*: concurrent updates and system failures can lead to inconsistent replicas. The replication system should offer mechanisms for both resolving conflicts and keeping consistency between multiple replicas and their updates.
- *Location and access transparency*: users should not need to know where the requested data are physically located and how to access.
- *Update efficiency*: Due to the enormous workload of replication, the update process should be multicast-based mechanisms and with less overhead for saving bandwidth.
- *Scalability*: the replication system should be scalable in terms of read/write operations in large number of replicas and especially at high update

rates. Note that decentralized mechanisms display better scalability but may cause problem of data consistency.

- *Mobility*: with the possible replication on mobile locations, the replication system should consider the uncertainty of connection conditions.
- *QoS-awareness*: the replication process may employ different update policies on different data source based on their service class in order to optimize the resource and performance.

Caching was traditionally applied to distributed file systems and its applications on networks incur new problems, e.g., where to place the cache and when to update it. Generally, caching and replication are complementary techniques for providing service availability and responsiveness. Caching directly targets minimizing the service delays by assuming that retrieving the required content from the cache, incurs less latency compared to getting it from the network. Caching may be viewed as a special case of replication when mirror sites store only partial content of the original site, and the request for content unavailable in cache will be redirected to the main site.

Client-side pre-fetching has recently gained much attention because of its potential performance benefits without requiring complicated changes or additions to Web services. The basic idea of pre-fetching stems from the fact that, after retrieving a page from server, a user usually spends some time viewing the page and, during this period, the generally idle network link can be used to fetch data in anticipation.

Server-side pushing, similarly to multicasting, saves the bandwidth usage by sending out the content just once instead of sending one copy to every single user. However, real pushing is difficult to implement due to the diversity of users' needs in terms of content, request time, as well as the difference of link capacities among users. For this reason, "client-pulling", i.e., the client requesting files periodically from the servers may be more practical for implementations [35].

(vi) Protocol memory and timer management

The system functions of the communication subsystems may affect the performance of upper protocol functions. For example, retransmission-based mechanisms or ordered delivery require efficient memory management of temporary data and fast index searching. For real-time applications with high requirement on data processing, e.g. video transcoding, protocol memory and timer management is a very important bridge between upper layer services and underlying hardware.

3.4.4 Network

To guarantee quality of service, networks carry the tasks of traffic shaping, policing, pricing, packet classifying, queueing and scheduling. In mobile and wireless environments, error control and ah hoc routing are also very important issues.

(i) Traffic shaping, policing and pricing

Traffic shaping and policing aim to minimize the rejection probability of traffic, which is essentially the same goal of rate control and congestion control mechanism. However, traffic shaping and policing choose a slightly different approach, i.e., by reducing the burstiness and self-similarity of traffic.

Traffic shaping can be employed at the ingress of the network or at intermediate nodes for per-hop based shaping, and it can be either flow aggregation based or an individual flow based. Two main shaping algorithms exist, namely Leaky Bucket (LB) based algorithms [36, 37] defined by ATM Forum and Token Bucket (TB) based algorithms [38, 39] defined by IETF respectively. The latter has been adopted as a reference algorithm for traffic conditioner in 3GPP [40]. However, for a CDMA-based system like UMTS, very little literature can be found on how traffic conditioning performs. Both algorithms regulate the bursty traffic in such a way that over a long term period the average allowed rate approaches the desired token rate r asymptotically and over a short time interval the burst size of the traffic is upper bounded by bucket size b. Traffic shaping will of course introduce traffic shaping delay which consumes part of the total end-to-end delay budget of a service, but the regulated flow is expected to have a lower packet loss ratio instead.

Traffic shaping may not guarantee the conformance between influx traffic and service requirements. *Traffic policing* function compares the conformance of the user data traffic with the QoS attributes of the corresponding service and takes corresponding actions, e.g., reject or penalize non-conformance flow. Traffic policing, in practice, is a network operator choice. More detailed description of the policy framework can be found in [41].

Unlike policing, *pricing* encourages an open 'don't-ask-just-do-it' approach, by charging the non-conformance packet or flow. If additional congestion avoidance mechanism pricing existing, policing can be removed from the data path, giving reduced latency and complexity [42].

(ii) Packet classifying, queueing and scheduling

Packet classifying, queueing and scheduling are critical components in providing service differentiation, or guarantee of end-to-end quality of service for multimedia services and applications in wireless packet networks. For an end-to-end system, transmission queues are deployed at source, destination and all the intermediate nodes along the transmission path. How to handle the received packets and schedule the transmissions at each queue will directly affect throughout, delay, jitter and even packet loss ratio of the corresponding traffic.

In general, packet classifying is required for service differentiation. It identifies the service class of received packets and hence enables different processes provided by queueing and scheduling algorithms. Nevertheless, packet classifying can be also used with resource reservation mechanism rather than queueing and scheduling to achieve the similar effect as service differentiation. For example, in the DiffServ (RFC2475) architecture, packet classifying is combined with service differentiation; while in the IntServ (RFC1633) architecture, packet classifying is combined with resource reservation. However, in both cases, packet classifying is a foundation for providing the concept of service classes.

Queueing and scheduling basically handles the temporal behavior of packets from the time they enter the input queue(s) of the corresponding node to the time they exit from the same node to the next hop or to the upper layer applications. The desirable features of a good queueing and scheduling algorithm includes [43, 44]:

- *Delay bound*: the algorithm must be able to provide bounded delay for individual sessions in order to support delay sensitive multimedia services.
- *Throughput*: the algorithm should be able to provide short-term throughput guarantee for error-free sessions and long-term throughput guarantee to all sessions.
- *Fairness and efficiency*: the algorithm should not penalize sessions that temporarily exceed their reserved channel bandwidths provided the bandwidth is unused. In the mean time, fairness must be ensured for the overuse among sessions.
- *Complexity and scalability*: the algorithm should be low-complexity in terms of enqueueing and/or dequeueing a packet, which is necessary for wireless networks with resource and energy constraints. In addition, the algorithm should operate efficiently as the number of sessions sharing the channel increases.

- *Energy consumption*: the algorithm should take into account the need to prolong the mobile terminal's battery life.
- *Robustness*: the algorithm should be robust to bursty errors in wireless channels and intelligently protect sessions with good channel conditions when relinquishing sessions experiencing bad channel conditions. Furthermore, the service degradation due to the bursty errors should be smoothed.
- *Isolation*: the algorithm should provide necessary mechanism for isolating the ill effects of misbehaving sessions from regular sessions.
- *Delay/bandwidth decoupling*: For most algorithms, the delay is tightly coupled with the reserved rate. However, it is desirable to decouple them because some high-bandwidth services may tolerate relatively large delays, e.g. web browsing and services.

Queueing and scheduling algorithms can be classified as *work-conserving* or *non-work-conserving* [19]. With a work-conserving algorithm, a server is never idle when there is a packet to send. Examples include Generalized Processor Sharing (GPS, also known as fluid fair queueing FFQ), Weighted Fair Queueing (WFQ, also known as Packet-by Packet GPS, PGPS), Worst-Case Fair Weighted Fair Queueing (WF^{2}Q), Virtual Clock (VC), Weighted Round Robin (WRR), Deficit Round Robin (DRR), and Self-Clocked Fair Queueing (SCFQ). In contrast, with a non-work-conserving algorithm, each packet is assigned, either explicitly or implicitly, an eligibility time. Even when the server is idle and packets are waiting in the queue, if no packets are eligible, none will be transmitted. Examples include Hierarchical Round Robin (HRR), Stop-and-Go Queueing (SGQ), and Jitter Earliest Due Date (Jitter-EDD). Non-work-conserving algorithms generally incur large delay than work-conserving algorithms, but they may be used for applications with strict energy and jitter constraints but low requirement on delay.

Recently developed algorithms for TDMA wireless networks include [43]: Channel State Dependent Packet Scheduling (CSDPS), an enhancement of CSDPS providing class based queueing (CBQ-CSDPS), Idealized Wireless Fair Queueing (IWFQ), a variant of IWFQ called Wireless Packet Scheduling (WPS), Channel Independent Fair Queueing (CIF-Q), Server Based Fairness Algorithm (SBFA), Wireless Fair Service (WFS), and Token Bank Fair Queueing Scheduling (TBFQ) [45]. Also, algorithms for CDMA networks include Dynamic Resource Scheduling (DRS), Wireless Multimedia Access Control Protocol with BER Scheduling (WISPER), and Wireless Adaptive Scheduling Algorithm (WASA) [46].

(iii) QoS-aware routing

QoS routing requires not only to find route(s) from a source to destination(s), but also route(s) that satisfies the end-to-end QoS requirements, e.g., throughput, delay and packet loss ratio.

In the design of QoS-aware unicast routing protocols, several issues need to be considered. First, in multi-hop mobile ad hoc networks, link conditions may be very different due to the physical distance, mobility, and the power limit of mobile stations. Second, due to wireless interference, newly established routes may deteriorate the performance existing routes. Therefore, it is natural to combine routing with admission control and resource reservation, to provide end-to-end QoS guarantee. Third, the ability of providing QoS is heavily dependent on characteristics of access technologies. For example, in TDMA-based ad hoc networks, bandwidth is divided into a number of time slots. Neighboring nodes cannot transmit in the same time slot. Therefore, QoS routing with admission control and resource reservation mainly focus on (i) how to reserve the required number of time slots in each hop that makes a multi-hop end-to-end route possible, and (ii) if it is impossible to find a single route that satisfies the bandwidth requirement, is multi-path routing possible. Another example, in CSMA/CA-based ad hoc networks, such as IEEE 802.11, the resource of channel bandwidth is contended among neighboring nodes. Due to the exponential backoff, throughput and delay of nodes in hotspots may change widely in time. Therefore, QoS routing may focus on (i) estimation of available bandwidth and (ii) load balancing. Existing routing protocols, such as DSR, AODV, and ZRP, were designed without explicitly considering aforementioned issues. Therefore, they cannot support QoS effectively.

Multimedia applications such as video on demand and video conferencing often require multicast services. Challenges on the design of QoS-aware multicast routing include:

- *Load-balancing*: in order to minimize the congestion, the new multicast member should be able to select the path based on the traffic characteristic such that it balances the load of nodes/links, or it has the path with the minimum load. Nevertheless, other QoS metric for path selection, e.g. delay and packet loss, may also be used.
- *Fault-tolerance*: the multicast path algorithm should be able to search for maximally disjoint (i.e., minimally overlapped) multiple paths such as the impact of link/node failures becomes significantly reduced, and

the use of multiple paths renders QoS services more robust in unreliable network conditions.

- *Dynamic update*: the performance of QoS routing can be seriously degraded if the state information cannot be updated in time. However, the network state changes dynamically due to the transient load fluctuation, connections in and out, and links up and down. These make the dynamic update of QoS routes extremely difficult.
- *Overhead*: the control overhead of the multicast protocol should be restrained.
- *Scalability*: the multicast routing should be scalable when the number of multicast members increases. Therefore, routing protocols that maintain global per-flow information may not be good option. Instead, routing protocols with hierarchical structures are desirable.

3.5 QoS Monitoring and Adaptation

Temporary QoS violations may be allowed in services with soft or statistical QoS guarantees. However, there is still the possibility, e.g., handoff and the change of link conditions, that the QoS violations cannot be recovered by itself without an additional control resolution. In this case, QoS adaptation mechanisms should be provided for necessary intervention. To accomplish QoS adaptation, QoS monitoring is also required to detect the violations. In general, Bochmann et al., pointed out that the design of QoS adaptation should be guided by the following premises [47]:

- QoS adaptation should be performed automatically without user intervention when possible.
- QoS violations should be recovered locally without the change of user configurations when possible.
- QoS adaptation should maintain the initial QoS agreements as long as possible, before any control action is taken.

QoS adaptation can be employed on individual media content (i.e., video, audio, etc.), individual QoS parameters (throughput, delay, jitter, etc.), or overall performance of a service. Recently, middleware design attracted a lot of attentions in providing an intermediate coordinator for QoS adaptation. Based on the controlled object, QoS adaptation can be divided into four categories:

- *Content-based adaptation*: control schemes adjust the media content parameters, such as video encoding bit-rate and frame rate.
- *Resource-based adaptation*: control schemes adjust resources reserved at the communication subsystem or at the network, e.g., to increase available bandwidth by expanding the flow window size or to reserve more communication subsystem resource like cache space.
- *Link adaptation*: many wireless access technologies, e.g., IEEE 802.11, allow multiple transmission rates by providing various modulation techniques or adjustable length of PN (Pseudo-Noise) code. Since transmission rates differ with the channel conditions, appropriate link adaptation and signaling mechanisms can maximize the throughput under dynamically changing channel conditions.
- *Scheduling-based adaptation*: control schemes adjust scheduling mechanisms in response to the change of network condition.
- *Routing-based adaptation*: control schemes change the route based on the detection of QoS violations.

4 Architectures

The basic elements of a QoS architecture include QoS specification, mapping, and one or more QoS mechanisms mentioned in Section 3.4. Although QoS adaptation is not mandatory, it is a highly desirable feature in wireless and mobile environments in which link conditions are constantly changing due to the mobility and interference.

Various architectures have been proposed for wired networks. Early frameworks for can be found in [8]. For IP-based Internet, IETF IntServ and DiffServ are two well known models. IntServ (RFC1633, 2205, 2201) aims to support QoS by per-flow or per-flow-aggregate based resource reservation, which lacks scalability due to the signaling overhead and complexity. MPLS (RFC3031) may be use with IntServ to support QoS on a per-user basis. DiffServ (RFC2475) provides a simple and scalable class-based per-hop-behavior (PHB) service differentiation. However, without the necessary mechanism of admission control and resource reservation, QoS violations are inevitable. Instead of deterministic QoS guarantees, only relative QoS guarantees can be provided. Both models are not complete satisfactory. Possible combination of both models has been proposed in [52], which achieves end-to-end QoS by exploiting IntServ in the access networks and DiffServ in the core. This allows scalability, due to the DiffServ aggregation, while keeping the advantages of end-to-end signaling by means of the RSVP protocol. Also, with the admission control mechanism added to DiffServ, deterministic QoS guarantees may be accomplished. ITU-T are also making effort to identify a set of generic QoS mechanisms and provide a structure for them, which is called IPQoS [48]. In this ongoing QoS architecture, three logical planes are initially identified, i.e., control plane, data plane, and management plane. Control plane contains mechanisms dealing with the pathways through which user data traffic travels. These mechanisms include admission control, resource reservation and QoS routing. Data plane contains mechanisms dealing with the user data traffic directly. The mechanisms include buffer management, congestion avoidance, packet marking, queueing and scheduling, traffic classification, shaping and policing. Management plane contains mechanisms dealing with the operation, administration, and management aspects of the user data traffic. These mechanisms include monitoring, policy, SLA, and traffic restoration. Also worth noting is that in addition to QoS mechanisms in three logical planes, issues of QoS signaling have been particularly addressed by ITU-T SG 11. A similar QoS framework can be also found in [61] with the emphasis on solving the problem of hyper handover, i.e. the handover between different administrative domains, access technologies, user terminals, or applications in a heterogeneous network.

Aforementioned architectures in general are designed for wired networks. Effort has been made to apply similar reservation-based or differentiation-based frameworks to the wireless domain. Besides the well known IEEE 802.11e service differentiation MAC scheme [18], many other QoS architectures, catering to various wireless access technologies, such as IEEE 802.16 WirelessMAN, MANET, 3G/4G IP-based wireless networks, can be found in literature:

- *Integrated QoS Architecture for GSM* [51]: It provided the QoS mapping between IP network and GPRS RLC (Radio Link Control) layer, and implemented the QoS routing and resource reservation mechanisms.
- *DiffServ-SPS-MPA (DiffServ-Streaming Proxy Server - Mobile Proxy Agent)* [49]: Verma and Barnes propose this DiffServ-based 3G architecture that uses a probing scheme to provide seamless QoS to streaming applications. The key elements of this architecture are a Streaming Proxy Server (SPS) and Mobile Proxy Agent (MPA). The

SPS functions as a transcoder that convert the incoming video into the format that is most suitable for appropriate QoS service class. The MPA carries the functions of QoS scheduling and multicast routing.

- *DiffServ-MIR-HMIP (DiffServ-Mobile IP Reservation Protocol - Hierarchical Mobile IP)* [50]: It implemented bandwidth allocation and reservation so as to provide users of a shared medium with a guaranteed bandwidth. The bandwidth control is done by token bucket based algorithm on outgoing interfaces of network elements.
- *QoS Architecture for 802.16* [59]: IEEE 802.16 is designed to provide broadband WirelessMAN (Metro Area Network) services. Based on traffic shaping, policing, priority scheduling and dynamic bandwidth allocation, this architecture aims to introduce various levels of QoS guarantees while still achieving high system utilization. It implements different scheduling strategies on different service categories. Wireless Fair Queueing (WFQ) is used for the high priority category to guarantee the strict delay requirement, while Weighted Round Robin (WRR) and FIFO scheduling policies are used for middle and low priority categories, respectively.
- *INSIGNIA* [62]: It was designed to support fast reservation, restoration and scalability in MANET with the in-band signaling and soft-state resource management. The framework of INSIGNIA is shown in Fig. 2.

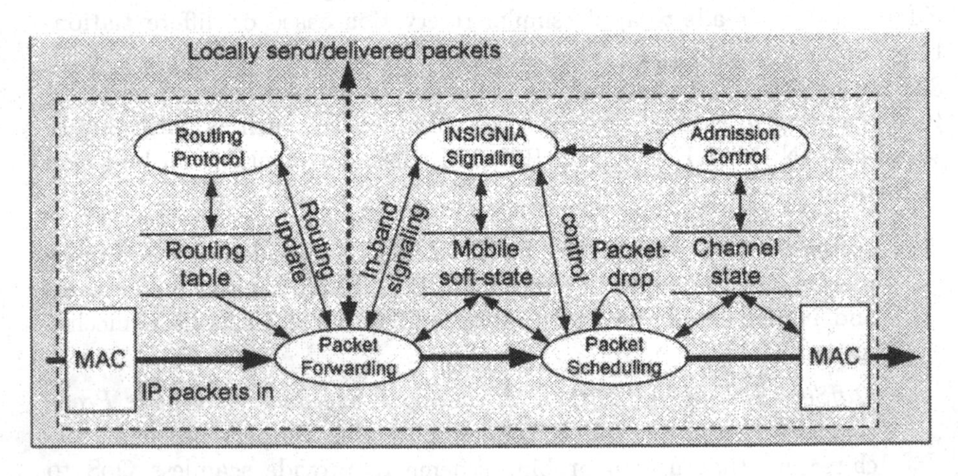

Figure 2: INSIGNIA QoS Framework

- *CADENUS Extension* [63]: It was proposed as a component-based QoS framework for heterogeneous networks with the high degree of complexity. Key components, named mediators, were designed to handle the mapping of service agreements and resource management.
- *FLO (Flexible Layer One)* [57]: By developing a flexible physical layer to GERAN (GSM/EDGE radio access network), operators can easily and quickly configure the coding rate, cyclic redundancy check (CRC) size, input block size, etc., at the physical layer to match the QoS requirements of the service to be carried at any particular time, which will improve the physical layer performance. FLO introduces the following advantages:
  - Physical layer flexibility
  - Significantly improved system capacity for carrying real-time packet services
  - Improved multiplexing ability
  - Provision for additional physical layer performance optimization
  - Simplified introduction of new services to the handset device
- *UMTS QoS Architecture* [53]: As shown in Table 3, UMTS supports four traffic types: conversational class, streaming class, interactive class, and background class. Key QoS entities include admission control module, bearer service manager, resource manager, traffic conditioner, and packet classifier, which are shown in Fig 3. Slight variations of UMTS QoS architecture can be found in [54, 55, 56].

Traffic parameters	Conversational class	Streaming class	Interactive class	Background class
Max. bit rate	x	x	x	x
Guaranteed bit rate	x	x		
Max. service data unit (SDU) size	x	x		
SDU format information	x	x		
SDU error ratio	x	x	x	x
Residual bit error ratio	x	x	x	x
Delivery of erroneous SDUs	x	x	x	x
Delivery order (y/n)	x	x	x	x
Transfer delay	x	x		
Traffic handling priority			x	
Admission/retention priority	x	x	x	x

Table 3: UMTS Traffic Classes and QoS Parameters

- *4G QoS Architecture* [58]: This architecture is set to support any type of user service with seamless mobility between different access technologies. The target technologies are Ethernet (802.3) for wireline access; 802.11 for wireless LAN access, and W-CDMA (the physical layer

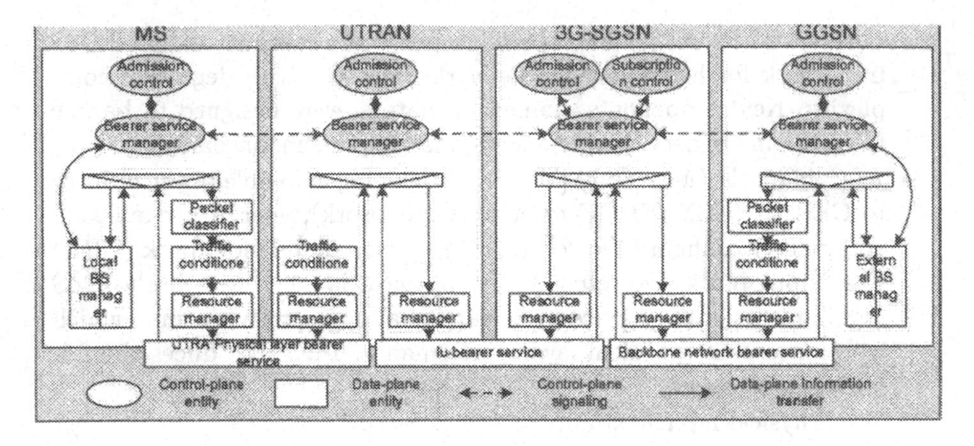

Figure 3: QoS Components in UMTS QoS Architecture

of UMTS) for cellular mobile access. The seamless inter-technology handoffs must be supported, therefore mobility and interoperability are key problems in this heterogeneous environment. Instead of dealing with intra-technology handoff or physical layer handoff, a network layer inter-technology handoff mechanism should be used in this architecture. Summarizing, illustrated in Fig 4, the key entities in this QoS architecture are:

- *Mobile Terminal (MT)*: the MT should be built in with enhanced IPv6 (RFC1883), able to perform marking according to user subscribed services and inter-operable between multiple access technologies; a network control plane, able to perform AAA registration / deregistration; an MT networking manager that will make the decision to execute handover and attach procedure based on information configured by user and received from the networking devices; and a radio convergence function (RFC) that interfaces with the radio layer in compliance with the IP layer QoS requirements.

- *Access Router (AR)*: which is the generic MT point of attachment to the network. On the top of enhanced IPv6 with IP Sec and DiffServ filtering, AR comprises of an AAA attendant that performs basic transport, authentication, registration, and security functions; a fast handoff module (FHO) with proper QoS signaling; a QoS attendant with interoperable functions of shaping, scheduling, and policing.

- *QoS Broker (QoSB)*: A QoSB is responsible for managing one or more ARs, controlling user access and access rights according to the information provided by the AAAC (Authentication, Authorization, Auditing and Charging) system.

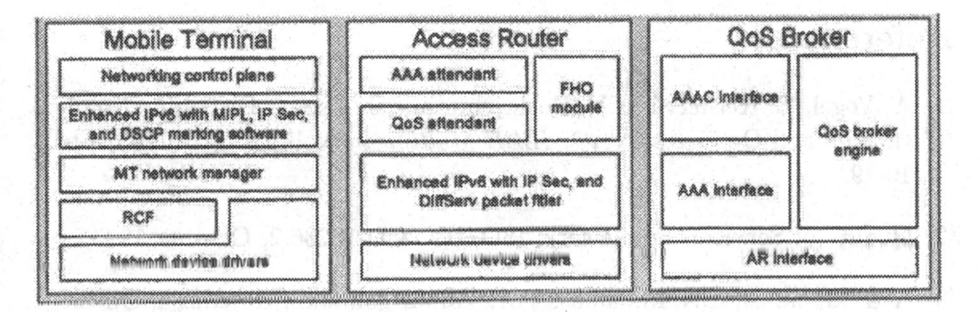

Figure 4: QoS Components in 4G QoS Architecture

5 Summary

The commercialization of multimedia services over wireless and mobile networks appears inevitable in coming years. This provides great opportunities, yet challenges for the research and development of QoS supports for various wireless access technologies.

In the short term, research is more likely to be focused on QoS design for specific multimedia applications over specific access technologies. In the long term, a seamlessly integrated QoS support for a wide range of customized multimedia services over various access technologies is the ultimate goal. This places significant challenges on new design issues, including e.g., interoperability, mobility management and vertical handoff between different wireless networks.

Although wireless technologies, such as 802.11, 802.16, Bluetooth and IP-based 3G cellular mobile networks, etc., target different markets, better QoS provisioning may make one technology more attractive and competitive. Therefore, QoS is moving towards the center stage of the development of next generation wireless networks. In this process, issues addressed in this article will play a critical role in the success of the next generation wireless multimedia services.

6 Acknowledgements

We would like to give our special thanks to Wenxiu Lin and Sue Jackson for their time and effort. This article would not have been possible without their generous support.

References

[1] A. Vogel, B. Kerhervé, G.V. Bochmann and J. Gecsei, Distributed multimedia and QoS - a survey -, *IEEE Multimedia* Vol. 2 No. 2 (1995) pp. 10-19.

[2] Quality of Service - Framework, ISO/IEC CD 13236.2, October 1995.

[3] A.G. Malamos, E.N. Malamas, T.A. Varvarigou and S.R. Ahuja, On the definition, modeling and implementation of Quality of Service (QoS) in distributed multimedia systems, *IEEE ISCC '99* (Egypt, Jul. 1999).

[4] ISO/IEC 14496-1, *Information Technology - Generic coding of audio-visual objects - Part 1: System.* Proposed Draft Amendment 1 (Seul, Mar. 1999).

[5] H. Zhu, M. Li, I. Chlamtac and B. Prabhakaran, Distributed Admission Control for IEEE 802.11 Ad Hoc networks, *TR UTD/EE/12/2003*, Univ. of Texas at Dallas.

[6] M. Li, H. Zhu, B. Prabhakaran and I. Chlamtac, End-to-end QoS Guarantee in Heterogeneous Wired-cum-Wireless Networks, *TR UTD/EE/12/2003*, Univ. of Texas at Dallas.

[7] K. Park, M. Sitharam and S. Chen, Quality of Service provision in noncooperative networks: heterogeneous preferences, multi-dimensional QoS vectors, and burs, *1st Intl. Conference on Information and Computation Economies* (1998).

[8] C. Aurrecoechea, A.T. Compbell and L. Hauw, A survey of QoS architectures, *Multimedia Systems* Vol. 6 No. 3 (Springer-Verlag, 1998) pp. 138-151.

[9] What is QoS?, www.getech.co.uk/products/sitara/whatisqos.htm

[10] ITU-T Rec. E.800, Terms and Definitions Related to Quality of Service and Network Performance Including Dependability (Aug. 1993).

[11] W.C. Hardy, *QoS Measurement and Evaluation of Telecommunications Quality of Service*, (Wiley, 2001).

[12] H. Zhu, I. Chlamtac and G. Zeng, Dynamic buffer management for multimedia synchronization in heterogeneous networks, *1st Intl. Workshop on Community Networks and FTTH/P/x* (Dallas, Oct. 2003).

[13] L.A. DaSilva, QoS mapping along the protocol stack: discussion and preliminary results, *IEEE International Conference on Communications* (ICC'00) Vol. 2 No. 18-22 (Jun. 2000) pp. 713-717.

[14] C. Schmidt and M. Zitterbart, Towards integrated QoS management, *5th IEEE Computer Society Workshop on Future Trends of Distributed Computing Systems* (Aug. 1995) pp. 94-101.

[15] *Draft ITU-T Recommendation H.263* (May 2, 1996).

[16] Joint Video Team (JVT) of ISO/IEC JTC1/SC29/WG11 and ITU-T SG16 Q.6, *Joint Committee Draft*, (May 2002).

[17] W. Zeng and J. Wen, 3G wireless multimedia: technologies and practical issues, in *Journal of Wireless Communications and Mobile Computing, Special Issue on Multimedia over Mobile IP* Vol. 2 No. 6 (Sep. 2002) pp. 563-572.

[18] H. Zhu and I. Chlamtac, Performance analysis for IEEE 802.11e EDCF service differentiation, to appear in *IEEE Trans. on Wireless Communications* (2004).

[19] H. Zhang, Service disciplines for guaranteed performance service in packet-switching networks, *Proc. IEEE* Vol. 83 No. 10 (Oct. 1995) pp. 1374-1396.

[20] H. Zhu, M. Li, I. Chlamtac and B. Prabhakaran, Survey of quality of service in IEEE 802.11 networks, to appear in *IEEE Wireless Communications*, Special Issue on Mobility and Resource Management, (2004).

[21] *SMIL*, available at http://www.w3.org/AudioVideo/.

[22] L. Delgrossi and L. Berger, *Internet stream protocol version 2 (ST2) protocol specification version ST2+*. Request for Comments RFC 1819 (August 1995) ST2 Working Group.

[23] S.N. Fabri and A.M. Kondoz, Provision of streaming media services over mobile networks, *2nd Intl. Conference on 3G Mobile Communication Technologies* (Conf. Publ. No. 477) (2001) pp. 104-108.

[24] P. V. Rangan, S. Ramanathan and S. Sampathkumar, Feedback techniques for continuity and synchronisation in multimedia information retrieval," *TOIS* Vol. 13 No. 2 (1995) pp. 145-176.

[25] A. Boukerche, S. Hong and T. Jacob, MoSync: a synchronization scheme for cellular wireless and mobile multimedia systems, *9th Intl. Symposium on Modeling, Analysis and Simulation of Computer and Telecommunication Systems* (2001) pp. 89-96.

[26] A. Hać and C.X. Xue, Synchronization in multimedia data retrieval, *International Journal of Network Management* Vol. 7 (1997) pp. 33-62.

[27] E. Biersack and W. Geyer, Synchronization delivery and play-out of distributed stored multimedia streams, *ACM/Springer-Verlag: Multimedia Systems* Vol. 7 (1999) pp. 70-90.

[28] H. Zhu, G. Zeng and I. Chlamtac, Control scheme analysis for multimedia inter- and intra-stream synchronization, *IEEE International Conference on Communications* (ICC'03) Vol. 1 (2003) pp. 7-11.

[29] H. Zhu, I. Chlamtac, J.A. Cobb and G. Zeng, SMART: A synchronization scheme for providing multimedia quality in emerging wireless Internet, *Vehicular Technology Conference* (VTC Fall'03) (2003).

[30] H. Balakrishnan, V.N. Padmanabhan, S. Seshan and R.H. Katz, A comparison of mechanisms for improving TCP performance over wireless links, *IEEE/ACM Transaction on Networking* (Dec. 1997).

[31] H. Balakrishnan, S. Seshan and R.H. Katz, Improving reliable transport and handoff performance in cellular wireless networks, *ACM Wireless Networks* Vol. 1 No. 4 (Dec. 1995).

[32] S. Biaz, M. Mehta, S. West and N.H. Vaidya, TCP over wireless network using multiple acknowledgements, *Technical Report 97-001*, Computer Science, Texas A&M University, (1997).

[33] H. Balakrishnan and R.H. Katz, Explicit loss notification and wireless web performance, *IEEE Globecom'98*.

[34] E. Ayanoglu, S. Paul, T.F. LaPorta, K.K. Sabnani and R.D. Gitlin, AIRMAIL: a link-layer protocol for wireless networks, *ACM/Baltzer Wireless Networks Journal* Vol. 1 (Feb. 1995) pp. 47-60.

[35] M. Conti, M. Kumar, S.K. Das and B.A. Shirazi, Quality of Service issues in Internet web services, *IEEE Transaction on Computer* Vol. 51 No. 6 (Jun. 2002) pp. 593-594.

[36] A. Elwalid and D. Mitra, Traffic shaping at a network mode: theory, optimum design, admission control, *IEEE Infocom'97* (1997) pp. 445-455.

[37] F.C. Harmantzis, D. Hatzinakos and I. Katzela, Shaping and policing of fractal a-stable broadband traffic, *Canadian Conf. on Elec. and Comp. Engin.* (CCECE'01) (May 2001) pp. 697-702.

[38] F.Y. Li and N. Stol, QoS provisioning using traffic shaping and policing in 3rd-Generation wireless networks, *IEEE WCNC'02* (2002) pp. 139-143.

[39] W.K. Wong, Y. Qian and V.C.M Leung, Scheduling for heterogeneous traffic in next generation wireless networks, *IEEE Globecom'00* Vol. 1 (Dec. 2000) pp. 283-287.

[40] 3GPP TS23.107v4.0.0, *QoS Concept and Architecture*, available at http://www.3gpp.org (Dec. 2000).

[41] IETF RFC 2753, *A framework for policy-based admission control*, R. Yavatkar, D. Pendarakis and R. Guerin (Jan 2000).

[42] B. Briscoe, M. Rizzo, J. Tassel and K. Damianakis, Lightweight policing and charging for packet networks, *IEEE OpenArch 2000* (Sep 1999).

[43] H. Fattah and C. Leung, An overview of scheduling algorithms in wireless multimedia networks, *IEEE Wireless Communications* (Oct. 2002) pp. 76-83.

[44] J.A. Cobb, M.G. Gouda and A. El-Nahas, Time-shift scheduling - fair scheduling of flows in high-speed networks, *IEEE/ACM Trans. on Networking* Vol. 6 No. 3 (Jun. 1998) pp. 274-285.

[45] W.K. Wong, H. Zhu and V.C.M. Leung, Soft QoS provisioning using Token Bank Fair Queueing Scheduling algorithm, *IEEE Wireless Communications* (Jun. 2003) pp. 8-16.

[46] P. Richardson, L. Sieh and A. Ganz, Quality of service support for multimedia applications in third generation mobile networks using adaptive scheduling, *Real-Time Systems* Vol. 21 (Kluwer Academic Publishers, 2001) pp. 269-284.

[47] A. Hafid and G.V. Bochmann, Quality-of-service adaptation in distributed multimedia applications, *Multimedia Systems*, Vol. 6 (Springer-Verlag, 1998) pp. 299-315.

[48] H.-L. Lu and I. Faynberg, An architecture framework for support of quality of service in packet networks, *IEEE Communications Magazine* (Jun. 2003) pp. 98-105.

[49] S. Verma and R. Barnes, A QoS architecture to support streaming applications in the mobile environment, *5th Intl. Symposium on Wireless Personal Multimedia Communications* Vol. 2 (Oct. 2002) pp. 514-520.

[50] G.L. Grand and E. Horlait, An end-to-end QoS architecture for mobile hosts, *IPDPS 2001* (San Francisco, CA, 2001).

[51] J. Mikkonen and M. Turunen, An integrated QoS architecture for GSM networks, *IEEE 1998 International Conference on Universal Personal Communications (ICUPC'98)* Vol. 1 (Oct. 1998) pp. 403-407.

[52] R. Mameli and S. Salsano, Use of COPS for Intserv operations over Diffserv: architecture issues, protocol design and test-bed implementation, *IEEE ICC 2001* Vol. 10 (Jun. 2001) pp. 3265-3270.

[53] R. Koodli and M. Puuskari, Supporting packet-data QoS in next generation cellular networks, *IEEE Communications Magazine* Vol. 39 No. 2 (Feb. 2001) pp. 180-188.

[54] L. Chen, H. Kayama, N. Umeda and Y. Yamao, Wireless QoS architecture and resource management for CDMA packet access multimedia cellular radio, *12th IEEE International Symposium on Personal, Indoor and Mobile Radio Communications* Vol. 1 (2001) pp. 64-68.

[55] S. Leroy, L Bos and J. De Vriendt, End-to-end UMTS quality of service architecture for the support of real-time multimedia in UMTS R5, *IEEE 3rd International Conference on 3G Mobile Communication Technologies* (Conf. Publ. No. 489) (May 2002) pp. 234-238.

[56] M. Ricardo, J. Dias, G. Carneiro and J. Ruela, Support of IP QoS over UMTS networks, *13th IEEE Intl Symposium on Personal, Indoor and Mobile Radio Communications* Vol. 4 (Sep. 2002) pp. 1909-1913.

[57] G. Platt, K. Pedersen and B. Sebire, Introducing the GERAN Flexible Layer One concept, *IEEE Wireless Commun.* (Jun. 2003) pp. 18-25.

[58] V. Marques, R.L. Aguiar, C. Garcia, J.I. Moreno, C. Beaujean, E. Melin and M. Liebsch, An IP-based QoS architecture for 4G operator scenarios, *IEEE Wireless Communications* (Jun. 2003) pp. 54-62.

[59] G. Chu, D. Wang and S. Mei, A QoS architecture for the MAC protocol of IEEE 802.16 BWA system, *IEEE Intl. Conf. on Communications, Circuits and Systems and West Sino Expositions* Vol. 1 (2002) pp. 435-439.

[60] H. Zhu, H. Wang, I. Chlamtac and Biao Chen, Bandwidth scalable source-channel coding for streaming video over wireless access networks, *1st Annual Wireless Networking Symposium* (WNCG'03) (Oct. 2003).

[61] X. Gao, G. Wu and T. Miki, QoS framework for mobile heterogenous networks, *IEEE ICC'03* Vol. 2 (May 2003) pp. 933-937.

[62] S.-B. Lee, G.-S. Ahn, X. Zhang and A.T. Campbell, INSIGNIA: an IP-based quality of service framework for mobile ad hoc networks, *Journal of Parallel and Distributed Computin* Vol. 60 No. 4 (2000) pp. 374-406.

[63] R. Chakravorty, I. Pratt, J. Crowcroft and M. D'Arienzo, Dynamic SLA-based QoS control for third generation wireless networks: the CADENUS extension, *IEEE ICC'03* Vol. 2 (May 2003) pp. 938-943.

[64] L. Georgiadis, P. Jacuet and B. Mans, Bandwidth reservation in multihop wireless networks: complexity and mechanisms, INRIA-RR-4876 (Jul. 2003).

[65] P. Gupta and P.R. Kumar, The capacity of wireless networks, *IEEE Trans. on Information Theory* Vol. 46 No.2 (Mar. 2000) pp. 388-404.

[66] J. Li, C. Blake, D.S.J. De Couto, H.I. Lee and R. Morris, Capacity of Ad Hoc wireless networks, *Mobile Computing and Networking* (2001) pp. 61-69.

[67] H. Schulzrinne, Operating system issues for continuous media, *Multimedia Systems* Vol. 4 (Oct. 1996) pp. 269-280.

Cross-Layer Architectures for Bandwidth Management in Wireless Networks

Klara Nahrstedt, Samarth H. Shah, and Kai Chen
Department of Computer Science
University of Illinois, Urbana, IL 61801
E-mail: {klara,shshah,kaichen}@cs.uiuc.edu

Contents

1 Introduction

1.1 Overview

In recent times, there has been a proliferation in the use of wireless communication all over the world. Several wireless networking technologies such as cellular networks, wireless local area networking (WLAN) and Bluetooth are commonly used in different environments. One problem spanning all types of wireless devices (e.g., cellular phones, PDAs, laptops, etc.) and networks, is the scarcity and variability of resources such as battery power, processor speed, memory, and wireless bandwidth. A major portion of research in wireless networking thus addresses the problem of how to effectively manage and optimally use these scarce and variable resources.

1.2 Focus

The topic of resource management in wireless networks is a very broad one. There are a number of research thrusts pertaining to different network technologies and different resources whose usage is to be optimized. Resource management is required for all types of wireless networks such as cellular networks, wireless LANs, mobile ad hoc networks and sensor networks. In this work, we focus on resource management schemes meant only for networks based on, or compatible with, the widely used and deployed IEEE 802.11 standard [1]. We consider Single-hop ad hoc ("pervasive" environments) [1], Access-point based [2, 3], as well as Multihop [4, 5, 6] network environments.

Furthermore, while there exists an extremely large body of work dealing with power control in wireless networks, and also work on processor-management, in this survey, the only resource whose management we address is *wireless bandwidth.*

Bandwidth, or channel capacity, in a wireless network is a resource shared by a number of wireless nodes in the vicinity of each other. The

[1]Some of the projects we refer to, e.g. [2, 3, 4], pre-date the standardization of the IEEE 802.11 protocol, but the approaches discussed are still applicable to it.

more time spent by a wireless node on the channel, the more data it can transmit and the greater its throughput. Thus, bandwidth management must be co-ordinated amongst the nodes that share the wireless channel. In the absence of bandwidth management, nodes sharing a wireless channel will attempt to completely capture it in order to maximize their throughput. Due to location dependent errors and due to interference, each node perceives a different channel capacity. The channel capacity perceived by each node also varies as a function of time. An application working over a wireless channel must be designed to dynamically adjust the quality of service (QoS) it provides its user(s). These factors make the problem of bandwidth management and allocation in wireless networks a challenging one.

1.3 Cross-Layer Design

In our study of bandwidth management architectures for different types of IEEE 802.11-based networks, we found one prominent common trait. All the major schemes that we focus on employ some kind of interaction, at each wireless network node, between the protocols at the different layers of the OSI protocol stack. This kind of interaction is not present in the protocol stack of a wireline network node.

There exist individual mechanisms at every layer of the protocol stack that assist in some way in controlling the usage of the wireless channel. Distributed fair scheduling [7, 8, 9, 10, 11, 12] and bandwidth monitoring and estimation [1, 5, 6] at the MAC layer, QoS routing at the network layer [13, 14, 4, 15], and rate adaptation [2, 6, 1, 5] at the transport and application layers are all examples of these individual mechanisms. The architectures we describe in detail in the following sections involve, with varying degree, *co-operation* between individual mechanisms at different layers of the protocol stack, usually through the sharing and exchange of state information between them. They hence deploy a cross-layer design for their services and protocols.

1.4 Outline

The rest of this survey is organized as follows. In the Section 2, we present a list of essential tasks in performing bandwidth management in wireless networks. In Section 3, we present brief summaries of the six cross-layer bandwidth management architectures we have chosen for the survey. We describe their salient characteristics, especially the internals of their respective essential tasks that we list in Section 2. Section 4 compares and contrasts the six architectures. We once again use the respective schemes' design of critical bandwidth management tasks as the criteria for their comparison. Section 5 lists some future directions and improvements that we might see in various tasks such as network-layer signaling, available bandwidth estimation, etc. Section 6 concludes this survey.

2 Essential Tasks in Bandwidth Management

In this section, we list the essential tasks in bandwidth management that are common to all the architectures we survey. This list of tasks forms a *methodology* for our study. In the subsequent sections, we examine in detail how the architectures go about accomplishing these tasks, compare and contrast these methods, and suggest improvements. Each of these tasks might occur at a different layer of the OSI protocol stack, and exchange information with the other tasks at other layers, thus bringing about cross-layer interaction.

1. *Available bandwidth estimation and monitoring:* Before distributing the channel capacity among the flows contending for it, a bandwidth management scheme must first have an estimate of the available bandwidth. This is likely to be less than the theoretical channel capacity (1, 2, 5.5 or 11 Mbps for IEEE 802.11 networks) due to the presence of location-dependent contention and physical errors. If these phenomena are rampant, the channel state is bad and available bandwidth is less. Most of the architectures described in this survey either assume or implement an available bandwidth estimation or channel quality monitoring module. Usually, the stimulus for rate adaptation arises from this module: when it detects a change in available bandwidth or channel quality, rate adaptation on the behalf of one or more flows must occur. Note that channel quality, and hence available bandwidth, are *location-dependent* features. Available bandwidth from the same source to different destinations can be different, depending on the location of the destinations.

2. *Signaling:* All the architectures we study employ a signaling protocol to carry reservation state over the network. The need for a signaling protocol in multi-hop bandwidth management is obvious, since

resources have to be reserved at each hop of the flow. However, a signaling protocol may be required even in single-hop wireless networks because the reservation state may be maintained centrally in such networks and a communication mechanism is required between the centralized reservation table and the mobile wireless hosts. The Utility-fair scheme [3] uses MAC-layer signaling while the BM [1] has its own signaling messages. The signaling protocol must be lightweight as it may have to be frequently employed due to bandwidth re-negotiations during the flow.

3. *Bandwidth allocation policy:* Once the available bandwidth is known and the requirements of the flows are known, the next step is to allot bandwidth amongst the competing flows. Some architectures simply adopt a first-come first-serve (FCFS) policy in which, if the bandwidth request is no more than the available bandwidth, the entire request is satisfied. Other architectures attempt to ensure fair sharing or price-based sharing of the available bandwidth. The bandwidth allocation policy comes into play at the time of bandwidth re-negotiation and for flow-dropping, in the absence of sufficient resources, also. Some schemes [4, 6] do not specify which flows must adapt their sending rate and which need not, when resources become scarce. In these schemes, fairness is hence impacted and providing even coarse rate or channel time guarantees becomes impossible.

4. *Reservation state:* Based on the bandwidth allocation policy, flows are admitted and allotted some portion of the available wireless channel bandwidth. Most of the architectures in this survey employ a *reservation-based*, as opposed to differentiated-services, approach to bandwidth management. All the architectures, except for SWAN [6] which keeps aggregate state, maintain per-flow reservation state. The state maintenance can be centralized or distributed. It is required at the time of admission control and rate allocation to know how the available bandwidth is being consumed by the existing flows. It is possible to keep per-flow reservation state only if the number of flows traversing a single wireless or ad hoc subnet is small. This is an implicit assumption of the schemes. In the Internet, it is impractical to maintain per-flow reservation state due to the presence of a large number of flows. SWAN eschews per-flow state not because storing it is costly (it is not costly, in fact, if the number of flows is small), but because

keeping the reservations coherent in a continuously changing environment is non-trivial. Other approaches also recognize this problem of keeping reservation state coherent. They work to keep the overhead of adaptation, which causes change in reservation state, minimal, while still attempting to optimize performance. SWAN's not maintaining per-flow state could result in decreased fairness and hence a degradation in quality of admitted flows. The guarantees provided by all the schemes are *channel-conditioned*, but those provided by SWAN are slightly weaker than those provided by other schemes, due to the absence of per-flow reservation state.

The architectures that maintain per-flow reservation state utilize *soft-state* reservations that need to be periodically refreshed. All reservation state must be maintained as soft-state in order to deal with drastic conditions (e.g., mobility, link breakage, device turn-off, etc.) that might cause a flow to terminate abnormally without even being able to properly release its reserved resources. Unless the resource reservations are time-out, the resources will be permanently lost to the network as a consequence of the abnormal termination of the flow.

5. *Rate control:* Once a flow has been allotted a certain portion of the available channel bandwidth, it must restrict its packet transmission rate to conform to its allotted share. This is essential to the co-operative sharing of the available bandwidth. Rate control may be done at various layers of the OSI protocol stack. Some schemes [5, 1] modify the application to adapt its quality and only generate packets at a rate no greater than the allotted bandwidth. Other possible rate-control and bandwidth share enforcement mechanisms are: employing a rate-based transport-protocol, priority-aware packet dropping, MAC-layer fair packet scheduling, and traffic shaping using token-bucket or leaky-bucket flow control.

6. *Application and flow adaptation:* A common feature of all the architectures is the presence of several adaptive levels at which the application is assumed to be able to operate. This is a vital requirement of the applications for them to be able to function uninterrupted in a continuously changing wireless environment. The stability of resource availability in wireline networks means that one-time admission control of a flow is often sufficient [1]. In a wireless network, the flows have to continuously adapt to the changing conditions. Even when

resource availability is low and flows are forced to lower their transmission rates, the application must still be able to provide the user *some* service, perhaps at a degraded quality.

The stimulus for flow adaptation may come from the available bandwidth monitor at the MAC layer, via network mechanisms such as ECN and in-band signals, or from the centralized reservation manager. The reaction to the stimulus could be a demand/release of bandwidth or a change in the packet transmission rate of the flow.

3 Description of Cross-Layer Architectures

This section contains brief descriptions of six cross-layer architectures that we have chosen to concentrate on for the purposes of our survey: TIMELY [2], Utility-fair [3], INSIGNIA [4], dRSVP [5], SWAN [6], and BM [1]. We present the architectures in the order in which they were published.

3.1 TIMELY

The TIMELY [2] adaptive resource management architecture considers a multi-cell wireless network that consists of a set of wireless access points connected to a wireline backbone, such as a cellular network or a group of inter-connected wireless LANs. The TIMELY architecture includes many components at different layers: 1) link layer scheduling, 2) resource reservation and advanced reservation, 3) resource adaptation, and 4) a priority-aware transport protocol. A key feature of the architecture is the coordination of resource adaptation among different layers, such that each layer may perform its task more intelligently and effectively.

TIMELY targets two problems in a multi-cell wireless network. First, wireless channel resources are highly dynamic. Second, resource reservation should be taken care of in advance before hand-off. To this end, TIMELY divides the bandwidth management task into three sub-tasks: 1) resource (or bandwidth) reservation for a flow; 2) advanced reservation before band-off; and 3) adaptation of the ongoing flows when resource variation occurs. Before starting, each flow specifies a range for each resource requested, such as low and high bandwidths, to the network, and the network performs admission control test over each link of the path in a centralized manner. The admission control test succeeds when the new flow can be admitted without violating the minimum rates of the ongoing flows. For best-effort traffic, the network has reserved certain portion of bandwidth for them and they do not need to go over the admission control procedure. The bandwidth of each link is assumed to be known by the network. In resource adaptation, TIMELY differentiates two types of flows. A *static* flow is a flow with both end-points residing in their respective cells over a threshold period of time, otherwise, it is a *mobile* flow. The goal of resource adaptation in TIMELY is to maximize the resource allocated to the static flows, and maintain only the minimum requested rate for the mobile flows (i.e, no adaptation), because the mobile flows are likely to be handed off to another cell. TIMELY's adaptation algorithm is based on the notion of network *revenue*. The network earns certain *admission fee* when it admits a flow, which is related to the flow's granted rate in such a way that, the marginal network revenue decreases for each unit of bandwidth beyond the flow's minimum requested rate. When the network changes a flow's rate, it has to pay an *adaptation credit* to the flow, or if the network drops the flow, it has to pay a larger *termination credit*. Because the network aims to maximize its long-term revenue, it has to make sure that the benefit it will receive by adapting or dropping a flow out-weights the credit it has to pay. As a result, the allocations for the ongoing flows are kept relatively stable. TIMELY relies on a conservative heuristic to select a set of flows to adapt, and decides how to adapt using a special weighted max-min algorithm [2]. In advanced reservation, TIMELY predicts the next hand-off cell(s) of a mobile host, and invokes the reservation procedure to reserve resource for the flow both in the predicted hand-off cell(s) and along the new path(s). As mentioned earlier, only the minimum rate is maintained for a hand-off flow. At the transport layer, each flow relies on a special adaptive protocol (called HPF) to interleave multiple packet sub-streams with different priorities in a single stream. When bandwidth reduction happens, only the most important sub-stream will be transmitted. HPF also relies on the resource adaptation signal from the network layer to react to dynamic resource changes.

3.2 Utility-Fair

The utility-fair adaptive service model [3] is a data link control model that accounts for the adaptation of wireless bandwidth variation, as well as application specific adaptation dynamics including adaptation time-scales and policies. It targets the data link layer in last-hop wireless systems, such as a wireless LAN, where an access point controls both uplink and downlink communications between the access point and a set of mobile devices.

In this service model, the bandwidth requirement of an application is represented by a *utility curve*, which maps the application's bandwidth into a utility (or satisfaction) level representing the application's perceived quality. Generally, an application requires a minimum level of bandwidth to operate, and is able to adapt to bandwidth variation beyond that. The minimum bandwidth requirement is served by the *sustained rate service* class of the data link layer. Beyond that, an application can choose from two adaptation classes. An *active adaptation service* class allows the application to control the application specific adaptation time-scale and policies. Adaptation time-scale is the smallest time that the application can successfully adapt to, and adaptation policy accounts for the granularity and amount of bandwidth variation that the application can tolerate during the adaptation process. The other adaptation class is the *passive adaptation service* which assumes that the application is able to adapt to any bandwidth variation at any time.

The data link control architecture employs a *centralized controller* at the access point, as well as a set of *distributed handlers* at each of the mobile devices. The central controller admits each application based on their sustained portion of bandwidth requirement, and allocates the rest of the channel bandwidth to the admitted flows according to their utility curves. It does *not* allocate the bandwidth equally to each application, since that will lead to different perceived quality levels for each application. Instead, the bandwidth allocation algorithm is based on the "utility-fair" criterion, which gives each application certain bandwidth such that all the applications perceive the *same* quality. The computation of utility-fair allocation is not complex when the utility curves are piece-wise linear [3]. After allocating the bandwidth for each application, the central controller advertises the allocation to those applications who require active adaptation service. The distributed handler, which acts as a proxy on behalf of the application, decides whether and how to accept the advertised bandwidth. For instance, the handler may implement a discrete adaptation policy to accept only certain level of bandwidth, or a hand-off adaptation policy to increase sending rate only after a hand-off. This allows great flexibility for the applications to program different adaptation dynamics. For those applications using the passive adaptation class, the central controller does not need to hear from them. After the bandwidth allocation process, the wireless access point enforces the allocated rates of different applications at the MAC layer.

3.3 INSIGNIA

INSIGNIA [4] is an end-to-end QoS framework that supports adaptive service in a multi-hop mobile ad hoc network. It targets adaptive real-time applications with two layers of media quality, i.e., a base layer and an enhanced layer. INSIGNIA provides fast, per-flow bandwidth reservation along a flow's path, and reacts quickly to route change by restoring the reservation status along the new path. It is designed to be light-weight and responsive to bandwidth variation and network topology changes.

The key part of INSIGNIA is an in-band signaling protocol coupled with a soft-state resource (bandwidth) management module at each router. In-band signaling means that the control information is carried with the data packets. Each data packet has a special IP header that contains the relevant INSIGNIA control information. When a flow needs to reserve bandwidth, it sets the *service mode* of the data packets to RES, which indicates that a reservation is being solicited, together with a MAX and a MIN bandwidth requests corresponding to its base and enhanced layers of traffic. At each router, a bandwidth management module decides whether the MAX or MIN requests can be granted, based on the current bandwidth of the wireless link and the set of admitted flows. If MAX or MIN bandwidth can be granted, the router sets an indicator in the special IP header, and keeps the flow's reservation status as a soft-state; if nothing can be granted, it changes the service mode of the packet from RES into BE, in order to notify the downstream routers that no reservation should be made for it. As a result, the bandwidth allocation policy at each router is first-come first-serve. After receiving the data packet, the receiver learns the reservation status from the IP header, and may notify the sender to scale up or scale down the media quality. The sender then controls its sending rate according to the quality permitted. After route change, INSIGNIA can quickly restore the reservation status along the new path when the data packets travel through it. Therefore, it is well suited for a dynamic, mobile, and variable bandwidth ad hoc network.

INSIGNIA's bandwidth management function is done independently at each router. It assumes that the router is able to obtain available bandwidth information from the link layer for admission control purpose, and that the link layer is able to provide QoS-driven access to the shared media for the adaptive and best-effort packets.

3.4 dRSVP

The dRSVP [5] protocol is a scheme to support per-flow dynamic QoS in a variable bandwidth network, such as a mobile network with wireless links. Although dRSVP is designed within the context of QoS, its core component is a bandwidth allocation and reservation algorithm that provides dynamic bandwidth guarantees for the passing flows at each router, therefore, it can also be considered as a bandwidth management scheme in wireless networks. dRSVP is a fully distributed scheme that can be applied to a multi-hop mobile ad hoc network, as well as a scaled down single-hop ad hoc network.

Similar to RSVP, dRSVP provides end-to-end bandwidth guarantees for a flow by reserving bandwidth along the flow's path. Unlike RSVP, a flow's bandwidth request in dRSVP is over a *range*, i.e., between a low and a high value, rather than using a single value, and the reserved bandwidth for the flow is at some point within the range. By using a range, the likelihood that a flow's bandwidth request can be maintained is increased, even when the wireless link's available bandwidth changes. In dRSVP, each router is assumed to be able to obtain the current available bandwidth from the link layer, and then allocates the bandwidth to the current flows, which is kept in soft-state, using a special bandwidth allocation algorithm. The bandwidth allocation algorithm divides up available bandwidth among the flows with consideration of the desired range for each flow, as well as their bottleneck allocations at other routers. Each router in dRSVP allocates and reserves bandwidth independently, and a flow's bottleneck reservation is carried to its upstream and downstream routers via a signaling protocol similar to RSVP. On start-up of a flow, it has to go through the admission control test which attempts to reserve bandwidth along the path. Once succeeded, the flow can start to send out packets according to the actual allocation it received, i.e., the sender assumes the responsibility of rate control. The sender also has to refresh its allocation state periodically with the routers, similar to the soft-state approach taken by RSVP. In dRSVP, each router is an independent component without any coordination between them, therefore, a sub-net link layer "bandwidth manager" may be needed to help coordinate their access to the shared wireless media.

3.5 SWAN

SWAN [6] is a distributed and stateless network model to support the delivery of real-time traffic over a multi-hop ad hoc network. It manages two different types of traffic: real-time and best-effort. Real-time traffic is guaranteed minimum delay at each node. SWAN does not maintain per-flow state at each node, and it requires only a best-effort 802.11 MAC layer without any QoS capability. Therefore, it is a simple and scalable proposal suitable for a large-scale ad hoc network.

SWAN has two important components: a *rate controller* at each node to restrict best-effort traffic, and a *sender-based* admission control procedure to admit a real-time flow. The rate controller is based on the observation that, MAC layer packet *delay* is a fairly good hint of the medium access contention. In order to support real-time traffic with minimum delay, the sending rate of *every* node in a neighborhood area should be throttled to avoid excessive contention. To this end, the rate controller obtains the packet delay measurement from the MAC layer, and throttles the best-effort traffic rate using a well-known AIMD (Additive Increase Multiplicative Decrease) algorithm normally found in TCP. This rate control algorithm is essentially a distributed bandwidth management technique where the nodes in a neighborhood area *coordinate* with each other to back-off their rates if necessary. When a real-time flow starts, it sends a *probing* packet to the destination, with a "bottleneck bandwidth" field in the packet. Each intermediate router decides an admissible rate for the flow by subtracting the current aggregate real-time traffic rate from the current available bandwidth, and updates the probing packet's "bottleneck bandwidth" field accordingly. On reception of the probing packet, the destination node extracts the bandwidth information and returns it to the sender. The sender can now start the real-time session with a rate within the admissible limit, hence rate control is the responsibility of each application.

When the current admitted real-time traffic can no longer be supported at a SWAN node, for instance, due to re-routing of some real-time flows or variation of wireless bandwidth, the node will notify the real-time flows by the way of marking their packets' explicit congestion notification (ECN) fields. On reception of such ECN marking, a real-time flow should stop sending packets and re-establish its admissible rate by probing the path again. To avoid synchronization effect, SWAN recommends randomly marking a subset of the real-time flows to alleviate congestion. As a result, no fairness can be guaranteed between flows. This is because SWAN keeps no flow state information and therefore has no way of knowing exactly which real-time flow to mark. By design, SWAN is a simple, distributed, and stateless architecture to support real-time traffic with coordinated bandwidth management techniques.

3.6 Bandwidth Management (BM)

The bandwidth management scheme proposed in [1] targets a wireless network where all the mobile nodes are within each other's transmission range, and packet transmission between them takes place in a *peer-to-peer* manner. This *single-hop ad hoc network* is a representation of many practical networking setups commonly found in smart-rooms, in-home networking, and hot-spot networks. The ad hoc mode in the IEEE 802.11 MAC protocol's DCF (Distributed Coordination Function) supports such a network.

A major challenge to support multimedia streaming over this network is that, wireless bandwidth cannot be guaranteed for each flow due to their unco-operative contention in accessing the shared medium. A flow's data rate may be highly fluctuating, and often cannot meet the minimum rate required in streaming the media. To this end, the bandwidth management (BM) scheme [1] proposes to include admission control and dynamic bandwidth management functions into the network. The key part of this scheme is a centralized bandwidth manager (BM) which runs at a node in the network rich in CPU and memory resources. The BM node can be a dedicated server which advertises its service to the adjacent mobile nodes as part of service discovery, or it can be dynamically elected using a simple leader election algorithm. Before starting a flow, the flow sends a message to the BM to request for certain *channel time proportion*, which is defined as the fraction of unit time for which the flow can have the channel to itself for transmission. Essentially, the resource to be shared in this network is the channel time, and the total channel time in the network is 100%. The reason a flow requests for channel time, instead of bandwidth, is that the *perceived bandwidth* of a source-destination pair may be different throughout the network due to location-dependent channel conditions, and such knowledge is only available at the source (or destination), not the BM. To this end, each node runs a *total bandwidth estimator* at its link layer, in order to measure the actual throughput that it can achieve between itself and the destination. The measured throughput, therefore, has taken into account the effect of channel contention as well as physical layer conditions. Using the estimated channel bandwidth, the flow is able to *map* its bandwidth requests, represented by a range of high and low bandwidths, into their corresponding channel time proportion requests.

The BM performs admission control for a new flow after receiving its requests. The admission control test successes when the new flow and the admitted flows can be granted at least their minimum channel time proportions. Beyond their minimum requests, the remaining channel time is allocated to them in a max-min basis. In other words, BM's bandwidth allocation is max-min fair with minimum rate guarantee. The admitted flows then control their transmission rates according to their granted channel time, so that co-operation between flows is achieved and the channel is fairly shared. The BM also provides dynamic bandwidth management for the set of admitted flows during their session life-time, for instance, when a flow perceives a different channel bandwidth at the link-layer bandwidth measurement, or when it needs to re-adjust its minimum or maximum requests due to change of traffic pattern such as in VBR flows. In this case, the flow sends a re-negotiation message to the BM to request a change. The BM then re-calculates the bandwidth allocation and distributes it to all the flows. Note that a flow may be cut out if the minimum requirement can no longer be supported. The resource reservation state is maintained at the BM as soft-state with a time-out period, but a flow may still send a tear-down message to explicitly release its allocated resource. By sharing the channel co-operatively, the BM scheme is able to better deliver multimedia traffic, using the commercially available best-effort IEEE 802.11 hardware without any QoS capability [1].

4 Comparison

In this section, we compare and contrast the various cross-layer architectures described in the previous section, with respect to how they perform their critical tasks. Figure 1 puts the comparison in a nutshell.

The second row of the table pertains to the adaptivity of the flow. Most of the architectures assume highly flexible flows that are adaptive over a continuous range of bandwidths. INSIGNIA is an exception: it assumes a flow can have only three bandwidth levels of operation. Mobile flows (flows in which at least one end-point is mobile) in TIMELY are pinned to their minimum bandwidth request.

The third row of the table pertains to the fairness in bandwidth allocation. Most schemes provide some notion of fairness or price-weighted allocation. However, INSIGNIA and SWAN are exceptions. Admission control is first-come first-serve (FCFS) in these schemes: flows are admitted while resources last. Which flows must adapt in response to available bandwidth variations, and how much they should adapt, is not deterministic.

The fourth row indicates how the decision to allocate bandwidth is made.

In the case of TIMELY, while the bandwidth allocation is decided by the weighted max-min servers, the reservation state is communicated via an RSVP-like signaling algorithm and maintained distributedly. The weighted max-min servers ensure co-ordination that is lacking in fully distributed allocation decisions. (See Section 5.) In fully distributed schemes, per-flow state is maintained at multiple nodes in the network. SWAN is an exception; it maintains aggregate flow state at each node in the network. In centralized schemes, per-flow state is maintained centrally. In both types of schemes, a signaling protocol is needed to communicate state information.

As evident from row 6 of the table, even some of the reservation-based schemes such as TIMELY and INSIGNIA provide differentiated services based on packet priorities. This is because reservations in a wireless environment are not hard, but are subject to dynamic variations in topology and channel characteristics. Packets are marked as low or high priority and when available bandwidth becomes unexpectedly scarce and rate control needs to be performed, preferably the lower priority packets are dropped.

5 Future Directions

In this section, we point out some of the weaknesses in the existing bandwidth management architectures and present some possible future directions of research in this area.

5.1 Improving Accuracy of MAC-layer Available Bandwidth Estimation

Most of the architectures studied in this survey involve a MAC-layer monitor to estimate the available bandwidth on each wireless interface. In actual fact, the available bandwidth is different over each wireless link, out of the same wireless interface. As illustrated in Figure 2, the available wireless bandwidth over node A's wireless interface to node C, 1.8 Mbps, is different from that over the same wireless interface but to node B, 1.4 Mbps. (Assume a 2 Mbps wireless channel.) This is because the level of contention and fading/interference experienced by packet transmissions is different over the two links. If the errors resulting from these phenomena greatly affect a wireless link, then a greater length of time must be expended towards sending a single IEEE 802.11 MAC frame, over the link, so the effective available bandwidth of the link (reciprocal of the time expended) is smaller.

Since the contention, fading and interference effects are different in different neighborhoods, different wireless links have different available bandwidths.

The MAC-layer monitors that estimate the available bandwidth out of each wireless interface usually *average* the available bandwidths to different neighbors to obtain a single value of available bandwidth. Thus, the available bandwidth out of node A's wireless interface is estimated to be 1.6 Mbps. This single value of available bandwidth may not be precise enough for accurate admission control.

In order to deal with this problem of inaccurate available bandwidth estimation in the face of location-dependent errors, the BM scheme [1], introduces per-*neighbor* available bandwidth estimation. It then uses the concept of *channel time proportion* (CTP) for more accurate admission control. In Figure 2, there are two flows requesting admission: the flow from A to B requests 1.2 Mbps and the flow from A to C requests 0.4 Mbps. Both flows are admissible if the single value of available bandwidth, 1.6 Mbps, is used, since their sum does not exceed the estimated available bandwidth out of A's interface.

If CTP is used, however, both flows cannot be simultaneously admitted. If the maximum available bandwidth from node A to node B is 1.4 Mbps, and a flow requests 1.2 Mbps, then it is, in effect, requesting $\frac{1.2}{1.4} = 85.7\%$ of unit time out of node A's wireless interface. The concept of channel time proportion may be better understood using frames per second as the unit of bandwidth, rather than bits per second. If 10 frames of some size can be sent from A to B and a flow requires a throughput of 8 frames of the same size per second, then in effect, it needs to spend eight-tenths of a second on A's wireless interface transmitting to B. In [1], frames per second throughput is normalized over different frame sizes to obtain a bits per second throughput over each wireless link.

The flow from A to B requests $\frac{1.2}{1.4} = 85.7\%$ of the channel time and the flow from A to C requests $\frac{0.4}{1.8} = 22.2\%$ of the channel time. Since the sum exceeds 100% of unit time out of node A's wireless interface, both flows are not simultaneously admissible.

Admission on the basis of a single available bandwidth value (1.6 Mbps in the example) can be considered a false admission [6]. Due to the presence of dynamic regulation of real time flows in SWAN to deal with false admissions, this architecture can mitigate the negative effects on performance, but fairness is nevertheless affected. Like in the BM scheme, future MAC-layer available bandwidth monitors will need to consider at flow admission time, that the available bandwidth out of the same interface is different for different neighbors, due to location-dependent errors. An increase or decrease in location-dependent errors may also trigger dynamic rate adaptation during the course of operation of a particular flow on an interface, after it has already been admitted.

Taking into consideration location-dependent errors still does not guarantee perfectly accurate admission control, although it does eliminate some false admissions. It is difficult to predict beforehand, the effect of admitting a new flow to a channel, on MAC transmission delays and effective throughput of existing flows. One of the most challenging problems is to determine a priori precisely how much of a deterioration will be produced in existing flows' quality by the admission of, say, a 200 kbps flow. Dynamic adaptation of the flow rate can help in mitigating this problem. If the admission of a new 200 kbps flow causes existing flows' available bandwidth to be adversely affected, they can perform dynamic rate adaptation in order to compensate.

5.2 QoS using IEEE 802.11e

The IEEE 802.11e draft specification is an upcoming effort to support QoS in wireless LANs [19]. It has two modes: the Enhanced Distributed Co-ordination Function (EDCF) mode and the Hybrid Co-ordination Function (HCF) mode. The EDCF mode is an improved version of the DCF mode of base IEEE 802.11, to provide QoS via differentiated service. QoS support is realized with the introduction of Traffic Categories (TCs). Upon finding the channel idle or after a collision occurs, higher priority TCs are likely to wait a shorter interval before attempting to transmit, while lower priority TCs wait longer. Thus higher priority TCs are likely to get first access to the channel.

The HCF mode is an improvement of the legacy IEEE 802.11 PCF mode. It supports both a contention-free period and a contention-period, as in the case of PCF. During the contention-free period channel access is solely through polling from the Hybrid Co-ordinator (HC). During the contention-period, channel access can be via listen-before-talk EDCF distributed co-ordination *as well as* through polling, when the HC's poll pulse wins the contention. The transition between the two periods is signaled using a beacon. A special time interval is set aside for mobile hosts to send the HC their resource requests, which the HC uses in determining polling frequency and length of transmit opportunities for the respective hosts, so that their requests can be satisfied. This solves the problem of unknown transmission times of polled stations in legacy IEEE 802.11 PCF.

The IEEE 802.11e EDCF could prove useful to the bandwidth management architectures which need packet classification and differentiation [2, 4, 6]. It remains to be seen whether vendors implement the IEEE 802.11e HCF mode, however. The legacy IEEE 802.11 PCF mode has problems with beacon delays that hinder its implementation, and this problem persists in IEEE 802.11e HCF also. If the IEEE 802.11e HCF is implemented, reservation-based bandwidth management schemes will be benefited because a protocol to make resource requests and a polling-based scheme for enforcement of allocated channel fractions will become available. However, a channel allocation policy that takes into account requests and computes allocated channel fractions will still need to be plugged in at the HC. A simple queue backlog based fair policy for this purpose is described in [20]. Other schemes, such as the max-min fair with minimum guarantees scheme of [1] could also be used. A mechanism will also be required to track variations in channel quality so that requests can be modified when this happens.

5.3 Network-Layer Improvements

One interesting feature of the bandwidth management schemes for multi-hop mobile ad hoc networks (MANETs) [5, 4, 6] is the *decoupling* of routing and resource management-related functions such as signaling, resource monitoring and resource reservation. In order to discover a route that supports the bandwidth requirements of the flow, the protocols rely on an underlying routing protocol, e.g. DSR [16], TORA [17] or AODV [18], to first discover all paths to the destination. They then check whether the resources available on each of these paths can support the bandwidth requirements of the flow, and determine the most optimal of the candidate routes. On the other hand, QoS-routing schemes [13, 14, 15] take network resource availability into account in the routing algorithm. They attempt to simultaneously optimize the resource-richness and path length of the route. While the latter approach may outperform the former, many researchers advocate the former because of its simplicity and flexibility: different resource management functions can be plugged into different routing protocols, rather than having an immutable resource management model built into the routing algorithm itself. The result of using such a simplified network-layer resource management scheme is that the routing protocol used is a simplistic fewest-hop shortest-path based one such as TORA, DSR or AODV. Future bandwidth management architectures for wireless networks might see the employment, at the network layer, of sophisticated congestion-aware routing algorithms and pre-emption of flows from routes by other flows. If a number of paths are available from source to destination, congestion-aware routing algorithms pick one with large available bandwidth. If another connection must use this, and only this, chosen route, the connection with multiple possible routes can be pre-empted from its chosen route onto another possible route. These more intelligent routing schemes will result in minimization of blocking factor and maximization of overall network throughput and lifetime.

5.4 Co-ordinated Bandwidth Allocation

In the fully distributed approaches described in Section 3, INSIGNIA, dRSVP, and SWAN, admission control is inaccurate due to the presence of *hidden flows* Figure 3 illustrates this problem. Node B is within the transmission range (solid circumference) of node A and there exists a 1 Mbps flow from A to B. Assume maximum channel capacity of 2 Mbps. Also assume that nodes perform unutilized channel capacity monitoring, prior to flow admission, as in INSIGNIA or SWAN. Node A is outside the interference range (dashed circumference) of node C, but node B is within C's interference range. Assume Now, if a 1.5 Mbps flow is admitted at node C, its transmissions from C will collide with node A's transmissions to node B. Hence, the throughput of the flow from node A to node B will be severely degraded.

There exists thus a problem of *co-ordination* between the flows in their admission control. The solution to this is that node C should be aware of the flow from node A to B and only admit flows requesting upto 1 Mbps channel bandwidth. Another possible solution is to have a subnet bandwidth manager, as suggested in [5], which can possess global knowledge of all flows and nodes in a neighborhood and co-ordinate the admissions at the respective nodes. The weighted max-min servers in TIMELY perform the co-ordination function. Obviously, both of these solutions violate the distributed and autonomous nature of the admission control in the respective schemes.

The regulation of the real-time flows in SWAN, to respond to mobility and false admission, also mitigates the problem of hidden flows. Admission in spite of the existence of a hidden flow can also be considered a false admission, although the term false admission is defined slightly differently in [6]. In the example of Figure 3, both the 1 Mbps and 1.5 Mbps flows will have to be regulated after they have both been admitted to deal with the problem of false admission. In [5], it is mentioned that a subnet bandwidth manager is required for co-ordination, but the details of its design are not provided. In the future, the design of an admission control scheme for multi-hop wireless networks will need to consider the problem of co-ordination between flows in a neighborhood that are hidden from each other to ensure accurate admissions and rate allocation.

6 Conclusion

Bandwidth is a scarce and variable resource in all types of wireless networks. Tracking the available bandwidth and having all applications using the wireless channel adapt their quality of service (QoS) to the available bandwidth is a complex procedure. It requires several adjustments at each layer of the OSI protocol stack. Moreover, the procedures at each layer of the stack must co-operatively interact with those at the other to ensure maximum satisfaction for users.

In this survey, we picked six approaches that each adopt such a *cross-layer* approach towards monitoring and adapting to the variable channel capacity. We briefly described the salient features of the six architectures and compared and contrasted the methods they employ at each layer of the protocol stack. We also give our views on future directions in cross-layer design for bandwidth management in wireless networks.

References

[1] S. Shah, K. Chen, and K. Nahrstedt, Dynamic bandwidth management in single-hop ad hoc wireless networks, *IEEE PerCom 2003* (Fort Worth, TX, Mar. 2003).

[2] V. Bharghavan, K. Lee, S. Lu, S. Ha, J. Li, and D. Dwyer, The TIMELY adaptive resource management architecture, *IEEE Personal Communication Magazine* Vol. 5 No. 4 (Aug. 1999).

[3] G. Bianchi, A. Campbell, and R. Liao, On utility-fair adaptive services in wireless networks, *IEEE/IFIP IWQoS 1998* (Napa, CA, May 1998).

[4] S. Lee, G. Ahn, X. Zhang, and A. Campbell, INSIGNIA: An IP-based quality of service framework for mobile ad hoc networks, *Journal of Parallel and Distributed Computing (Academic Press), Special Issue on Wireless and Mobile Computing and Communications* Vol. 60 No. 4 (Apr. 2000).

[5] M. Mirhakkak, N. Shult, and D. Thomson, Dynamic bandwidth management and adaptive applications for a variable bandwidth wireless environment, *IEEE JSAC* Vol. 19 No. 10 (Oct. 2001).

[6] G. Ahn, A. Campbell, A. Veres, and L. Sun, SWAN: service differentiation in stateless wireless ad hoc networks, *IEEE InfoCom 2002* (New York, NY, Jun. 2002).

[7] B. Bensaou, Y. Wang, and C. Ko, Fair medium access in 802.11 based wireless ad hoc networks, *IEEE MobiHoc 2000* (Boston, MA, Aug. 2000).

[8] D. Eckhardt and P. Steenkiste, Effort-Limited Fair (ELF) scheduling for wireless networks, *IEEE InfoCom 2000* (Tel Aviv, Israel, Mar. 2000).

[9] V. Kanodia, C. Li, A. Sabharwal, B. Sadeghi, and E. Knightly, Distributed multi-hop scheduling and medium access with delay and throughput constraints, *ACM MobiCom 2001* (Rome, Italy, Jul. 2001).

[10] H. Luo, S. Lu, and V. Bharghavan, A new model for packet scheduling in multihop wireless networks, *IEEE MobiCom 2000* (Boston, MA, Aug. 2000).

[11] H. Luo, P. Medvedev, J. Cheng, and S. Lu, A self-coordinating approach to distributed fair queuing in ad hoc wireless networks, *IEEE Infocom 2001* (Anchorage, AK, Apr. 2001).

[12] N. Vaidya, P. Bahl, and S. Gupta, Distributed fair scheduling in a wireless LAN, *ACM MobiCom 2000* (Boston, MA, Aug. 2000).

[13] R. Sivakumar, P. Sinha, and V. Bharghavan, CEDAR: a core-extraction distributed ad hoc routing algorithm, *IEEE JSAC* Vol. 17 No. 8 (Aug. 1999).

[14] S. Chen and K. Nahrstedt, Distributed quality-of-service routing in ad-hoc networks, *IEEE JSAC* Vol. 17 No. 8 (Aug. 1999).

[15] S. Shah and K. Nahrstedt, Predictive location-based QoS routing in mobile ad hoc networks, *IEEE ICC 2002* (New York, NY, Apr. 2002).

[16] D. Johnson and D. Maltz, Dynamic source routing in ad hoc wireless networks, *Mobile Computing* (Chapter 5, , Kluwer Academic Publishers, 1996) pp. 153-181.

[17] V. Park and S. Corson, A highly adaptive distributed routing algorithm for mobile wireless networks, *IEEE InfoCom 1997* (Kobe, Japan, Apr. 1997).

[18] C. Perkins and E. Royer Ad-hoc on-demand distance vector routing, *IEEE WMCSA 1999* (New Orleans, LA, Feb. 1999).

[19] S. Mangold, S. Choi, P. May, O. Klein, G. Hiertz, and L. Stibor, IEEE 802.11e wireless LAN for quality of service, *European Wireless* (Florence, Italy, Feb. 2002).

[20] P. Garg, R. Doshi, R. Greene, M. Baker, M. Malek, and X. Cheng, Achieving higher throughput and QoS in 802.11 wireless LANs, *IEEE IPCCC* (Phoenix, AZ, Apr. 2003).

Resource Management and Connection Admission Control in Wireless Networks

Tuna Tugcu
Georgia Tech Savannah
School of Electrical and Computer Engineering
Georgia Institute of Technology
Savannah, GA 31407
E-mail: `Tuna.Tugcu@gtsav.gatech.edu`

Contents

1 Introduction

Resource management is the key factor that determines the performance of wireless networks since the radio spectrum is the scarce resource. Coexistence of multiple wireless networks with different technologies limits the radio spectrum available for new wireless networks. Besides the shortage of radio resources, mobility of the users and the physical properties of radio channels make resource management in wireless networks more challenging than wireline networks. On the other hand, the success of wireless telephone networks has increased both the number of mobile users and their expectations from wireless networks to a level comparable to wireline networks. Efficient resource management elevates system throughput to support the increasing wireless population while satisfying user expectations.

Connection Admission Control (CAC) process follows the paging process and helps resource management by deciding which connection requests are admitted into the network. Since each connection consumes resources, admission of a request has considerable effects on other connections. Therefore, while admitting a request, CAC must ensure that enough resources can still be allocated for the existing connections.

CAC schemes in wireless networks differ from their counterparts in wireline networks in various aspects. Since the radio resources constitute the bottleneck, CAC schemes in wireless networks generally focus more on the management of radio resources rather than the resources in the wireline backbone. Admission of a connection in one cell affects other connections both in that cell and the surrounding cells. Radio transmission due to one connection causes interference on other connections. Such interference is a limiting factor on network capacity, especially for interference-limited networks like CDMA. The most significant difference between CAC schemes for wireless and wireline networks is, however, support for mobility. A wireless network must be able to convey the connection of an *active* user (i.e., a user with an ongoing connection) from one cell to the other when the user moves between cells. To avoid service interruption during such handoff operations, the network needs to employ prioritization or reservation schemes to keep connection dropping and blocking rates at reasonable levels. Once a connection request is admitted, the network must do its best to keep the connection alive.

CAC also plays a significant role in providing *Quality of Service (QoS)*. Especially, future telecommunication networks like 3G and *Next Generation Wireless Systems (NGWS)* aim providing integrated services such as voice, high bandwidth data, and multimedia with QoS support. CAC must consider QoS requirements during connection setup and handoffs. In the case of NGWS, QoS requirements must be translated between the subsystems for vertical handoffs. CAC needs to work hand in hand with resource management to achieve this task.

In this chapter, we study how resource management and connection admission are performed in wireless networks. Following the discussion of issues common to both wireline and wireless networks, we explain how CAC is performed in wireless networks. We handle the topics of radio spectrum assignment and connection admission control together. Though not among the topics of this chapter, we briefly introduce QoS issues due to its relationship with CAC. This discussion is followed by admission control in NGWS. We finally conclude by summarizing the important points in admission control.

2 Resource Management and Connection Admission Control

The performance of a wireless network is bounded by the efficiency of resource management. The resources are consumed by the connections that are admitted into the network by the CAC scheme. To better understand how resource management is performed, we need to study radio spectrum assignment to the cells, admission of new and handoff connections into the network, and QoS provisioning together.

2.1 Managing the Radio Spectrum

Wireless networks are composed of two parts: the radio network and the wireline backbone. The capacity in the wireline backbone can be increased if necessary. Therefore, the backbone does not constitute the bottleneck. The capacity of the radio network, on the other hand, is dictated by the radio spectrum allocated to the wireless network. Coexistence of multiple generations of wireless networks limits the frequency band available for new wireless networks, making the radio spectrum the scarcest resource in the network. It is the task of CAC and resource management schemes to administer this scarce resource efficiently to maximize user satisfaction and network throughput.

Radio resource assignment schemes for FDMA-based networks can be categorized under three groups [1]:

- *Fixed Channel Allocation (FCA):* In FCA schemes, the service provider partitions the available radio spectrum into several frequency bands. Then, the frequency bands are assigned to cells according to a reuse plan. Since the frequency assignment is fixed, the network cannot adapt to fluctuations in connection traffic. Though it is possible to develop several frequency assignment plans for different traffic scenarios, the network is still vulnerable to unexpected traffic loads. Channel borrowing is another approach to cope with fluctuations, but adjacent channel interference limits the use of this approach. Due to ease of implementation, FCA schemes are widely used.

- *Dynamic Channel Allocation (DCA):* In DCA schemes, all available radio channels are gathered in a central pool, and the channels are allocated to cells on a need basis. Although DCA schemes adapt to changing traffic patterns better than FCA schemes, their complexity and centralized mechanism are the major drawbacks. Furthermore, under high traffic load conditions, the performance of DCA schemes deteriorate.

- *Hybrid Channel Allocation (HCA):* HCA schemes combine FCA and DCA schemes to overcome the problems in both approaches. The set of channels is split into fixed and dynamic subsets. The channels in the fixed subset are preferred as much as possible. The dynamic subset is used to adapt to fluctuations.

As opposed to FDMA-based networks, frequency planning is not a problem in CDMA networks. However, inner- and outer-cell interference are the key issues in admission control [2, 3, 4, 5]. Admitting a connection request in one cell has adverse effects on the capacity of that and the surrounding cells. An interference-based approach should be taken for admission control and possible reservations (discussed in Section 2.2) in CDMA networks.

2.2 Tradeoff Between Connection Blocking and Connection Dropping Rates

CAC for wireline networks varies from CAC for wireless networks in several aspects. In wireline networks, a conservative approach is taken in the admission of new connections to ensure that the ongoing connections get enough resources while reducing the *connection blocking rate*, the ratio of rejected new connection requests. However, the situation is more complex in the case of wireless networks due to the mobility of the users. In addition to connection blocking rate, the CAC scheme in wireless networks must also consider the *connection dropping rate* (also called *forced connection termination rate*), the ratio of ongoing connections that are forcefully terminated due to insufficient radio resources.

Connection blocking and dropping rates are the most tangible criteria from the perspective of the mobile user. In the literature, it is widely accepted that connection dropping is more annoying than connection blocking [4, 6, 7] and more important in determining user satisfaction. Since it is possible that a user in one cell may visit any one of the neighbor cells, each ongoing connection exerts pressure on the surrounding cells. In order to reduce connection dropping rate, CAC must consider the pressure from surrounding cells while making the admission decision for a new connection request. In other words, during the admission decision for a new connection, CAC must put aside some resources for possible handoff connections from the surrounding cells. Therefore, the CAC scheme running in a cell cannot work independent of the surrounding cells.

Prioritization and queuing are two straight forward solutions for lowering the connection dropping rate [8]. In prioritization, handoff connections are given higher priority compared to new connections. However, this approach results in the starvation of new connections in cells intersected by the highways. Since the number of handoff requests to these cells is will be very high, stationary users in such cells will not be able to get enough radio resources. Queuing, on the other hand, is not practical since the time required to complete connection establishment or handoff is very short compared to connection duration.

A more effective way to combat connection dropping rate is dedicating some resources, called *guard channels,* specifically to handoff connections. Since the guard channels are dedicated handoff connections, new connection requests will not be granted if all channels except the guard channels are busy. Therefore, determining the optimum number of guard channels is a crucial issue for the network performance. If the number of guard channels is too high, many new connection attempts will fail although there are free channels. On the contrary, if the number of guard channels is too low, many handoff events will fail resulting in a high forced connection termination rate. Thus, there is a tradeoff between connection blocking and connection dropping. The decision for the optimum number of guard channels is both time and space dependent. The number of guard channels depends on the location of the cell, random events like traffic congestion, accidents, or festivals. Therefore, assigning a predetermined number of guard channels to each cell increases dropping rate in some cells and blocking rate in others.

The number of guard channels can be adjusted dynamically to cope with varying connection traffic using reservations. The basic idea is to estimate the cells that are on the future path of the user and reserve resources on those cells. Since the user's path is independent of the planning of the spectral resources, it is not possible to exactly know the set of cells he will visit. However, a good estimation can be made by considering the fact that users move toward a destination rather than making random moves. Therefore, the path that a user follows can be modeled as the concatenation of multiple line segments. A reservation area may be formed by considering this estimated path. The performance of the network depends on how close the estimation is to the actual path of the user. Though not guaranteed, it will be very likely that the user will remain in this reservation area in the near future. If the reservation area is very large, it is very likely that the actual path of the user is covered by the reservation area and the connection survives while new connections in the cells in the reservation area suffer. On the other hand, if the reservation area is too small to cover the actual path, the connection may be dropped when the user leaves the reservation area.

A reservation area that covers the user's path for the lifetime of the connection cannot be constructed during connection setup. There are several reasons:

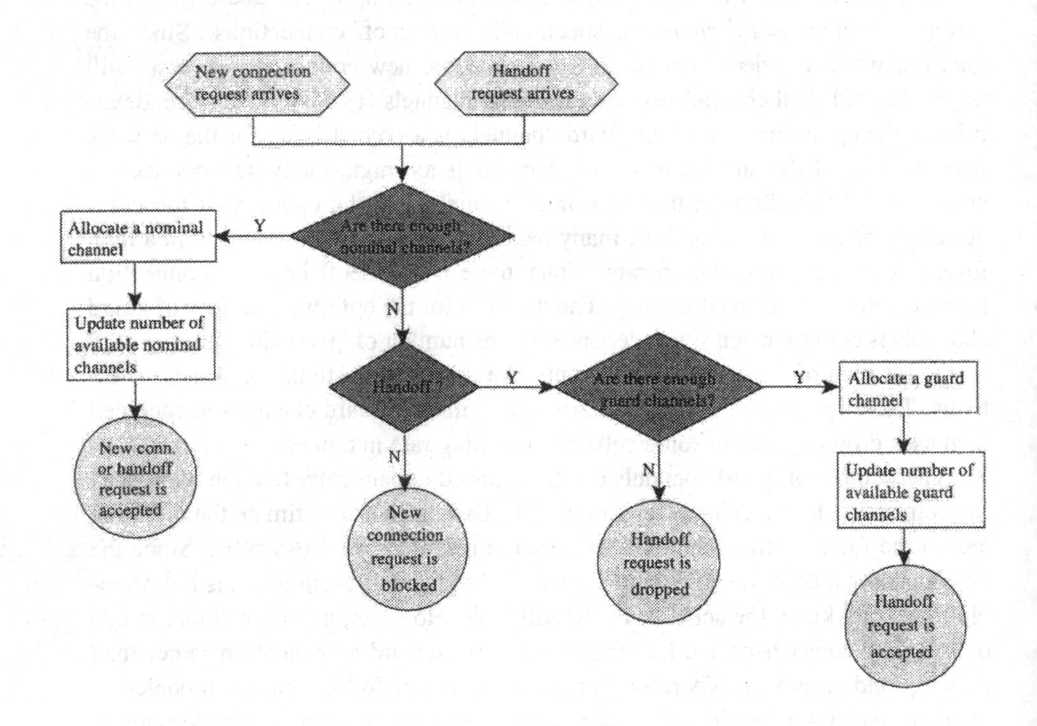

Figure 1: Generic CAC with Reservations

- The lifetime of the connection is not known in most cases.
- The reservation area will be very large.
- Uncertainty will be higher.
- Reservations in the cells that are likely to be visited close to the end of the connection will be reserved early unnecessarily.

Due to these reasons, the reservation area must be constructed to cover a short period and be advanced gradually. This will make the reservation area smaller and more accurate. Thus, less resources will be locked for handoff connections, resulting in lower connection blocking rate. A generic algorithm for CAC with reservations is given in Figure 1.

In the literature, most of the previous work on guard channels assigns a fixed number of guard channels to each cell [1, 2, 3, 9]. As stated above, fixed number of guard channels is vulnerable to fluctuations in new connection generation and handoff rates. Methods for variable number of guard channels have been proposed in [4, 10, 11, 12, 13, 14]. The work in [10] and [11] is for TDMA/FDMA networks. The air interface is not specified in [12]. Only the work in [4, 13, 14] propose variable number of guard channels for CDMA networks. In [15], channels are borrowed from stationary calls participating in handoffs in order to allocate them to handoff requests by moving mobiles.

2.3 Resource Management in UMTS

In the case of *Universal Mobile Telecommunications System (UMTS)*, resource management functions are mainly performed by the *Radio Resource Control (RRC)*. RRC manages the signaling between the mobile terminals and *UMTS Terrestrial Radio Access Network (UTRAN)* at the network layer. The basic function of RRC is to manage connection establishment (and re-establishment if necessary), termination, and maintenance. RRC performs admission control for (possibly multiple) connection to the mobile terminal by selecting the parameters that describe the radio link. For each connection admitted, RRC allocates the radio resources required by the QoS specifications. The connection is terminated either by an explicit request from the *Non-Access Stratum (NAS)* or due to a connection failure. RRC is also capable of re-establishing the connection if it is lost. If a connection is terminated, all resources allocated to the connection are released. It is the duty of RRC to execute the necessary signaling to the mobile terminal about the allocation of radio resources.

The mobile terminal can be in one of two modes: *idle* or *connected*. When powered on, a mobile terminal searches for the current cell by scanning for the

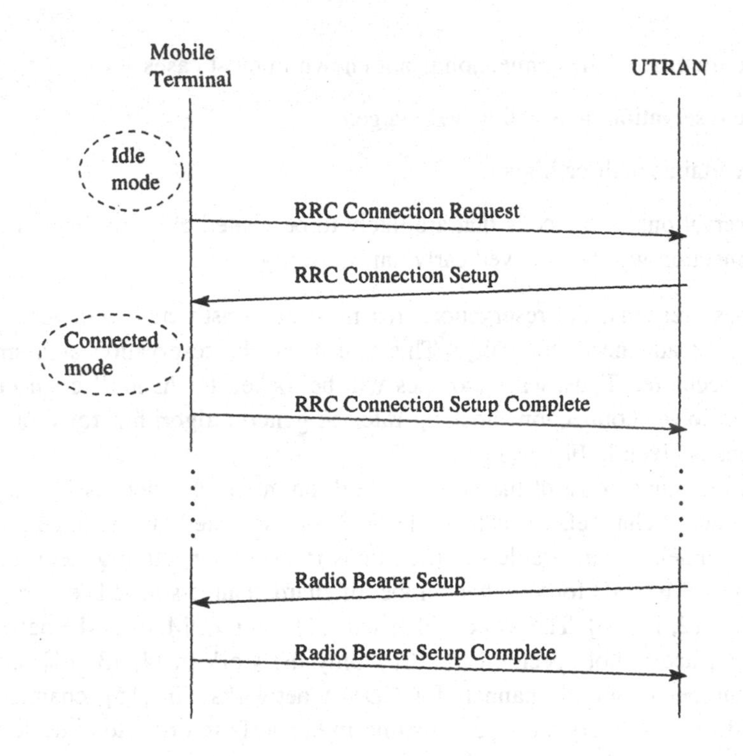

Figure 2: Connection Establishment

Broadcast CHannel (BCH). When a BCH is discovered, the mobile terminal camps on the corresponding cell and goes to idle mode. The mobile terminal keeps on listening to the BCH even in the idle mode to detect any change of cell. The mobile terminal establishes an RRC connection with the UTRAN in order to communicate. The connection establishment procedure for RRC and radio bearer is given in Figure 2.

The mobile terminal transmits its signal measurements to UTRAN. RRC decides which values are measured at what time, and how they are reported. RRC utilizes these measurements to perform the mobility functions such as handoff, cell selection, and paging area update operations. It selects and reselects the cell according to the cell selection criteria and the idle mode measurements. RRC also ensures that the required QoS is provided to the mobile terminal.

RRC interacts with the *BMC (Broadcast/Multicast Control)* to execute broadcast and multicast requests from the network. If a message is to be broadcasted to a group of cells, RRC configures the BMC for cell broadcast services. It also allocates the required resources for the broadcast. Finally, RRC also executes functions

Table 1: QoS Attributes

Category	Attributes	Units
Bandwidth-related	Average data rate	bps
	Peak data rate	bps
	Minimum acceptable data rate	bps
	Maximum burst size (max number of bits sent at peak rate)	bits
Delay-related	Maximum delay	msec
	Maximum jitter (variation in delay)	msec
Reliability of connection	BER (Bit Error Rate)	-
	FER (Frame Error Rate)	-
	Maximum loss ratio (undelivered frames/received frames)	

like power control, data protection and encryption.

2.4 Quality of Service

For voice networks, the resources required for each connection simply constitute a channel. Since all channels have the same bandwidth, the number of channels is equivalent to the number of connections in the network. The picture becomes more complex when data and multimedia connections are considered. Such connections have different expectations from the network such as specific bandwidth, limited delay, jitter, and error rate. These expectations define the *Quality-of-Service (QoS)* required for the connection. QoS requirements are specified by the source of the connection during connection request and negotiated with the network.

CAC tries to maximize the number of connections admitted into the network while maintaining the QoS requirements of existing connections. Therefore, CAC must avoid congestions in advance and come up with a Boolean decision, accept or reject, for each request. Connections are generally defined using the attributes in Table 1.

Typically, a CAC scheme that provides QoS provisioning is composed of three basic components: traffic descriptors, admission criteria, and measurements (Figure 3). A traffic descriptor is a set of parameters that characterize a traffic source. A typical traffic descriptor is a token bucket, composed of a token fill rate r and a token bucket size b. A source described by such a token bucket transmits at most

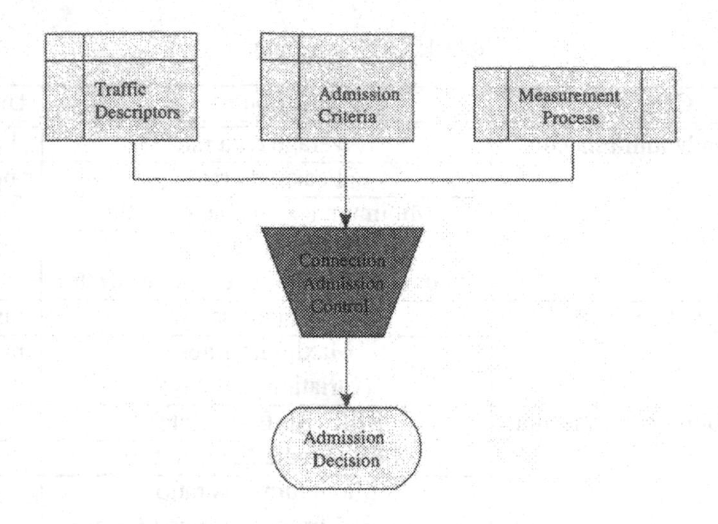

Figure 3: Components of CAC

$r \times t + b$ bytes. Admission criteria are the rules used by the CAC scheme to make the decision. CAC must consider the effects of the new connection on the existing connections. To make the decision, CAC needs an accurate measure of the amount of congestion and the amount of resources in the network. Measurements can be made using a time window, point samples, or exponential averaging [16, 17, 18].

In order to make the admission decision, CAC considers the following:

- Network measures (e.g., multipath, path loss, interference, etc.)
- Traffic characteristics of all existing connections.
- QoS requirements of all existing connections.

As mentioned in Section 2.2, handoff connections must be prioritized over new connections to lower connection dropping rate. If prioritization is provided by means of reservations, the resources required by the reservations should also be considered.

The network must monitor the traffic sent over each admitted connection using a *traffic policer* and ensure that the connection does not violate the QoS specifications. On the mobile terminal side, the data must be pushed into the network according to the QoS specifications using a *traffic shaper* so that frames are not dropped by the policer in the network. Flows from multiple queues are placed in frames and transmitted over the same physical channel using a *scheduler*. If the quality of the radio resources deteriorates, QoS should also be degraded gradually. In the case of multimedia communications, feedback must be provided to the encoder to reduce the offered data rate.

In UMTS, there are four QoS classes:

- The *conversational* is for real-time applications like telephony, voice over IP and video conferencing. It is strictly delay sensitive.
- The *streaming* class is for real-time video streams with a limited delay variation.
- The *interactive* class is for applications that request data from remote servers. E.g.: Remote access, web browsing, etc.
- The *background* class is for applications that are not time sensitive. E.g.: E-mail access, file download, etc.

The QoS parameters used for describing these traffic classes are:

- maximum bit rate,
- guaranteed bit rate,
- delivery order,
- maximum packet length,
- delivery of erroneous packets,
- residual bit error rate,
- packet error ratio,
- transfer delay
- traffic handling priority,
- allocation/retention priority,
- source statistics descriptor,
- SDU format information.

For a connection that requires QoS provisioning, the application layer service attributes are translated to radio access bearer service attributes. The RRC of the mobile terminal communicates with the RRC of UTRAN to setup the radio bearer as in Figure 2.

3 Connection Admission Control in the Next Generation

Several factors compel wireless networks to achieve a performance comparable to the wireline networks. Some of these factors can be listed as:

- Recent advances in technology (especially in signal processing, battery, and screen design).
- Increase in the population of the mobile society.
- Decrease in the prices of wireless devices and service.
- Introduction of new, exciting services.

Next Generation Wireless Systems (NGWS) aim providing high bandwidth access anytime/anywhere to satisfy user expectations. Various types of services, including multimedia with different levels of QoS requirements, will be provided independent of the location and speed of the user. Though existing wireless networks are also designed with these objectives in mind, they fail to satisfy all of the requirements simultaneously due to constraints like global coverage, indoor/outdoor communications, and frequent handoffs. Wireless LANs (WLANs), Personal Communication Systems (PCS), satellite systems, and their future generations together with new wireless networks like 4G Mobile are candidate subsystems for NGWS (Figure 4). These subsystems and the new technologies to come will serve collaboratively in order to provide high bandwidth access everywhere.

The air interface of all of these subsystems are different. The NGWS architecture is independent from the details of individual subsystems. While the current systems are based on FDMA, TDMA, or CDMA, new subsystems are likely to use different technologies. One of the important issues is that the signal quality of a radio channel is both time and frequency dependent. The effective data rates of two channels with the same bandwidth may be different due to frequency selective fading of the radio signal. Furthermore, the effective data rate of the same channel also varies in time. To compensate for these problems, *Orthogonal Frequency Division Multiplexing (OFDM)* based air interface is considered for future wireless networks [20]. The details of OFDM-based air interface are out of the scope of this chapter.

Since the service areas of the subsystems overlap, the mobile terminals will have access to multiple subsystems simultaneously. However, NGWS must select one of the subsystems that can accommodate the connection request according to service class, QoS specifications, accessibility and availability of the subsystems, and user preferences. The selection of the subsystem is a critical factor in the

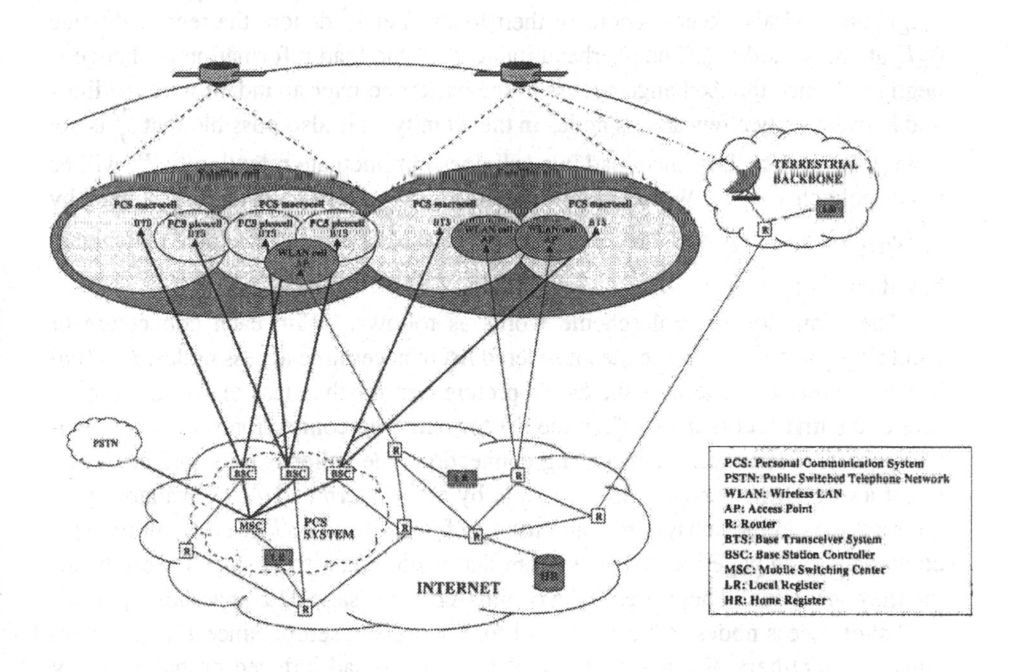

Figure 4: NGWS Architecture

performance of the overall NGWS. Trivial solutions like selecting the subsystem with best signal level will result in the accumulation of the connections in some of the subsystems. Such an accumulation will cause blocking of service for mobiles that cannot access lightly loaded subsystems, resulting in higher outage rate, lower throughput, and unstable service throughout the NGWS [19].

To explain the global admission control scheme that manages the resources in all subsystems, we first introduce the notation. Let b_i^s denote the ith access node (e.g., base station, etc.) of subsystem s, and c_i^s denote the capacity of b_i^s. Each access node b_i^s periodically transmits its load information, l_i^s, to all of its neighbors, and also keeps record of their loads. Let $l_j'^t$ denote the recorded value of l_j^t at access node b_i^s. The overhead induced by the load information exchange is negligible since the exchange occurs in the backbone over abundant wireline links and between only a few access nodes in the vicinity. It is also possible that $l_j'^t$ is not exactly up-to-date, but since load in a cell does not fluctuate relentlessly, $l_j'^t$ will be reasonably close to l_j^t. We denote the new load of b_i^s after request rq is accepted by $\widehat{l}_i^s(rq)$. We also denote the recorded load of b_j^t at b_i^s after rq is accepted (calculated based on $l_j'^t$) by $\widehat{l'}_j^t(rq)$.

The admission control scheme works as follows. With each connection or handoff request rq, we associate an ordered list of accessible access nodes, $\mathcal{L}_{ac}(rq)$, in which ordering criteria is the user's preferences for the class of rq. $\mathcal{L}_{ac}(rq)$ is sent to the first access node, b_i^s, in the list for outgoing connection setup and handoff requests. However, for incoming connections the caller is a remote node that is not aware of the subsystems accessible by mobile terminal MT, availability of the resources in the subsystems, and user preferences for MT. Furthermore, MT could have been paged over a subsystem that is not likely to be used for establishing the connection. Therefore, in the paging reply message MT specifies $\mathcal{L}_{ac}(rq)$, the list of access nodes that can be used for connection setup. Since $\mathcal{L}_{ac}(rq)$ contains the identifiers of a few access nodes, the overhead induced on paging reply message is not significant. The caller sends the connection setup request to the first access node, b_i^s, in $\mathcal{L}_{ac}(rq)$. The algorithm of the CAC scheme for NGWS is given in Figure 5.

4 Conclusion

The performance of a wireless network is bounded by the efficiency of resource management. The resources are consumed by the connections that are admitted into the network by the CAC scheme. Assignment of the radio resources to the cells, admission of new and handoff connections into the network, and QoS provi-

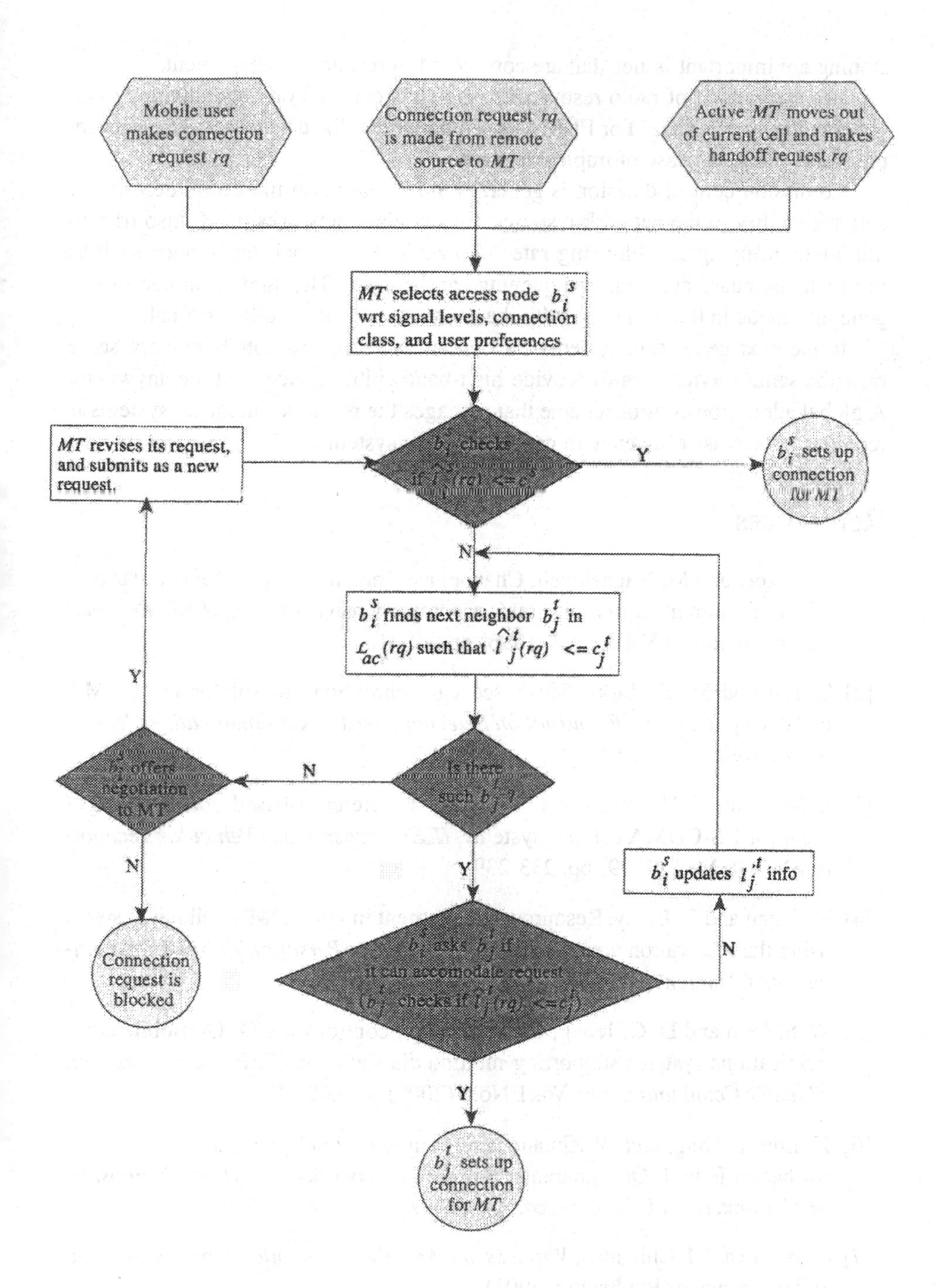

Figure 5: CAC Algorithm for NGWS

sioning are important issues that are considered in resource management.

The assignment of radio resources to the cells depends on the technology that governs the air interface. For FDMA-based networks, fixed schemes are generally preferred due to the ease of implementation.

Admission control decision is generally made based on the traffic descriptors and availability of the network resources. In wireless networks, CAC also tries to minimize dropping and blocking rates. To achieve this goal, techniques such as reservations, guard channel, and queuing can be used. The admission decision is generally made in the backbone with the assistance of the mobile terminal.

In the next generation systems, several subsystems will collaboratively serve over the same service area to provide high-bandwidth service anytime/anywhere. A global admission control scheme that manages the resources in all subsystems is required to increase efficiency in next generation systems.

References

[1] I. Katzela and M. Naghshineh, Channel assignment schemes for cellular mobile telecommunication systems: a comprehensive survey, *IEEE Personal Communications* Vol.3 No.3 (1996) pp. 10-31.

[2] Z. Liu and M. E. Zarki, SIR-based call admission control for DS-CDMA cellular systems, *IEEE Journal on Selected Areas in Communications* Vol.12 No.4 (1994) pp. 638-644.

[3] S. M. Shin, C.-H. Cho, and D. K. Sung, Interference-based channel assignment for DS-CDMA cellular systems, *IEEE Transactions Vehicular Technology* Vol.48 No.1 (1999) pp. 233-239.

[4] T. Tugcu and C. Ersoy, Resource management in DS-CDMA cellular systems using the reservation area concept, 4th *European Personal Mobile Communications Conference* (2001).

[5] W. S. Jeon and D. G. Jeong, Call admission control for CDMA mobile communications systems supporting multimedia services, *IEEE Transactions on Wireless Communications* Vol.1 No.4 (2002) pp. 649-659.

[6] X. Luo, I. Thng, and W. Zhuang, A dynamic channel prereservation scheme for handoffs with QoS guarantee in mobile networks, 4th *IEEE Symposium on Computers and Communications* (1999).

[7] Y.-B. Lin and I. Chlamtac, *Wireless and Mobile Communications,* (New York, Wiley Computer Publishing, 2001).

[8] G. Senarath and D. Everitt, Performance of handover priority and queuing systems under different handover request strategies for microcellular mobile communication systems, *IEEE Vehicular Technology Conference* (1995) pp. 897-901.

[9] B. Gavish and S. Sridhar, Threshold priority policy for channel assignment in cellular networks, *IEEE Transactions on Computers* Vol.46 No.3 (1997) pp. 367-370.

[10] D. A. Levine, I. F. Akyildiz, and M. Naghshineh, A resource estimation and call admission algorithm for wireless multimedia networks using the shadow cluster concept, *IEEE/ACM Transactions on Networking* Vol.5 No.1 (1997) pp. 1-12.

[11] C. Oliviera, J. B. Kim, and T. Suda, An adaptive bandwidth reservation scheme for high-speed multimedia wireless networks, *IEEE Journal on Selected Areas in Communications* Vol.16 No.6 (1998) pp. 858-873.

[12] Y. C. Kim, D. E. Lee, B. J. Lee, Y. S. Kim, and B. Mukherjee, Dynamic channel reservation based on mobility in wireless ATM networks, *IEEE Communications Magazine* Vol.37 No.11 (1999) pp. 47-51.

[13] J. Hou, Y. Fang, and A. N. Akansu, Mobility-based channel reservation scheme for wireless mobile networks, *IEEE Wireless Communications and Networking Conference* (2000).

[14] Y. Ma, J. J. Han, and K. S. Trivedi, Call admission control for reducing dropped calls in CDMA cellular systems, *IEEE Computer Communications* Vol.25 No.7 (2002) pp. 689-699.

[15] D.-J. Lee and D.-H. Cho, Performance analysis of channel-borrowing handoff scheme based on user mobility in CDMA cellular systems, *IEEE Transactions on Vehicular Technology* Vol.49 No.6 (2000) pp. 2276-2285.

[16] S. Jamin, P. Danzig, S. Shenker, and L. Zhang, A measurement-based admission control algorithm for integrated services packet networks (Extended version), *ACM/IEEE Transactions on Networking* Vol.5 No.1 (1997) pp. 56-70.

[17] D. Tse and M. Grossglauser, Measurement-based call admission control: analysis and simulation, *IEEE INFOCOM* (1997) pp. 981-989.

[18] S. Floyd, Comments on measurement-based admissions controlled-load service, *ACM Computer Communications Review* (1996).

[19] T. Tugcu and F. Vainstein, Mathematical foundations of resource management in next generation wireless systems, *IEEE International Symposium on Personal, Indoor and Mobile Radio Communications* (2003).

[20] S. Coleri, M. Ergen, A. Puri, and A. Bahai, "Channel estimation techniques based on pilot arrangement in OFDM systems, *IEEE Transactions on Broadcasting* Vol. 48 No. 3 (2002) pp. 223-229.

[21] D. J. Goodman, *Wireless Personal Communications Systems,* (Massachusetts, Addison-Wesley, 1997).

[22] X. Qiu, K. Chawla, J. Chuang, and N. Sollenberger, Network-assisted resource management for wireless data networks, *IEEE Journal on Selected Areas in Communications* Vol.19 No.7 (2001) pp. 1222-1234.

[23] C. Posner and R. Guerin, Traffic policies in cellular radio that minimize blocking of handoffs, *Proceedings of the 11th ITC* (1985).

[24] D. Hong and S. Rappaport, Traffic modeling and performance analysis for cellular mobile radio telephone systems with prioritized and nonprioritized handoff procedures, *IEEE Transactions on Vehicular Technology* Vol.35 No.3 (1986) pp. 7792.

[25] A. S. Acampora and M. Naghshineh, Control and quality of service provisioning in high-speed microcellular networks, *IEEE Personal Communications* Vol.1 No.2 (1994) pp. 36-43.

[26] A. Sutivong and J. M. Peha, Novel heuristics for call admission control in cellular systems, *IEEE* 6th *International Conference on Universal Personal Communications* (1997) pp. 129-133.

[27] N. D. Tripathi, J. H. Reed, and H. F. Vanlandingham, Handoff in cellular systems, *IEEE Personal Communications* Vol.5 No.6 (1998) pp. 26-37.

Real-Time Guarantees in Wireless Networks

Shengquan Wang, Ripal Nathuji, Riccardo Bettati, and Wei Zhao
Department of Computer Science
Texas A&M University, College Station, TX 77843
E-mail: {swang,rnathuji,bettati,zhao}@cs.tamu.edu

Contents

1 Introduction

The convenience of wireless communications has led to an increasing use of wireless networks for both civilian and mission critical applications. Many of these kinds of applications require delay-guaranteed communications. In the following, we describe approaches to providing delay-guaranteed services in wireless networks.

A significant amount of work has been done on *real-time* communication over *wired* networks [3, 11, 13, 14, 20, 21]. *Wireless* networks are sufficiently different from their wired counterparts so that technologies developed for wired networks cannot be directly adopted: in most wired network models for real-time systems (and many other systems requiring quality of service support), the communication links are assumed to have a fixed capacity over time. This assumption may be invalid in wireless (radio or optical) environments, where link capacities can be temporarily degraded due to fading, attenuation, and path blockage. For example, in a digital cellular radio transmission environment, radio wave reflection, refraction, and scattering, may cause the transmitted signal to reach the receiver by more than one path. This gives rise to the phenomenon known as multipath fading [16]. Also, mobile terminals exhibit time variations in their signal level due to motion [19]. These characteristics of wireless links all result in performance degradation. In order to improve the performance of wireless links, error control schemes are used. Common error control methods used in wireless communications include forward error correction (FEC), automatic repeat request (ARQ) and their hybrids [2, 5].

The difficulty of provisioning real-time guarantees in wireless networks stems from the need to explicitly consider both the channel transmission characteristics and the error control mechanisms put in place to alleviate the channel errors. To describe the underlying communication infrastructure in terms of channels and protocols, appropriate models are needed. There is a large volume of literature dealing with the representation and analysis of channel models, and most of these models directly characterize the fluctuations of signals and provide an estimate of the performance characteristics, such as symbol error rate vs. signal-to-noise ratio [22]: The classical two-state Gilbert-Elliott model [4, 6] for burst noise channels, which characterizes error sequences, has been widely used and analyzed. In [18], a multiple-state quasi-stationary Markov channel model is used to characterize the wireless nonstationary channel. This model was developed based on experimental measurements of several real channels. In [19, 24], a finite-state Markov channel was described that has multiple states representing the reception at different signal-to-noise levels. A fluid version of the Gilbert-Elliott model was used in [12] to perform analysis of delay and packet-discard performance as well as the effective capacity for QoS support over a wireless link with ARQ and FEC. In the following, we will describe a very general framework to analyze delay peformance on wireless links. We will illustrate it with the example of a Rayleigh fading channel model with hybrid ARQ/FEC error control. The wireless link will be modeled as a fluid version of Finite-State Markov model. It is important to note that the framework presented here is by far not limited to this particular channel (Rayleigh with ARQ/FEC), but can be applied to many other models as well.

In order to provide real-time guarantees, one needs both an appropriate description of the underlying wireless links and a traffic model, which is the description of the workload carried on links. The traffic model in turn depends upon the desired service requirements. Real-Time communication service requirements can be guaranteed in two forms: deterministic services and statistical services. *Deterministic services* require that the delay and delivery guarantees are satisfied for all packets. This provides a very simple model to the application. However, they tend to heavily overcommit resources because the resource management subsystem must assume a worst-case scenario. In real systems, this frequently results in significant portions of network resources being wasted. *Statistical services* allow packets to be occasionally dropped or excessively delayed. Statistical services thus significantly increase the efficiency of network usage by allowing increased statistical multiplexing of the underlying network resources. A number of approaches have been presented in the literature to provide statistical guarantees for deterministically constrained traffic streams. We will make use of *rate-variance envelopes* [11]. The resulting resource management framework for statistical services will have two main benefits: first, it allows for a reduction of the resource commitment through statistical multiplexing of the traffic. Second, it allows for seamless integration of stochastic link models, largely independently of the detailed nature of the specific model.

In the following, we focus on providing end-to-end delay guarantees via connection admission control mechanisms that make sure that end-to-end delay requirements are not violated – both for new and existing connections – after a new connection has been admitted. We adopt two different admission control mechanisms, which differ by the point in time during which they perform explicit delay computations. We first introduce *Delay-Based Admission Control* (DBAC) as a straightforward approach to providing real-time guarantees. In DBAC, the delay tests for the admission decisions are done at connection establishment time. The necessary delay computations are based on run-time flow information. This allows DBAC to achieve high resource utilization, at the cost of computational overhead at runtime. The second scheme we consider is *Utilization-Based Admission Control* (UBAC). In UBAC, safe resource utilization bounds are computed at each wireless link *during system (re-)configuration time*, i.e., before runtime. Run-time admission control is then reduced to a simple utilization-based test along the path of the flow: As long as the utilization of the links along the path of a flow is not beyond the safe utilization bound computed offline, the probabilistic performance guarantee can be met. We will show in our experimental evaluation that UBAC's admission probability is comparable to that of the much more expensive DBAC scheme.

In Section 2, we describe the wireless link model. The underlying network and traffic models for this study are introduced in Section 3. In Section 4, we develop a delay analysis methodology to provide real-time guarantees. Admission control mechanisms are presented in Section 5. In Section 6, we provide extensive experimental data to illustrate the QoS performance. A summary is given in Section 7.

2 Models of Wireless Networks and Links

2.1 Overview

We consider a wireless network that consists of a number of wireless links, each of which connects two wireless nodes. This kind of networks is used widely in mission critical systems ranging from terrestrial-based infrastructures to satellite environments. Fig. 1 shows an example of a wireless system that falls into our network model.

To guarantee an end-to-end delay, delay characteristics on each wireless link need to be analyzed. Thus, in the rest of this section, we will mostly discuss models related to wireless links in our networks. Underlying wireless links are physical *wireless channels*. For the purpose of delay guarantees, a wireless channel model describes the channel error statistics and its effect on channel capacity. A large number of such models have been described and evaluated in the literature, based on the Rayleigh Fading Channel, or (by adding a line-of-sight component) the Rician Fading Channel [16]. Typical channel error statistics models, such as the binary symmetric channel, are

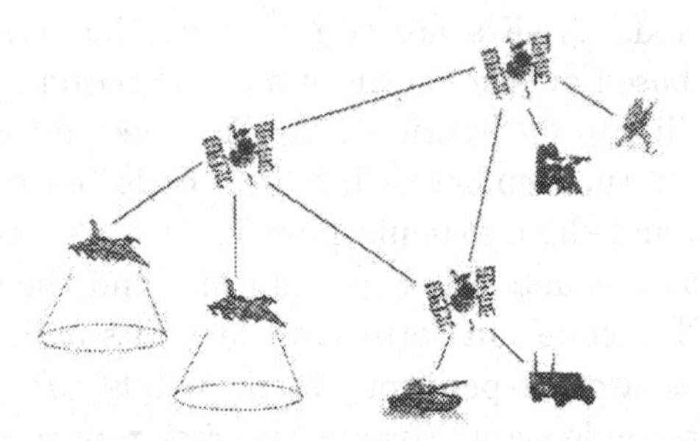

Figure 1: An Ground-space-ground Wireless Communication System

modeled as finite-state Markov models, and can be used to represent time-varying Rician (and other) channels in a variety of settings [1, 7, 8].

In addition to the physical channel, the formulation of a *link model* has to account for error control schemes used at the link layer. In the following, we first consider the framework of the wireless link, and then lay out a more detailed description of our Markov link model. The framework and description will largely follow the approach presented by Krunz and Kim in [12]. We will extend their two-state Markov model to a more general finite-state Markov model. Finally we derive the stochastic service curve that will be used in the delay analysis.

2.2 Framework of a Wireless Link

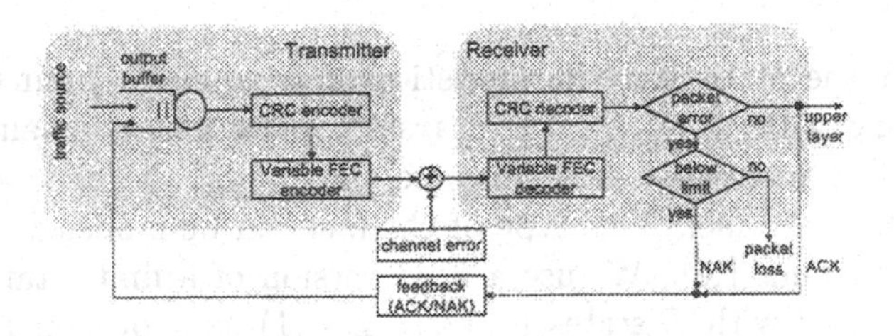

Figure 2: Wireless Link Framework [12]

We consider a hybrid ARQ/FEC error control scheme (Fig. 2) and assume a *stop-and-wait* (SW) scheme for ARQ: the sender transmits a codeword to the receiver and waits for an acknowledgement. If a positive acknowledgement (ACK) is received, the sender transmits the next codeword. If a negative acknowledgement (NAK) is received, however, the same codeword is retransmitted. NAK's are triggered at the receiver by an error detector, typically based on some form of a cyclic redundancy check.

The FEC capability in the hybrid ARQ/FEC mechanism is characterized by three parameters: the number of bits in a code block (n), the number of payload bits (k), and the maximum number of correctable bits in a code block (r). Note that n counts the k payload bits and the extra parity bits. Assuming that a FEC code can correct up to r bits and that bit errors in a given channel state are independent, the probability $P_{nc}(p)$ that a packet contains a non-correctable error, given a bit error rate p, is given by [12]

$$P_{nc}(p) = \sum_{j=r+1}^{n} \binom{n}{j} p^j (1-p)^{n-j}. \tag{1}$$

To account for the FEC overhead, the actual average service capacity observed at the output of the buffer is $C \cdot \frac{k}{n}$, where C is the maximum capacity for the wireless channel.

2.3 Markov Link Model

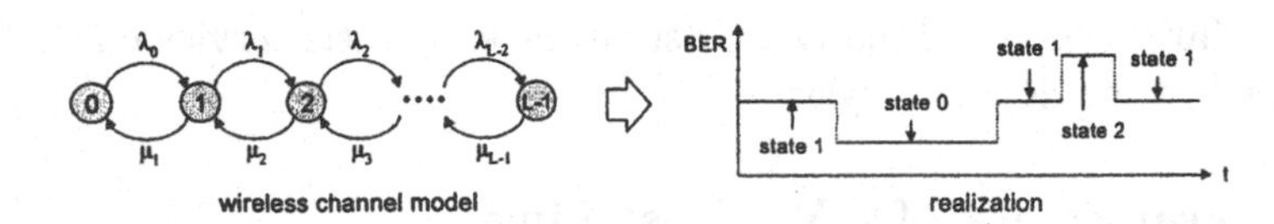

Figure 3: Fluid Version of Finite-State Markov Model of a Wireless Channel

Although the statistical charateristics of a wireless channel can significantly vary with time, the basic system parameters remain constant over short time intervals. Therefore we can model the channel to be a quasi-stationary channel. This type of channel can be modeled with finite-state Markov chains [18]. We use a fluid version of a finite-state Markov-Modulated model with L states $(0, 1, \ldots, L-1)$ as shown in Fig. 3 [12]. The bit error rates (BER) during State i are given by p_i, where we assume $0 \leq p_0 < p_1 < \cdots < p_{L-1} \leq 1$. The durations in State i before being transitioned to State $i+1$ and $i-1$ are exponentially distributed with means $\frac{1}{\lambda_i}$ and $\frac{1}{\mu_i}$, respectively. We assume that the transitions only happen between adjacent states.

It is generally difficult to get analytically tractable results that accurately represent the behavior of ARQ and FEC and map the channel model into the respective link model. To solve this, the authors in [12] assume that the packet departure is described by a fluid process with an average constant service capacity that is modulated by the channel state (Fig. 4). Each state

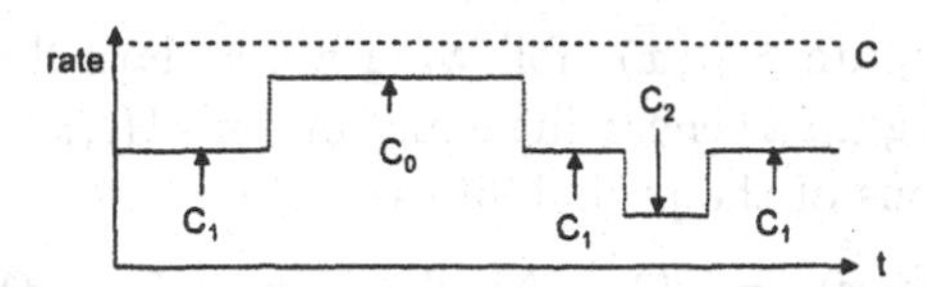

Figure 4: Approximate Model of a Wireless Link

i then gives rise to a stationary link-layer service capacity C_i, which takes packet re-transmissions into account. The total time needed to successfully deliver a packet, conditioned on the channel state, follows a geometric distribution. Let N_{tr} denote the number of retransmissions (including the first transmission) until a packet is successfully received. For the given packet error probability p_i of the channel in State i, the expected value for N_{tr} is $E[N_{tr}] = \frac{1}{1-p_i}$ [1]. Thus, C_i can be written as [12]

$$C_i = C \cdot \frac{k}{n} \cdot (1 - P_{nc}(p_i)). \tag{2}$$

As the state transition rates of the channel are not affected by ARQ or FEC, the result is a Markov-modulated model with L state $(0, 1, \ldots, L-1)$ with link capacity C_i associated with State i.

2.4 Stochastic Service Curve of a Wireless Link

In order to determine the performance guarantees that can be given by a wireless link, we must describe the amount of service that the link can provide. For this we make use of so-called *service curves.* In the following we show how we derive the service curve from a given link model.

The *stochastic service curve* $S(t) = \int_0^t C(\tau)d\tau$ is defined as the traffic amount that can be served during time interval $[0, t]$ by the wireless channel, where $C(\tau)$ is the capacity at time τ. Correspondingly, we define $S_i(t)$ as the traffic amount that can be served during time interval $[0, t]$ with the system in State i at time t, $F_i(t, x)$ and $F_S(t, x)$ as the cumulative probability distribution of $S_i(t)$ and $S(t)$, respectively. We denote π_i as the probability

[1]If we predefine a limit N_l on number of retransmissions, $E[N_{tr}] = \frac{1-p_i{}^{N_l}}{1-p_i}$ [12].

that the link is in State i at any time when the system is steady, and we then have

$$F_S(t,x) = \sum_{l=0}^{L-1} \pi_i F_i(t,x). \tag{3}$$

We need to compute $F_i(t,x)$: following a standard fluid approach [17], we proceed by setting up a generating equation for $F_i(t,x)$ at an incremental time Δt later in terms of the probabilities at time t.

$$\begin{aligned} F_i(t+\Delta t, x) &= (\lambda_{i-1}\Delta t)F_{i-1}(t, x - C_{i-1}\Delta t) \\ &\quad +(1-(\mu_i+\lambda_i)\Delta t)F_i(t, x - C_i\Delta t) \\ &\quad +(\mu_{i+1}\Delta t)F_{i+1}(t, x - C_{i+1}\Delta t), \end{aligned} \tag{4}$$

as $i = 1, \ldots, L-2$, and

$$\begin{aligned} F_0(t+\Delta t, x) &= ((1-\lambda_1)\Delta t)F_0(t, x - C_0\Delta t) \\ &\quad +(\mu_1\Delta t)F_1(t, x - C_1\Delta t), \end{aligned} \tag{5}$$

$$\begin{aligned} F_{L-1}(t+\Delta t, x) &= (\lambda_{L-2}\Delta t)F_{L-2}(t, x - C_{L-2}\Delta t) \\ &\quad +((1-\mu_{L-1})\Delta t)F_{L-1}(t, x - C_{L-1}\Delta t). \end{aligned} \tag{6}$$

Both sides are divided by Δt in the above equations. As $\Delta t \to 0$, we have

$$\begin{aligned} &\frac{\partial F_i(t,x)}{\partial t} + C_i \frac{\partial F_i(t,x)}{\partial x} = \\ &\quad \lambda_{i-1}F_{i-1}(t,x) - (\lambda_i + \mu_i)F_i(t,x) + \mu_{i+1}F_{i+1}(t,x), \end{aligned} \tag{7}$$

as $i = 1, 2, \ldots, L-2$, and

$$\frac{\partial F_i(t,x)}{\partial t} + C_i \frac{\partial F_i(t,x)}{\partial x} = \begin{cases} -\lambda_i F_i(t,x) + \mu_{i+1}F_{i+1}(t,x), & i = 0 \\ \lambda_{i-1}F_{i-1}(t,x) - \mu_i F_i(t,x), & i = L-1 \end{cases}. \tag{8}$$

The initial conditions are $F_i(0,x) = \begin{cases} 0, & x \le 0 \\ 1, & x > 0 \end{cases}$ for $i = 0, 1, \ldots, L-1$. The partial differential equations can be rewritten in matrix form as follows:

$$\frac{\partial \mathbf{F}}{\partial t} + \mathbf{C}\frac{\partial \mathbf{F}}{\partial x} = \mathbf{QF}, \tag{9}$$

where $\mathbf{F} = (F_0(t,x), F_1(t,x), \ldots, F_{L-1}(t,x))^{\perp}$, $\mathbf{C} = \mathrm{diag}(C_0, C_1, \ldots, C_{L-1})$ and

$$\mathbf{Q} = \begin{pmatrix} -\lambda_0 & \mu_1 & \cdots & 0 \\ \lambda_0 & -(\lambda_0+\mu_1) & \cdots & 0 \\ \vdots & \vdots & \vdots & \vdots \\ 0 & 0 & \cdots & \mu_{L-1} \\ 0 & 0 & \cdots & -\mu_{L-1} \end{pmatrix}. \tag{10}$$

The above linear first-order hyperbolic PDEs can be solved numerically, and the $F_i(t,x)$'s can be computed. Furthermore, if we define $\boldsymbol{\pi} = (\pi_0, \pi_1, \ldots, \pi_{L-1})^{\perp}$, the π_i's in (3) are given by

$$\boldsymbol{\pi} = \boldsymbol{\pi}\mathbf{Q}, \quad \text{and} \quad |\boldsymbol{\pi}| = 1. \tag{11}$$

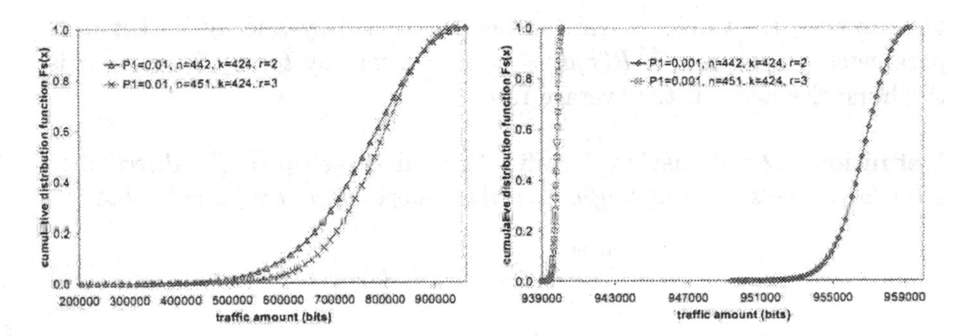

Figure 5: The Stochastic Service Curve for a Wireless Link

Fig. 5 shows simulated data for the distribution of $S(t)$ for a two-state Markov model, where we specify $C = 2$ Mbps, $\lambda_0 = 10, \lambda_1 = 30, p_0 = 10^{-6}$. We vary the BER p_1 and the code parameters (n, k, r). The data illustrates that BER and coding substantially affect the service distribution.

In Section 4, we will illustrate how the service distribution $F_S(t,x)$ can be used to perform statistical delay analysis. In the next section, we will introduce the traffic model that is used to derive the other important component required for delay analysis: *traffic arrival.*

3 Traffic Model

We model the *traffic arrival* for a flow as a stochastic arrival process $\mathcal{A} = \{\mathcal{A}(\tau), \tau \geq 0\}$, where random variable $\mathcal{A}(\tau)$ denotes the incoming traffic amount of the flow at a link server during time interval $[0, \tau]$. The arrival process $\mathcal{A}$ is stationary and ergodic. Since $\mathcal{A}(\tau)$ is stationary, $\mathcal{A}(\tau+t)-\mathcal{A}(\tau)$ possesses the same probability distributions for all τ. Therefore, the random variable $R(t) = \frac{\mathcal{A}(t_0+t)-\mathcal{A}(t_0)}{t}$ exists, which is callled the *stochastic traffic arrival rate.* The traffic arrival can be bounded either deterministically or stochastically by the traffic arrival envelope as follows:

Definition 3.1 (Deterministic Traffic Arrival Envelope) *The function $b(t)$ is called the deterministic traffic arrival envelope of the traffic arrival with rate $R(t)$ if*

$$\int_{t_0}^{t_0+t} R(\tau)d\tau \leq b(t), \tag{12}$$

for any $t_0, t \geq 0$.

For example, the traffic arrival can be constrained by a leaky bucket with parameters (σ, ρ) as $\int_{t_0}^{t_0+t} R(\tau)d\tau \leq \sigma + \rho \cdot t$, for any $t_0, t \geq 0$, where σ is the burst size and ρ is the average rate.

Definition 3.2 (Statistical Traffic Arrival Envelope) *The distribution $B(t)$ forms the statistical traffic arrival envelope of the traffic arrival $\mathcal{A}$ if*

$$\int_{t_0}^{t_0+t} R(\tau)d\tau \preceq_{st} B(t), \tag{13}$$

for any $t_0, t \geq 0$, where $X \preceq_{st} Y$ means $\Pr\{X < Z\} \leq \Pr\{Y < Z\}$.

We will be describing traffic arrivals using the rate-variance envelope [11], which describes the variance of the traffic arrival rate during a time interval. The rate-variance envelope is used as a simple way to capture the second-moment properties of temporally correlated traffic flows. It is a key factor for computing delay violation probabilities.

4 Statistical Delay Analysis in a Wireless Network

In this section, we will perform the delay analysis that is needed in order to provide end-to-end guarantees. Our analysis will be based on the service description (service curves) introduced in Section 2 and the workload description (traffic arrival) discussed in Section 3.

A probabilistic real-time guarantee can be defined as a bound on the probability of exceeding a deadline, i.e., $\Pr\{D > d\} \leq \epsilon$, where the delay D suffered by a packet is a random variable, d is the given deadline, and ϵ is the given violation probability, which is generally small.

We consider networks that use static-priority schedulers at the network nodes, as opposed to previous work considering FIFO buffers [12]. For *wired* networks with static priority scheduling, we addressed the issue of how to provide statistical real-time guarantees in [20], based on Knightly's earlier work in [11]. Define C as the capacity of a link and G_i as a group of flows that are served by the link at priority i. Assume $b_{i,j}(t)$ and $B_{i,j}(t)$ to be the deterministic and statistical bound, respectively, for the traffic arrival for the individual flow $j \in G_i$. Then the *delay violation probability* $\Pr\{D_i \geq d_i\}$ for a random packet with priority i at the output link can be bounded by

$$\Pr\{D_i \geq d_i\} \leq \max_{t<\beta_i} \Pr\{B^*(t+d_i) \geq C \cdot (t+d_i)\}, \tag{14}$$

where $B^*(\cdot)$ is the amount of aggregated traffic of same and higher priorities:

$$B^*(t+d_i) = \sum_{q=1}^{i-1} \sum_{j \in G_q} B_{q,j}(t+d_i) + \sum_{j \in G_i} B_{i,j}(t), \tag{15}$$

and β_i is a bound on the busy interval and is defined as follows:

$$\beta_i = \min\{t > 0 : \sum_{q=1}^{i} \sum_{j \in G_q} b_{q,j}(t) \geq C \cdot t\}. \tag{16}$$

The above formula cannot be applied directly for wireless links however, as their capacities vary over time. Fortunately, as the following observation shows, it is not difficult to integrate stochastic arrivals and a stochastic service curve to compute delay violation probabilities: consider a wireless link with a static-priority scheduler and maximum capacity C. Let $C(t)$ be the available capacity for traffic as a function of time. Thus $C - C(t)$ is the *unavailable* capacity of link at time t. We can equivalently model this system if we define a *virtual traffic arrival* with instantaneous capacity $C - C(t)$ to a link with constant capacity C, by requiring that this virtual traffic is given strictly highest priority during scheduling. Packet delays for real traffic in the original system are identical to delays in this virtual-traffic model. In particular, if the wireless link has a stochastic service curve $S(t)$, then the equivalent virtual traffic on the wireless link has the stochastic envelope $B'(t) = C \cdot t - S(t)$. This gives raise to the following theorem:

Theorem 4.1 *Consider a wireless link with a static-priority scheduler and stochastic service curve $S(t)$. Assume $B_{i,j}(t)$ is the statistical bound for the traffic arrival of the individual flow $j \in G_i$. Then, the delay violation probability for a random packet with priority i can be bounded by*

$$\Pr\{D_i \geq d_i\} \leq \max_{t>0} \Pr\{B'(t+d_i) + B^*(t+d_i) \geq C \cdot (t+d_i)\}, \tag{17}$$

where $B'(t) = C \cdot t - S(t)$, *and* $B^*(t + d_i)$ *is defined in Equation (15).* [2]

We make the following observations about Theorem 4.1: First, $B'(t + d_i)$ and $B^*(t + d_i)$ are independent. Given their distribution functions, the distribution function of the summation $B^*(t + d_i) + B'(t + d_i)$ can be obtained by their direct convolution. Second, the distribution of $B'(t + d_i)$ can be directly obtained from $S(t)$, which we in turn derived in Section 2. Note that (17) holds for any $S(t)$, no matter what specific wireless link model is chosen. The main challenge for statistical delay analysis is how to obtain the distribution function of $B'(t + d_i)$, i.e., how to clearly describe the traffic arrival envelope. It is inherently very difficult for the network to enforce or police the stochastic properties of traffic streams. Consequently, if a particular application does not conform to the chosen stochastic model, no guarantees can be made. Moreover, if admitted to the network, such a non-conforming stream could adversely affect the performance of other applications if it is statistically multiplexed with them. Therefore, we must find a means to describe the non-conforming traffic so that we can perform delay analysis.

We will use the approach previously developed in [20] for the statistical delay analysis. We start by representing the input traffic flows as a set of random processes. Traffic policing ensures that these processes are independent. If we know the mean value and the variance of each individual traffic random variable, and the number of flows is large enough, then by the *Central Limit Theorem* we can approximate the random process of the set of all flows combined. The Central Limit Theorem states that the summation of a set of independent random variables converges in distribution to a random variable that has a *Normal Distribution* [3]. In the following, we illustrate how using *rate-variance envelopes*, the mean rate and the rate-variance of each individual flow can be determined by deterministic traffic models.

The rate-variance envelope $RV(t) = var(R(t))$ describes the variance of the arrival rate for the incoming flow over an interval of length t [11]. We assume that a flow of priority i is controlled by a leaky bucket with burst size σ_i and average rate ρ_i at each router. Assume that Flow j in the group

[2] Here the maximum busy interval is canceled out due to the possibly unconstrained stochastic service curve. The virtual traffic may produce an infinite-length maximum busy interval. So the delay violation probability may appear to be loose. In our simulation data, we find that the maximum value will be achieved for relatively small values of t, therefore, the bound is tight.

[3] In [10], the author experimentally found the normal distribution approximation to be highly accurate in predicting the performance of a buffered priority multiplexer.

of flows G_i has mean rate $\phi_{i,j}$ and rate-variance envelope $RV_{i,j}(t)$. With application of a Gaussian approximation over intervals, $B^*(t+d_i)$ in (17) can be approximated by a normal distribution $N(\phi_i(t), RV_i(t))$ [11], where

$$\phi_i(t) = (t+d_i)\sum_{q=1}^{i-1}\sum_{j\in G_q}\phi_{q,j} + t\sum_{j\in G_i}\phi_{i,j}, \tag{18}$$

$$RV_i(t) = (t+d_i)^2\sum_{q=1}^{i-1}\sum_{j\in G_q} RV_{q,j}(t+d_i) + t^2\sum_{j\in G_i} RV_{i,j}(t). \tag{19}$$

Given the deterministic traffic arrival envelope $b_{i,j}(t) = \sigma_i + \rho_i t$, for any flow j in G_i, we can easily obtain mean rate $\phi_{i,j}$ for each individual flow, and an adversarial mode is chosen for obtaining the rate-variance envelope $RV_{i,j}(t)$ [11]. [4] We obtain the *mean rate* and the *rate-variance envelope* as follows:

$$\phi_{i,j} = \rho_i, \tag{20}$$

$$RV_{i,j}(t) \leq \frac{\rho_i\sigma_i}{t}. \tag{21}$$

In summary, this leads to the following lemma:

Lemma 4.2 *Define $n_q = |G_q|$, $q = 1, 2, \ldots, i$. With application of a Gaussian approximation over intervals, $B^*(t+d_i)$ can be bounded by a normal distribution $N(\phi_i(t), RV_i(t))$, i.e.,*

$$\Pr\{B^*(t+d_i) < x\} \leq \Phi\left(\frac{x - \phi_i(t)}{\sqrt{RV_i(t)}}\right), \tag{22}$$

where

$$\phi_i(t) = (t+d_i)\sum_{q=1}^{i-1} n_q\rho_q + t n_i\rho_i, \tag{23}$$

$$RV_i(t) \leq (t+d_i)\sum_{q=1}^{i-1} n_q\rho_q\sigma_q + t n_i\rho_i\sigma_i, \tag{24}$$

and

$$\Phi(a) = \frac{1}{\sqrt{2\pi}}\int_{-\infty}^{a}\exp\left(-\frac{x^2}{2}\right)dx. \tag{25}$$

[4] In adversarial mode, the traffic arrival process conforms to a binomial distribution, where the rate-variance envelope is upper bounded.

The distribution function of the summation $B^*(t+d_i)+B'(t+d_i)$ can be obtained by convolution. Define this distribution function as $F_B(t+d_i, x)$. Then, the delay violation probability can be upper-bounded with utilization as shown in the following theorem:

Theorem 4.3 *Consider a wireless link with a static-priority scheduler and stochastic service curve $S(t)$. Assume the same traffic envelope as in Theorem 4.1. The delay violation probability for a random packet with priority i is bounded by*

$$\Pr\{D_i \geq d_i\} \leq 1 - \min_{t>0} F_B(t+d_i, C \cdot (t+d_i)). \qquad (26)$$

We have now derived the statistical delay formula for a single wireless link. Based on this result, we obtain the end-to-end delay violation probability along each path as follows: Given the delay violation probability ϵ_i and the end-to-end deadline d_i along route $\mathcal{R}$, we can partition d_i into $\{d_i^k : k \in \mathcal{R}\}$, and the delay guarantee is met when [9]

$$\Pr\{D_i^{e2e} > \sum_{k\in\mathcal{R}} d_i^k\} \leq 1 - \prod_{k\in\mathcal{R}}(1 - \Pr\{D_i^k > d_i^k\}) \leq \epsilon_i. \qquad (27)$$

This bound on the end-to-end real-time guarantee gives rise to several possible approaches to admission control and connection establishment. In the next section, we will describe two such approaches.

5 Admission Control Mechanisms

Recall that it is the admission control mechanism that decides if a new connection can be admitted. The decision is based upon whether the end-to-end delay requirements can be met for both newly arriving and existing connections. Admission control mechanisms for systems with static-priority schedulers differ to a great extent by the point in time during which the delay computations for the admission tests are performed. In this section, we describe two such approaches that perform the delay computation previously described. The first of these approaches executes delay computations at run-time during connection establishment, while the second approach utilizes off-line computations during system (re-)configuration and so simplyfies the computation needed at run-time.

5.1 Delay-Based Admission Control (DBAC)

Delay-based admission control (DBAC) is a mechanism that makes admission decisions by analyzing the system state at run-time. In particular, DBAC performs delay calculations at flow establishment time using run-time information about flows to determine both whether the introduction of the new flow causes existing flows to miss their real-time requirements and whether guarantees can be provided for the new flow. If both of these conditions are met, the flow is accepted, otherwise it is rejected by the admission controller. Note that in systems with multiple priorities, only flows of equal and lower priorities need to be checked.

We employ the equations from previous sections to calculate the delay-violation probabilities required in the DBAC algorithm.Though the DBAC algorithm is conceptually simple, it is computationally complex. Note that an increased delay on nodes on the intended route for the new flow causes long end-to-end delays on routes that cross the new flow. Therefore, for every admission control decision, the admission controller must perform delay violation probability calculations for both the intended flow route, and possibly several other routes in the system as well. Each of these computations requires solving the system of partial differential equations (9) that model the wireless links, as well as an expensive convolution operation in (26) to derive the end-to-end probability. It is also necessary for the admission controller to be aware of the network topology and routing. Thus DBAC becomes extremely costly as the system scales in size [5].

The computational overhead of DBAC motivates the need to develop a system that significantly reduces computational complexity, without sacrificing system efficiency. In the following section we describe how admission control overhead at run-time can be reduced by performing some computation ahead of flow establishment.

5.2 Utilization-Based Admission Control (UBAC)

With UBAC, we decrease the admission control overhead by reducing the amount of computation that needs to be performed at run-time. For this to be possible, we assume that each link server reserves a certain percentage of capacity for every particular traffic priority. It is the responsibility of the admission control module to ensure that the capacity usage of individual

[5] Note that the overhead only occurs at flow establishment time, not during packet forwarding time.

traffic priorities does not exceed the reserved portion. This is necessary to provide isolation among different traffic priorities and hence to guarantee end-to-end delays to the flows.

As opposed to run-time calculations per flow, UBAC requires off-line delay computations per traffic priority to obtain what we call a *safe utilization bound*. Since flow population information is unavailable for off-line calculations, we must obtain a *flow-population-insensitive* statistical delay formula, which can be used to compute the safe utilization bound. During run-time, UBAC checks whether the link utilization allocated to each traffic priority (this allocated utilization should not exceed the safe utilization bound) is not exceeded. The total number n_i of flows of priority i on a link is therefore subject to the following constraint:

$$n_i \leq \frac{\alpha_i}{\rho_i} C, \tag{28}$$

where α_i is the ratio of the link capacity allocated to traffic of priority i, and ρ_i is the average rate of priority i traffic. With this constraint, the mean rate and the rate-variance can be upper-bounded as follows:

$$\phi_i(t) = (t + d_i) \sum_{q=1}^{i-1} \alpha_q C + t\alpha_i C, \tag{29}$$

$$RV_i(t) = (t + d_i) \sum_{q=1}^{i-1} \alpha_q \sigma_q C + t\alpha_i \sigma_i C. \tag{30}$$

Correspondingly, Lemma 4.2, Theorem 4.3 and Equation (27) can be re-formulated using the flow-population insensitive definition for the new $\phi_i(t)$ and $RV_i(t)$ given above. Thus we observe that the benefit of UBAC over DBAC is the former's ability to perform admission control without heavy run-time computations. As our performance evaluation will illustrate, UBAC is still able to provide comparable resource efficiency.

6 Performance Evaluation

In this section, we evaluate the performance of the two approaches discussed in the previous sections. The simulated wireless network could be representative of an ground-space-ground wireless communication system (Fig. 1). We allow any pair of nodes in the network to establish a real-time priority connection (voice in this case). All traffic is routed along the shortest-path route. In our wireless link model, we assume that all links in the network have a maximum capacity of 2 Mbps. Links follow a two-state Markov model as previously defined. In the simulation, we specify the link parameters as follows: $\lambda_0 = 10, \lambda_1 = 30, p_0 = 10^{-6}$, and we vary the bit error rate (BER) p_1 for State 1 (BAD state). We also adopt five different Bose-Chaudhuri-Hocquenghem (BCH) [15] coding schemes for FEC. We assume that requests for real-time flow establishment form a Poisson process, and that flow lifetimes are exponentially distributed with an average of 180 seconds. [6]

In obtaining our results, we are interested in two metrics: i) *WCAU* – The *worst-case achievable utilization* is the maximum link utilization that can be safely allocated to real-time traffic in UBAC; ii) *Admission Probability* – This is the probability that a flow can be admitted without violating delay guarantees. Both metrics reflect on the efficient use of network resources.

We find that the conclusions we draw based on the cases described here generally hold for other cases we have evaluated.

6.1 WCAU Comparison

The underlying network topology in the WCAU experiment is the network shown in Fig. 1, where nodes communicate through a space-based reachback network. We vary the link characteristics by varying the bit error rate (BER) p_1 for State 1 (BAD state). We also consider five different BCH coding schemes with increasing level of correctability (i.e., different (n, k, r) [12]. In our traffic model, we assume that all traffic belongs to a single real-time priority. We simulate voice traffic, with bursts $\sigma = 640$ bits, and average rate $\rho = 32000$ bps. We assume that the end-to-end deadline is 15 ms. The end-to-end deadline violation probability is either 10^{-6} or 10^{-3}.

The WCAU can be computed by Equation (27) that we obtained in the previous section using simple binary search. The results of our WCAU experiments are shown in Fig. 6. The following observations can be made from these results: 1). *Sensitivity of WCAU to channel coding:* Our results show the performance tradeoff of using various channel codes. Codes that provide greater error correction decrease the amount of actual traffic included in packets. For low error rates, this capability is not worthwhile, as shown in Fig. 6, since error correction is rarely useful, and in fact decreases the overall achievable utilization. 2). *Sensitivity of WCAU to BER:* As the BAD-state BER p_1 is increased from 0.001 to 0.01, the WCAU decreases

[6] A real system would support best-effort traffic as well. Since this traffic would not affect the results of this evaluation, we omit it from our experiments.

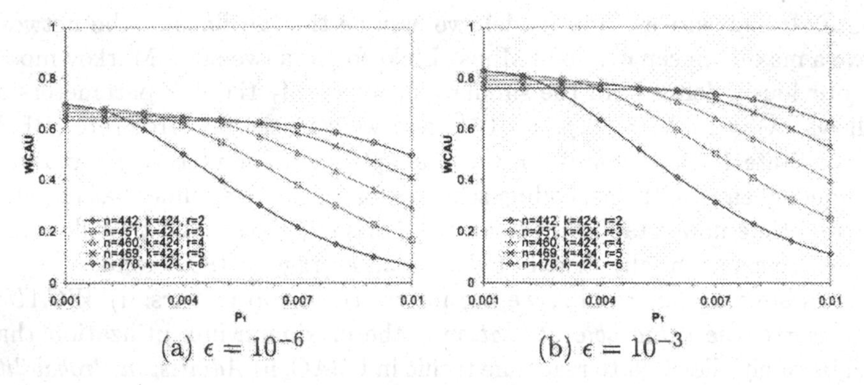

(a) $\epsilon = 10^{-6}$ (b) $\epsilon = 10^{-3}$

Figure 6: WCAU Comparison

for all cases. These results support the intuition that, as the error probability of the network increases, the amount of capacity that can be supported for real-time traffic should decrease. 3). *Sensitivity of WCAU to deadline violation probability:* As expected, the WCAU increases when the deadline violation probability is decreased. In other words, allowing higher loss probabilities creates additional available utilization for real-time traffic.

6.2 Admission Probability Comparison

In addition to the topology (Fig. 1) used in the last sub-section (called *Net 1* in this context), we use a random network topology (generated with GT-ITM [23] using the Waxman 2 method) with the same number of total nodes, which we refer to as *Net 2*. We use this randomly generated topology in order to support the fact that our results are not dependent upon a particular topology. We fix bit error rate (BER) p_1 for State 1 (BAD state) $p_1 = 0.001$ and choose BCH coding scheme with parameters $(n = 442, k = 424, r = 2)$. The end-to-end deadline violation probability is 10^{-6}.

We simulate the case when there is only a single real-time priority in the network with same parameters σ, ρ, d as the first simulation. We also simulate the case when there are two real-time priorities in the network to see how multiple priorities affect the admission probability. In this case, we choose additional higher-priority traffic as follows: $\sigma = 1280$ bits, $\rho = 64000$ bps, $d = 0.005$ s. The capacity is allocation with ratio $\alpha_{high} : \alpha_{low} = 1 : 3$.

As expected, in both cases, the admission probability decreases with increasing flow arrival rate. The substantial conclusion we draw from these

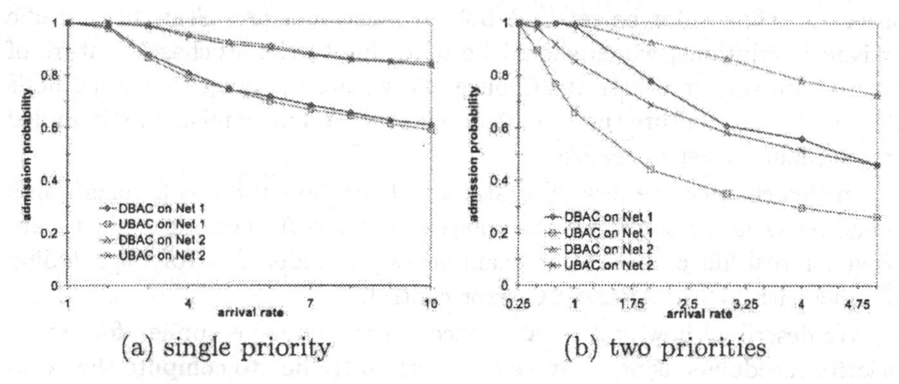

(a) single priority (b) two priorities

Figure 7: Admission Probability Comparison

results is with regard to the relationship between UBAC and DBAC: It is clear from Fig. 7(a) that in the single-priority case UBAC is in fact able to provide the same efficiency with regard to network resource allocation as DBAC. This result is significant because it means that the efficiency of DBAC can be provided with low run-time overhead by using UBAC. Thus costly run-time delay computations can be removed without sacrificing performance. From Fig. 7(b), we find that DBAC obtains more gains in terms of admission probability than UBAC when there are multiple priorities. This can be attributed to the fact that the pre-allocation of capacity in UBAC disables the capacity sharing between the traffic with different priorities, so that the overall achievable utilization is decreased. Therefore, DBAC achieves much higher admission probabilities than UBAC in the multiple-priority case.

7 Conclusions

The statistical nature of service provided by wireless links inherently precludes deterministic delay guarantees. Means must therefore be used that allow definition and enforcement of *statistical* guarantees. This requires a service description that captures the stochastic characteristic of the service provided. We present *statistical service curves* and show how they accurately represent the service provided by wireless links in a tractable manner.

In end-to-end systems, where traffic traverses more than one link, the effect on traffic as it traverses multiple links must be captured in order to formulate traffic arrival characterizations at downstream links, and in order to define enforcement and policing mechanisms. Statistical traffic arrival descriptions, which should be used due to the stochastic nature of service, are very impractical to police. So we are studying to use methods that accurately capture the statistical behavior of deterministically bounded traffic (much easier to realize).

In this chapter, we described statistical service curves as a generic link characterization framework. We showed how this framework could be applied for real-life models. For example, we considered a Rayleigh fading channel with hybrid ARQ/FEC error control.

We described how statistical service curves could be applied for static-priority schedulers using what we call "virtual traffic" to compute the available service to real-time traffic. We did not assume a particular traffic pattern; instead we used rate-variance envelopes, a simple and general traffic characterization. Such envelopes described the variance of the flow rates as a function of the interval length. This methodology made our approach applicable to any particular situation.

We evaluated two different admission control mechanisms (delay-based and utilization-based) and illustrated performance trade-offs between resource utilization and overhead. For each admission control mechanism, statistical delay analysis was performed correspondingly. We described the benefits and drawbacks of both mechanisms and also quantified performance characteristics with simulation data.

References

[1] P. Bello, Aeronautical channel characterization, *IEEE Transactions on Communications* Vol. 21 No.5 (May 1973) pp. 548-563.

[2] J. B. Cain and D. N. McGregor, A recommended error control architecture for ATM networks with wireless links, *IEEE Journal on Selected Areas in Communications* Vol. 15 No. 1 (Jan. 1997) pp. 16-28.

[3] A. Dailianas and A. Bovopoulis, Real-time admission control algorithms with delay and loss guarantees in ATM networks, In *Proceedings of IEEE Infocom* (Toronto, Canada, Jun. 1994).

[4] E. O. Elliott, Estimates of error rates for codes on burst-noise channels, *Bell Syst. Tech. J.* Vol. 42 No. 9 (Sep. 1963) pp. 1977-1997.

[5] R. Fantacci, Queuing analysis of the selective repeat automatic repeat request protocol wireless packet networks, *IEEE Transactions on Vehicular Technology* Vol. 45 No. 2 (May 1996) pp. 258-264.

[6] E. N. Gilbert, Capacity of a burst-noise channel, *Bell Syst. Tech. J.* Vol. 39 No. 8 (Sep. 1960) pp. 1253-1265.

[7] J. Hagenauer and W. Papke, Data transmission for maritime and land mobile using stored channel simulation, In *Proceeding of IEEE Veh. Technol. Conference* (San Diego, CA, USA, 1982).

[8] R. W. Huck, J. S. Butterworth, and E. E. Matt, Propagation measurements for land mobile satellite services, In *Proceeding of IEEE Veh. Technol. Conference* (Toronto, Canada, 1983).

[9] E. Knightly, H-bind: A new approach to providing statistical performance guarantees to VBR traffic, In *Proceedings of IEEE INFOCOM* (San Francisco, CA, USA, Mar. 1996).

[10] E. Knightly, Second moment resource allocation in multi-service networks, In *Proceedings of ACM Sigmetrics* (Seattle, WA, Jun. 1997).

[11] E. Knightly, Enforceable quality of service guarantees for bursty traffic streams, In *Proceedings of IEEE Infocom* (San Francisco, CA, USA, Mar. 1998).

[12] M. Krunz and J. G. Kim, Fluid analysis of delay and packet discard performance for QoS support in wireless networks, *IEEE Journal on Selected Areas in Communications* Vol. 19 No. 2 (Feb. 2001) pp. 384-395.

[13] J. Kurose, On computing per-session performance bounds in high-speed multi-hop computer networks, In *Proceedings of ACM Sigmetrics* (Newport, RI, Jun. 1992).

[14] J. Liebeherr, D.E. Wrege, and D. Ferrari, Exact admission control in networks with bounded delay services, *IEEE/ACM Transactions on Networking* Vol. 4 No. 6 (Dec. 1996) pp. 885-901.

[15] S. Lin and Jr. D. J. Costello, *Error Control Coding: Fundamentals and Applications* (Prentice Hall, New Jersey, 1983).

[16] J. G. Proakis, *Digital Communications*, (McGraw-Hill, New York, 1995).

[17] M. Schwartz, *Broadband Integrated Networks* (Prentice Hall, New Jersey, 1996).

[18] B. Vucetic, An adaptive coding scheme for time-varying channels, *IEEE Trans. Commun.* Vol. 39 (May 1991) pp. 653-663.

[19] H. S. Wang and N. Moayeri, Finite-state Markov channel - a useful model for radio communication channels, *IEEE Transactions on Vehicular Technology* Vol. 45 No. 2 (Feb. 1995) pp. 258-264.

[20] S. Wang, D. Xuan, R. Bettati, and W. Zhao, Differentiated services with statistical real-time guarantees in static-priority scheduling networks, In *Proceedings of IEEE RTSS* (London, UK, Dec. 2001).

[21] S. Wang, D. Xuan, R. Bettati, and W. Zhao, Providing absolute differentiated services for real-time applications in static-priority scheduling networks, In *Proceedings of IEEE Infocom* (Anchorage, Alaska, USA, Apr. 2001).

[22] D. Wu and R. Negi, A wireless channel model for support of quality of service, In *Proceedings of IEEE GLOBECOM* (San Antonio, TX, USA, Nov. 2001).

[23] E. W. Zegura et al., Modeling topology of large internetworks, http://www.cc.gatech.edu/projects/gtitm/.

[24] Q. Zhang and S. A. Kassam, Finite-state Markov model for Rayleigh fading channels, *IEEE Transactions on Communications* Vol. 47 No. 11 (Nov. 1999) pp. 1688-1692.

Fair Real-Time Scheduling over a Wireless LAN

Insik Shin, Sanjeev Khanna, and Insup Lee
Department of Computer Information and Science
University of Pennsylvania, Philadelphia, PA 19131
E-mail: {ishin,sanjeev,lee}@cis.upenn.edu

Contents

References

1 Introduction

Wireless communication technology has gained widespread acceptance in recent years. The IEEE 802.11 standard [10] has led wireless local area networks (LANs) into greater use. Beyond the original bandwidth of 2 Mbps at the wireless medium, the IEEE 802.11a [11] and 802.11b [12] supplements increase the bandwidth to 54 Mbps and 11 Mbps, respectively. With such high bandwidth, the demand for supporting multiple time-sensitive high-bandwidth traffic applications, such as video-on demand and interactive multimedia, in wireless LANs has been increasing. Thus, it is important to provide fair bandwidth allocation to multiple real-time traffics in a wireless LAN.

Many real-time packet scheduling and fair packet scheduling algorithms have been developed for wired networks. However, it is not clear how well these algorithms work for wireless networks, since wireless channels are subject to unpredictable *location-dependent* and *bursty* errors. In the presence of such unpredictable errors, real-time traffic applications can not fully utilize channel bandwidth assigned to them. A real-time traffic application may fail to send or receive some of its real-time packets in time due to channel errors. When experiencing packet losses above some specified thresholds, the application undergoes degraded quality of service (QoS).

In this paper, we investigate the problem of fair scheduling of real-time packets with deadline constraints over wireless LAN, aiming at both achieving fair degradation and maximizing the system throughput. To achieve fairness for the applications that naturally want to minimize the degree of their gradated QoS respectively, we choose the scheduling objective of minimizing the maximum degree of the degraded QoS among all applications. Simultaneously, we try to maximize the overall system throughput. We show that for the problem of throughput maximization alone, there exists a simple online algorithm that achieves a performance ratio of two with respect to the optimal. However, for the problem of achieving fairness, we show that no online algorithm can guarantee a bounded performance ratio with respect to the optimal. Thus, no online algorithm can ensure a bounded performance ratio for the combined objectives. In contrast to the online scheduling problem with unpredictable channel errors, if all the errors are assumed to be known in advance, we show that there is a polynomial time offline scheduling algorithm that optimally achieves our scheduling objectives.

We then study four online algorithms and evaluate their performance using simulations. The first two are EDF (Earliest Deadline First) and GDF (Greatest Degradation First) that consider only one aspect of our scheduling goal respectively. EDF is naturally suited for maximizing throughput while GDF seeks to minimize the maximum degradation. The next two are algorithms, called EOG (EDF or GDF) and DDT (Deadline and Degradation Together), that consider the two aspects of our scheduling goal. EOG simply considers the deadline constraint or the degradation constraint at a time, whereas DDT considers the both constraints at the same time. Our simulation results show that the DDT is as good as the EDF for maximizing the throughput and that it is the best for minimizing the maximum degradation.

This paper is organized as follows. Section 2 provide related works. Section 3 describes our scheduling model, including parameters and objectives. Section 4 provides theoretical results mentioned above. Section 5 presents the four online scheduling algorithms and their examples. Section 6 introduces the error handling mechanisms that can be used by online scheduling algorithms. Section 7 presents the comparative evaluation of the four online algorithms based on simulation results. Section 8 summarizes the paper and identifies future work.

2 Related Work

There has been previous work on providing QoS guarantees over wireless links using call admission and scheduling [4, 5, 6]. While this previous work focused on providing service guarantees, it did not consider the issue of fairness or degradation. Several approaches have been introduced to deal with real-time task scheduling problem considering degraded QoS in an overload situation, where the system cannot meet the deadlines of all tasks. The notion of (m, k)-firm deadlines was introduced in [9] to represent less stringent guarantees for temporal constraints. In the (m, k)-firm deadlines model, it is adequate to meet the deadline constraints of m out of any k consecutive instances of a task in the overload condition. The *imprecise computation* [7, 16] and *IRIS (Increased Reward with Increased Service)* [8, 2] models were proposed to provide minimal quality of service in the overload condition. In these models, each task consists of a mandatory subtask and an optional subtask. The mandatory subtask should be completed before its deadline in order to provide minimal quality of service. The optional subtask can be executed before its deadline in order to enhance the quality of service after the completion of its corresponding mandatory subtask if there are enough resources in the system that are not committed to execute mandatory subtasks of other tasks. The execution of an optional subtask is associated with error in the imprecise computation model or reward in the IRIS model. The longer the optional subtask executes, the smaller the error, or the higher the reward. A typical scheduling objective is to ensure that the mandatory parts of all tasks are completed by their deadlines while the total errors/rewards in the system are minimized/maximized. These models addressed real-time scheduling problem with QoS degradation issues while optimizing a system-wide performance measure (error or reward). However, the issue of fairness was not addressed in these models.

Significant research efforts have been made to adapt packet fair scheduling algorithms to a wireless domain taking care of the wireless channel error characteristics [13, 14, 15]. These packets, however, are assumed to have no deadline constraints. There has been previous study on the issue of temporal fairness in periodic real-time scheduling. The notion of *pfairness* was introduced in [3] to minimize the length of time during which a task is denied service. This fairness notion enforces scheduling each periodic task proportional to its temporal property, which is the ratio of its execution time requirement to its period. This pfairness notion is different from our fairness notion in that the pfairness focuses on the temporal fairness for a single task, whereas we focus on the fairness in reducing the QoS degradations among multiple tasks (flows). Another study [18] is comparable with ours in that the both studies consider the same fairness notion of minimizing the maximum QoS degradation in real-time scheduling. This study used a finite range (window) to keep track of packet losses of real-time multimedia streams and employed a scheduling policy that mainly worked on the greatest degradation (loss) first basis. However, our study is distinguishable with this study in that we consider an optimization issue of a system-wide performance measure such as the system throughput as well as the fairness issue, whereas this study considers only the fairness issue.

3 Scheduling Model and Goals

3.1 System Model

We consider real-time traffic as *isochronous* (or synchronous) traffic that consists of message streams that are generated by their sources on a continuing basis and delivered to their respective destinations on a continuing basis. Such traffic includes periodic and sporadic messages that are characterized by stringent timing constraints. Periodic packets are generated at regular time intervals, and sporadic packets are generated at irregular intervals. The timing constraint of such a message is represented by a *deadline* that is the instant of time by which its delivery is required to be completed. The deadline is said to be *hard* if failure to meet it is considered as a fatal fault. The deadline is said to be *soft* if it is undesirable to miss it but a few misses of the deadlines are tolerable. We refer to a message stream as a flow and a message instance as a packet.

In this paper, we will only consider the scheduling of periodic packets with soft deadlines. Examples of such packets include constant bit-rate (CBR) digitized voice and video data packets. We model a periodic soft real-time flow, f_i, as (p_i, e_i), $1 \leq i \leq N$. This means that each packet of flow f_i is generated with a period of p_i, each packet must be delivered to the destination within p_i units of time from its generation or arrival at the source; otherwise, the packet is lost, and a packet loss rate of up to e_i is acceptable. For simplicity, we assume that all the packets of f_i are of the same size L.

We consider scheduling such real-time flows over a packet-switched wireless network that consists of multiple cells. Each cell is assumed to consist of a base station (BS) and multiple mobile hosts (MHs). Each flow is either an uplink (from a MH to the BS) or a downlink (BS to MH). The BS performs the scheduling of real-time packet deliveries using a polling scheme like one of the standard IEEE 802.11 PCF (Point Coordination Function) protocol [10]. In such a polling scheme, the BS polls MHs according to its polling scheduling algorithm and an MH being polled is allowed to transmit its data packet. The acknowledge of the packet transmission is piggybacked in the next poll packet. The source of each flow f_i should notify the BS of a tuple (p_i, e_i) information prior to being scheduled. Each flow f_i is associated with a channel, ch_i, which is in one of two states, namely, *error state* or *error-free state*, at any time instant. When channel ch_i is in error state, flow f_i experiences a channel error. A packet delivery of flow f_i fails if the flow perceives the channel error at any time instant during the packet delivery.

3.2 Scheduling Objectives

We want to measure the performance of scheduling algorithms along two dimensions: from the points of view of the system and the user.

From the viewpoint of the system, it would be desirable to deliver as many packets as possible over all flows. We define the system throughput, T_{sys}, to be the fraction of delivered packets as follows:

$$T_{sys} = \frac{\sum_{i=1}^{N} M_i^a}{\sum_{i=1}^{N} M_i}, \quad (1)$$

where let M_i be the number of packets that flow f_i was supposed to deliver and let M_i^a be the number of packets that flow f_i actually successfully delivered.

From the viewpoint of the user, each user wants all periodic packets to be delivered on time. However, unpredictable wireless channel errors may cause the user with real-time traffic to receive degraded quality of services due to packet losses. We define a variable, called the degradation value, ϵ_i, to represent the degree of the degradation in quality of service. In other words, the degradation value is the distance between the desired quality of service and the quality of service being served. Each user naturally wants to minimize ϵ_i for its f_i. In order to be fair to all flows, we would like to minimize the maximum degradation value among all the flows. When ϵ_{max} is the maximum degradation value among all the flows, the smaller ϵ_{max} an scheduling algorithm generates, the fairer we consider the algorithm is. We consider a packet loss rate as a QoS parameter and define the degradation value ϵ_i for each flow f_i to be the distance between its acceptable packet loss rate e_i and the actual packet loss rate as follows:

$$\epsilon_i = 1 - \frac{M_i^a}{M_i} - e_i. \tag{2}$$

In the presence of unpredictable errors, the problem is to determine which packets need to be scheduled, and in what order, so as to (a) minimize the maximum degree of degraded service among all flows, and (b) maximize the overall system throughput subject to (a). In other words, our scheduling objective is to determine the smallest ϵ_{max} for which we can ensure that $M_i^a \geq \lceil M_i(1-e_i-\epsilon_{max})\rceil$ and maximize $\sum_{i=1}^{N} M_i^a$.

4 Theoretical Results

We now establish some theoretical results that give an insight into the difficulty of finding good schedules in presence of errors. We start with some simple results concerning the worst-case behaviour of online algorithms for our problem. We measure the performance of any online algorithm in the competitive analysis framework of Sleator and Tarjan [17]. We say that an online algorithm is c-competitive if on any input sequence, it is guaranteed to produce a solution that is at least $1/c$ times as good as an optimal solution. Thus a 1-competitive algorithm gives essentially an optimal solution itself.

Proposition 1 *For any $\delta > 0$, there does not exist a $(2-\delta)$-competitive online algorithm for throughput maximization even when the system contains only two hosts.*

Proof. Consider a system with only two hosts, each one having one packet to transmit at some time t, with deadline $t+2$. If the online algorithm chooses the first time slot to schedule MH_1, an adversary generates an error in time slot 2 for MH_2, causing the online algorithm to loose its packet. On the other hand, if the online algorithm first schedules the packet from MH_2, the adversary generates an error for MH_1 at time $t+1$. In either case, an optimal algorithm could have scheduled 2 packets. This process can be repeated indefinitely and thus the proposition follows. □

In fact, the lower bound above is tight for throughput maximization.

Proposition 2 *There exists a 2-competitive online algorithm for throughput maximization.*

Proof. Consider the following online algorithm: at any time t, schedule any available packet from an error-free mobile host. Let P_{OPT} and P_{ON} denote the set of packets that are scheduled by the optimal offline algorithm and the online algorithm respectively. Moreover, let $P_{Both} = P_{OPT} \cap P_{ON}$ and $P_{Miss} = P_{OPT} \setminus P_{ON}$. Consider a packet $m \in P_{Miss}$ scheduled by the optimal at some time t. Since the online algorithm did not schedule this packet, it must be the case that it scheduled some other packet at time t. Therefore, $|P_{Miss}| \leq |P_{ON}|$. Putting together, $|P_{OPT}| = |P_{Both}| + |P_{Miss}| \leq 2 * |P_{ON}|$. □

It fact, it is easy to see that any greedy online algorithm is 2-competitive for throughput maximization. However, once we take fairness into account, the situation becomes intractable.

Proposition 3 *For any $c \geq 1$, there does not exists a c-competitive algorithm for minimizing ϵ_{max} even when the system contains only two hosts.*

Proof. Consider the same scenario as described in the proof of Proposition 1. Assume that $\epsilon_1 = \epsilon_2 = 0$. Clearly, on the input sequence described there, an optimal algorithm achieves $\epsilon_{max} = 0$. On the other hand, since the online algorithm can be forced to miss half the packets, at least one of the flows has a degradation of $1/2$. Thus for any online algorithm, ϵ_{max} can be forced to be $1/2$. The ratio of these two quantities is unbounded and the proposition follows. □

Given the difficulty of the online case even for the simpler goal of throughput maximization, a natural question to ask is if the problem remains intractable when the errors are all known in advance (i.e., an offline setting). We next show that this case is essentially equivalent to the maximum flow problem, well-known to be solvable efficiently in polynomial time (see [1], for instance).

Theorem 4.1 *There is a polynomial time offline algorithm that determines an optimal fair schedule with maximum throughput.*

Proof. We will reduce our problem to the maximum flow problem. For any $\epsilon > 0$, let $M_i^\epsilon = \lceil M_i(1 - e_i - \epsilon) \rceil$ where $1 \leq i \leq N$. Roughly speaking a schedule is said to be an (ϵ, α)-*schedule* if (i) it schedules at least M_i^ϵ packets for each flow i, and (ii) it sends at least $(\sum_{i=1}^{N} M_i^\epsilon) + \alpha$ packets overall. Our goal is to determine a pair (ϵ^*, α^*) such that (i) for any $\epsilon > \epsilon^*$ there does not exist an $(\epsilon, 0)$-schedule, and (ii) for any $\alpha > \alpha^*$, there does not exist an (ϵ^*, α)-schedule. We will construct an instance of the maximum flow problem for every candidate pair (ϵ, α) and output the solution corresponding to the best such pair found.

Let T denote $\sum_{i=1}^{N} M_i$, the total number of packets. It is easy to verify that there are only $O(\log T)$ candidate pairs need to be considered. We first use binary search over the set $\{1/T, 2/T, \ldots, 1\}$ to identify the smallest ϵ for which an $(\epsilon, 0)$-schedule exists. Then another binary search on the set $\{1, 2, ..., T\}$ determines the largest α for which this schedule stays feasible. Thus from here on, we assume without loss of generality that we know the optimal values of ϵ and α.

We construct a directed graph $G = (V, E)$ as follows. The vertex set of G contains two special vertices, namely a source s and a sink t, vertices $s_1, ..., s_N$ for each of the N flows, a vertex w, vertices $u_1, ..., u_p$ corresponding to the packets generated by the various flows, and vertices $v_1, ..., v_T$ corresponding to the various time slots that are available for scheduling packets. There is a directed edge from a vertex u_i to a vertex v_j if the packet corresponding to u_i could be scheduled at the jth time slot (i.e., the corresponding flow is in good state and the deadline of the packet has not yet expired). There is an edge from any vertex s_i to a vertex u_j if u_j corresponds to the packet generated by the ith flow. There are edges from each vertex v_j to the vertex t. All edges described thus far have a capacity of 1 on each edge. Finally, there are edges from the source vertex s to each of the s_i's as well as to the vertex w. The vertex w is connected to each u_i's by an edge of capacity 1. The directed edge (s, s_i) has capacity equal to M_i^ϵ and the edge (s, w) has capacity equal to α.

It is now an easy consequence of our construction that there exists an s-t flow of value $\sum_{i=1}^{N} M_i^\epsilon + \alpha$ if and only if there exists an (ϵ, α)-schedule. □

5 Online Scheduling Algorithms

This section presents online scheduling algorithms and their examples.

5.1 Online Scheduling Algorithms

We will denote by m_i^k the kth packet of a flow f_i. Each packet m_i^k is associated with an arrival (generation) time $A(m_i^k)$, at which it is ready to be transmitted.

Since each flow f_i is periodic, $A(m_i^k) = A(m_i^{k-1}) + p_i$. Each packet m_i^k is also associated with a deadline $d(m_i^k)$, beyond which m_i^k cannot be scheduled for transmission. $d(m_i^k)$ is simply updated such that $d(m_i^k) = A(m_i^k) + p_i$. A packet is automatically dropped (lost) when its deadline expires (i.e., m_i^k is dropped at $d(m_i^k)$). For simplicity, we abbreviate m_i^k as m_i whenever there is no confusion.

The scheduling goal of our scheduling model is to minimize ϵ_{max} and to maximize T_{sys} subject to the minimum ϵ_{max}. Under online scheduling, however, it is not feasible to maximize T_{sys} after finding the minimum ϵ_{max}. Hence, we slightly modify our scheduling goal for online scheduling algorithms such that our goal is (1) to minimize ϵ_{max} and (2) to maximize T_{sys} simultaneously. We consider the following online scheduling algorithms in order to achieve our scheduling goal.

5.1.1 Deadline First: EDF

This EDF (Earliest Deadline First) algorithm considers the deadline of packets first and then consider the degradation values of packets for tiebreaking. In other words, at time t, a packet m_i is scheduled if $A(m_i) \leq t < d(m_i)$, and $d(m_i)$ is the minimum among all the packets. A packet with a higher degradation value wins a tie. This algorithm is known to provide the maximum overall system throughput when there is no channel error. However, since this algorithm may starve some flows, it is not expected to perform well in minimizing ϵ_{max}.

5.1.2 Degradation First: GDF

This GDF (Greatest Degradation First) algorithm considers the degradation values of packets first and then the deadlines of packets for tiebreaking. At time t, a packet m_i is scheduled if $A(m_i) \leq t < d(m_i)$ and $\epsilon_i = \epsilon_{max}$. A packet with an earlier deadline wins a tie. Intuitively, this algorithm is expected to result in minimizing ϵ_{max}. However, it is not suitable for maximizing the overall system throughput. For instance, we have two packets, m_i and m_k, at time t such that $d(m_i) = t + 1$ and $d(m_k) = t + 2$. We can schedule the two packets if m_i is scheduled at t and m_k is scheduled at $t + 1$. If $\epsilon_k > \epsilon_i$, this algorithm schedules m_k at t. Then, it has no packet to schedule at $t + 1$ since m_i is dropped at $t + 1$.

5.1.3 Deadline or Degradation First : EOG

The EDF and the GDF are expected to perform well in one aspect of our scheduling goal respectively. However, they are not suitable to achieve the two aspects of our scheduling goal simultaneously. Thus, we consider a hybrid algorithm, EOG (EDF or GDF), that attempts to combine positive aspects of the EDF and the GDF in expectation of performing well in the both aspects of our scheduling goal. At any scheduling decision time, if there is a packet whose deadline expires soon, packets are scheduled on the EDF basis. Otherwise, packets are scheduled on the GDF basis. At time t, if there is a packet m_i satisfying both $A(m_i) \leq t$ and $d(m_i) = t + 1$, m_i is scheduled. Otherwise, it chooses a packet m_i satisfying $A(m_i) \leq t < d(m_i)$ and $\epsilon_i = \epsilon_{max}$. Since this algorithm is a hybrid of the EDF and the GDF, it is expected to provide an average performance of the both.

5.1.4 Deadline and Degradation Together : DDT

In addition to the EOG algorithm that considers one aspect of our scheduling goal at a time, we introduce an algorithm that takes into account the two aspects of our scheduling goal simultaneously. This algorithm has an acceptance test for new arrived packets with the following principle: when the packets in a scheduling queue are to be scheduled, a new packet m_i is accepted into the scheduling queue if and only if no packet m_k is removed out of the scheduling queue due to the insertion of m_i into the queue, where $\epsilon_k > \epsilon_i$. We define the notion of virtual deadline for a packet. When a packet m_i arrives, its virtual deadline $v(m_i)$ is initially set as its absolute deadline $d(m_i)$. Then, the acceptance test is performed as follows (its pseudo code is shown in Figure 1):

- If there is no packet m_k in the scheduling queue such that $v(m_i) = v(m_k)$, the packet m_i is accepted into the scheduling queue.

- If there is a packet m_k in the scheduling queue such that $v(m_i) = v(m_k)$ and $\epsilon_k \leq \epsilon_i$, the packet m_i is accepted into the scheduling queue and a new acceptance test is initiated for m_k.

- If there is a packet m_k in the scheduling queue such that $v(m_i) = v(m_k)$ and $\epsilon_k > \epsilon_i$, $v(m_i)$ decreases by one. At time t, if $v(m_i) \geq t$, this acceptance test continues. Otherwise, the packet m_i is not accepted into the scheduling queue.

At any scheduling decision time t, this algorithm chooses a packet from the scheduling queue according to the EDF algorithm. That is, it chooses a packet m_i if $A(m_i) \leq t < v(m_i) \leq d(m_i)$ and $d(m_i)$ is the minimum among all the packets in the scheduling queue. Since this algorithm takes into consideration the deadlines and the degradation values of the flows simultaneously, it is expected to perform well in the two aspects of our scheduling goal.

```
Procedure AcceptanceTest(packet m) {
    /* This procedure performs an acceptance test on packet m based on its virtual deadline */
    /* Let Q be the queue of packets to be scheduled */

    /* a virtual deadline is initially set as an absolute deadline */
    v = GetAbsoluteDeadline(m);
    while (v ≥ t) {                          /* t is the current time */
        k = GetPacketOutOfQueue (v);         /* get a packet k that has a virtual deadline v */
        if (k = NULL) {                      /* if there is no packet k that has virtual deadline v */
            InsertQueue(Q, m);               /* packet m is accepted into the scheduling queue Q */
            return;
        }
        else if (m.ε > k.ε) {                /* m.ε - degradation value of packet m*/
            InsertQueue(Q, m);               /* packet m is accepted into Q */
            /* remove packet k out of Q, and then rename it as m in order to re-calculate
               its virtual deadline */
            m = RemovePacketOutOfQueue (Q, k);
        }
        v = v - 1;        /* decrease the virtual deadline by one */
    }
    /* packet m is dropped, since its virtual deadline is calculated as ahead of the current time */
    DropPacket(m);
}
```

Figure 1: DDT Acceptance Test Pseudo Code

5.2 Example of Online Scheduling Algorithms

We present an example to illustrate the four online scheduling algorithms Suppose that five packets arrive at time t as shown in Table 1. The schedules generated by the algorithms are shown in Table 2. For instance, EDF would schedule m_1 at time $t + 1$ since m_1 has the greatest degradation value among the packets with the earliest deadline. In addition to packets to be scheduled, Table 2 also shows the virtual deadlines of the packets for DDT. DDT would also schedule m_1 at time $t + 1$ since m_1 has the earliest deadline among the packets whose virtual deadlines are greater than t. In this example, EDF and DDT schedule four packets, while GDF and EOG schedule only three packets.

Packet m_i	$d(m_i)$	ϵ_i
m_1	$t+2$	0.08
m_2	$t+2$	0.04
m_3	$t+3$	0.10
m_4	$t+3$	0.06
m_5	$t+4$	0.12

Table 1: Packets in Example

Algorithm	$t+1$	$t+2$	$t+3$	$t+4$
EDF	m_1	m_2	m_3	m_5
GDF	m_5	m_3	m_4	
EOG	m_5	m_1	m_3	
DDT ($v(m_i)$)	m_1 ($t+2$)	m_3 ($t+3$)	m_4 ($t+1$)	m_5 ($t+4$)

Table 2: Schedules in Example

6 Wireless Error Handling

The presence of errors raises the issue of how to handle the situation where a packet was scheduled, but its delivery failed due to a channel error. This section presents a wireless error handling mechanism that can be used by any online scheduling algorithm.

In handling such a packet that experienced an error, one way is to drop the packet and the other way is to reschedule the packet at some time later. The packet drop would not be good if the packet can be rescheduled for a successful transmission later. Considering the bursty nature of wireless channel errors, rescheduling the packet immediately would not be good either. It is desirable to delay rescheduling the packet to sometime later such that it can escape an error burst and can be delivered on time. Since errors are unpredictable, however, it is not clear how long rescheduling the packet should be delayed. We call this rescheduling delay *backoff time*.

The backoff time b_i of each flow f_i needs to be defined long enough to escape a potential bursty error and short enough for the packet to be scheduled prior to the expiration of its deadline. When a packet m_i experienced an error at time t, we consider the following ways to define b_i such that m_i becomes ready for retransmission at $t + b_i$:

- *constant* backoff: b_i is fixed to a constant integer.

- *random-exp* backoff: b_i is randomly chosen in $[1, BW]$, where BW exponentially increases when the same packet experiences another channel error.

- *half-deadline* backoff: b_i is set to $(t + d(m_i))/2$.

7 Simulation Results

This section evaluates the performance of the four online algorithms based on simulation results.

7.1 Simulation Environments

For evaluating the performance of the four online algorithms as a polling scheduling algorithm, we consider a polling scheme such as the IEEE 802.11 PCF protocol. The base station polls a mobile host using a poll control frame, and the polled mobile host is allowed to transmit its data frame. During simulations, we used a logical time unit such that the data frame transmission time is 1-time-unit long. The total simulation time was 50,000 time units for each simulation run. The main simulation parameters manipulated were:

- *Backoff Scheme:* when a flow experiences a transmission failure, its backoff time is determined according to one of the following backoff schemes: constant, random-exp, and half-deadline.

- *Polling Overhead (PO):* for a polling scheme such as the IEEE 802.11 PCF protocol, the polling overhead is defined as the ratio of the poll control frame transmission time to the data frame transmission time. The polling overhead is one of the following values: 0.01, 0.05, and 0.1.

- *The Number of Nodes (N):* we used the parameter of the number of nodes N to characterize the simulation workload. The number of nodes is one of the following values: 1, 2, 4, 8, 16, 32, and 64.

- *Flow Set:* each node has a single periodic flow $f_i(p_i, e_i)$. The system utilization factor is defined as $\sum 1/p_i$. When the system utilization factor U is low, the four online scheduling algorithms perform indistinguishable. Thus, the system utilization factor is randomly chosen in the range [0.95, 0.99] to clearly distinguish the performance of the four algorithms. Then, each p_i is determined in the range [5, 100] according to the system utilization factor. Each e_i is randomly chosen in the range [0.01, 0.05].

- *Maximum Error Duration (ED) and Error Rate (ER):* we used the following two parameters to characterize unpredictable location-dependent bursty wireless channel errors: maximum error duration and error rate. During a single simulation run, each channel has multiple independent error-state intervals to capture the location-dependent error property. The maximum error duration is the upper-limit of a single error-state interval. The duration of each error-state interval is randomly determined in the range between 1 and the maximum error duration to capture the bursty error property. The maximum error duration is one of the following values: 1, 2, 4, 8, 16, 32, and 64 time units. The error rate is defined as the ratio of the total error-state duration to the total simulation time. All channels have the same error rate for each simulation run. The error rate is one of following values: 0.05, 0.1, 0.15, and 0.2.

We have two simulation measures. The first measure is the system throughput T_{sys} which is one aspect of our scheduling goal. The second measure is the maximum degradation value ϵ_{max} among all flows, which is the other aspect of our scheduling goal.

7.2 Simulation Results

For each value of the number of nodes, we created 30 different flow sets, and for each pair of maximum error duration and error rate, we created 10 different wireless channel error scenarios. We define a *simulation case* as a combination of the following four simulation parameters: the polling overhead PO, the number of nodes N, the maximum error duration ED, and the error rate ER. For a single simulation case, we performed 300 simulation runs. At 95% confidence level, all differences to the average simulation results were smaller than or equal to 0.021 in terms of T_{sys} and 0.078 in terms of ϵ_{max} for each simulation case.

Figure 2 compares the performance of three backoff schemes in terms of T_{sys} and ϵ_{max}. With the following simulation parameters: $N = 32$, $ER = 10\%$, $PO = 0.05$, and $U \in [0.95, 1.00]$, we simulated four scheduling algorithms under three backoff schemes and then plotted the average values of the four scheduling algorithms' results. In Figure 2, we can see that the half-deadline backoff scheme outperforms the other two schemes in terms of both the system throughput and the maximum degradation value. In the both figures for T_{sys} and ϵ_{max}, we can see that the performance stays stable when the maximum error duration is between 1 and 16, but begins worse sharply from the maximum error duration of 16 or 32. We believe this can be explained as follows. When 32 flows were created for the system utilization factor between 0.95 and 1.00, the mean period for the 32 flows was close

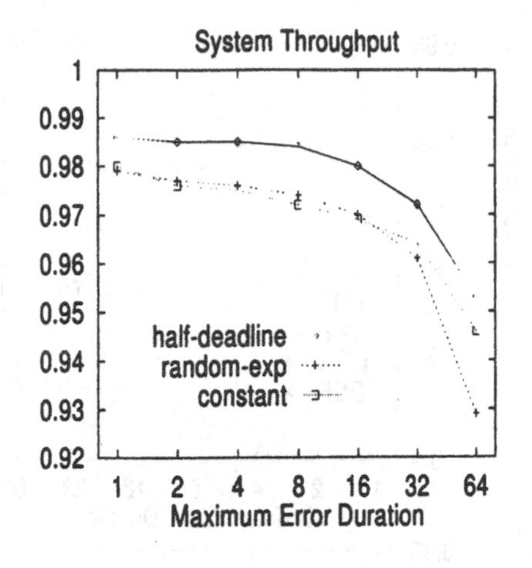

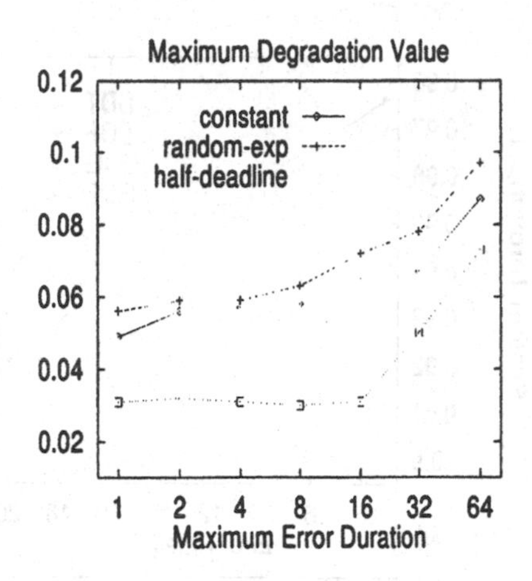

Figure 2: Backoff Scheme Comparison

to 32. According to our error model, the average error duration is a half maximum error duration. Thus, in case of the maximum error duration smaller than 16, the flows that experienced an error would have mostly enough time to recover the error by re-transmitting after the error without a deadline miss. When the maximum error duration begins longer than 16, however, there would be an increasing number of flows that experienced an error but had no chances to re-transmit between the end of the error and a deadline. Given these simulation results showing that the half-deadline scheme is the best among the three backoff schemes, we use only the half-deadline backoff scheme for the other simulations.

To compare the four scheduling algorithms in terms of the system throughput T_{sys} and the maximum degradation value ϵ_{max}, we performed simulations for each possible simulation case of (ER, ED, N, PO) and obtained the average results for each simulation case. Figure 3 plots the average system throughput and Figure 4 plots the average maximum degradation values of the four scheduling algorithms as a function of each simulation parameter, respectively.

Figure 3 compares the four scheduling algorithms in terms of the system throughput T_{sys}. According to Figure 3, we can see that EDF and DDT provide nearly the same system throughput, overlapping each other in the figures, and outperform the other two algorithms across simulation parameters. It is expected since EDF is known as an optimal online scheduling algorithm in maximizing the system throughput and DDT uses EDF in scheduling packets that passed admission tests. The performance gap between EDF/DDT and EOG/GDF increases, as the error rate increases and the number of nodes increases. When the maximum error duration increases, however, the performance gap decreases. As shown in Fig-

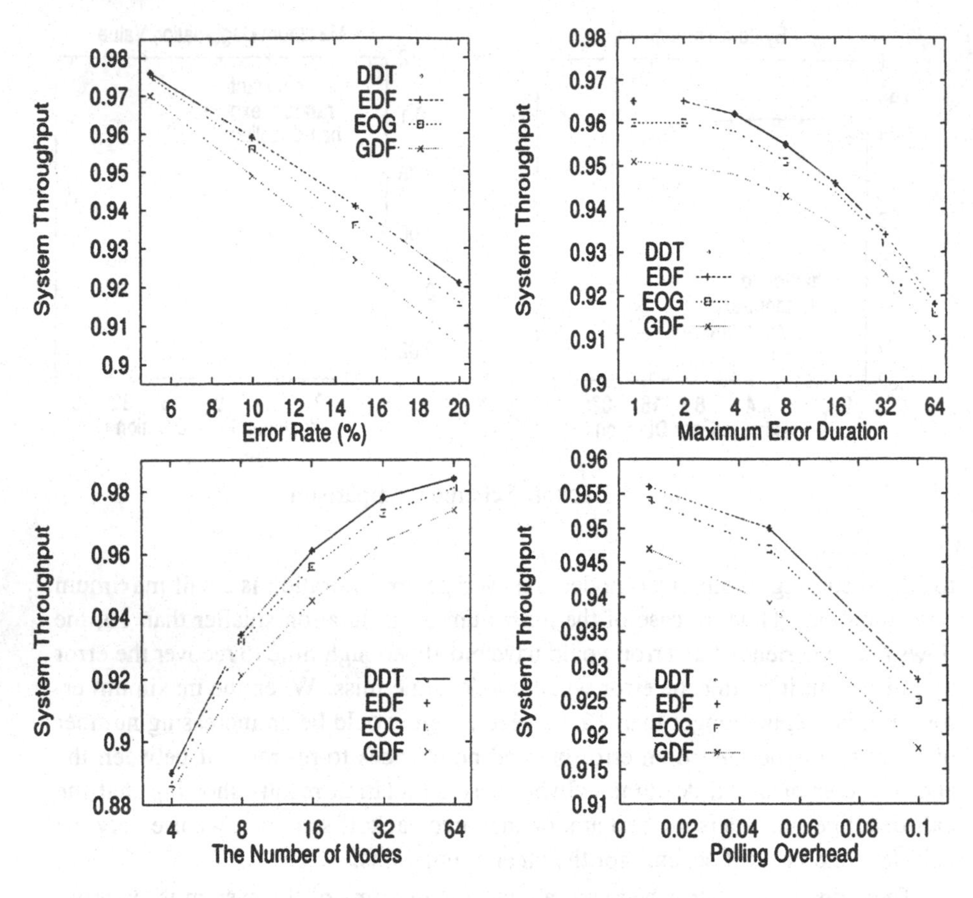

Figure 3: Online Algorithm Comparison: System Throughput

ure 2, when the maximum error duration increases, many flows begin to lose retransmission chances prior to deadlines and thus the system throughput decreases leaving smaller chances for different algorithms to show their different scheduling behaviors. Unlike the other three simulation parameters, the polling overhead is not a considerable factor to affect the system throughput gaps among the scheduling algorithms.

Figure 4 compares the four scheduling algorithms in terms of the maximum degradation value ϵ_{max}. It is shown that DDT and EOG show very close performances in minimizing ϵ_{max}, nearly overlapping in the figures, and outperform the other two algorithms. Even though GDF is focused on minimizing ϵ_{max}, it is interesting that DDT and EOG perform better than GDF. We believe that this phe-

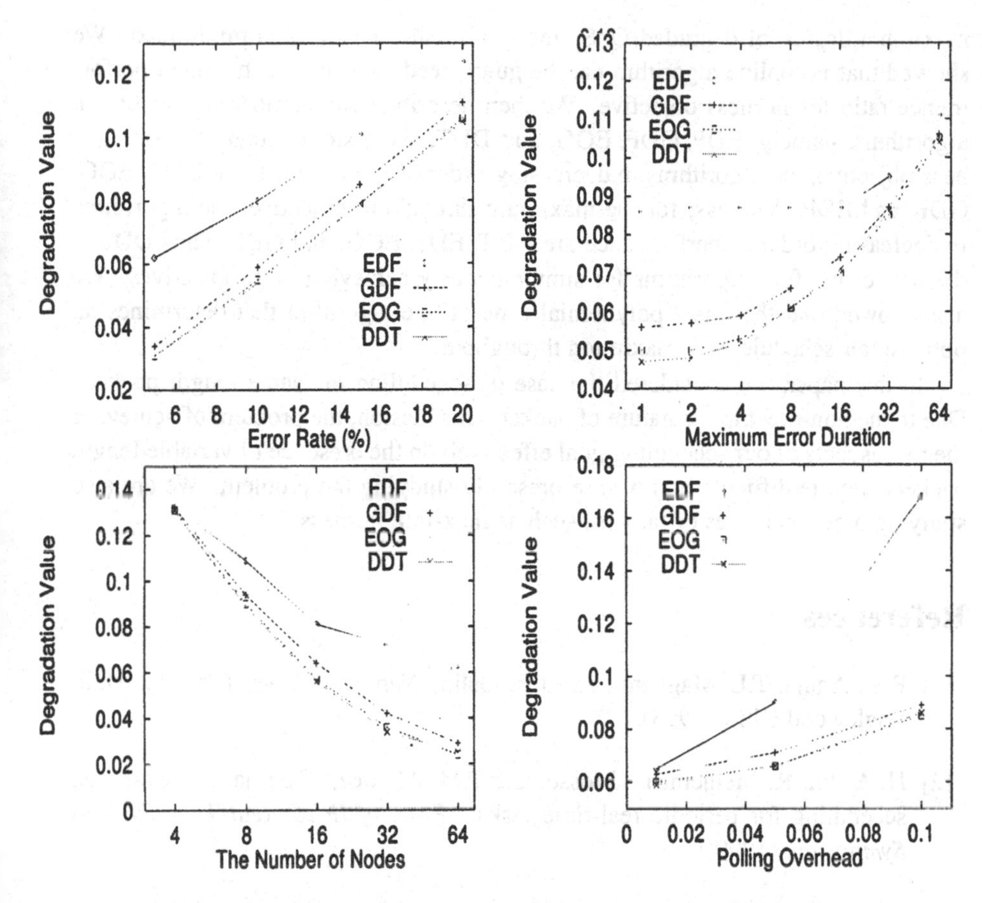

Figure 4: Online Algorithm comparison: Maximum Degradation

nomenon can be explained by the fact that DDT and EOG achieve higher system throughput than GDF does. As one might expect, EDF performs very poorly in comparison to the other algorithms. While the performance gap between EDF and the other algorithms is very sensitive to the maximum error duration, the number of nodes, and the polling overhead, it is relatively insensitive to the error rate.

8 Conclusion

We studied the problem of scheduling packets with deadlines in wireless network with unpredictable channel errors. We considered scheduling objectives of achieving fairness and maximizing the overall throughput. Fairness is to ensure that the maximum degree of degraded QoS among all real-time flows is minimized. We showed that no online algorithm can be guaranteed to achieve a bounded performance ratio for fairness objective. We then described and compared four online algorithms, namely, EDF, GDF, EOG, and DDT, using simulations. For the fairness objective, the algorithms in decreasing order of performance are DDT, EOG, GDF, and EDF. Whereas, for the maximum throughput objective, the algorithms in decreasing order of performance are DDT, EDF, EOG, and GDF. Thus DDT is the best of the four algorithms for simultaneously achieving both objectives. We also showed that there is a polynomial time offline algorithm that determines an optimal fair schedule with maximum throughput.

In this paper, we considered the case of scheduling the same length packets. Due to the non-preemptive nature of packet transmission, the problem of achieving the two aspects of our scheduling goal effectively in the presence of variable length packets is quite difficult, and we are presently studying the problem. We are also studying other measures of fairness such as max-min fairness.

References

[1] R.K. Ahuja, T.L. Magnanti, and J.B. Oslin, *Network Flows*, (Prentice-Hall, Englewood Cliffs, 1993).

[2] H. Aydin, R. Melhem, D. Mosse, and P.M. Alvarez, Optimal reward-based scheduling for periodic real-time tasks, *Proc. of IEEE Real-Time Systems Symposium* (1999).

[3] S. Baruah, N. K. Cohen, C. G. Plaxton, and D. A. Varvel, Propotionate Progress: a Notion of Fairness in Resource Allocation, *Algorithmica* Vol. 15 No. 6 (1996) pp. 600-625.

[4] J. Capone and I. Stavrakakis, Delivering diverse delay/dropping qos requirements in a TDMA environment, *Proceedings of ACM/IEEE MOBICOM conference* (Sep. 1997).

[5] C. Chang, K. Chen, M. You, and J. Chang, Guaranteed quality-of-service wireless access to ATM networks, *IEEE Journal of Selected Areas in Communications* Vol. 15 No. 1 (1997).

[6] S. Choi and K. G. Shin, A unified wireless LAN architecture for real-time and non-real-time communication services, *ACM Transactions on Networking* (Feb. 2000).

[7] J.-Y. Chung, J. W.-S. Liu, and K.-J. Lin, Scheduling periodic jobs that allow imprecise results, *IEEE Transactions on Computers* Vol. 19 No. 9 (1990).

[8] J. K. Dey, J. Kurose, and D. Towsley, On-line scheduling policies for a class of IRIS (Increasing Reward with Increasing Service) real-time tasks, *IEEE Transactions on Computers* Vol. 45 No. 7 (1996) pp. 802-813.

[9] M. Hamdaoui and P. Ramanathan, A dbynamic priority assignment technique for streams with (m, k)-firm deadlines, *IEEE Transactions on Computers* Vol. 44 No. 12 (1995) pp. 1443-1451.

[10] IEEE, *IEEE Std 802.11 - Wireless LAN Medium Access Control (MAC) and Physical layer (PHY) Specifications* (1997).

[11] IEEE, *IEEE Std 802.11a - Wireless LAN Medium Access Control (MAC) and Physical layer (PHY) Specifications: High Speed Physical Layer(PHY) in the 5 GHz Band* (1999).

[12] IEEE, *IEEE Std 802.11b - Wireless LAN Medium Access Control (MAC) and Physical layer (PHY) Specifications: High Speed Physical Layer(PHY) in the 2.4 GHz Band* (1999).

[13] S. Lu, V. Bharghavan, and R. Srikant, Fair scheduling in wireless packet networks, *IEEE/ACM Transactions on Networking* Vol. 7 No. 4 (1999).

[14] S. Lu, T. Nandagopal, and V. Bharghavan, A wireless fair service algorithm for packet celluar networks, In *Proceedings of ACM/IEEE MOBICOM Conference* (Oct. 1998).

[15] T. Ng, I. Stoica, and H. Zhang, Packet fair queueing algorithms for wireless networks with location-dependent error, In *Proceedings of ACM/IEEE MOBICOM Conference* (Oct. 1998).

[16] W.-K. Shih and J. W.-S. Liu, On-line scheduling of imprecise computations to minimize error, In *Proc. of IEEE Real-Time Systems Symposium* (Dec. 1992).

[17] D.D. Sleator and R.E.Tarjan, Amortized efficiency of list update and paging rules, *Communications of the ACM* Vol. 28 No. 2 (1985) pp. 202-208.

[18] R. West, K. Schwan, and C. Poellabauer, Scalable scheduling support for loss and delay constrained media streams, *Proc. of IEEE Real-Time Technology and Applications Symposium* (Dec. 1998).

Inter-Domain Radio Resource Management for Wireless LANs

Yasuhiko Matsunaga
Internet Systems Research Laboratories
NEC Corporation
E-mail: y-matsunaga@bl.jp.nec.com

Randy H. Katz
Computer Science Division
University of California, Berkeley
E-mail: randy@cs.berkeley.edu

Contents

1 Introduction

Efficient radio resource usage has always been a primary concern in wireless communication systems, such as wireless LANs (WLANs) and cellular systems. Demands for mobile communications have been increasing rapidly over the last decade although the frequency range suitable for personal communications is scarce.

To date, radio resource management (RRM) researches have focused on intra-domain issues, particularly radio resource usage optimization by dynamic channel allocation [1, 2] or load balancing [3] inside a single wireless network operator. This is a valid assumption in most cellular systems where each spectrum range is exclusively licensed to a specific wireless operator. On the contrary, WLANs use unlicensed frequency bands shared by various public and private systems, and their radio resources are managed independently [4]. As the number of WLAN systems proliferates and overlap, lack of radio resource management coordination can cause significant performance degradation due to inter-domain interference. Moreover, the partitioning of frequencies independent of actual user/traffic density leads to congestion in the successful operators' network, while less popular operators retain unused excess capacity.

To overcome this situation, we propose a radio resource broker that enables coordinated radio resource management among WLAN systems. The radio resource broker (RRB) collects radio link configuration and statistics from different domains, and optimizes radio resource usage by changing frequency channel or transmission power at access points. The RRB also performs network-initiated handover for dynamic load balancing across domains.

Such a cooperative approach is suitable for environments like multi-tenant buildings and university campuses where there is a strong incentive to use spectrum resources efficiently and fairly. Even for competing public hotspot WLAN operators, the RRB offers an approach for maximizing the performance of otherwise uncoordinated independent network deployments.

Although our current research is targeted for WLANs, the basic principles of radio resource redistribution can be applied to other wireless systems. For example, some European cellular network operators intend to share the same frequency pool as well as radio access network equipment [5]. The notion of an RRB is also useful for cellular operators who share radio resources, and can help enable the spectrum underlay market being investigated by the spectrum policy task force of FCC [6].

In this paper, we first review the current status of regulation and standardization activities on sharing radio resources in Section 2. Then we introduce our inter-domain radio resource management architecture, functions of the RRB, and radio resource usage optimization and redistribution procedures performed at the RRB in Section 3. Because the workload characteristics affect traffic engineering performance, we measure the WLAN load statistics in a university campus environment and derive an empirical WLAN workload model in Section 4. Using this model, we evaluate the performance of proposed inter-domain RRM scheme by simulation in Section 5. We show the optimality of radio resource usage versus channel stability, and demonstrate the effectiveness of radio resource redistribution through dynamic channel compensation and network-initiated load balancing. Finally, we list the related research works in Section 6 and conclude the paper in Section 7.

2 Regulations and Standardization Activities

This section gives background information on regulation and standardization related to sharing radio resources among different domains in unlicensed and licensed bands.

2.1 Unlicensed Band

Unlicensed band devices are subject to control output power less than the limit specified by local regulations. The power cap of unlicensed devices is usually set much smaller than that of the licensed band devices to mitigate interference. Most current WLAN systems conform to IEEE 802.11a/b/g standards and operate in either 2.4 GHz or 5 GHz band. These bands are getting congested due to the prolific growth of WLAN systems. The fundamental solution to avoid congestion is to allocate additional spectrum. Since there is no extra frequency space available in 2.4 GHz band, regulatory bodies have been examining the feasibility of allocating additional spectrum in 5 GHz band. In the FCC regulatory domain, 5.15-5.25 GHz (UNII-1), 5.25-5.35 GHz (UNII-2), and 5.725-5.875 GHz (UNII-3) bands are already allocated for WLANs. In Europe, 5.15-5.35 GHz and 5.47-5.725 GHz bands are available for WLANs, provided that the devices are capable of Dynamic Frequency Selection (DFS) and Transmit Power Control (TPC) to avoid harmful co-channel operation with radar systems [7]. In Japan, 5.15-5.25 GHz, 4.90-5.00 GHz, and 5.030-5.091 GHz bands are allocated for WLANs (The latter two bands are dedicated for public use, and the access points must be licensed. The maximum output power limits of the unlicensed and licensed WLAN stations are 10 mW and 250 mW, respectively.) Current unlicensed band allocations in the FCC, European and Japanese regulatory domains are summarized in Table 1. Furthermore, it was agreed on global basis in WRC-03 [8] that 5.15-5.25 GHz, 5.25-5.35 GHz and 5.47-5.725 GHz bands will be assigned for WLAN devices with interference mitigation techniques such as DFS.

Regulatory Domain	Frequency Band	Maximum Output	Common Unlicensed Devices
FCC	2.4-2.4835GHz	1W	WLAN, Bluetooth, Cordless Telephone, Microwave Oven
	5.15-5.25GHz	50mW	WLAN (indoor)
	5.25-5.35GHz	250mW	WLAN
	5.725-5.825GHz	1W	WLAN, Cordless Telephone
Europe	2.4-2.4835GHz	100mW (EIRP)	WLAN, Bluetooth, Microwave Oven
	5.15-5.35GHz	200mW (EIRP)	WLAN (indoor)
	5.47-5.725GHz	1W (EIRP)	WLAN
Japan	2.4-2.497GHz	10mW/MHz (EIRP)	WLAN, Bluetooth, Microwave Oven
	4.9-5.0GHz	10mW	WLAN (public use)
	5.03-5.091GHz	10mW	WLAN (public use)
	5.15-5.25GHz	10mW/MHz (EIRP)	WLAN (indoor)

Table 1: Unlicensed Band Allocations for Wireless LANs

With regard to standardization, IEEE standard 802.11 [9] specifies a CSMA/CA mechanism for a medium access control among WLAN devices. When a WLAN device execute CSMA/CA, it senses the medium before transmitting a packet over the air. If it encounters another device already transmitting, it will randomly back off and wait to reattempt the transmission. Besides the basic CSMA/CA mechanism, a protocol for TPC and DFS was standardized in the 802.11h task group [10]. 802.11h is primarily intended for 5 GHz WLANs to coexist with radars in Europe, but it can also be used in rest of the world. Benefits of TPC and DFS are auto-configuration, interference reduction, uniform utilization of available channels, and link reliability improvement. An outline of TPC and DFS procedures specified in the 802.11h standard is as follows.

1. The access point advertises a country code and a local power constraint to the stations in beacon and probe response frames. The station adjusts transmit power according to the country's regulatory requirement and local constraint.

2. The station associates with the access point, notifying its transmit power ranges and supported channels.

3. The access point and the station may exchange 'TPC Request' and 'TPC Report' frames to estimate link margin and to control power level precisely.

4. The access point and the station may exchange 'Measurement Request' and 'Measurement Report' messages to estimate interference levels in the region. Before sending a 'Measurement Request' message, the access point may send a 'Quiet' message to the stations to suppress packet transmission during the measurement period.

5. Based on the measurement results, the access point may decide to switch to the new channel, and inform the associated stations. The decision algorithm is beyond the scope of the standard,

While such TPC and DFS mechanisms provide an accurate estimation of interference levels, they do not distinguish the source WLAN domains from which interference comes. Therefore 802.11h itself does not enable fair radio resource allocation among WLAN domains.

2.2 Licensed Band

In licensed bands, it is common that two or more services share the same spectrum. Each service is classified as primary or secondary for the spectrum, and stations of secondary services must to give way to stations of primary services by mitigating mutual interference. When several services share the same bands, they often must coordinate, or carefully select frequency, location and power. Trade associations and consulting engineers typically perform this coordination on behalf of licensees [11].

Recently, FCC adopted spectrum leasing rules that enables licensees to lease their rights in any amount of spectrum and for any period during the term of the license easily [12]. Such secondary spectrum market is expected to provide more efficient and dynamic use of spectrum resources, and promote the growth of new innovative services. Similar arguments have been made by the government of United Kingdom in 2002 [13].

At the time of writing (Feb. 2004), 3GPP (3rd Generation Partnership Project) is investigating the signaling architecture of RAN (Radio Access Network) sharing across different mobile network operators [14, 15]. RAN sharing is attractive for mobile operators to share the heavy deployment costs for mobile networks, especially in the roll-out phase. It is expected that the operators do not only share the radio network elements, but may also share the radio resources, e.g. the operators' licensed 3G spectra.

3 Inter-Domain Radio Resource Management

3.1 Radio Resource Broker Concept

Since our goal is to allocate shared radio resources fairly across multiple WLAN domains, we need to have a trusted third party agent who is independent from each domain's financial interests. Figure 1 depicts the role of a proposed radio resource broker (RRB), which acts as a point of radio resource coordination across domains. In Figure 1, the geographical coverage of WLAN domains A, B, and C are overlapped. The RRB collects measured radio resource usage statistics from access points and mobile clients in each domain by SNMP (Simple Network Management Protocol) [16]. It is a reasonable assumption that the WLAN access points can be monitored and controlled remotely because most enterprise-class access points support SNMP. SNMP version 3 also provides user authentication and message encryption functions [17]. Prior to the measurement, the RRB should know the IP address and generalized region location of each access point to create the coverage overlap map. To collect statistics from individual mobile clients and control them, we need a new client Management Information Base (MIB) for radio resource management. The RRB then analyzes the measured data and provides feedback to optimize the radio resource usage and to redistribute the radio resources fairly across domains. Feedback includes optimal channel and power allocation for the access points, and network-initiated inter-domain handover for mobile clients. The details of the feedback mechanisms are described in the following sections.

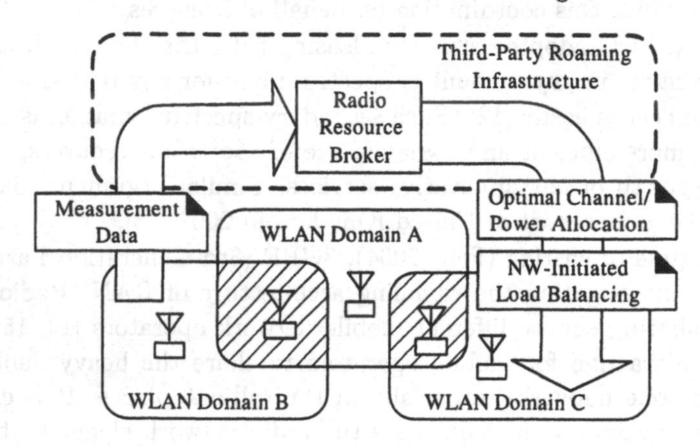

Figure 1: Radio Resource Broker.

3.2 Resource Usage Optimization

In this section, we describe the radio resource usage optimization framework based on an integer programming formulation. The RRB attempts to optimize each domain's radio resource usage by minimizing the network cost function $\mathrm{F}(x)$ under the restrictions of available channels and power levels. The objective function and the constraints are:

$$\min \quad \mathrm{F}(x) = \sum_{i=1}^{N_x} \Phi(\rho(i, Channel(x), Power(x))) \tag{1}$$

$$\begin{aligned} \text{s.t.} \quad \rho(i, Channel(x), Power(x)) &\leq 1 \\ Channel(x) &\in Set\{C_{x1} \cdots C_{xj}\} \\ Power(x) &\in Set\{P_{x1} \cdots P_{xk}\} \end{aligned} \tag{2}$$

where N_x is the number of cells in domain x, $\Phi(y)$ is the link cost function, and $\rho(i, Channel(x), Power(x))$ is the link utilization at cell i in domain x. The first constraint means the link utilization must be less than 1. The numbers of available channels and power levels are set by the RRB, and remain constant during each optimization process. It is well known that load balancing can be an effective allocation strategy, when the cost is convex as a function of the allocated loads [3]. Therefore we adopted a piecewise linear convex link cost function $\Phi(y)$ similar to the one used in the traffic engineering research paper [18].

$$\Phi(y) = \begin{cases} 1 & \text{for} \quad 0 \leq y < 1/3 \\ 5 & \text{for} \quad 1/3 \leq y < 2/3 \\ 50 & \text{for} \quad 2/3 \leq y < 3/4 \\ 500 & \text{for} \quad 3/4 \leq y < 4/5 \\ 5000 & \text{for} \quad y \geq 4/5 \end{cases} \tag{3}$$

$\rho(i, Channel(x), Power(x))$, the link utilization function of cell i, is defined as the sum of own-cell utilization and co-channel utilization by neighbor cells. Strictly, the cell utilization should consider link adaptation because the wireless medium usage depends on the link speed (e.g. 1, 2, 5, 11 Mb/s for 802.11b) used at the time of packet transmission. If per link-speed statistics are not available from the access points, the utilization can be approximated by calculating the ratio of the measured load to the maximum throughput.

$$\begin{aligned} &\rho(i, Channel(x), Power(x)) \\ &= \text{(cell } i\text{'s utilization)} + \text{(co-channel utilization by neighbor cells)} \\ &= \text{(cell } i\text{'s utilization)} \\ &\quad + \sum_{x=1}^{N_d} \sum_{j=1, j \neq i}^{N_x} [\delta_{ij} \bullet \text{(cell } j\text{'s utilization)} \bullet S_{ij}/S_i] \end{aligned} \tag{4}$$

where

$$\delta_{ij} = \begin{cases} 1 & \text{if cell } i\text{'s channel} = \text{cell } j\text{'s channel} \\ 0 & \text{otherwise} \end{cases} \tag{5}$$

N_d is the number of domains, S_i is cell i's service area, and S_{ij} is the overlapped region of cell i's service area and cell j's interference area. We assume every cell uses non-overlapped channels, and clients are uniformly distributed and associate with the nearest access point for the sake of simplicity. The service area and the interference area shapes are approximated as circles. The radius of the interference area is where received signal level equals the carrier sense threshold. In general, the interference area is much larger than the service area [19]. The overlapped region S_{ij} can be calculated analytically under the condition that each area is represented by a circle. To calculate the overlapped region S_{ij}, we assumed indoor path loss model recommended in ITU-R P.1238-2 [20],

$$L[dB] = 20 \log_{10} f + N \log_{10} d + L_f(n) - 28 \tag{6}$$

where L is the total loss, f is the frequency in MHz, N is the distance power loss coefficient, d is the separation distance between the access point and the client terminal, and L_f is the floor penetration loss factor. For example, the N value of 30 is suggested for 2.0 GHz in an office environment. Since radio propagation can be very complex in indoor and metropolitan environments, one can elaborate the calculation of S_i and S_{ij} by making use of three-dimensional radio propagation prediction tools [21].

3.3 Radio Resource Redistribution

After the RRB performs radio resource optimization for each domain described in the Section 3.2, it compares each domain's radio resource usage and redistributes those resources. The algorithm is shown in Figure 2. The calculation is executed for all administrative domains sharing radio resources in the region of interest. Examples include an office building, a university campus, and an airport.

The RRB calculates each domain's network cost with (F2(x)) and without (F1(x)) assuming the presence of other domains. F2(x) is identical to the network cost function as defined in Equations (1), (3), (4) and (5). F1(x) differs from F2(x) in that F1(x) omits per-domain summation of co-channel utilization by neighbor cells in Equation (4). Then the RRB derives the cross-domain impairment I(x) by subtracting F1(x) from F2(x). The cross-domain impairment relative to the average of impairment of all (N_d) domains is accumulated as a credit. Intuitively, a WLAN domain with larger credit than others has been experiencing interference more severe than others. The RRB allocates more radio resources to the domains with a larger credit by loosening channel or power constraints in Equation (2). It also allocates less radio resources to the domains with a smaller credit by tightening channel or power constraints, or initiating an inter-domain handover to a less-congested domain.

Note that the radio resource redistribution and resource usage optimization processes are separated. Allocating more radio resources to the domain with larger credit can degrade the global performance, because it is suffering

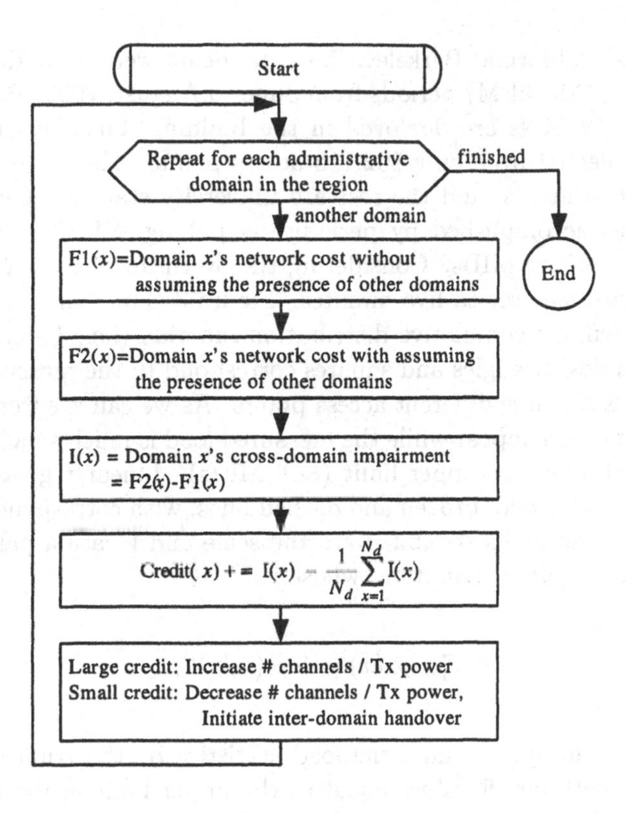

Figure 2: Radio Resource Redistribution Flowchart.

more interference from other domains. The RRB optimizes radio resource usage under the constraint set by radio resource redistribution process.

4 Workload Characterization

4.1 Measurement

Since the workload greatly affects resource allocation control performance, we need an accurate WLAN workload model with which to assess the effectiveness of our RRB architecture. To the authors' knowledge, previous WLAN measurement studies [22, 23, 24] show various aspects of WLAN statistics but none of them gave a realistic workload model suitable for simulations. Therefore we collected load distribution statistics from WLAN access points located at the Computer Science Division's building at the

University of California, Berkeley. Measurements were done during weekday daytime (9AM-6PM) periods from June to August, 2003. Both 802.11a and 802.11b WLANs are deployed in the building, but the measurement data were collected from only 802.11b access points. The majority of users are graduate students, and the rest are university staff and faculty. Measurement was accomplished by periodically polling WLAN access points' interface and bridge MIBs. Considering the overhead of measurement, the polling interval was set to five minutes. Figure 3 shows the log-log plot of the complementary cumulative distribution function (i.e., 1 - c.d.f.) versus the load. Circles, triangles and squares correspond to the measured WLAN load statistics at three different access points. As we can see from Figure 3, linear relationships appear while the measured load is much smaller than the MAC layer throughput upper limit ($\sim$ 6 Mb/s). Linear regression results are also plotted in solid, broken and dashed lines, with corresponding Pareto distribution parameters. α and β are the scale and location parameters of a generalized Pareto distribution, whose c.d.f. is

$$F_{pareto}(x) = 1 - (\alpha/x)^{\beta}. \tag{7}$$

Therefore we can approximate the load statistics by the truncated Pareto distribution, with cutoff values equal to the upper limit of the MAC layer throughput. Figure 4 shows the measured burst duration statistics of the same WLAN access points. We used a burst threshold of 100 kb/s to define ON/OFF states. Observed ON/OFF duration statistics also showed heavy-tailed behavior, and OFF durations were generally longer than ON durations. The distributions of ON and OFF states are also approximated by the Pareto distribution.

4.2 Empirical WLAN Workload Model

Following the measurement results in the previous section, we derived an empirical WLAN workload model as shown in Figure 5. The per-cell workload is modeled as a two-state Markovian arrival process. The load is generated only in the ON state, and the load distribution is a truncated Pareto distribution. The ON-state duration, and the OFF-state duration also follow Pareto distribution. We used this empirical WLAN workload model in the simulations for performance evaluation.

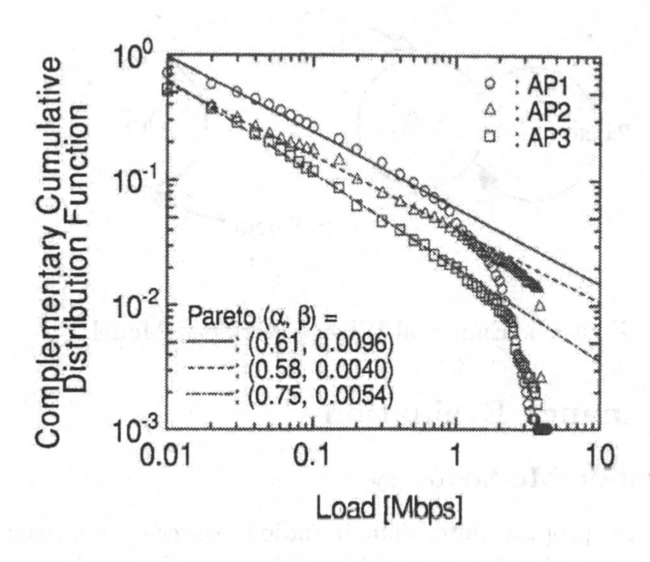

Figure 3: Measured WLAN Load Statistics.

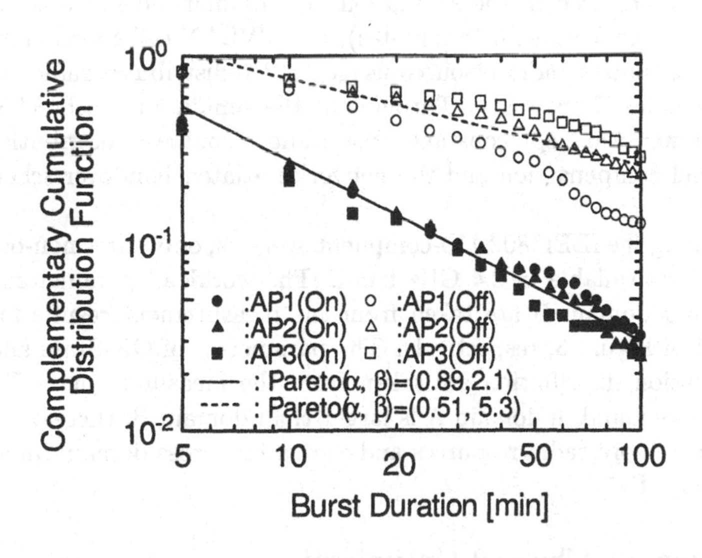

Figure 4: Measured WLAN Burstiness Statistics.

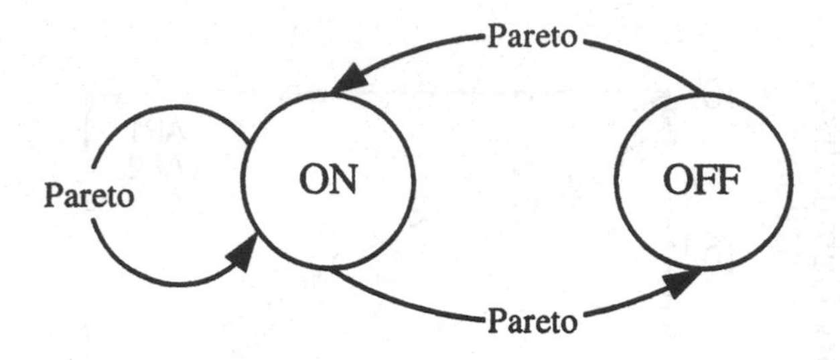

Figure 5: Empirical WLAN Workload Model.

5 Performance Evaluation

5.1 Simulation Methodology

We confirmed the proposed inter-domain radio resource optimization and redistribution schemes by simulation. Figure 6 shows the cell layout and the initial channels of access points. Two WLAN domains are co-located and sharing the radio resources. Each access point is separated by 30 m, and domain A's access points and domain B's access points are placed alternately. The parameters used in the simulation are summarized in Table 2. After each measurement interval (5 minutes), each WLAN cell's load is updated, the RRB optimizes radio resource usage, and redistributes radio resources among domains if necessary. Throughout the simulation, each cell's transmission power was kept constant. For radio resource redistribution, only the channel compensation and the network-initiated handover schemes are evaluated.

Assuming the IEEE 802.11b-complient systems, only three non-overlapping channels are available in 2.4 GHz band. The workload parameters for domain A and domain B are taken from the measurement results for AP 1 and AP 3 in Figure 3, respectively. The parameters of ON-State and OFF-State duration distribution are taken from the measurement in Figure 4. The traffic demand in domain A is larger than domain B, therefore domain A consumes more radio resources and incurs less cross-domain impairment than domain B.

5.2 Dynamic Channel Optimization

Dynamic channel allocation helps to reduce the probability of congestion among multiple cells by assigning different channels to congested cells. Fig-

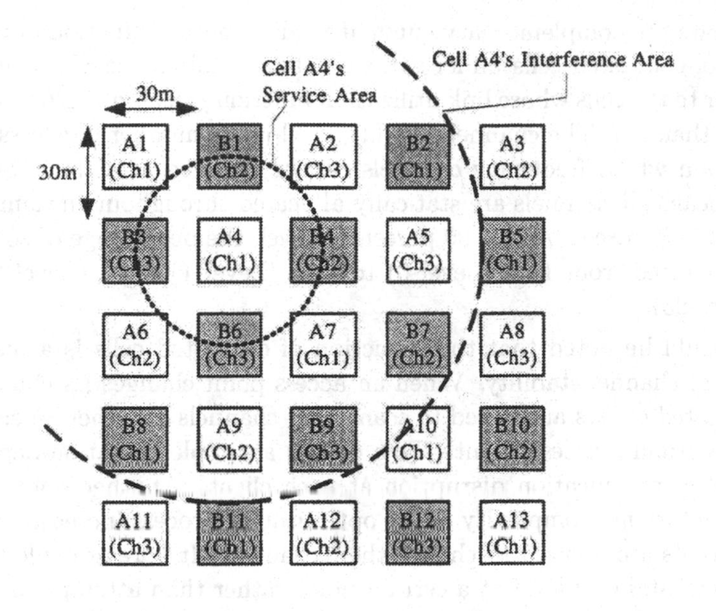

Figure 6: Cell Layout.

	Domain A	Domain B
Number of APs	13	12
Measurement Interval	5 minutes	
Number of Channels	3	
Cell Radius	30 m	
Tx Power	15 dBm	
Radio Frequency	2.4 GHz	
Carrier Sense Threshold	-82 dBm	
ON-State Load α	0.61	0.75
ON-State Load β	0.0096	0.0054
ON-State Load Cutoff	6.0 Mb/s	6.0 Mb/s
ON-State Duration α	0.89	
ON-State Duration β	2.1	
OFF-State Duration α	0.51	
OFF-State Duration β	5.3	

Table 2: Simulation Parameters.

ure 7 shows the complementary cumulative distribution function of the percentage of congested cells under various channel stability levels. Congested cells refer to the cells whose link utilization function $\rho(i, Channel(x), Power(x))$ is larger than 0.8. The channel stability level is the number of access points per domain whose frequency channels can be changed at a time. Stability level 0 means all channels are statically allocated throughout the simulation as shown in Figure 6. At the 99 percentile level, the percentage of congested cells is reduced from 14% (Level 0) to 13% (Level 1), 10% (Level 2), and 7.6% (Level3).

It should be noted that the reduction of congested cells is achieved at the cost of channel stability. When an access point changes its channel, all its associated clients are forced to scan other channels and then re-associate with the original access point. This causes an implementation-dependent temporal communication disruption at each client. Another point is that the computational complexity of the optimization process increases as more access points are allowed to change their channels. It is reasonable to limit the channel stability level at a certain point rather than attempt to achieve optimal usage by changing channels frequently. Channel stability level 2 is used in the simulations in the following sections.

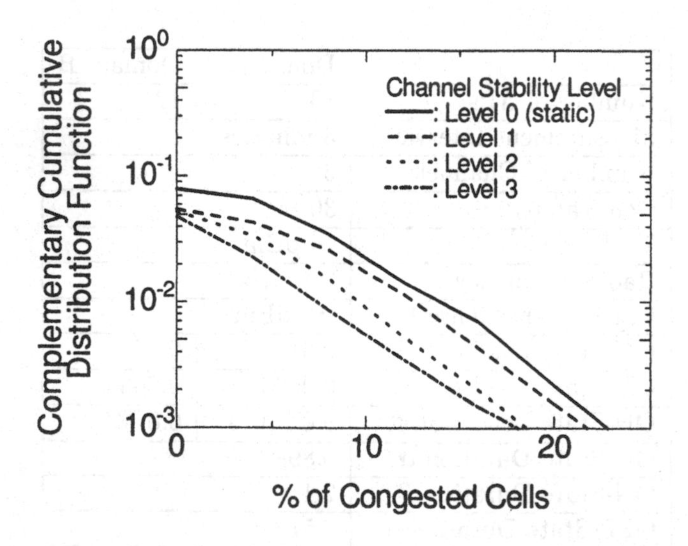

Figure 7: Distribution of Congested Cells under Various Channel Stability Levels.

5.3 Radio Resource Redistribution

Figure 8 shows the effect of radio resource distribution by channel compensation. The solid line shows the time-evolution characteristics of domain B's credit without any radio resource redistribution mechanism. As domain A's traffic demand is larger than domain B's, domain B suffers more cross-domain impairment than domain A, leading to domain B's credit growing infinitely as the time elapses. On the other hand, the dashed line shows that domain B's credit remains within $\pm$10,000 when the channel compensation is enabled. This means the radio resources are fairly redistributed from domain A to domain B by the channel compensation scheme in the long run.

We also simulated a network-initiated load balancing. Figure 9 shows the domain B's credit value over time with load balancing. Both load balancing with and without channel compensation provide a good convergence of domain B's credit around zero for the short and long time scales. The difference in convergence time between Figure 8 and Figure 9 is due to network-initiated load balancing. Load balancing immediately reduces congestion by relocating clients to a less congested domain. On the other hand, channel compensation rebalances cross-domain impairment by assigning fewer channels to the domain with the smaller cross-domain impairment. Since only a limited number of cells can change channels at any given time, it takes more time to redistribute radio resources than to perform client load balancing.

6 Related Work

Intra-domain radio resource optimization is a classical research topic, and much works have been done for circuit-switched cellular network. Dynamic channel allocation was investigated extensively in 1990's, see Katzela and Naghshineh's review paper [1] for a variety of channel allocation algorithms. Alanyali and Hajek [3] showed a simple least loaded cell selection policy can achieve optimal load balancing result for large call arrival rates. Recently dynamic channel allocation and network-initiated load balancing for WLANs were also investigated [25] and actually implemented by several vendors [26, 27]. On the other hand, there are few researches on inter-domain radio resource management. Zhuang et al. [28] presented the architecture of multi-domain resource reservation for end-to-end QoS guarantee in UMTS, but they assumed each domain's resource is separate and is managed independently. To the authors' knowledge, there is no previous work in the fair allocation of shared radio resources among different domains.

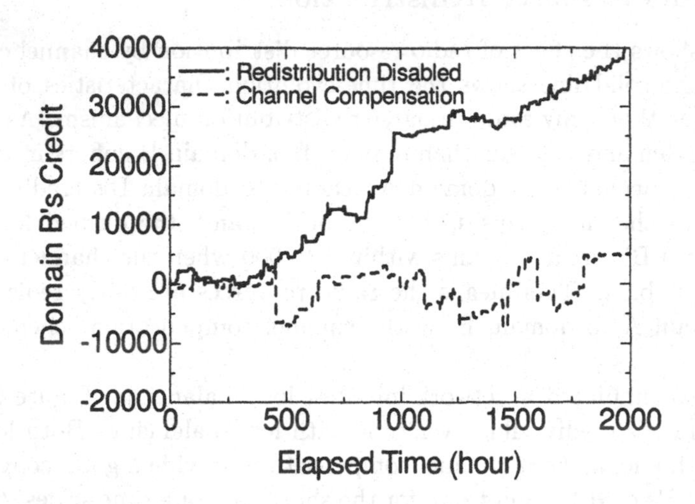

Figure 8: Radio Resource Redistribution by Channel Compensation.

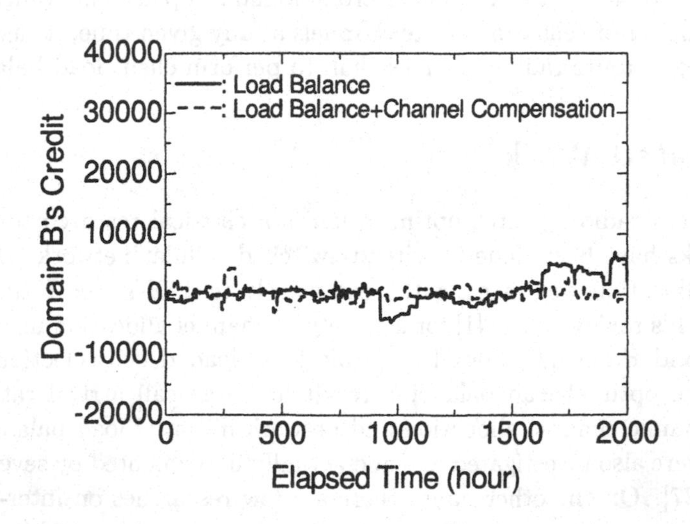

Figure 9: Radio Resource Redistribution by Network-Initiated Cross-Domain Load Balancing.

7 Conclusion

In this paper, we presented a framework of fair-sharing the radio resources among different administrative domains. Based on the measurement data collected from the WLAN access points and stations, the RRB optimizes and redistributes the radio resources among the domains by dynamic channel compensation, power control, or network-initiated load balancing. To characterize the cell-level workload, we collected statistics from a campus WLAN and derived an empirical two-state Markov model. Using that workload model, we confirmed our proposed schemes by simulating radio resource redistribution between two domains with different traffic demands. Simulation results show each domain's credit (cross-domain impairment) is kept within an allowable level in the long run by using either channel compensation or load balancing schemes. Thus the fair allocation of radio resources among domains is possible. We also showed that network-initiated load balancing is effective for redistributing radio resources in a short time span.

8 Acknowledgments

The authors would like to thank Fred Archibald for allowing to collect campus WLAN statistics. The authors also would like to thank Kazuhiro Okanoue and anonymous reviewers for many valuable comments. Yasuhiko Matsunaga was supported by NEC Corporation as a visiting industrial fellow at University of California, Berkeley from Dec. 2002 to Dec. 2003. Funding for this work was provided in part by California MICRO program, with matching support from Cisco, HP, Nortel, and NTT Communications.

References

[1] I. Katzela, M. Naghshineh, Channel Assignment Schemes for Cellular Mobile Telecommunication Systems : A Comprehensive Survey, *IEEE Personal Communications* Vol.3 No.3 (1999) pp. 10-31.

[2] J. Zander, S-L. Kim, M. Almgren, O. Queseth, *Radio Resource Management*, (Artech House, 2001).

[3] M. Alanyali, B. Hajek, On Simple Algorithms for Dynamic Load Balancing, *IEEE Infocom* (1995) pp. 230-238.

[4] N. P. Reid, R. Seide, *Wi-Fi (802.11) Network Handbook*, (McGraw-Hill, 2002).

[5] Pyramid Research, *3G Network Sharing: The European Case for UMTS*, (2001).

[6] Spectrum Policy Task Force, *http://www.fcc.gov/sptf/*

[7] European Radiocommunications Committee, On the harmonised frequency bands to be designated for the introduction of High Performance Radio Local Area Networks (HIPERLANs), *ERC/DEC/(99)23*, (1999).

[8] ITU-R, Use of the bands 5150-5250 MHz, 5250-5350 MHz and 5470-5725 MHz by the mobile service for the implementation of wireless access systems including radio local area networks, *Resolution [COM5/16] (WRC-03)*, (2003).

[9] IEEE Std 802.11, Part 11: Wireless LAN Medium Access Control (MAC) and Physical Layer (PHY) Specifications, (1999).

[10] IEEE Std 802.11h-2003, Spectrum and Transmit Power Management Extensions in the 5 GHz band in Europe, (2003).

[11] B. Knobb, *Wireless SPECTRUM Finder*, (McGraw-Hill, 2001).

[12] FCC 03-113, Report and Order and Further Notice of Proposed Rulemaking, (2003).

[13] M. Cave, Review of Radio Spectrum Management, *http://www.spectrumreview.radio.gov.uk/*, (2002).

[14] 3GPP TR 22.951, Service Aspects and Requirements for Network Sharing, (2003).

[15] 3GPP TR 23.851(draft), Network Sharing; Architecture and Functional Description, (2003).

[16] IETF RFC 3411, An Architecture for Describing Simple Network Management Protocol (SNMP) Management Frameworks, (2002).

[17] IETF RFC 3414, User-based Security Model (USM) for version 3 of the Simple Network Management Protocol (SNMPv3), (2002).

[18] B. Fortz, M. Thorup, Internet Traffic Engineering by Optimizing OSPF Weights, *IEEE Infocom* (2000) pp. 519-528.

[19] K. Ishii, T. Ohsawa, Theoretical Analysis of Throughput Performance for Wireless LAN System in Multi-cell Environment, *Trans. IEICE* Vol. J83-B No. 3 (2000) pp. 267-275.

[20] ITU-R P. 1238-2, Propagation data and prediction methods for the planning of radio communication systems and radio local area networks in the frequency range of 900 MHz to 100 GHz, (2001).

[21] Y. Watanabe, H. Furukawa, K. Okanoue, S. Yamazaki, RADIOSCAPE: System Design Tool for Indoor Wireless Communications via the Internet, *IEEE ICWLHN* (2001).

[22] D. Kotz, K. Essien, Analysis of a Campus-wide Wireless Network, *ACM Mobicom* (2002).

[23] M. Balazinska, P. Castro, Characterizing Mobility and Network Usage in a Corporate Wireless Local-Area Network, *ACM MobiSys* (2003).

[24] A. Balachandran, G. M. Voelker, P. Bahl, and V. Rangan, Characterizing User Behavior and Network Performance in a Public Wireless LAN, *ACM SIGMETRICS* (2002).

[25] A. Balachandran, G. M. Voelker, and P. Bahl, Hot-Spot Congestion Relief in Public-area Wireless Networks, *IEEE WMCSA* (2002).

[26] Airespace, *http://www.airespace.com/*

[27] Cisco Systems Inc., Data Sheet for Cisco Aironet 350 Series Access Point, (2001).

[28] W. Zhuang, Y. S. Gan, Q. Gao, K. J. Loh, K. C. Chua, Multi-Domain Policy Architecture for IP Multimedia Subsystem in UMTS, *Network Control and Engineering for QoS, Security and Mobility*, (Kluwer Academic Publisher, 2002) pp. 27-38.

High Performance Broadband Wireless Infrastructure Extension Networks

Rene L. Cruz, Arvind Santhanam, and Bongyong Song
Department of Electrical and Computer Engineering
University of California, San Diego, CA 92093
E-mail: cruz@ece.ucsd.edu, arvind@cwc.ucsd.edu, bosong@ucsd.edu

Contents

1 Introduction

Over the last decade there has been accelerated progress in the wireless communications arena. The convergence of many services including Voice, Text Messaging, Internet Access and embedment of digital cameras, MP3 players and GPS among others, all packed in a compact handheld device has provided the common man with the ability to communicate and entertain himself around the world in an untethered manner. The convergence of Voice and Data access wirelessly has resulted in a surge in demand for such services. With only a limited spectrum available to service providers, this sharp increase in demand has imposed greater stress on the service providers resources. The cost of wireless access remains high and supports low bandwidth communications. Nevertheless, the economics of wireless networking make it one of the most promising and viable technology.

Today's wireless networks can broadly classified into two classes, namely, cellular networks and wireless LANs (WLANs, e.g. IEEE 802.11). Cellular networks provide quality voice communications and low bandwidth data access over vast geographic spaces. Wireless LANs are somewhat antipodal in nature to cellular networks in that they can enable high speed data communications (upto 54 Mbps using 802.11g), but are greatly constricted in terms of spatial coverage (range of 100 meters). Another contrasting feature between cellular systems and WLANs is in the way these networks are planned. While cellular networks are carefully planned and managed in a centralized manner in any given area, WLANs lack proper network planning and spawn themselves in a erratic manner. However, both these networks share a common feature in that they are 'single-hop' networks. The communication between the mobile user and the AP is through a direct communication link. Moreover, since the AP is co-located with existing wireline infrastructure, the reach of today's networks is highly limited.

1.1 A Broadband Wireless Network

There is an earnest desire among service providers and consumers alike to marry the best of the above wireless systems, namely, the vast coverage of cellular networks with the promised bandwidth of WLANs. This can be enabled by using the concept of multihop networking. Each AP in a multihop network serves all the users (possibly mobile) within a certain *footprint*. The area of each footprint is determined by the peak power at which the AP/user can transmit and the traffic intensity to and from the other nodes in its vicinity. All APs may share a common frequency band

and can communicate with each other through wireless communication links. The function of each AP is to ensure that the aggregate traffic generated by the users it services is transmitted to a gateway node, a node that is connected to the high bandwidth wireline infrastructure. Symmetrically, the gateway node must ensure transfer of all traffic destined for a certain user is forwarded to the AP that services the user.

Each AP is capable of acting as a router node in that, it can relay traffic of any other AP to the desired destination. The data rate of a link decreases at least as fast as the inverse square of distance, and therefore it becomes efficient for an AP to use its neighboring APs as repeaters and transmit information to its desired destination, particularly when the destination node is quite far away. By wirelessly communicating traffic to the gateway, APs no longer need to be co-located with wireline infrastructures thereby increasing their reach significantly. We define such a network of APs as a Broadband wireless network. The Access Points are static nodes with relatively stable channels between them. Moreover, since they are unlikely to be energy limited (but would be peak power constrained though), sophisticated resource allocation algorithms can be employed to support high data rates. Since mobile users are likely to be in close proximity of an AP, the bandwidth they receive would be comparable to that achieved by WLANs today thus enabling popular applications like streaming media, large file downloads, online gaming etc.

The structure of a broadband access network as we have defined it here is more complex than that of traditional 'single-hop' wireless networks as wireless LANs and cellular networks. However, this increase in complexity of the network architecture enables the possibility of realizing significant increases in coverage, throughput, energy efficiency, thereby translating to a lower cost per bit of information communicated for the end user. On the flip side, broadband wireless access networks as we have defined here are simpler than many "ad-hoc" wireless networks that have been proposed in the literature, whereby end-users are responsible for forwarding traffic of other users. In broadband wireless access networks, users need only directly communicate with a nearby AP.

In this chapter, we shall outline an approach which minimizes the average power consumption of the interconnected AP network to transport desired traffic. First let us look at how wireless networks compare with alternate technologies, specifically wireline technologies like Ethernet, DSL, Cable Modem etc.

1.2 Comparative Advantages of Wireless Networks

First of all, the amount of time required to set up a wireless infrastructure network is relatively small, in the order of weeks or smaller. Small-scale wireless installations such as indoor 802.11 LANs are almost as simple as a plug-and-play system. Because of the relative ease in deployment of wireless networks, they can be built to scale gracefully with the short-term forecasted demand, thereby making them economically beneficial for service providers. A wireline network, on the other hand, may take years to deploy, since cabling tends to be a rather tedious, daunting and expensive operation. Lack of cabling makes wireless networks fairly inexpensive to deploy, particularly so in metropolitan areas where denizens are increasingly mobile but bandwidth starved.

Cognizant of the surging demand in urban areas and the great promise offered by the burgeoning WiFi technology, regulatory bodies worldwide are opening up of several frequency bands, increasing the amount of bandwidth available in wireless networks. The frequency bands 2.4-2.484 GHz, 5.15-5.35 GHz and 10-66 GHz are some examples for IEEE802.11b, IEEE802.11a, and IEEE802.16 technologies respectively. High bandwidth metropolitan area networks (MANs) are currently being architected and are likely to be deployed in the coming years. The IEEE802.16 protocol has been designed to standardize the communications between wireless interfaces, and promises bandwidths as high as 155 Mbits/sec between point-to-point links over long distances in urban areas. Standardization of communication protocols (IEEE802.11, IEEE802.16, Bluetooth etc.), competition among manufacturers of wireless components and the resulting economies of scale are playing a pivotal role in drastically reducing the cost of Access Points (APs), a key component in a wireless infrastructure. Propelled by the availability of low cost hardware, many companies worldwide are venturing in building wireless networks in cities and townships.

A common problem with many currently deployed technologies is that the algorithms implemented in these networks are highly inefficient and fail to scale over large geographic areas, translating to higher cost for the end-user. It is therefore imperative that the planning and design of a broadband wireless network be done in a systematic and efficient manner. In this article, we outline an approach to provide Quality of Service in broadband wireless networks.

1.3 Technical Challenges facing Wireless Networks

Today's cellular data networks are capable of supporting low bandwidth data access over large geographic areas but are relatively costly compared to their wireline counterparts for the same bandwidth. At the other end of the spectrum lie Wireless LANs which provide high bandwidth access (upto 54 Mbps using 802.11g) at virtually no cost but are excessively constrained in terms of spatial coverage. In current wireless LANs, the user needs to be within the vicinity (at most 150 meters) of an IEEE802.11b Access Point (AP) to communicate. Moreover, since wireless LANs are generally setup without any major global coordination and share the same unlicensed spectrum, they are prone to mutual signal interference from neighboring WLANs as well as other systems as Bluetooth that share the same frequency bands.

In addition to these problems, the quality of a wireless connection is much inferior to a wireline connection because of random disturbances over the wireless channel caused by reflections and scattering of the signal. The unpredictability in a wireless channel, even between static wireless nodes with relatively stable channels, arises due to the limited coordination amongst transmissions in a given geographical area, resulting in high signal interference between transmissions. The *hidden node* and *exposed node* problems [6] encountered in IEEE802.11b networks and the *near-far* problem in CDMA networks are primarily responsible for the low operational efficiency of the respective networks in terms of energy and throughputs. These problems are attributed to the poor coordination amongst transmissions and their respective signal powers, something that is characteristic in all wireless networks today. Current wireless networks operate at high levels of inefficiency and that translates to a high cost (in dollars) per unit data transferred over the wireless media.

1.4 Approaches to Increase Spectral Efficiency: Link Scheduling and Power Control

A broadband wireless network can overcome many of these problems that beset current wireless systems today. Recent research [9] [5] [2] [1] has shown that great increases in capacity of wireless systems are possible provided coordination amongst APs is done in a judicious manner. Indeed, coordination is necessary because of multi-access interference between signal transmissions. Due to the limited availability of the frequency spectrum, communicating links in the network must share it, and contend for this resource. When two links transmit at the same time in a given geographic space over the same frequency band, they interfere with each other, limiting data throughput. Scheduling links in a wireless network becomes imperative in order to realize high throughputs for all the links in the network. Since multi-access interference between links is determined by the transmit power of the respective transmitters (scaled by the path-loss), high throughputs can be achieved by proper choice of transmit powers among the transmitting nodes. The problem of transmission scheduling to maximize network throughput was first studied by Ephremides and Troung in [4]. The authors computed a link scheduling policy subject to the constraint that the signal interference for each link scheduled was acceptably low. They showed that this problem was NP-complete, and developed a heuristic with a polynomial time-complexity to solve it. Their scheduling algorithm was somewhat conservative, since they limited the signal interference to a fairly low value. Recently, Cruz and Santhanam [2] generalized the problem in [4] by explicitly accounting for multi-access signal interference. Specifically, they developed a link scheduling and power control algorithm that maximizes the weighted sum of average link rates in a network, when the devices are subject to average and peak power constraints. Toumpis and Goldsmith [9] have also studied the advantages of joint routing, link scheduling, power control and successive interference cancellation to solve a similar problem.

Wireless devices are generally powered by batteries and therefore would like to consume as little power as possible to transmit their information. Even in the case of wireless devices connected to a constant source of power, minimizing energy consumption is vital so that one may realize high data rates. Although health issues associated with RF transmissions are a controversial matter, nevertheless it is always desirable to limit transmission energy as much as possible. Recently, Elbatt and Ephremides [5] proposed a scheduling algorithm and power control strategy that minimizes the total power consumption in the network, when each link in the network is guaranteed a desired data rate. The heuristic a selects candidate subsets of concurrently active links, and applies the power control algorithm developed in [7] in order to find the minimal power vector for each candidate subset. The objective of their optimization problem is to maximize network throughput by supporting as many links as possible with a given fixed SIR, while remaining energy efficient. They propose a heuristic to identify large subsets of concurrently active links that can be efficiently supported, and evaluate the throughput of the heuristic via simulation for a uniform traffic demand on links.

Subsequent to their work, Cruz and Santhanam [3] proposed a systematic optimization for a problem formulation that is motivated by similar objectives. In particular, the authors compute subsets of simultaneously active links, herein called *transmission modes*, as well as the associated transmission powers and transmission schedules, in order to minimize the total *average* transmission power expended across the network. The constraints of this optimization problem are that each link in the network must have an average data rate no less than a given prescribed value, and a peak transmission power for each node.

1.4.1 Hierarchical Approach for Large Wireless Networks

Signal interference issues are paramount in wireless networks, and much benefit can accrue from coordinating transmissions as mentioned above. It is however, impossible to control transmissions at a fine-grained level over vast geographic spaces. A general approach in large wireless networks is to use a 'divide and conquer' approach. The main idea of this approach is to partition the network into *clusters*, where each cluster contains a small number of links (interconnecting a subset of APs) that are spatially close to each other. Links in a cluster are scheduled at a fine scale of time and this is done somewhat independently of links in other clusters. At the top level of the hierarchy, scheduling is done at the cluster level to determine which clusters are active and for how long they are active. If the clusters that are simultaneously active are 'far enough', then the signal interference emanating from one cluster can be approximated as static ambient noise at another. This 'external interference' at a cluster can then be used dynamically to compute the appropriate resource allocation strategies in each cluster. This approach is outlined in greater detail in [18] to minimize the total power consumption in a wireless network to support a given number of end-to-end wireless connections, each requiring an average bandwidth guarantee. Similar approaches are used in GSM cellular networks, where a 'cell', hexagonal in shape, is assigned a frequency band and has 6 hexagonal cells surrounding it. The 6 neighboring cells are not allowed to use the same frequency band as the central cell, mandating a need for at least 7 frequency bands. The major motivation for such a frequency allocation is to ensure that the inter-cell interference is acceptably low.

1.5 Multihop Networking using Multiple Antennas

An important attribute in the problems mentioned thus far is that they assume the use of antennas of fixed radiation pattern as the physical interface over the wireless channel. The function of antennas is simplified by assuming omni-directional antennas, and in this case the amount of interference is mostly controlled by power control and scheduling. However, in recent years a more proactive approach to combating multi-access interference has made significant progress. Transmitters and receivers are equipped with multiple antenna arrays and their beam patterns are dynamically configured in order to maximize the network capacity by amplifying signal power and suppressing interference. This approach is often called the Multiple-Input and Multiple-Output (MIMO) technique and was initially studied by Telatar [8], who realized that a single MIMO radio link can deliver multiple parallel data streams which are mutually non-interfering and the capacity can be greatly increased. The number of parallel data pipes linearly increases with the size of the array. When it comes to multiple MIMO links, transmit and receive beam patterns are adjusted so that the interference between transmissions is nullified or at least significantly suppressed. Also, the link gains achievable over a point-to-point link scales *linearly* as the product of the transmitter and receiver antennas. Suppressed interference and increased link gain enable wireless links to operate like wireline connections and support very high data rates, ushering in the age of wireless broadband networking. Song and Cruz [1] extended the problem formulated in [3] to include multiple antennas at the transmitter and receiver of each node. They propose a joint beamforming, power control and link scheduling algorithm that minimizes the total average power consumption to support desired rates along communication links. While beamforming, power control and scheduling need to be jointly optimized, their approach decouples these mechanisms, greatly reducing the computational complexity. This problem and the approach are explained at high level in this article.

The rest of this article is organized as follows. Section 2 describes a general system model applicable to multi-antenna wireless networks and a set of assumptions made. Section 3 formulates the problem and outlines our approach to solve it. Section 4 presents a numerical example of a broadband wireless access network in a metropolitan area, e.g. downtown San Diego. Finally, we conclude this article with some closing remarks in Section 5.

2 System Model

Consider a set of wireless nodes $\mathcal{N} = \{1, \ldots, N\}$ constructing a multi-hop wireless network. A single radio frequency band is shared by multiple nodes and the channel access is coordinated by a coordinator station, which could be one of the N nodes. We assume that the all the necessary information

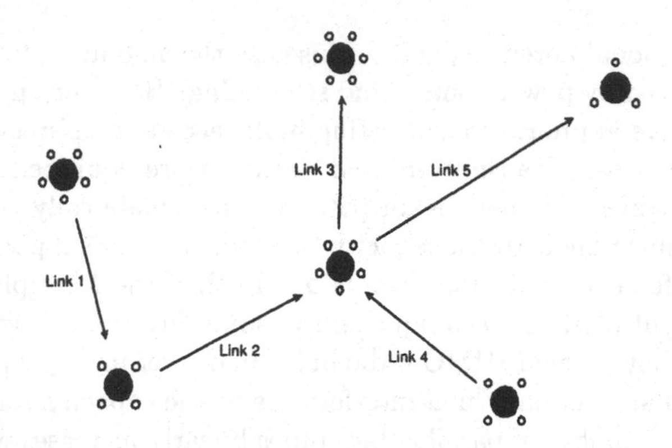

Figure 1: A MIMO multi-hop wireless network composed of 6 nodes and 5 directional links (N=6, L=5). Each node is equipped with different number of antennas.

is timely and reliably delivered to the coordinator. Wireless communication is facilitated by a multiple-input and multiple-output (MIMO) antenna system. At each node, antenna arrays with Q and P elements are used for transmission and reception, respectively. An example of a MIMO multi-hop wireless network composed of 6 nodes and 5 directional links is shown in Figure 1. Notice that a transmitter can support multiple links simultaneously by transmitting signals to different destinations using different beams. Similarly, at the receiver, there can be multiple concurrent incoming data streams from different sources. A corresponding MIMO tranceiver architecture is illustrated in Figure 2. A transmit/receive array processor produces different transmit/receive array weight vectors for different links. As a result, any communication link has a corresponding transmit weight vector and a receive weight vector, and there may be multiple such vectors at each node depending on the number of links emanating/incident from/to the node. While, in general, Q and P can vary for each node, for simplicity of exposition we assume identical numbers of transmit/receive antennas at each node.

A subset $\mathcal{L} = \{1, \ldots, L\}$ among the possible $N(N-1)$ directional links constitutes a network topology and $\mathcal{L}$ is assumed to be prespecified from the decision made by a network routing algorithm. We denote the transmitter of link l as T_l and the receiver as R_l. In a broadband wireless access network, we envision that the APs are stationary in space.Thus, we assume a static network in which the placement of nodes and the link topology do not change with time. For simplicity of exposition, we divide time into slots,

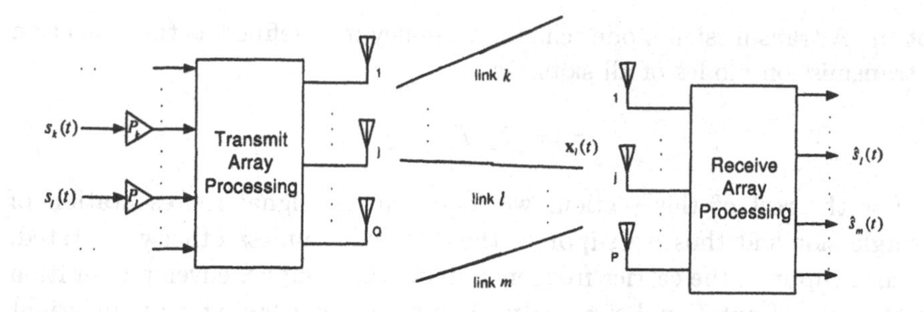

Figure 2: Tranceiver architecture in a power controlled MIMO multi-hop wireless network.

each of equal duration and indexed by the positive integers. Transmissions begin and end on slot boundaries. All radio resource management schemes considered in this chapter - beamforming, power control and link scheduling - operate in units of slot. Thus, beam patterns and transmit powers are kept constant for the duration of a slot.

We assume that each node is capable of power control. Let P_i^{max} be the maximum power for node i. Let $\mathcal{L}_i$ be a subset of $\mathcal{L}$ which consists of links originating at node i. Each node must conform to the peak transmission power constraint in every slot. We thus have

$$0 \leq \sum_{l \in \mathcal{L}_i} P_l(m) \leq P_i^{max} \text{ and } 0 \leq P_l(m) \quad \forall\, i,\, m \tag{1}$$

where $P_l(m)$ denotes the transmit power of link l in slot m.We assume that $P_l(m)$ can take on arbitrary values that satisfy(1). The network power vector in slot m defined $\mathbf{P}(m) \triangleq [P_1(m), \ldots, P_L(m)]^T$ describes the network power control status in slot m. The superscript T denotes the transpose operation.

We denote the transmit array weight vector and the receive array weight vector of link l in slot m as $\mathbf{v}_l(m)$ and $\mathbf{u}_l(m)$, respectively. Each element of a weight vector is a complex number that represents a magnitude multiplier and phase shift associated with the corresponding antenna element, which is electronically configurable for each slot.We assume that the l_2-norm of a weight vector is always kept to be unity, i.e. $\|\mathbf{v}_l(m)\| = 1$ and $\|\mathbf{u}_l(m)\| = 1$ for all l and m. By collecting all the transmit weight vectors, we construct the network transmit weight matrix in slot m defined as $\mathbf{V}(m) \triangleq [\mathbf{v}_1(m) \ldots \mathbf{v}_L(m)]$. Similarly, the network receive weight matrix in slot m is defined as $\mathbf{U}(m) \triangleq [\mathbf{u}_1(m) \ldots \mathbf{u}_L(m)]$.

The set $T_m = \{\mathbf{P}(m), \mathbf{V}(m), \mathbf{U}(m)\}$ precisely describes the network-wide radio resource utilization in slot m and we call it a *transmission mode* in slot m. A transmission mode scheduling policy π is defined as the collection of transmission modes of all slots, i.e.

$$\pi = \{T_1, T_2, \ldots\}. \tag{2}$$

For the rest of this section, we focus on the signal representation of a single slot and thus we suppress the slot index unless otherwise stated. We also suppress the carrier frequency using the complex envelope notation for the sake of notational simplicity. Assuming a flat-fading, time-invariant channel environment, the channel between T_i and R_j is described by a $P \times Q$ matrix $\mathbf{H}_{ji}$ where q-th column of $\mathbf{H}_{ji}$ represents the channel responses from the q-th transmit antenna to the antenna array at the receiver. Then the $P \times 1$ received signal vector at the node R_l is given by

$$\begin{aligned} \mathbf{x}_l(t) &= \sum_{i \in \mathcal{L}} \sqrt{P_i} \mathbf{H}_{li} \mathbf{v}_i s_i(t) + \mathbf{n}_l(t) \qquad (3) \\ &= \sqrt{P_l} \mathbf{H}_{ll} \mathbf{v}_l s_l(t) + \sum_{i \neq l} \sqrt{P_i} \mathbf{H}_{li} \mathbf{v}_i s_i(t) + \mathbf{n}_l(t) \end{aligned}$$

where $s_l(t)$ is the transmitted signal and $\mathbf{n}_l$ is $P \times 1$ thermal noise vector at R_l. If the channel is characterized by a set of independent Rayleigh fading, each element of a channel matrix is a circularly symmetric complex Gaussian random variable. In a line of sight (LOS) channel environment, another extreme scenario having no fading at all, the channel matrix can be described as

$$\mathbf{H}_{ji} = \sqrt{g_{ji}} \tilde{\mathbf{H}}_{ji} = \sqrt{g_{ji}} \mathbf{a}_P(\theta^t_{ji}) \mathbf{a}^T_Q(\theta^r_{ji}) \tag{4}$$

where g_{ji} denotes the common channel gain, $\mathbf{a}_P(\cdot)$ represents the *array manifold vector* describing the relative phase delay between P array elements associated with the angle of interest, and θ^t_{ji} (θ^r_{ji}) is the angle to the transmitter T_i (to the receiver R_j) at node R_j (T_i). One commonly used array configuration is the standard uniform linear array (ULA) and its geometry is illustrated in Figure 3. The array mainfold vector for a standard ULA with P elements, for instance, is given by $[\ldots e^{j(n-\frac{P-1}{2})\pi\cos(\theta)} \ldots]^T, 0 \leq n \leq P-1$. Here, we model a signal as a narrowband plane wave.

The output of the combiner at node R_l is given by

$$\hat{s}_l(t) = \mathbf{u}_l^H \mathbf{x}_l(t) \tag{5}$$

where the superscript H denotes the conjugate transpose operation. In determining the array weight vectors, our major concern is the signal to

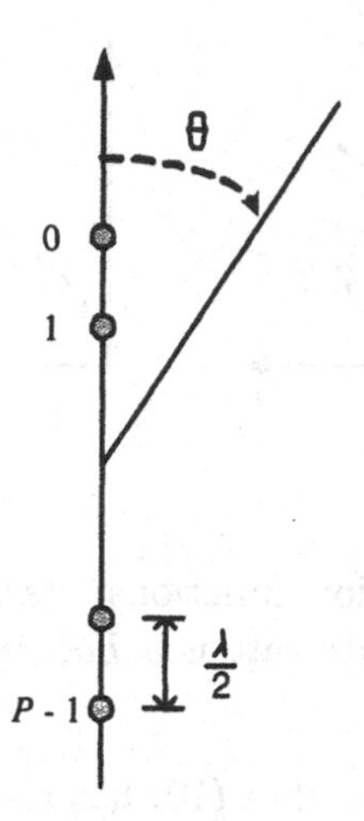

Figure 3: Antenna geometry of a standard uniform linear array (ULA) of size P. Neighboring antenna elements are spaced by half wavelength.

interference plus noise ratio (SINR). Assuming the wide sense stationarity of signals, the output power of an array for link l is given by

$$\begin{aligned} E\{|\, \hat{s}_l(t)\, |^2\} &= E\{\mathbf{u}_l^H \mathbf{x}_l(t)\mathbf{x}_l^H(t)\mathbf{u}_l\} \\ &= \mathbf{u}_l^H E\{\mathbf{x}_l(t)\mathbf{x}_l^H(t)\}\mathbf{u}_l \\ &= \mathbf{u}_l^H \mathbf{\Phi}_l^x \mathbf{u}_l. \end{aligned} \tag{6}$$

If we further assume that signals are uncorrelated and zero mean, we obtain the spatial covariance matrix of the received signal as

$$\begin{aligned} \mathbf{\Phi}_l^x &= P_l \mathbf{H}_{ll}\mathbf{v}_l\mathbf{v}_l^H\mathbf{H}_{ll}^H + \sum_{i\neq l} P_i \mathbf{H}_{li}\mathbf{v}_i\mathbf{v}_i^H\mathbf{H}_{li}^H + \sigma_{n_l}^2 \mathbf{I} \\ &= \mathbf{\Phi}_l^s + \mathbf{\Phi}_l^{in}, \end{aligned} \tag{7}$$

where $\mathbf{\Phi}_l^s = P_l\mathbf{H}_{ll}\mathbf{v}_l\mathbf{v}_l^H\mathbf{H}_{ll}^H$ represents the covariance matrix of the signal and $\mathbf{\Phi}_l^{in} = \sum_{i\neq l} P_i\mathbf{H}_{li}\mathbf{v}_i\mathbf{v}_i^H\mathbf{H}_{li}^H + \sigma_{n_l}^2\mathbf{I}$ is that of interference and noise. The SINR of link l at the array output can be written as

$$\Gamma_l = \frac{\mathbf{u}_l^H \mathbf{\Phi}_l^s \mathbf{u}_l}{\mathbf{u}_l^H \mathbf{\Phi}_l^{in} \mathbf{u}_l} \tag{8}$$

$$= \frac{P_l \mathbf{u}_l^H \mathbf{H}_{ll}\mathbf{v}_l\mathbf{v}_l^H\mathbf{H}_{ll}^H \mathbf{u}_l}{\sum_{i\neq l} P_i \mathbf{u}_l^H \mathbf{H}_{li}\mathbf{v}_i\mathbf{v}_i^H\mathbf{H}_{li}^H \mathbf{u}_l + \sigma_{n_l}^2 \mathbf{u}_l^H \mathbf{u}_l} \tag{9}$$

$$= \frac{P_l G_{ll}}{\sum_{i\neq l} P_i G_{li} + \sigma_{n_l}^2}, \tag{10}$$

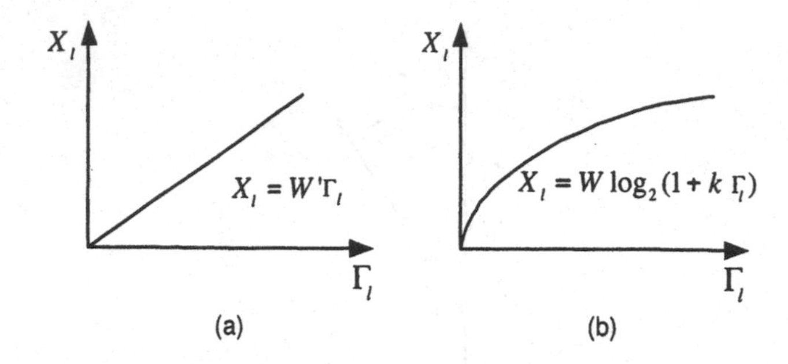

Figure 4: Two link adaptation functions. (a) The data rate is a *Linear* function of SINR. (b) The data rate is a *Logarithmic* function of SINR.

where $G_{ji} = |\mathbf{u}_j^H \mathbf{H}_{ji} \mathbf{v}_i|^2$. Note that (10) has the form of SINR formula used in conventional omnidirectional antenna networks. We refer to G_{ji} as the *effective* link gain between T_i and R_j since it reflects the array gain as well as the channel gain. Obviously, this quantity can vary from one slot to another even in a static channel environment. The effective link gain is controlled by the array processing at both transmitters and receivers.

Many advanced wireless systems employ link adaptation technique to increase the system capacity [10][11]. The data rate of a link is dynamically adjusted based on the link condition. While real world systems provide only a discrete set of data rates, continuous rate models are useful for theoretical analysis. In general, the data rate of a link is a monotonically non-decreasing function of the corresponding SINR and *linear* and *logarithmic* link adaptation functions are commonly used in the literature. These models are depicted in Figure 4. In a logarithmic model, the data rate for link l is given by [12][13]

$$X_l = W \log(1 + k\Gamma_l). \tag{11}$$

Note that this model stems from the Shannon capacity formula. When the interference is Gaussian and $k = 1$, this model becomes the Shannon formula. The parameter k represents the constant gap from the capacity limit and $k = 0.16$ when uncoded M-QAM (M-ary quadrature amplitude modulation) is used [13]. Hence, we can assume $0.16 \leq k \leq 1$ for a practical communication systems.

In this chapter, we assume a *linear* link adaptation in which the transmission rates are adapted in proportion to the SINR. In this model, the data rate for link l is given by

$$X_l = W'\Gamma_l, \tag{12}$$

for some constant W'. Although this model is only valid for infinite bandwidth channels from the viewpoint of information theory [14], it is commonly used for CDMA systems where data rate is primarily adjusted by a spreading factor [15][16]. Also, this is a fairly appropriate model for low SINR networking [17] as the logarithm in Shannon capacity formula is well approximated by a linear function. Equation (9) and (12) state that the data rates of all links are completely determined by the choice of transmit powers, transmit array weight vectors and receive array weight vectors.

Using (10) and (12), the achieved data rate for link l is

$$X_l = W' \frac{P_l G_{ll}}{\sum_{i \neq l} P_i G_{il} + \sigma_{n_l}^2}. \tag{13}$$

The long-term average rate of link l is then defined as

$$X_l^{avg} = \lim_{S \to \infty} \frac{1}{S} \sum_{m=1}^{S} X_l(m). \tag{14}$$

Let C_l be a given minimum acceptable average data rate of link l. Then we must have

$$X_l^{avg} \geq C_l \text{ for all } l \in \mathcal{L}. \tag{15}$$

We define the required minimum average rate vector as $\mathbf{C} = [C_1, C_2, \ldots, C_L]^T$. The average power consumed by the transmitter of link l, P_l^{avg}, is given by

$$P_l^{avg} = \lim_{S \to \infty} \frac{1}{S} \sum_{m=1}^{S} P_l(m). \tag{16}$$

Using (14) and (16), we can define the network average rate vector and the network average power vector as $\mathbf{X}^{avg} = [X_1^{avg}, \ldots, X_L^{avg}]^T$ and $\mathbf{P}^{avg} = [P_1^{avg}, \ldots, P_L^{avg}]^T$, respectively.

3 Optimal Policy

Our goal of this study is to find an optimal policy that minimizes the total power consumption for a given set of capacity requirement $\mathbf{C}$. We present a duality approach for solving this problem. This approach helps us to decouple beamforming, power control and link scheduling.

3.1 Problem Formulation : Primal Problem

To achieve the most energy efficient networking, the cost function to minimize is given by

$$h(\mathbf{P}^{avg}) \triangleq \sum_{l \in \mathcal{L}} P_l^{avg}. \tag{17}$$

We minimize $h(\mathbf{P}^{avg})$ for a given $\mathbf{C}$ by assigning an optimal transmission mode to each slot. This primal problem is succinctly stated as

$$\min_{\pi}\{h(\mathbf{P}^{avg})\} \quad \text{subject to (1) and (15),} \tag{18}$$

where the minimization is done over all possible policies defined as (2). Note that this optimization involves, in each slot, choosing optimal power level and optimal transmit array weights for each transmitter and determining corresponding optimal receive array weights for each receiver. Our approach below reduces the problem to a convex optimization problem over a *single* slot.

3.2 Outline of our approach

As mentioned earlier, we solve the primal problem using a duality approach. Specifically, we associate a dual variable with each average rate constraint and construct a dual objective function by adding the constraints (scaled by their dual variable) to the primal objective function and taking the minimum over the allowable set of signal powers and array weight vectors. For the exact details of our approach, refer to [3] and [1].

The dual problem is completely solvable for omnidirectional antenna systems (where $P = Q = 1$) as the dual objective function can be written as the minimum of a finite number of affine functions [3]. The dual problem involves maximization of the dual objective function subject to non-negativity constraints on the dual variables. The solution to the dual problem provides us with a set of $L+1$ (or less) transmission modes that are used in the optimal policy. We further show that without loss of generality it can be assumed that the instantaneous signal power of transmission for any link l in an optimal policy is either 0 or P_i^{max}. To completely specify the optimal policy, we must determine the fraction of time for which each transmission mode is active. Each of the $L+1$ transmission modes (power vectors) achieves a certain instantaneous data rate on each link. The optimal policy time-shares these rate vectors in a way that the average rate constraints are satisfied on each link. Using the property that, in the optimal transmission schedule, each link meets its minimum rate guarantee with *equality*, we are able to construct a set of L linear equations (one for each link) with the time-fractions as the variables. An additional equation ensures that the time-fractions add up to unity. These $L+1$ equations are linearly independent and are guaranteed to have a solution by the principle of *strong duality*. We determine the optimal time-fractions for the $L+1$ transmission modes by solving the set of equations through a simple matrix inversion.

When it comes to MIMO systems, the solution to the dual problem is still unsolved. However, an optimal policy for a MIMO system shares a lot of properties with that of an omnidirectional antenna system [1]. For a linear rate adaptation model as in (12), a simple 2 level (On-Off) power control suffices for the most energy efficient networking and every optimal policy is composed of no more than $L+1$ transmission modes. While optimal MIMO beamforming is still an open problem, it is shown that the beamsteering at the transmitter and the MVDR beamformer at the receiver is optimal for LOS channels based on the *link optimal criteria*. This criteria achieves the highest SINR of a link when all other links are freezed. When this beamforming algorithm is used, the above mentioned procedure for finding optimal omnidirectional policies can be used as (10) and (13) are as in the omnidirectional case.

4 Numerical Examples

In this section, we apply our algorithm to a broadband wireless infrastructure network as might be deployed in a metropolitan city. Access points (APs) are placed at street corners, atop buildings, and are connected in multi-hop fashion to a central AP which is connected to the high-speed wireline network. This architecture is depicted in figure 5. The network is connected in a grid arrangement and is comprised of 9 Access Points (represented by elevated discs in figure 5) and 24 uni-directional links between the APs. This architecture is symbolically illustrated in Figure 6. A blank circle represents a regular AP not having direct access to wired infrastructure and the dark circle denotes the central AP connected to the Internet. Note that only inbound links are represented. Outbound links have opposite directions even though they are not illustrated.

It is assumed that a single wireless channel with bandwidth $W = 20$Mhz is shared by all nodes for both transmission and reception purpose, i.e. TDD system. The constant in the capacity model (12) is assumed to be $W' = \frac{W}{ln2}$. The LOS MIMO channel with a free-space pathloss governs the propagation

Figure 5: Configuration of a broadband wireless access network.

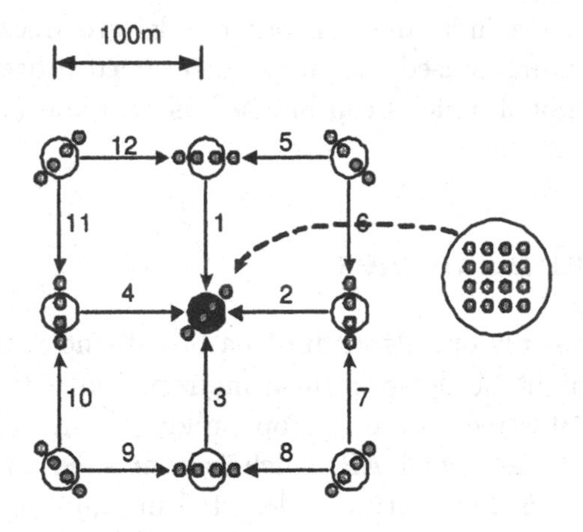

Figure 6: Configuration of a broadband wireless access network. The central AP has a square array of size 16 which is perpendicular to the ground ($P = Q = 16$). Other access points have a standard ULA of size 4 ($P = Q = 4$).

law of all links. We design the network assuming that each AP generates an identical amount of traffic C. Hence inner layer links (1, 2, 3 and 4) have the capacity requirement of $2C$ and outer layer links have only $\frac{C}{2}$. All nodes have identical peak transmission power constraints of 0.5 Watt and suffer from ambient noise with equal power for each node. The links operate at SNR = -10dB. Here, SNR represents the signal to ambient noise ratio

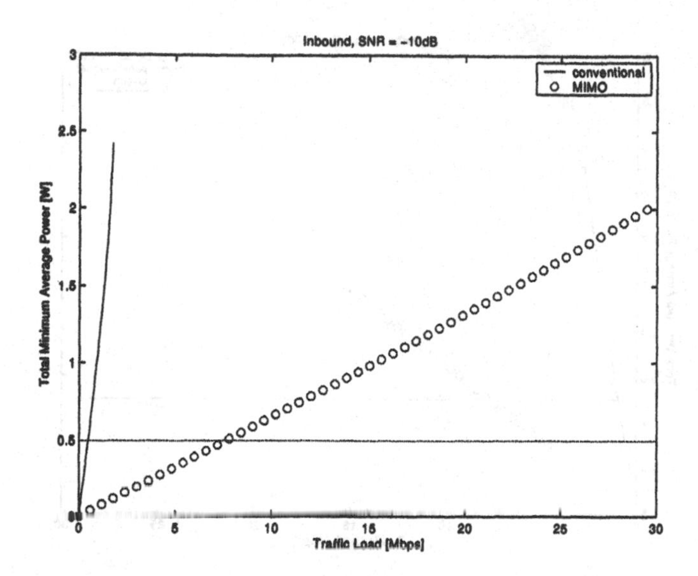

Figure 7: Inbound performance of a HCS. The total minimum average power $H(\mathbf{C})$ is plotted as a function of the amount of traffic C that each node generates.

when omnidirectional antennas are used at both ends. The actual SNR with antenna arrays is boosted up by the amount of the array gains. Each node has a standard ULA of size four serving as both a transmit array and a receive array ($P = Q = 4$) except for the central AP. The central AP is equipped with a square array of size 16 for both transmission and reception ($P = Q = 16$) in order to support higher capacity demands in inner layer links.

We first consider individual optimizations of inbound and outbound networks and then consider joint optimization of the entire network. Figure 7 illustrates the minimum required average transmit power of the inbound network as a function of the traffic load per access point, C. We refer to this function as $H(\mathbf{C})$. The performance of a conventional omnidirectional antenna network having a single transmit antenna and dual receive antenna system is depicted together for comparison. In this case, all links are 1×2 single-input and multi-output (SIMO). Our algorithm suggests that that TDMA policies are best in both cases up to the traffic load corresponding to the total minimum average power of 0.5 Watt; a TDMA policy is being used in region below the horizontal line which is the limit of a TDMA policy. At this limit, only one link is on at a time and there is no idle period,

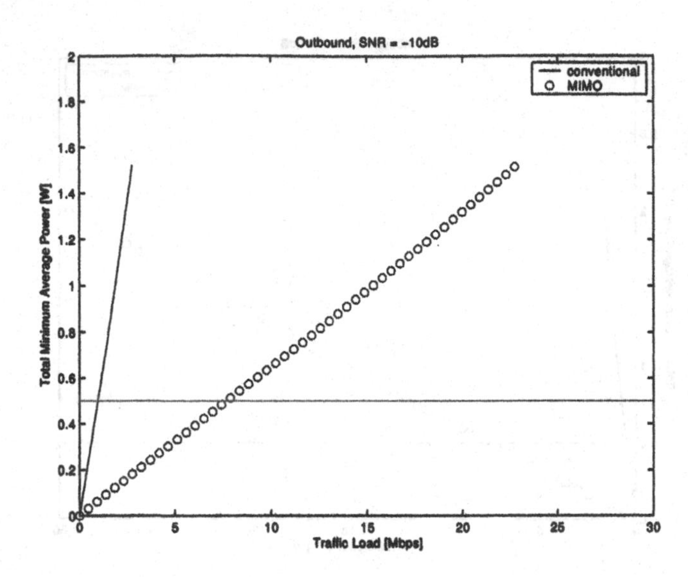

Figure 8: Outbound performance of a HCS. The total minimum average power $H(\mathbf{C})$ is plotted as a function of the amount of traffic intended to each node.

hence the total minimum average power is $H(\mathbf{C}) = 0.5$ Watt. Intuitively, a TDMA policy is optimal up to the TDMA limit as the co-channel interference is avoided. Beyond the limit of a TDMA policy, however, simultaneous transmissions occur and links suffer and links suffer from co-channel interference in addition to the ambient noise. Hence the energy efficiency of transmissions which is observed by the slope of $H(\mathbf{C})$ is degraded in case of a conventional system (the slope gets steeper as the traffic load increases). Surprisingly, the slope remains unchanged when MIMO is applied. The interference nulling capability of the arrays keeps the slope of $H(\mathbf{C})$ constant. Note that the fixed slope demonstrates the optimality of the policies found by our algorithm, since our algorithm is known to be optimal up to the TDMA limit and $H(\mathbf{C})$ is a convex function of $\mathbf{C}$. The maximum capacity of 29.47 Mbps is obtained using eight transmission modes each of them containing four active links. Since four links are active at their peak power all the time, the average power consumption is 2 Watts (0.5+0.5+0.5+0.5) at the maximum operating point. Optimal modes are chosen among all possible 322 modes. Interestingly, in this policy, a single receiver (the central AP) is required to simultaneously receive signals from two transmitters using the space division multiple access (SDMA) principle.

Figure 8 depicts the performance of outbound links (the central AP to other APs). An identical amount of traffic C is destined from the access point to each other node. The maximum outbound capacity is found to be 22.77Mbps which is approximately 6.7Mbps smaller than the inbound capacity. This is inherently due to the topology we have chosen, since the central AP has four outgoing links, while other APs have a smaller number of outgoing links, and the peak transmission power for each AP is the same. This results in lower outbound capacity than inbound capacity. Thirteen ($=L+1$) transmission modes are chosen among all possible 189 modes to achieve the maximum capacity. It is found that 3.03 links are simultaneously active on average at the maximum operating point. As a result, the corresponding average power consumption is 1.5 Watts.

Unfortunately, the capacity requirement might be asymmetric the other way, i.e. an outbound intensive capacity requirement. This is the case when the web download traffic is prevailing. One way we can support this asymmetric traffic request is to time share two optimal policies corresponding to the inbound and the outbound. This time sharing policy can be represented as $\pi_{ts} = \xi\pi^*_{out} + (1-\xi)\pi^*_{in}$ where π^*_{out} and π^*_{in} are, respectively, the outbound policy and the inbound policy we have found previously, and ξ ($0 \leq \xi \leq 1$) is the time fraction used for the outbound policy.[1] This policy supports the maximal outbound capacity of 22.77ξ Mbps and the maximal inbound capacity of $29.47(1-\xi)$ Mbps. Let $r = \frac{C_{out}}{C_{in}}$ be the outbound to inbound traffic ratio. Then the maximal outbound capacity is given by $22.77 \times \frac{29.47r}{22.77+29.47r}$. The total average power consumption is given by $1.5\xi + 2(1-\xi)$ Watts.

We can come up with a better policy by jointly scheduling outbound links and inbound links. We consider a larger network composed of 24 links (12 inbound links + 12 outbound links) and find the best policy. In general, a joint scheduling policy achieves higher capacity than that of a time sharing policy as more links can be scheduled simultaneously. For example, the maximal outbound capacity of the joint scheduling policy for $r = 3$ is 19.5 Mbps whereas the time sharing policy supports only 18.1 Mbps. The joint scheduling policy is composed of 25 ($=L+1$) transmission modes including those modes scheduling inbound links and outbound links simultaneously. This improvement is achieved by searching optimal transmission mode from a larger space which is composed of 2286 modes. In real world networks, this capacity and computational complexity tradeoff should be determined by various factors such as the network size, available processing power, etc.

[1] This policy provides the equivalent capacity to the spectrum partitioning policy where the total bandwidth W is is partitioned into two chunks, ξW and $(1-\xi)W$.

A clustering approach designed to reduce computational complexity in large networks is described in [18], which is akin to the decoupling approach for inbound and outbound traffic that we have described here.

5 Closing Remarks

Wireless networks have traditionally acted as a mere supplement to high bandwidth wireline networks. The traffic demand supportable in a static, multi-hop wireless network using IEEE 802.11b Access Points is extremely low [6] compared to wireline technologies as DSL and Cable Modem. This is mainly due to lack of coordination among link transmissions in a given geographic space that results in high signal interference between them. The adverse effect of multi-access interference can be limited by judiciously scheduling links and by proper choice of their associated signal powers of transmission. By using multiple antenna elements at the transmitter and receiver nodes and adjusting their beam patterns appropriately, signal interference can be significantly reduced. The data rates achievable by joint beamforming, power control and link scheduling are so high, they have made it possible for a wireless network to be an attractive alternative to existing wireline Internet technologies. In this chapter we have assumed that channel conditions are relatively stable. We have not explicitly considered the problem of channel estimation in a network context, which is an interesting area for further research. In practice, channel conditions may change, and it is important to consider the the overhead of channel estimation, computation of a resource allocation policy, and dissemination of the computed policy. Even if this overhead is significant, as long as the channel conditions remain stable for sufficiently long, the approach outlined here may yield significant performance improvements relative to other approaches to wireless networking where signal transmissions are largely uncoordinated. When channel conditions change on a very fast time scale, the approach outlined here may still be valuable, if appropriate link margins are factored in.

References

[1] B. Song, R. L. Cruz, Joint beamforming, power control and link scheduling in multi-hop wireless networks, *Proceedings of MilCom 2003* (Oct. 2003).

[2] R. L. Cruz, A. V. Santhanam, Optimal link scheduling and power control in multi-hop wireless networks, *Proc. IEEE Globecom 2002.*

[3] R. L. Cruz, A. V. Santhanam, Optimal routing, link scheduling and power control in multi-hop wireless networks, *Proc. IEEE Infocom 2003.*

[4] A. Ephremides, T. V. Truong, Scheduling broadcasts in multihop radio networks, *IEEE Trans. on Communications* Vol 38 (Apr. 1990).

[5] T. Elbatt, A. Ephremides, Joint scheduling and power control for wireless ad-oc networks, *Proc. IEEE Infocom 2002.*

[6] S. Xu, T. Saadawi, Revealing the problems with 802.11 MAC protocol in multi-hop wireless ad hoc networks, *Journal of Computer Networks* Vol. 38 No. 4 (Mar. 2002).

[7] G. Foschini, Z. Miljanic, A simple distributed autonomous power control algorithm and its convergence, *IEEE Trans. Veh. Technol.* Vol. 42 No. 4 (Nov. 1993) pp. 641-646.

[8] I. E. Telatar, Capacity of multi-antenna Gaussian channels, *European Transactions on Telecommunications* Vol. 10 No. 6 (Nov./Dec. 1999) pp.585-595.

[9] S. Toumpis, A. J. Goldsmith, Capacity regions for wireless adhoc networks, *International Symposium on Communication Theory and Applications* (Apr. 2001).

[10] S. Nanda, K. Balanchandran, and S. Kumar, Adaptation techniques in wireless packet data services, *IEEE Commun. Mag.* Vol. 38 (Jan. 2000).

[11] F. J. Block and M. B. Pursley, An adaptive- transmission protocol for direct-sequence spread-spectrum packet radio networks, *Proc. IEEE MILCOM* (2002).

[12] X. Qiu and K. Chawla, On the performance of adaptive modulations in cellular systems, *IEEE Trans. Commun.* Vol. 47 No. 6 (Jun. 1999) pp. 884-985.

[13] S. Catreux, P. F. Driessen, and L. J. Greenstein, Data throughputs using multiple-input multiple-output (MIMO) techniques in a noise-limited cellular environment, *IEEE Trans. Wireless Commun.* Vol. 1 No. 2 (Apr. 2002).

[14] T. M. Cover and J. A. Thomas, *Information Theory*, (Whiley and Sons, New York, 1991).

[15] F. Berggren, S. Kim, R. Jantti and J. Zander, Joint power control and intracell scheduling of DS-CDMA nonreal time data, *IEEE J. Selected Areas Commun.* Vol. 19 No. 10 (2001) pp. 1860-1869.

[16] S. Ramakrishna and J. M. Holtzman, A scheme for throughput maximization in a dual-class CDMA system, *IEEE J. Selected Areas Commun.* Vol. 16 No. 6 (Aug. 1998) pp. 830-844.

[17] Timothy J. Shepard, A channel access scheme for large dense packet radio networks, *Proc. ACM SIGCOMM Conference (SIGCOMM'96)* (Aug. 1996) pp. 219-230.

[18] R. L. Cruz and A. V. Santhanam, Hierarchical link scheduling and power control in multihop wireless networks, *Allerton Conference 2002* (Monticello, IL, Oct. 2002).

Truthful Computing in Wireless Networks

Xiang-Yang Li and WeiZhao Wang
Department of Computer Science
Illinois Institute of Technology, Chicago, IL 60616
E-mail: xli@cs.iit.edu, wangwei4@iit.edu

Contents

1 Introduction

1.1 Ad Hoc Wireless Networks

Wireless network has received significant attention over past few years due to its potential applications in various situations such as battlefield, emergency relief and environmental monitor, etc. Unlike wired networks and cellular networks which have fixed infrastructures, wireless *ad hoc* network enjoys a more flexible composition. Each mobile node has a transmission range and energy cost. A node v can receive the signal from another node u iff node v is within node u's transmission range. We assume that when u sends a packet, it consumes node u some energy and it does not cost node v any energy to receive it. If the receiving node is not within the sender's transmission range, then it must choose some intermediate nodes to repay the message. So unlike the wired networks, all nodes in the wireless ad hoc network should be able to act as a router. On the other aspect, the wireless node usually uses omni-directional antenna, which means that it can use a broadcasting-like manner to distribute the message to all nodes within its transmission range. We consider a wireless ad hoc network $G = (Q, E)$ consisting of a node set Q with $|Q| = n$ distributed in a two-dimensional plane, and directed edge $e = uv \in E$ if v can receive signal from u directly.

There are two different category of wireless ad hoc nodes: *fixed* transmission range and *adjustable* transmission range. For *fixed* transmission range nodes, their transmission range have been fixed and can't be adjusted afterward. So there is a directed arc from u to v if node v is in the transmission range of node u. Here the transmission cost depends on node u regardless of the distance between two nodes. Thus it can be considered as a *node weighted graph*. If all nodes' transmission range is the same, by properly scaling, we can assume all nodes have transmission range 1. Thus, wireless topology can be modeled by a *Unit Dish Graph(UDG)*. The wireless nodes of second category have *adjustable* transmission range: they can adjust their transmission range when necessary. Thus the cost to send a packet from node u to v not only depends on u but also depends on the geometry distance of u and v. For example, under most power attenuation model, the power needed to support a link uv is $|uv|^{\beta}$, where $\beta \in [2,5]$ depends on the transmission environment. We call this graph *edge weighted graph*. The weight of an edge uv is the power needed to support the communication between u and v.

1.2 Why Truthful Computing

Many existing works in wireless ad hoc networking assume that each individual wireless node (possibly owned by individual selfish users) will follow prescribed protocols without deviation. However, each user may modify the behavior of an algorithm for self-interested reasons.

Consider a user in a campus environment equipped with a laptop. The user might expect that his battery-powered laptop will last without recharging until the end of the day. When he participates in various ad hoc networks, he will be expected to relay traffic for other users. If he accepts all relay requests, his laptop might run out of energy prematurely. Therefore, to extend his laptop's lifetime, he might decide to reject all relay requests. If every user argues in this fashion, then the throughput that each user receives will drop dramatically. For some extreme cases, those students who needn't access the network even wouldn't care about the existence of the whole wireless *ad hoc* network. Clearly, they won't relay any messages at all. Thus, a stimulation mechanism is required to encourage users to provide service to other users.

Throughout this chapter, we address these stimulation mechanism that stimulates every individual node following prescribed protocols without deviation, which also known as *truthful computing*.

1.3 Approaches and Challenges

There are generally two ways to implement the truthful computing: *credit based method* and *incentive based method*. The first category used various non-monetary approaches including auditing, system-wide optimal point analysis and special hardware. The basic idea of credit based method is that all nodes will cooperate in order to achieve the system optimal performance, and the overall system optimum will in turn benefit the individual node. Some methods falling in this category can be found in [1, 2, 3, 4, 5, 6, 7]. The drawback of this method is that overall system optimum doesn't necessarily guarantee the individual optimality. Thus the nodes still have the incentive to deviate from their normal activity. The second method borrowed some ideas from the micro-economic and game-theoretic world, which involves the monetary transfer. The key result of this category is that all nodes won't deviate from their normal activities because they will benefit most when they reveal their true cost, even knowing all other nodes' true costs. We can thus achieve the optimal system performance.

In wireless ad hoc network environment, it is very expensive for one node to run the centralized algorithm for all nodes. Thus we often need to design some distributed algorithms or even localized algorithms. But one difficult problem has risen in distributed truthful computing environment: the algorithm is running on the selfish-node, is it a paradox asking the node itself to truthfully compute its own payment? On the other hand, there are some questions in the wireless ad hoc networks that are computationally intractable, so can we design some approximation method without losing the truthfulness of the mechanism?

The rest of the chapter is organized as follows. In Section 2, we review the credit based methods proposed in the literature used for truthful computing in wireless ad hoc networks. In Section 3, we discuss in detail the incentive based methods used for unicast and multicast in wireless ad hoc networks. In Section 4, we discuss the truthful computing at other layers of the wireless ad hoc networks such as MAC, TCP, and application layer. We conclude this Chapter in Section 5.

2 Credit Based Methods

Credit based methods have been widely proposed to solve the selfishness in wireless ad hoc networks for several years. Most of them are based on the simulation and are heuristics. They usually lack of formal analysis and theoretical proof, but several of them work well in the real world.

In [4], nodes, which agree to relay traffic but do not, are termed as misbehaving. They used *Watchdog* and *Pathrater* to identify misbehaving users and avoid routing through these nodes. *Watchdog* runs on every node keeping track of how the other nodes behave; *Pathrater* uses this information to calculate the route with the highest reliability. Notice that this method ignores the reason why a node refused to relay the transit traffics for other nodes. A node will be wrongfully labeled as misbehaving when its battery power cannot support many relay requests and thus refused to relay. It also does not provide any incentives to encourage nodes to relay the message for other nodes.

In [2], Buttyan *et al.* focused on the problem how to stimulate selfish nodes to forward the packets for other nodes. Their approach is based on a so called *nuglet counter* in each node. A node's counter is decreased when sending its own packet, and is increased when forwarding other nodes' packet. All counters should always remain positive. In order to protect the proposed mechanism against misuse, they presented a scheme based on a trusted and tamper resistant hardware module in each node, which generates cryptographically protected security headers for packets and maintains the nuglet counters of the nodes. They also studied the behavior of the proposed mechanism analytically and by means of simulations, and showed that it indeed stimulates the nodes for packet forwarding.

In [6], they still use a nugget counter to store the nuglets and besides that they use a fine which decreases the nugget counter to prevent the node from not relaying the packet. They use Packet Purse Model to discourage the user to send useless traffic and overload the network. The basic idea presented in [6] is similar to [2] but different in the implementation.

In [7], two acceptance algorithms are proposed. These algorithms are used by the network nodes to decide whether to relay traffic on a per session basis. The goal of them is to balance [1] the energy consumed by a node in relaying traffics for others with energy consumed by other nodes to relay its traffic and to find an optimal trade-off between energy consumption and session blocking probability. By taking decisions on a per session basis, the per packet processing overhead of previous schemes is eliminated. In [1], a distributed and scalable acceptance algorithm called GTFT is proposed. They proved that GTFT results in Nash equilibrium and the system converges to the rational and optimal operating point. We emphasize, however, that all the above algorithms are based on heuristics and lack a formal framework to analyze the optimal trade-off between lifetime and throughput. More importantly, they assumed that each path is h hops long and the h relay nodes are chosen with equal probability from the remaining $n-1$ nodes, which is

[1]It is impossible to strictly balance the number of packets a node has relayed for other nodes and the number of packets of this node relayed by other nodes since, in a wireless ad hoc network, majority of the packet transmissions are relayed packets. For example, consider a path of h hops. $h-1$ nodes on the path relay the packets for others. If the average path length of all routes is h, then $1-1/h$ fraction of the transmissions are transit traffics.

unrealistic.

In [8], Salem *et al.* presented a novel charging and rewarding scheme for packet forwarding in multi-hop cellular networks. In their network model, there is a base-station to forward the packets. They use symmetric cryptography to cope with the lying. To count several possible attacks, it precharges some nodes and then refunds them only if a proper acknowledgment is received. Their basic payment scheme is still based on nuglets.

In [3] Jakobsson *et al.* described an architecture for fostering collaboration between selfish nodes of multi-hop cellular networks. Based on this architecture, they provided mechanisms based on per packet charge to encourage honest behavior and to discourage dishonest behavior. In their approach, all packet originators attach a payment token to each packet, and all intermediaries on the packet's path to the base station verify whether this token corresponds to a special token called *winning ticket*. Winning tickets are reported to nearby base stations at regular intervals. The base stations, therefore, receive both reward claims (which are forwarded to some accounting center), and packets with payment tokens. After verifying the validity of the payment tokens, base stations send the packets to their desired destinations, over the backbone network. The base stations also send the payment tokens to an accounting center. Their method also involves some traditional security method including auditing, node abuse detection and encryption etc.

Generally speaking, these methods need some extra equipment, including special hardware, which is not very realistic under certain situation. In addition some methods assume that every node will enjoy better performance if the whole system's performance increases, but it is easy to construct some counter cases. One of these counter cases is the TCP/IP's congestion control scenery. Nodes using TCP/IP protocols will decrease their packet sending rate when they encounter some packet loss or timeout, so the overall system can survive the network congestion. Considering some malicious users, if they don't decrease their sending rate even they meet packet loss or time-out, they will enjoy a much faster sending rate than other nodes which conform to the rule, in the meanwhile the overall system performance will decrease sometime. Thus, we will concentrate much on incentive based methods instead of credit based methods in the following sections, which has also been studied extensively in wired networks and economics recently.

3 Incentive Based Method

3.1 Mechanism Design

In designing efficient, centralized (with input from individual agents) or distributed algorithms and network protocols, the computational agents are typically assumed to be either *correct/obedient* or *faulty* (also called adversarial). Here agents are said to be *correct/obedient* if they follow the protocol correctly; agents are said to be *faulty* if (1) they stop working, or (2) they drop messages, or (3) they act arbitrarily, which is also called *Byzantine failure*, i.e., they may deviate from the protocol in arbitrary ways that harm other users, even if the deviant behavior does not bring them any obvious tangible benefits.

In contrast, economists design market mechanisms in which it is assumed that agents are *rational*. The rational agents respond to well-defined incentives and will deviate from the protocol only if it improves their gain. A rational agent is neither correct/obedient nor adversarial.

Notice that, besides correct/obedient, adversarial, and rational agents, there is another set of agents, called *irrational*, which behave strategically but do not follow a behavior modeled by the mechanism designer. They behave irrationally with respect to the mechanism, e.g., they may have utility functions depending on more than just their own preferences. Another example is that some agents may be unable to act rationally if the strategy calculation is too expensive.

In this chapter, we always assume that the agents are rational. In addition, the mechanism used in this chapter is not computationally expensive.

A standard economic model for analyzing scenarios in which the agents act according to their own self-interest is as follows.

1. There are n agents. Each agent i, for $i \in \{1, \cdots, n\}$, has some private information t^i, called its *type*. The type t^i could be its cost to forward a packet in a network environment or its willing payment for a good in an auction environment. The type vector $t = (t^1, t^2, \cdots, t^n)$ of these agents is called a *profile*.

2. Each agent i has a set of strategies A^i that it can choose from. For each strategy vector $a = (a^1, \cdots, a^n)$, i.e., agent i plays strategy $a^i \in A^i$, the mechanism computes an *output* $o = o(a^1, \cdots, a^n)$ and a *payment* vector $p = (p^1(a), \cdots, p^n(a))$. Here the payment $p^i(a)$ is the money given to each participating agent i under strategy vector a. If $p^i(a) <$

0, it means that the agent has to pay $-p^i(a)$ to participate in the action.

3. Agent i has preference given by a valuation function v^i that assigns a real number $v^i(t^i, o)$ to each possible output o. Here, we assume that the valuation of an agent does not depend on other agents' types. Everything in the scenario is public except the type t^i, which is known to agent i only.

4. For Agent i's *utility* is $u^i = v^i(t^i, o) + p^i$. By assumption of rationality, agent i always tries to maximize its utility u^i.

3.2 Truthful Mechanism Design

A mechanism is *strategy-proof* or *Truthful* if the types are part of the strategy space A^i and each agent maximizes its utility by reporting its type t^i as input *regardless* of what other agents do. We will focus our attention on the truthful mechanism in the rest of our chapter.

The following are some natural constraints which any truthful mechanism must satisfy, before that we introduce a notation that will be used very often in the following sections.

Let a^{-i} denote the vector of strategies of all other agents except i, i.e., $a^{-i} = (a^1, a^2, \cdots, a^{i-1}, a^{i+1}, \cdots, a^n)$. Let $a|^i b = (a^1, a^2, \cdots, a^{i-1}, b, a^{i+1}, \cdots, a^n)$, i.e., each agent $j \neq i$ uses strategy a^j and the agent i uses strategy b.

1. **Incentive Compatibility (IC)**: For strategy-proof mechanism, the payment function should satisfy the incentive compatibility, i.e., for each agent i,

$$v^i(t^i, o(a|^i t^i)) + p^i(a|^i t^i) \geq v^i(t^i, o(a|^i a^i)) + p^i(a|^i a^i).$$

In other words, revealing the type t^i is the *dominating strategy*. If the payment were computed by a strategyproof mechanism, agent i would have no incentive to lie about its type because its overall utility would be no greater than it would have been if he had told the truth.

2. **Individual Rationality (IR)**: It is also called Voluntary Participation. Every participating agent must have non-negative utility, i.e.,

$$v^i(t^i, o(a|^i t^i)) + p^i(a|^i t^i) \geq 0.$$

Notice that here an agent is guaranteed to have non-negative utility if it reports its type truthfully no matter what other agents do.

3. **Polynomial Time Computability (PC)**: All computation, the computation of the output and the payment, is done in polynomial time.

3.3 VCG Based Mechanism

Arguably the most important positive result in mechanism design is what is usually called the generalized Vickrey-Clarke-Groves (VCG) mechanism by Vickrey [9], Clarke [10], and Groves [11]. The VCG mechanism applies to mechanism design maximization problems where the objective is to maximize the sum of all agents' valuations and the set of possible outputs is assumed to be finite.

A maximization mechanism design problem is called *utilitarian* if the function $g(o,t)$(also called *objective function*) to be maximized satisfies $g(o,a) = \sum_i v^i(a^i, o)$. A direct revelation mechanism $m = (o(a), p(a))$ belongs to the VCG family if (1) the output $o(a)$ computed based on the type vector a maximizes the objective function $g(o,a) = \sum_i v^i(a^i, o)$, and (2) the payment to agent i is

$$p^i(a) = \sum_{j \neq i} v^j(a^j, o(a)) + h^i(a^{-i}).$$

Here $h^i()$ is an arbitrary function of a^{-i} and different agent could have different function $h^i()$ as long as it is defined on a^{-i}. It is proved by Groves [11] that a VCG mechanism is truthful. Green and Laffont [12] proved that, under mild assumptions, VCG mechanisms are the only truthful implementations for utilitarian problems.

An output function of a VCG mechanism is required to maximize the objective function. This makes the mechanism computationally intractable in many cases. Notice that replacing the optimal algorithm with non-optimal approximation usually leads to untruthful mechanisms. In their seminal paper on algorithmic mechanism design, Nisan and Ronen [13] add computational efficiency to the set of concerns that must be addressed in the study of how privately known preferences of a large group of selfish agents can be aggregated into a "social choice" that results in optimal allocation of resources.

Similar to the *utilitarian* mechanism design problem, a maximization mechanism design problem is called *weighted utilitarian* if there exists positive real numbers $\beta_1, \cdots, \beta_n$ such that the objective function is $g(o,a) = \sum_i \beta_i \cdot v^i(a^i, o)$. A direct revelation mechanism $m = (o(a), p(a))$ belongs to the *weighted VCG family* if (1) the output $o(a)$ computed based on the type vector a maximizes the objective function $g(o,a)$, and (2) the payment to agent i is $p^i(a) = \frac{1}{\beta_i} \sum_{j \neq i} \beta_j \cdot v^j(a^j, o(a)) + h^i(a^{-i})$. Here $h^i()$ is an arbitrary function of a^{-i}. It is proved by Roberts [14] that a weighted VCG mechanism is truthful.

3.4 Network Model

We consider a set $Q = \{q_0, q_1, \cdots, q_{n-1}\}$ of n wireless nodes. Here q_0 is used to represent the access point (AP) of the wireless network to the wired network if it presents. Let $G = (Q, E)$ be the directed communication graph defined by Q, where E is the set of links (q_i, q_j) such that the node q_i can communicate directly to the node q_j. We assume that the graph G is node bi-connected. In other words, we assume that, the remaining graph, by removing any node q_i and its incident links from the graph G, is still connected. The bi-connectivity of the communication graph G will prevent the monopoly on the network as will see later in addition to provide fault tolerance.

We also assume that each wireless node has an omni-directional antenna and a single transmission of a node can be received by *any* node within its vicinity, i.e., all its neighbors in G. A node q_j can receive the signal from another node q_i if node q_j is within the transmission range of the sender q_i. Otherwise, they communicate through multi-hop wireless links by using some intermediate nodes to relay the message. Consequently, each node in the wireless network also acts as a router, forwarding data packets for other nodes. We assume that each wireless node q_i has a fixed cost c_i of relaying/sending a data packet to any (or all) of its outgoing neighbors. This cost c_i is a private information, only known to node q_i. In the terminology of economic theory, c_i is the type of node q_i. All n nodes together define a cost vector $c = (c_0, c_1, \cdots, c_{n-1})$, which is the profile of the network G. Based on this network model, we will address two important routing problems–Unicast and Multicast in the following two subsections.

3.5 Unicast

3.5.1 Statement of Problem

If a node q_i wants to send data to the access point q_0, typically, the path with minimum total relaying cost from node q_i to node q_0 under profile c is used to route the packets. We call this path Least Cost Path (LCP) and denote it as $\mathbf{L}CP(c, i, 0)$. Consider a (directed) path $\Pi(i, 0) = q_{r_s}, q_{r_{s-1}}, \cdots q_{r_1}, q_{r_0}$ connecting node q_i and node q_0, i.e., $q_{r_s} = q_i$ and $q_{r_0} = q_0$, and node q_{r_j} can send signal directly to node $q_{r_{j-1}}$. The cost of the path $\Pi(i, 0)$ is $\sum_{j=1}^{s-1} c_{r_j}$, which excludes the cost of the source and the target nodes.

To stimulate cooperation among all wireless nodes, node q_i pays some nodes of the network to forward the data for node q_i to the access point. Thus, each node q_j declares a cost d_j, which is its claimed cost to relay the packets. Note that here d_j could be different from its true cost c_j. Then node q_i computes the least cost path $\mathbf{LCP}(d, i, 0)$ according to the declared cost vector $d = (d_0, d_1, \cdots, d_{n-1})$. For each node q_j, node q_i computes a payment $p_i^j(d)$ according to the declared cost vector d. The *utility*, in standard economic model, of node q_j is $u_j = p_i^j(d) - c_j$. We always assume that the wireless nodes are rational: it always tries to maximize its utility $u_j = p_i^j(d) - c_j$.

We assume that the cost c_i is based on per packet or per session, whichever is appropriate. If the cost is per packet and a node q_i wants to send s packets to the access point q_0 in one session, then the actual payment of q_i to a node q_k will be $s \cdot p_i^k$.

If the payment scheme is not well-designed, a node q_j may improve its utility by lying its cost, i.e., declares a cost d_j such that $d_j \neq c_j$. Our objective is then to design a payment scheme such that each node q_j has to declare its true cost, i.e., $d_j = c_j$, to maximize its utility. Using the standard assumption from economic model, we assume that the wireless nodes do *not* collude with each other to improve their utility.

3.5.2 Pricing for Unicasting

For unicast problem, the output function $o(c)$ is just the LCP connecting q_i and q_0. The valuation $v^j(c_j, o(c))$ of a node q_j on the output $o(c)$ is $-c_j$ if node q_j is on the path and 0 otherwise. In other words, if node q_j is on the path, then node q_j will incur a cost c_j to carry the transit traffic for node q_i. We require that the pricing mechanism be strategyproof and nodes carrying no transit traffic receive no payment. Node q_i always prefers to find a path that maximizes the total valuation of all nodes, i.e., to find a path with the minimum total cost. In other words, given a path $\Pi(c, i, 0)$, the objective function is $\sum_{j \in \Pi(c,i,0)} c_j = \sum x_j(c, i, 0) \cdot c_j$, where $x_j(c, i, 0) = 1$ if node j belongs to the path $\Pi(c, i, 0)$ and $x_j(c, i, 0) = 0$ otherwise. The payment $p_i^k(c)$ to a node q_k on the LCP from q_i to node q_0 by node q_i is

$$p_i^k(c) = x_k(c, i, 0)c_k + [\sum_{j=1}^{n-1} x_j(c|^k\infty, i, 0)c_j - \sum_{1}^{n-1} x_j(c, i, 0)c_j]. \qquad (1)$$

Here $\sum_{j=1}^{n-1} x_j(c|^k\infty, i, 0)c_j$ is the cost of LCP without q_k and $\sum_{j=1}^{n-1} x_j(c, i, 0)c_j$ is the cost of LCP using q_k.

This payment $p_i^k(c)$ can then be interpreted as follows: the payment to a node q_k in the LCP equals to c_k, plus the improvement of the least cost path from q_i to q_0 due to the existence of node q_k. Notice that if node q_k does not belong to the least cost path, clearly, its presence does not improve the cost of LCP, thus its payment is 0. From now on, we use the term *q_k-avoiding-path* to refer to a path that does not pass through node q_k, and denote the least cost such path by $\mathbf{L}CP^{-k}(c, i, 0)$. Let $c(i, 0)$ be the cost of $\mathbf{L}CP(c, i, 0)$ and $c^{-k}(i, 0)$ be the cost of the least cost *q_k-avoiding-path* $\mathbf{L}CP^{-k}(c, i, 0)$. Notice this payment falls into the VCG mechanism, so it is strategy-proof.

3.5.3 The Distributed Algorithm

In the previous subsection, we presented a strategyproof pricing mechanism for unicast routing. Now we focus our attention on how to compute this price p_i^k in a distributed manner.

The algorithm has two stages. First, all nodes together find the Shortest Path Tree T rooted at q_0. Second, every node q_i computes its payment p_i^k in a distributed manner which is based on the algorithm in Feigenbaum *et. al* [15].

In the first stage, the shortest path tree T rooted at q_0 can be easily implemented using Dijkstra's algorithm, so we omit the details of the implementation here. In the second stage, based on the tree T found in the first stage, every node knows its parent and children in tree T. Initially at node q_i, entry p_i^k is set to ∞, if q_k is on $\mathbf{L}CP(c, j, 0)$; otherwise, p_i^k is set to 0. Every node now broadcasts its p_i^k to its neighbors. When a node q_i receives an updated price from its neighbor q_j, it updates the price entries as follows:

1. If q_j is the parent of q_i, node q_i updates
$$p_i^k = \min(p_i^k, p_j^k) \quad \text{if } q_k \in \mathbf{L}CP(c, i, 0).$$
2. If q_i is the parent of q_j, node q_i updates
$$p_i^k = \min(p_i^k, p_j^k + c_i + c_j) \quad \text{if } q_k \in \mathbf{L}CP(c, i, 0).$$
3. If nodes q_i and q_j are not adjacent in tree T, for every $q_k \in \mathbf{L}CP(c, i, 0)$, node q_i updates
$$p_i^k = \min(p_i^k, p_j^k + c_j + c(j, 0) - c(i, 0)) \quad \text{if } q_k \in \mathbf{L}CP(c, i, 0),$$

$$p_i^k = \min(p_i^k, c_k + c_j + c(j,0) - c(i,0)) \text{ if } q_k \notin \mathbf{LCP}(c,i,0).$$

Whenever any entry p_i^k changes, the entry p_i^k is sent to all neighbors of q_i by node q_i. When the network is static, the price entries decrease monotonically and converge to stable values after finite number of rounds. Notice that, here we assume that all nodes will forward these control messages, used to calculate the payment later, for free.

3.5.4 Truthful Implementation

The algorithm presented in the previous section is simple and efficient, but notice that this algorithm relies on the selfish node q_i to calculate the payment p_i^k for itself, which cannot prevent node q_i from manipulating the calculation in its favor. In [15], the authors pointed out that if agents *are required to sign all of the messages that they send and to verify all of the messages that they receive from their neighbors*, then the protocol can be modified so that all forms of cheating are detectable. Notice that even using this approach, all nodes must keep a record of messages sent to and received from its neighbors so that an audit can be performed later if a disagreement happens. Further more, their method only applies to the *edge weighted graph*. Thus, in this section, we focus our attention on how to design a new distributed algorithm based on the previous algorithm to guarantee truthful price calculation. The following method is a review of approach [16].

While it is quite obvious to conceive that the node q_i has the incentive to not correctly calculate its payment p_i^k in the second stage, it is not so straightforward to notice that the node q_i also has the incentive to lie about his shortest path even in the first stage. We give an example to show that even we can guarantee that the node q_i calculates his payment correctly in the second stage, it is still necessary for us to worry about nodes' lying in the first stage.

In Figure 1, the shortest path between q_0 and q_1 should be $q_0q_2q_3q_4q_1$, it is easy to compute the payments of node q_1 to nodes q_2, q_3, and q_4 are both 2. Thus, its overall payment to send a packet to q_0 is 6. However, if node q_1 lies that it is not a neighbor of q_4, then the shortest path becomes $q_0q_5q_1$. Now node q_1 only needs to pay q_5 to send the packet and the payment is 5. Consequently, node q_1 benefits from lying about the connection of the network. This rises from the fact that the least cost path doesn't necessary to be the path that you pay the least. This example also shows that *there is no truthful mechanism for directed edge weighted graph* when we assume that the nodes are agents since each node can choose which links to report to minimize its total payment.

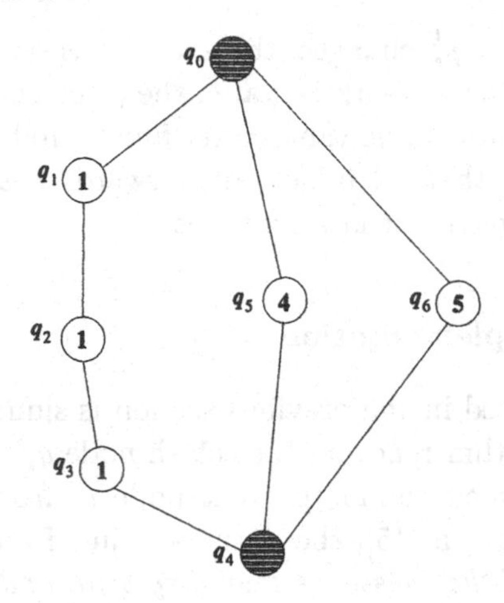

Figure 1: The node has the incentive to lie about its shortest path

We then modify the first stage of the algorithm as follows.

Algorithm 1 ***Modified Distributed Algorithm***

First Stage:

1. For every node q_i, it has two entries: $Dis(q_i)$ which stores the shortest distance to q_0 and its corresponding first hop neighbor $FH(q_i)$ on the least cost path. Initially, if q_0 is q_i's neighbor then set $Dis(q_i)$ to 0 and $FH(q_i)$ to q_0; else set $Dis(q_i)$ to ∞ and $FH(q_i)$ to NULL. Every node broadcasts its information to its neighbors.

2. For every node q_i, when it receives a broadcasting information from its neighbor q_j, first it compares $Dis(q_i)$ with $Dis(q_j) + c_j$: if $Dis(q_i) > Dis(q_j) + c_j$ then sets $Dis(q_i)$ to $Dis(q_j) + c_j$ and $FH(q_i)$ to q_j. After that it compares q_i and $FH(q_j)$:

   (a) Case 1: q_i is not $FH(q_j)$. If $Dis(q_i) + c_i < Dis(q_j)$ then node q_i contacts q_j directly using a reliable connection, asking q_j to update $Dis(q_j) = Dis(q_i) + c_i$ and $FH(q_j) = q_i$. After updating, node q_j should rebroadcast this information.

   (b) Case 2: q_i is $FH(q_j)$. If $Dis(q_i) + c_i \neq Dis(q_j)$ then node q_i contacts q_j directly using reliable connection, asking q_j to update

$Dis(q_j) = Dis(q_i) + c_i$ and $FH(q_j) = q_i$, after that q_j should rebroadcast this information.

Second Stage:

1. When q_i receives a broadcasting information p_j^k from its neighbor q_j, it updates the p_i^k using the payment updating algorithm presented in subsection 3.5.3. Additionally, if q_i triggers the change for p_j^k, it should recalculate p_j^k for node q_j using the payment updating algorithm in previous section to verify it. If it is not correct, then node q_i notifies node q_j and other nodes.

2. For every node q_i, when its entry for p_i^k changes, it not only broadcasts the value of p_i^k, but also broadcasts the information of the node that triggers this change.

It has been shown in [16] that the above approach can prevent nodes from misreporting its link information and its cost, and miscalculating the payment.

3.5.5 Collusion

Using the standard assumption from economic model, we assumed that the wireless nodes do *not* collude to improve their utility. But in practical situation, the collusion could happen very often and much disaster than the single node lying case. For example, if two nodes q_{k_1} and q_{k_2} know that the removal of them will disconnect some nodes from the access point, then these two nodes can collude to declare arbitrarily large costs and charge a monopoly price together. Notice that, by declaring much higher costs together, one node's utility may decrease, but the sum of their utilities is guaranteed to increase (thus, they share the increased utilities). So the collusion of nodes discussed here is different from the traditional *group strategyproof* concept studied in [17, 18]. A pricing mechanism is said to be group strategyproof in [17, 18] if any subset of agents colludes, then each agent of this subset cannot improve its utility without decreasing the utility of some other agent. Clearly, this formulation of group strategyproofness cannot capture the scenario we described before. We say that a mechanism is *k-agents strategyproof* if, when any subset of k agents colludes, the *overall* utility of this subset is made worse off by misreporting their types; a mechanism is *true group strategyproof* if it is *k-agents strategyproof* for any k. Clearly, we cannot design a *true group strategyproof* mechanism for the unicasting routing problem studied here: if all nodes but node q_i collude and declare arbitrarily high cost, then node q_i has to pay a payment arbitrarily higher than the actual payment it needs to pay if these nodes do not collude. Thus, it is interesting to design some mechanism that is *k-agents strategyproof* for some small integer k. Clearly, a *k-agents strategyproof* mechanism exists only if the underlying network topology is at least $k+1$ node connected.

For *k-agents strategyproof* problem, it is usually divided into two general categories: anonymous *k-agents strategyproof* problem and specific *k-agents strategyproof* problem. For the first category, what we only know is that k agents will collude, but we don't know exactly which the k agents are. It was conjectured that ever for $k = 2$, there are no strategy-proof mechanism, which means that two nodes can collude together and ask for arbitrary high price. So usually, we focus our attention on the second category. A simple case is that if we know exact which k agents probably will collude. Similar to finding the q_k-avoiding path, we can find a path that avoid these k nodes. It is easy to verify this method is strategy-proof. More sophisticated cases are these collusion nodes have some special property, for example, they should be neighbors in the wireless ad hoc. It is an open problem whether can design a *k-agents strategyproof* when knowing that possible colluding nodes are neighbors of each other.

3.5.6 Over Payment

Remember that the payment of a node q_i to a node q_k on the $\mathbf{L}CP(c,i,0)$ is

$$c_k + [\sum_{j=1}^{n-1} x_j(c|^k\infty, i, 0)c_j - \sum_{j=1}^{n-1} x_j(c, i, 0)c_j].$$

Clearly, node q_i overpays the nodes on the $\mathbf{L}CP(c,i,0)$ to make sure that they will not lie about their costs. The overpaid value is the value of $\sum_{j=1}^{n-1} x_j(c|^k\infty, i, 0)c_j - \sum_{j=1}^{n-1} x_j(c, i, 0)c_j$. In theory, it is not difficult to construct a network example such that the over-payment of a node q_i could be arbitrarily large. But in practice, after conducting extensive simulations when the cost of each node is chosen independently and uniformly from a range and the network topology is a random graph, we find out that the over-payment is small compared to the cost of LCP.

The metrics of the overpayment used in our simulations are *Total Overpayment Ratio* (TOR), *Individual Overpayment Ratio* (IOR), and *Worst Overpayment Ratio* (WOR). The TOR of a graph is defined as $\sum_i p_i / \sum_i c(i,0)$, i.e., the total payment of all nodes over the total cost of all LCPs. The IOR of a graph is defined as $\frac{1}{n}\sum_i p_i/c(i,0)$, i.e., the average overpayment ratio over all n nodes. The worst overpayment ratio is defined as $\max_i p_i/c(i,0)$, i.e., the maximum overpayment ratio over all n nodes. Remember that here p_i is the total payment of node q_i to all nodes on the LCP from q_i to q_0 and $c(i,0)$ is the total cost of nodes on the LCP from q_i to q_0. We found that the IOR and TOR are almost the same in all our simulations and they take values around 1.5. In all of our simulations, the average and the maximum are taken over 100 random instances.

In the first simulation, we randomly generate n nodes uniformly in a $2000m \times 2000m$ region. The transmission range of each node is set as $300m$. The cost of each node q_i to forward a packet to another node q_j is $\|q_iq_j\|^\kappa$ where κ varies between 2 and 2.5. The number of nodes in our simulations varies among 100, 150, 200, $\cdots$, 500. Figure 2 (a) illustrates the difference between IOR and TOR when graph model is UDG and $\kappa = 2$. We found that the values of IOR and TOR are almost the same and both of them are stable when the number of nodes increases. Figure 2 (d) illustrates the overpayment with respect to the hop distance to the source node. The average overpayment ratio of a node stays almost stable regardless of the hop distance to the source. The maximum overpayment ratio decreases when the hop distance increases, which is because large hop distance to the source node will smooth off the oscillation of the relay cost difference: for node closer to the source node, the second shortest path could be much larger than the shortest path, which in turn incurs large overpayment; for node far away from the source, the second shortest path has total cost almost the same as the shortest path, which in turn incurs small overpayment. Keep in mind that the IOR and TOR indeed increase when the hop distance to the source increases. Figure 2 (b) and (c) illustrate the overpayment for UDG graph when $\kappa = 2$ and $\kappa = 2.5$ respectively.

In our second set of simulations, we vary the transmission range of each wireless node from $100m$ to $500m$, and the cost $c_{i,j}$ of a node q_i to send a packet to another node q_j within its transmission range is $c_1 + c_2\|q_iq_j\|^\kappa$, where c_1 takes value from 300 to 500 and c_2 takes value from 10 to 50. The ranges of c_1 and c_2 we used here reflect the actual power cost of a node to send data at $2Mbps$ rate in on second. When node q_j is not within the transmission range of node q_i, cost $c_{i,j}$ is set to ∞. Figure 2 (e) and (f) illustrate the overpayment for random graph when $\kappa = 2$ and $\kappa = 2.5$ respectively.

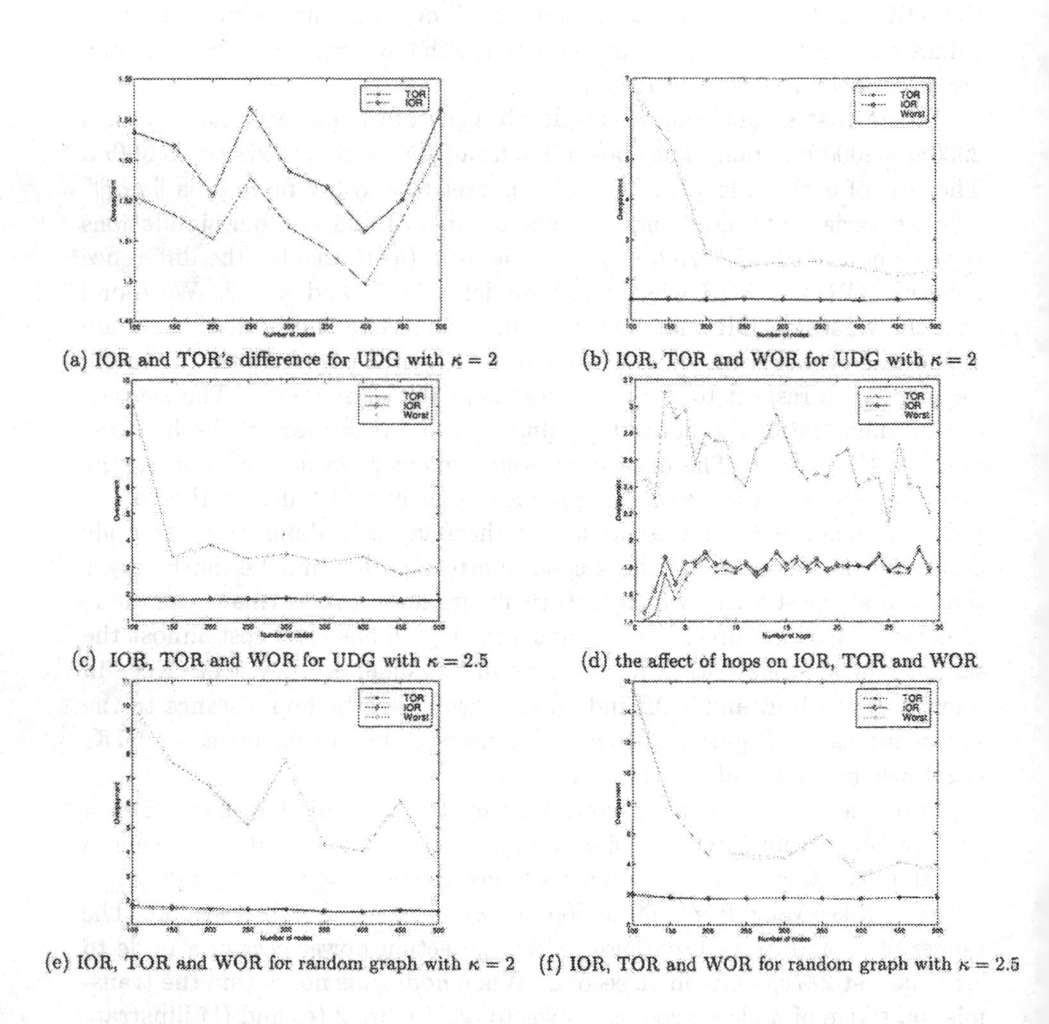

(a) IOR and TOR's difference for UDG with $\kappa = 2$ (b) IOR, TOR and WOR for UDG with $\kappa = 2$

(c) IOR, TOR and WOR for UDG with $\kappa = 2.5$ (d) the affect of hops on IOR, TOR and WOR

(e) IOR, TOR and WOR for random graph with $\kappa = 2$ (f) IOR, TOR and WOR for random graph with $\kappa = 2.5$

Figure 2: Overpayment ratios IOR, TOR and WOR for UDG and random graphs.

3.6 Multicast

3.6.1 Statement of Problem and Related Works

Assume that there is a set of users $R \subset Q$ that wants to receive information from the access point q_0. Each receiver node q_i from R has a valuation $v_i \geq 0$ of receiving the information, and which is the actual payment the node is willing to pay to receive the information. In addition, each node $q_i \in Q$ in the network has a cost c_i to forward data packets for other node.

Assume that each node q_i declares a valuation $w_i \geq 0$ if it receives the information and a cost $d_i \geq 0$ for forwarding the data to the access point q_0. The access point q_0 will then decide a subset of nodes R' to receive the information, compute a multicast tree spanning this set of nodes R', and compute a payment p_i for each node q_i. Notice that, if $p_i > 0$, then we say that the access point *pays* the node q_i for forwarding the data packets; if $p_i < 0$, then we say that the access point *charges* the node q_i for receiving the data packets. Clearly, we should have non-receiver relay nodes get positive payments. Notice that a receiver node may misreport its valuation of the data also.

The above multicast question is different from the question studied by Feigenbaum *et. al* [19]. They assumed that there is a multicast infrastructure, given any set of receivers $R \subset Q$, connects the source node to the receivers. Additionally, for each user $q_i \in R$, they assumed a *fixed* path from the source to it, determined by the multicast routing infrastructure. Then for every set R of receivers, the delivery tree is merely the union of the fixed paths from the source to the receivers R. They also assumed that there is a link cost associated with each communication link in the network and the link cost is *known* to everyone. For each receiver q_i, there is a valuation v_i that this node values the reception of the data from the source. This information v_i is only known to q_i. Node q_i will report a number v'_i, which is the amount of money he/she is willing to pay to receive the data. The source node then selects a subset $R' \subset R$ of receivers to maximize the difference $\sum_{i \in R'} v'_i - C(R')$, where $C(R')$ is the cost of the multicast tree $T(R')$ to send data to all nodes in R'. The approach of fixing the multicast tree is relatively simple to implement but could not model the greedy nature of all wireless nodes in the network since it requires that the link costs of the tree are known to every node.

3.6.2 Computational Hardness and Strategy-proof Hardness

The Prize Collecting Steiner Tree problem (PCST) is closely related to the problem of finding a maximum efficiency multicast tree. Given a graph $G = (Q, E)$, a cost vector c for all nodes, a valuation function v for all nodes, and a subset of receiver nodes r, the objective of PCST is to find a tree T, which minimizes $PC(T) = c(T) + v(\overline{T})$, where $v(\overline{T})$ is the sum of valuations of nodes not in the tree T. Let $V = \sum_{i \in N} v(i)$. For any tree T, we have $PC(T) = V + c(T) - v(T)$. Since $v(T) - c(T)$ is the revenue of the network by performing the multicast using the tree T, minimizing $PC(T)$ is equivalent to maximizing the total revenue of the multicast. It is well known that although the PCST problem can be approximated within $2 - 1/n$, but the revenue maximizing multicast problem is NP-hard, and it cannot be approximated within any constant factor.

For some specific problems, there exist some constant approximation algorithms. If q_0 should send data to all nodes in R regardless of the value of w_i, the problem become the Minimum Steiner Tree Problem: finding a minimum cost tree spanning all receivers. It is well-known that finding the minimum cost Steiner tree the both general node weighted and edge weighted graphs are NP-Hard, and even in the Euclidean or rectilinear metrics [20]. There are several polynomial time approximation algorithm presented in [21, 22, 23, 24].

In [25], they gave an approximation algorithm with the best known approximation ratio approaching $1 + \frac{ln3}{2} \approx 1.55$. However, this heuristic may be not practical for ad hoc wireless networks due to its implementation complexity. Takahashi and Matsuyama [22] gave a simple 2-approximation algorithm for Steiner minimum tree in edge weighted graphs. This algorithm maintains a tree T which initially contains only the source node. At each iterative step, the tree T is grown by one path from T of least cost can reach a destination not yet in T. Such path can be found by collapsing the entire tree T into one artificial node and then applying the single-source shortest-path algorithm. This procedure is repeated until all required nodes are included in T. This algorithm can be regarded as an adaptation of the Prim's algorithm for MST.

Even we can find the polynomial time constant approximation algorithm, it is still difficult to find a strategy-proof mechanism based on this constant approximation algorithm easily. In [13], they have already pointed out that *Replacing the optimal algorithm with a non-optimal approximation usually leads to untruthful mechanisms.* For example, if we using Takahashi and Matsuyama's 2 approximation algorithm and VCG mechanism to calculate the payment: denote T^{-p_k} as the tree without the node p_k and $W(T)$ as the sum of the node weight of this tree, then p_k's payment is $W(T) - W(T^{-p_k}) + c_k$. Figure 3 gives an example that a node may lie its cost to improve its utility. Here, q_0 is the access point and r_1, r_2 are receiving nodes. It is easy to calculate that $W(T) = 8$ and $W(T^{-p_k}) = 5$. Thus, the payment to node q^k is $5 - 8 + 4 = 1$, which is less than q^k's true cost 4. This violates the individual rationality (IR).

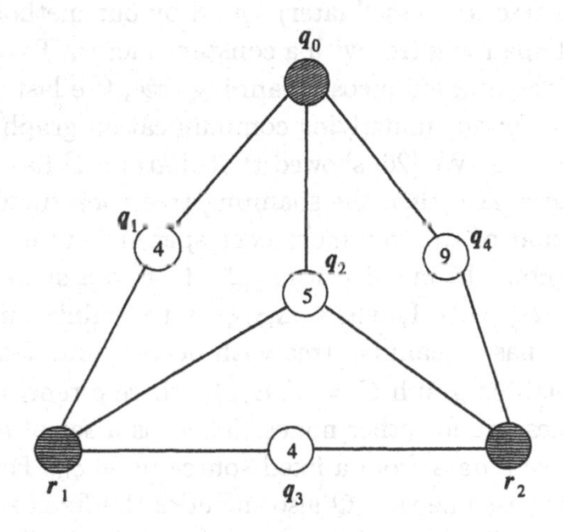

Figure 3: Non-optimal approximation usually leads to untruthfulness

Given any algorithm that approximates the minimum cost spanning tree with a factor α, we may design a payment function

$$\alpha \cdot W(T) - W(T^{-p_k}) + c_k.$$

However, it is unknown whether this payment function will satisfy the IR property. We suspect so and believe that counter-examples can be constructed such that $W(T^{-p_k}) = \alpha \cdot W(OPT^{-p_k})$ and $W(OPT^{-p_k}) > W(T)$. Notice that the first condition can be satisfied since we only have an α-approximation algorithm, and the second condition can be satisfied easily if $W(OPT) = W(T)$ since $W(OPT^{-p_k}) \geq W(OPT)$. In other words, we need design an example such that the *alpha*-approximation algorithm produces the best solution with node q_k and produces the worst solution without node q_k.

3.6.3 Node Weighted Receiving Relay Free UDG Graph Cases

In this subsection, we study a special case of multicast routing and propose an optimal computable truthful method. We assume that (1) it is node weighted and with $c_i = 0$ if $p_i \in Q$, i.e., all receiver nodes will relay the message for free; (2) all receiver nodes must receive the data; (3) the graph is a UDG graph. The truthfulness of our mechanism actually does not depend on the third assumption. The third assumption only guarantees that the spanning tree (discussed later) found by our method approximates the minimum cost spanning tree with a constant factor. To achieve constant approximation of the minimum cost spanning tree, the last assumption can actually be relaxed to: the underlying communication graph G has a degree bounded spanning tree. We [26] showed that if graph G has a spanning tree with bounded degree Δ_T, then the spanning tree constructed for multicast is Δ_T-approximation of the minimum cost spanning tree. Given a graph G, there is a polynomial time algorithm [27] to find a spanning tree whose degree is at most $\Delta_{OPT} + 1$, where Δ_{OPT} is the minimum degree bound such that graph G has a spanning tree with degree bounded by Δ_{OPT}.

Consider a weighted graph $G = (V, E, c)$, where c represents the cost of a node relaying message for other nodes. There is a set of receivers $Q \subset V$ that want to receive a data from a fixed source node q_0. For the simplicity of notation, we also assume that Q also includes the fixed source node. We then present the algorithm to construct a tree spanning all receivers and its cost is no more than 5 times of the minimum cost.

Algorithm 2 ***Reduction MST Algorithm***

1. First, we calculate the pairwise shortest path $LCP(q_i, q_j, c)$ between any two nodes in $q_i, q_j \in Q$ when the node costs vector is c. Construct a complete graph $K(Q, E')$ using Q as its vertices, and edge q_iq_j corresponding to $LCP(q_i, q_j, c)$, and its weight is the cost of $LCP(q_i, q_j, c)$ in G.

2. Calculate the minimum spanning tree on $K(Q, E')$. The resulting tree is denoted as $RMST(G)$.

For convenience of our analysis, we assume that no two nodes in $G(V, E, c)$ have the same cost, and also there are no two paths in $G(V, E, c)$ with the same length. Dropping this assumption doesn't change the result of our analysis.

Theorem 3.1 *[26] The $RMST(G)$ is a Δ_T-approximation of the minimum cost tree spanning all receivers if G has a spanning tree of degree bound Δ_T.*

Corollary 3.2 *[26] The $RMST(G)$ is a 5-approximation of the minimum cost tree spanning all receivers if G is a unit disk graph.*

Based on the $RMST(G)$ constructed by Algorithm 2, we [26] designed a truthful mechanism for calculating the payment. Before presenting the payment definition, we define some terms first.

If we change the cost of a node $v_k \in V$ to d_k, we denote the new graph as $G|^k d_k$. If we remove one vertex v_k from G, we denote the resulting graph as $G \backslash v_k$. If we apply Algorithm 2 on a graph G after removing a node v_k, we denote the resulting MST as $RMST(G\backslash v_k)$. Obviously, $RMST(G\backslash v_k) = RMST(G|^k \infty)$.

Given a spanning tree T, and a pair of nodes p and q on T, clearly there is a unique path connecting them. We denote such path as $\Pi_T(p,q)$, and the edge with the maximum length on this path as $LE(p,q,T)$. For simplicity, we use $LE(p,q,c)$ to denote $LE(p,q,RMST(G))$ and use $LE(p,q,c|^k d_k)$ to denote $LE(p,q,RMST(G|^k d_k))$.

Now we present the truthful mechanism to calculate the payment.

1. First each node $v_k \in V$ is required to report a cost, say d_k.

2. For every node $q_k \in V \backslash Q$ in G, first calculate $RMST(G)$ and $RMST(G|^k \infty)$ according to the nodes' declared costs vector d.

3. For any edge $e = q_i q_j \in RMST(G)$ and any node $v_k \in LCP(q_i, q_j, c)$, we define the payment to node v_k based on the virtual link $q_i q_j$ as

$$p^k(q_i q_j) = |LE(q_i, q_j, d|^k \infty)| - |LCP(q_i, q_j, d)| + d_k.$$

   Here $|\Pi|$ denotes the total cost of a path Π. If a node v_k is not on $LCP(q_i, q_j, c)$, then the payment $p^k(q_i q_j)$ to node v_k based on the virtual link $q_i q_j$ is 0. If the path $LCP(q_i, q_j, c)$ is not used in $RMST(G)$, then the payment to any node on path $LCP(q_i, q_j, c)$ based on edge $q_i q_j$ is also 0. The final payment to node v_k based on RMST is

$$p^k(d) = \max p^k(q_i q_j).$$

It is proved in [26] that this algorithm is not only truthful, but also it is optimal regarding the individual payment among all these truthful mechanisms based on the spanning tree $RMST$. For details of this algorithm please refer to [26].

3.6.4 Sharing Cost and Payment

Under some circumstance, if we have fixed the architecture of the multicast tree, it is also not trivial to design a *reasonable* cost-sharing mechanism to determine which users receive the transmission and how much they are charged. Here *reasonable* means at least:

1. Receivers cannot be charged more than what they are willing to pay.
2. The transmission costs of shared network links cannot be attributed to any single receiver.
3. The source node would not broadcast if the total payment received from the receiving node is less than what it should pay the relaying nodes (or links).

This problem has been formalized as the multicast cost-sharing problem(MCSP): For a graph $G = (V, E)$ and a tree T spanning the receiving nodes R, each receiving node has a utility u_i for receiving the information and known only to itself, so it can declare his utility as $u_i' \neq u_i$. Every internal node in the tree has a cost c_i to relay the data, so the access point q_0 should pay these nodes for relaying transit traffic. We let $x_i \geq 0$ denote how much user i is charged and σ_i denote whether user i receives the transmission; $\sigma_i = 1$ if the user receives the multicast transmission, and $\sigma = 0$ otherwise. We use u to denote the input vector $(u_1, u_2, \cdots, u_{|R|})$. The mechanism M is then a pair of functions $M(u) = (x(u); \sigma(u))$. The receiver set for a given input vector is $R(u) = \{i | \sigma_i = 1\}$. A user's individual welfare is given by $w_i = \sigma_i u_i - x_i$, The cost of the tree $T(R(u))$ reaching a set of receivers $R(u)$ is $W(T(R(u))) = \sum_{v_i \in T(R(u))} c_i$, and the overall welfare, or net worth, is $NW(R(u)) = \sum_{i \in R(u)} u_i - w(T(R(u)))$.

The goal of MCSP problem is to find a strategy-proof mechanism $M(u)$ subject to:

1. **No Positive Transfers (NPT)**, which means that the mechanism cannot pay receivers to accept the transmission,
2. **Voluntary Participation (VP)**, which means that no receiver can be forced to pay more than what it is willing to pay.
3. **Efficiency**: a configuration that will maximize $NW(R(u))$.

It may also has some additional desirable properties like:

1. **Consumer Sovereignty:** A receiver is always able to guarantee acceptance of the information if his price is increased to a sufficiently large value.

2. **Budget Balance:** the amount paid by the receiver exactly equals the cost of transmission.

The multicast cost-sharing problem has been studied extensively in recent years, first from a networking perspective in [28], then from a mechanism-design perspective [29], and most recently from an algorithmic perspective [19], [30], [31], [32].

Two mechanisms can be used to solve this problem: marginal cost (MC) and Shapley value (SH). MC mechanism satisfies strategy-proof, NPT, VP, CS and can be computed by a simple, distributed algorithm that uses only two modest-sized messages per link of the multicast tree in [19], [30], but one drawback is that it is not budget balanced, which means that sometimes it will have a budget surplus and sometimes it will have a budget deficiency. Shapley Value is to share the node q_i's cost within all its downstream receiving nodes. It is budget balanced and group-strategyproof and, among all mechanisms with these two properties, minimizes the worst-case welfare loss. But the SH method has a bad network complexity, computing the SH mechanism requires, in the worst case, that $O(|P|)$ bits be sent over $O(|N|)$ links, where P is the set of potential receivers, and N is the set of tree nodes.

The other interesting question is when the multicast infrastructure is not fixed, as the questions studied in previous subsection, and the receivers have to pay some other nodes to get data from the source. The mechanism described in the previous subsection provides a payment scheme to relay nodes such that they will not lie about their relay costs, but did not specify how the payment will be shared by all the receivers. We would like to design a payment sharing method such that it is better for each individual receivers to use multicast than to use unicast individually. In other words, the payment shared by a receiver node q_k should be no more than the total payment of receiver q_k when it uses unicast to connect with the source node.

4 Other Problems

Besides wireless ad hoc network, game theory has been used extensively in computer science, we briefly discuss some of the applications in this section.

4.1 Non-cooperation of Topology Control

In wireless ad hoc networks, usually the nodes can adjust their transmission ranges to achieve some desired properties, which is known as the topology control. Several topology control issues in a non-cooperative environment have been addressed before in [33]. In order to meet the connectivity requirement, we are given node pairs $(s_1, t_1), \cdots, (s_k, t_k)$, and each s_i needs to connect to t_i. Each node s_i has to choose a radius so that it gets t_i while keeping the radius as small as possible. If the radius of s_i cannot reach the target t_i, it relies on some other nodes s_j to relay the message and it is assumed that these nodes on the chosen path will relay. Notice that node s_i's only purpose is to connect to t_i, so it wouldn't care about whether other nodes can connect to their destinations or not. But the complication of the problem comes from the fact that the path connecting the source and target may contain several intermediate nodes. If a node enlarges its transmission range to connect more nodes, it is possible that it will have more choices for the intermediate nodes, which will in turn result in a smaller overall energy expenditure. Modeling the cost of a radius vector $\overline{r}$ for all nodes as $C(\overline{r}) = \sum_v r_v^\alpha$ where α is a constant between 2 and 5, they define the utility as $U(v) = f_{\overline{r}}(v) - r_v^\alpha$ where $f_{\overline{r}}(v)$ denoted the number of vertices w that v can reach. Their goal is to find a Nash Equilibria in this game, but unfortunately the existence of the Nash Equilibria and even approximate one is not guaranteed, the figure below is a graph falling into this category.

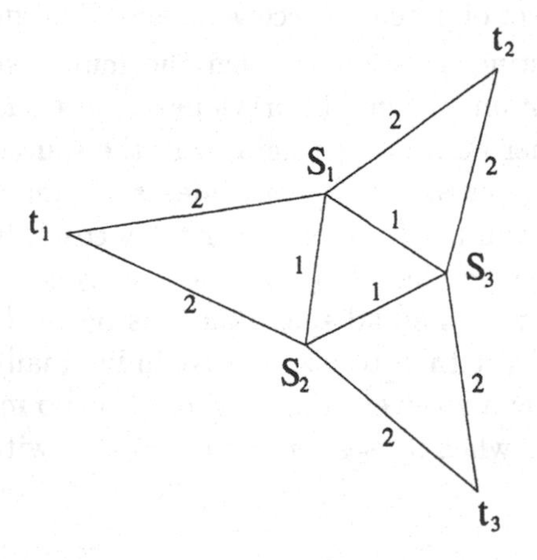

Figure 4: No Nash Equilibria

There are lots of other issues left untouched in this category, including min-power assignment problem, k-connectivity (node or edge) problem, undirected path problem and connected dominate set problem.

4.2 Incentives for Cooperation in Peer-to-Peer Networks

Peer-to-peer (P2P) file-sharing systems combine sophisticated searching techniques with decentralized file storage to allow users to download files directly from one another. The first mainstream P2P system, Napster, attracted public attention for the P2P paradigm as well as tens of millions of users for itself. Now P2P networks become a new platform for distributed applications, allowing users to share their computational, storage, and networking resources with their peers to the benefit of every participant. Most p2p system designs focus on traditional computer science problems including scalability, load-balancing, fault-tolerance, and efficient routing. While many poor to poor systems have implicitly assumed that peers will altruistically contribute resources to the global pool and assist others, recent empirical studies have shown that a large fraction of the participants engage in *freeriding* [34], [35]: 20% to 40% of Napster and almost 70% of Gnutella peers share little or no files [1, 2]. So it has been wildly acknowledged that the P2P file systems should take the user incentives and rationalities into consideration. So some recent literatures have been focus on how to solve this issue.

In [36], they define a game that models the file sharing scenario of P2P networks: n agents $a_1, ..., a_n$ participate in the system. Each agent a_i's strategy, denoted $S_i = (\sigma, \delta)$, consists of two independent actions: σ describes what proportion to share with other users, with σ_0(none), σ_1 (moderate) or σ_2 (heavy); δ determine how much to download from the network in each period. Each user can choosing between three levels: δ_0 (none), δ_1 (moderate) or δ_2 (heavy). Using this model, they propose several payment schemes: Micro-Payment Mechanisms, Quantized Micro-Payment Mechanism. Both schemes use *Nash Equilibrium* to find the agents' rational strategies and involve monetary transfer.

In [37], they use a rating scheme whereby a user is given a level in the P2P system to alleviate the free-rider problem. They discuss several issues of how the rating system should be, and based on the idea that when a user a receives a request from a user b, it uses user a's reputation to decide whether he will provide the service or not to. They give two distributed rating scheme to incentivize cooperation: *Structured Verification Scheme* (SVS) and *Lightweight Unstructured Verification Scheme* (LUVS). In SVS Scheme, every user i should have a supervisor and the supervisor should be chosen so there won't be easy collusion. (They use the ring structure *Chord* [38] to achieve that). User b gets user a's information and updates user a's reputation involving user a's supervisor. In LUVS scheme, every user will keep a list of the customers she has served and a list of servers from whom she has been served, along with the details of the transactions. Now, when a user a wants some service from a user b, a sends the list of its customers along with the request. If user b decides to verify a's rating, it samples a subset of a's customer list to confirm if they have received the claimed service from a or not. If most members in this sample say a yes, b trusts a and provides service depending on the rating of a. But they also pointed out that these schemes may suffer from the collusion problem.

4.3 Resource Allocation

With the advances in computer and networking, especially the Internet, thousands of heterogeneous computers have been interconnected to provide a great resource including computing, storage, etc. Because of the heterogeneity of these resources, it is challenging to design an architecture to accommodate all. Further more, in many cases, the resources are owned by multiple organizations and it is often required to allocate the limited resource to maximize the total user satisfaction (called *Social Choice*).

Resource allocation problem has been studied in human economies for a long time. In economic model of a computer system, the consumers are applications such as web clients, computational tasks, multimedia entertainment consumers, and ISP users. The suppliers are these computer systems who control the resources like CPU time, memory, cache, disks, network bandwidth, etc. Suppliers control access to their resource via prices, and consumers buy resources from the suppliers to satisfy their needs. The price is decided by a way similar to that in economic world: the demand and supply curve.

4.4 Cooperation in MAC Layer

Wireless MAC protocols such as IEEE 802.11 use cooperative contention resolution mechanisms for sharing the channel, it is usually based on a fully distributed mechanism to control the access to the network. For example, in the CSMA protocol, a wireless node senses whether the channel idle every DIFS (Distributed Inter Frame Space) seconds. If the channel is idle, then the node transmits the frame. Otherwise, it does a binary back-off with NAV (Network Allocation Vector) seconds which specifies the minimal time of deferral. This mechanism guarantees the fair use of the bandwidth if all nodes conform to this protocol.

But some nodes can modify their behavior in order to gain some advantage over other nodes. In stead of doing a binary back-off with NAV when collision is detected, some nodes can keep the DISF a constant in order to gain more chances to transmit a package. If one node or a few nodes play this trick, other nodes will suffer from unfair bandwidth share. The more disastrous scenery happens when the majority of the nodes or all nodes play this trick, which would result in a situation that every node (including these nodes playing this trick) share a much lower bandwidth than when no nodes play this trick. Thus, some truthful methods must be designed to deal with cooperation issue in the MAC layer also.

4.5 Cooperation for TCP/IP on End-node

For years, the conventional wisdom has been that the continued stability of the Internet depends on the widespread deployment of "socially responsible" congestion control. But what about if network end-points behaved in a selfish manner? Under this assumption, each flow attempts to maximize the throughput it achieves by modifying its congestion control behavior. In [39], they use a combination of analysis and simulation to determine the Nash Equilibrium of this game. They also addressed the efficiency of the network operating at these Nash equilibria.

Unlike the MAC layer scenery where the protocol contained in some hardware, which are very hard to manipulate, TCP/IP stacks in some operating systems like Linux, FreeBSD, is very easy to modify. So it is more desirable to design some truthful methods to prevent the misbehavior from happening in the TCP/IP layer.

5 Conclusion

In this Chapter, we assume that all wireless nodes are possibly owned by individual users and the users are able to modify the algorithms deployed on them for the sake of their own interests. We also assume that each wireless node has a cost to forward the data for other nodes and a node will only relay the data if it got a payment to cover its relay cost. We studied how the source node can design a payment scheme to all relay nodes such that the relay nodes have to report its cost truthfully to maximize their profits both for the case of unicast and the case of multicast. We also discussed the selfishness of wireless nodes in other layers including MAC layer, TCP/IP layer, and application layer.

References

[1] V. Srinivasan, P. Nuggehalli, C. F. Chiasserini, and R. R. Rao, Cooperation in wireless ad hoc networks, *IEEE Infocom* (2003).

[2] L. Buttyan and J. Hubaux, Stimulating cooperation in self-organizing mobile ad hoc networks, *ACM/Kluwer Mobile Networks and Applications(MONET)* Vol. 8 No. 5 (Oct. 2003).

[3] M. Jakobsson, J.-P. Hubaux, and L. Buttyán, A micro-payment scheme encouraging collaboration in multi-hop cellular networks, *Proceedings of Seventh International Financial Cryptography Conference (FC)* (2003).

[4] S. Marti, T. J. Giuli, K. Lai, and M. Baker, Mitigating routing misbehavior in mobile ad hoc networks, *Proceedings of MobiCom* (2000).

[5] L. Blazevic, L. Buttyan, S. Capkun, S. Giordano, J. P. Hubaux, and J. Y. Le Boudec, Self-organization in mobile ad-hoc networks: the approach of terminodes, *IEEE Communications Magazine* Vol. 39 No. 6 (Jun. 2001).

[6] L. Buttyan and J. P. Hubaux, Enforcing service availability in mobile ad-hoc WANS, *Proceedings of IEEE/ACM Workshop on Mobile Ad Hoc Networking and Computing (MobiHOC)* (2000).

[7] V. Srinivasan, P. Nuggehalli, C. F. Chiasserini, and R. R. Rao, Energy efficiency of ad hoc wireless networks with selfish users, *European Wireless Conference 2002 (EW2002)* (2002).

[8] N. B. Salem, L. Buttyan, J. Hubaux, and M. Kakobsson, A charging and rewarding scheme for packet forwarding in multi-hop cellular networks, *Proceedings of IEEE/ACM Workshop on Mobile Ad Hoc Networking and Computing (MobiHOC)* (2003).

[9] W. Vickrey, Counterspeculation, auctions and competitive sealed tenders, *Journal of Finance* (1961) pp. 8-37.

[10] E. H. Clarke, Multipart pricing of public goods, *Public Choice* (1971) pp. 17-33.

[11] T. Groves, Incentives in teams, *Econometrica* (1973) pp. 617-631.

[12] J. Green and J. J. Laffont, Characterization of satisfactory mechanisms for the revelation of preferences for public goods, *Econometrica* (1977) pp. 427-438.

[13] N. Nisan and A. Ronen, Algorithmic mechanism design, *ACM Symposium on Theory of Computing* (1999).

[14] K. Roberts, The characterization of implementable choice rules, in J. J. Laffont (ed.) *Aggregation and Revelation of Preferences*, Papers presented at the 1st European Summer Workshop of the Econometric Society, (North-Holland, 1979), pp. 321-349.

[15] J. Feigenbaum, C. Papadimitriou, R. Sami, and S. Shenker, A BGP-based mechanism for lowest-cost routing, *Proceedings of the 2002 ACM Symposium on Principles of Distributed Computing* (2002).

[16] Y. Wang X.Y. Li, W.Z. Wang and O. Frieder, Truthful low-cost routing in selfish ad hoc networks, submitted for publication (2003).

[17] K. Jain and V. V. Vazirani, Group strategyproofness and no subsidy via LP-duality, (2002).

[18] H. Moulin and S. Shenker, Strategyproof sharing of submodular costs: Budget balance versus efficiency, *Economic Theory* (2002), Available in preprint form at http://www.aciri.org/shenker/cost.ps.

[19] J. Feigenbaum, C. H. Papadimitriou, and S. Shenker, Sharing the cost of multicast transmissions, *Journal of Computer and System Sciences* Vol. 63 No. 1 (2001) pp. 21-41.

[20] M. R. Garey and D. S. Johnson, The rectilinear steiner problem is NP-complete, *SIAM Journal of Applied Mathematics* Vol. 32 (1977) pp. 826-834.

[21] P. Berman and V. Ramaiyer, Improved approximations for the steiner tree problem, *Journal of Algorithms* Vol. 17 (1994) pp. 381-408.

[22] H. Takahashi and A. Matsuyama, An approximate solution for the steiner problem in graphs, *Math. Jap.* Vol. 24 (1980) pp. 573-577.

[23] A. Zelikovsky, An 11/6-approximation algorithm for the network steiner problem, *Algorithmica* Vol. 9 (1993) pp. 463-470.

[24] A. Zelikovsky, Better approximation bounds for the network and euclidean steiner tree problems, Technical report cs-96-06, University of Virginia (1996).

[25] G. Robins and A. Zelikovsky, Improved steiner tree approximation in graphs, *Proceedings of ACM/SIAM Symposium on Discrete Algorithms* (2000) pp. 770-779.

[26] W.Z. Wang and X.-Y. Li, Truthful low-cost multicast in selfish networks, submitted for publication (2003).

[27] B. Raghavachari, Algorithms for finding low degree structures, in D. Hochbaum (ed.) *Approximation Algorithms for NP-Hard Problems*, (PWS Publishers Inc., 1997) pp. 266-295.

[28] S. Herzog, S. Shenker, and D. Estrin, Sharing the "cost" of multicast trees: an axiomatic analysis, *IEEE/ACM Transactions on Networking* Vol. 5 No. 6 (1997) pp. 847-860.

[29] H. Moulin and S. Shenker, Strategyproof sharing of submodular costs: budget balance versus efficiency, *Economic Theory* Vol. 18 (2001) pp. 511-533.

[30] J. Feigenbaum, A. Krishnamurthy, R. Sami, and S. Shenker, Hardness results for multicast cost sharing, Techical Report YALEU/DCS/TR1232 (Jun. 2002), http://ftp.cs.yale.edu/pub/TR/tr1232.ps.

[31] M. Adler and D. Rubenstein, Pricing multicast in more practical network models, *Proceedings of the 13th Symposium on Discrete Algorithms* (ACM Press/SIAM , New York/Philadelphia 2002) pp. 981-990.

[32] K. Jain and V. Vazirani, Applications of approximation to cooperative games, *Proceedings of the 33rd Symposium on the Theory of Computing* (ACM Press, New York 2002) pp. 364-372.

[33] L. Anderegg and S. Eidenbenz, Ad hoc-vcg: a truthful and cost-efficient routing protocol for mobile ad hoc networks with selfish agents, *Proceedings of the 9th Annual International Conference on Mobile Computing and Networking* (ACM Press 2003) pp. 245-259.

[34] J. Sweeny, An experimental investigation of the free-rider problem, *Social Science Research* Vol. 2 (1973).

[35] G. Marwell and R. Ames, Experiments in the provision of public goods: I. resources, interest, group size, and the free-rider problem, *American Journal of Sociology* Vol. 84 (1979).

[36] P. Golle, K. Leyton-Brown, I. Mironov, and M. Lillibridge, Incentives for sharing in peer-to-peer networks, *Lecture Notes in Computer Science* Vol. 2232 (2001).

[37] D. Dutta, A. Goel, R. Govindan, and H. Zhang, The design of a distributed rating scheme for peer-to-peer systems, *Proceedings of the Workshop on the Economics of Peer-to-Peer System* (2003).

[38] E. Adar and B. Huberman, Free riding on Gnutella, (2000).

[39] A. Akella, S. Seshan, R. Karp, S. Shenker, and C. Papadimitriou, Selfish behavior and stability of the internet: a game-theoretic analysis of TCP, *Proceedings of the 2002 Conference on Applications, Technologies, Architectures, and Protocols for Computer Communications* (ACM Proc. 2002) pp. 117 130.

Resource Allocation of Spatial Time Division Multiple Access in Multi-hop Radio Networks

Peter Värbrand and Di Yuan
Department of Science and Technology
Linköping Institute of Technology, Sweden
E-mail: petva@itn.liu.se, diyua@itn.liu.se

Contents

1 Introduction

A multi-hop radio network is characterized by a set of wireless networking units setting up a temporary communication infrastructure. Communication links are established in an ad hoc fashion, that is, two units establish a link if the signal strength between them is sufficiently high. A unit can, in addition to sending and receiving data of its own, forward data packets for other units. This is referred to as the multi-hop functionality. Multi-hop radio networks is a type of ad hoc networks which, in general, refer to wireless networking environments that lack permanent infrastructure.

One important application area of multi-hop radio networks is military command and control systems for communications between a command center and units spread out in some terrain, as well as communications between the units. For such systems, solutions based on some permanent infrastructure are not feasible. Because a multi-hop radio network can be deployed instantly, it provides an excellent solution candidate in this application.

Rational design of multi-hop radio networks, as well as ad hoc network in general, is a challenging task [18]. Here, we consider resource management for access control. The goal of resource management is to organize the transmission activities of the radio units, such that the network performance with respect to resource efficiency is optimized.

Access control is closely related to the radio resource of the physical layer. Generally speaking, not all units in a multi-hop radio network will be able to access the medium simultaneously, as the amount of physical resource (usually a frequency spectrum) is limited. Typically, the signal from one unit to another appears as interference for the other units. Having too many units transmitting at the same time will therefore cause a high level of interference.

A simple solution to organize transmission activities is time division multiple access (TDMA). In this scheme, transmission is organized using time slots. Every unit (or, sometimes, every link) is assigned a time slot, which is dedicated to this unit. A unit or a link may not be active except in its own slot. Several techniques exist for providing synchronization in TDMA. One particular solution for outdoor, tactical military communication is a satellite system (such as GPS).

Though simple to implement, a serious drawback of TDMA is its poor utilization of the resource. Resource utilization can be considerably improved using (controlled) simultaneous transmissions. In particular, units that are spatially separated can simultaneously transmit, provided that the interference level due to the transmissions is acceptable. Access control schemes based on this principle are referred to as spatial time division multiple access, or STDMA [28].

STDMA is a promising access control scheme for multi-hop radio networks. STDMA is more suitable for implementing quality-of-service requirements than contention-based schemes, such as those derived from carrier-sense multiple access (CSMA). In addition, from the resource utilization viewpoint, STDMA is competitive in comparison to some other schemes for access control, including CSMA and code division multiple access (CDMA). Simulations reported in [21] show that STDMA is superior to a CSMA scheme and a time-hopping CDMA scheme in maximizing network throughput. For low-traffic scenarios, STDMA also outperforms the other two schemes in terms of delay.

In STDMA, access control is implemented using a transmission schedule. The schedule consists of a number of time slots, and specifies which networking units are allowed to transmit in each of the time slots. The length of the schedule (or, the number of the time slots) equals the length of a data frame. The schedule is used repeatedly for every data frame. Like TDMA, STDMA is a synchronized protocol.

STDMA, in contrast to TDMA, needs an algorithm for generating the transmission schedule. The level of resource utilization depends greatly on the scheduling algorithm. Ideally, the algorithm should be able to compute an optimal STDMA schedule (for a given objective), and recompute the schedule whenever a change in the network topology occurs. Secondly, the algorithm should be sufficiently fast for being used on-line. Moreover, the algorithm should be distributed. Such an algorithm is unlikely to be found and implemented, because of the complexity of STDMA scheduling. Previous research effort devoted to STDMA scheduling has therefore focused on fast, heuristic algorithms. However, without having access to optimal solutions, it is difficult to judge the quality of heuristic solutions.

In this chapter, we report some recent advances in studying STDMA scheduling from a mathematical programming perspective. The goal is to derive the maximum achievable performance of STDMA in terms of resource utilization. A couple of assumptions are made for this purpose. First, it is assumed that the schedule is computed off-line (i.e., there is no time constraint for computation). Secondly, the computation is performed with the full knowledge of the interference. Even under these two assumptions, solving the STDMA scheduling problem is still far from trivial. If, however, we can solve the STDMA scheduling problem under these two assumptions to optimality, or, alternatively, obtain very sharp bounds to optimum, then we will be able to benchmark other heuristic algorithms as well as to assess the true potential of STDMA.

We consider two distinct resource management objectives. The first objective is to find a minimum-length schedule in which every unit (or every link) receives at least one slot. The second objective arises in traffic-sensitive scheduling, and amounts to finding a schedule that maximizes the network throughput for a given traffic distribution. (We will show later that these two objectives are very closely related to each other.) We discuss mathematical programming models and methods that can be used to derive very sharp bounds for both objectives. Moreover, we present a simple but effective greedy algorithm, which, in our numerical experiments, has a near-optimal performance.

The remainder of this chapter is organized as follows. We present the networking model in Section 2, and define four optimization problems in Section 3. Section 4 reviews some previous work in the area. Mathematical models and solution methods are discussed in Sections 5 and 6, respectively. Numerical results are presented in Section 7. In Section 8 we draw some conclusions and provide some perspectives related to future research.

2 Networking Model

2.1 Notation and Assumptions

We will use the following networking model (e.g., [19, 20]) for subsequent discussions. A multi-hop radio network is modeled using a directed graph $G = (N, A)$, where N is a set of nodes representing radio units, and A is a set of directed links. A link (i, j) exists from i to j if the signal-to-noise ratio (SNR) is above a threshold, that is, if

$$SNR(i,j) = \frac{P_i}{L_b(i,j)N_r} \geq \gamma_0. \quad (1)$$

In (1), P_i is the transmission power of node i, $L_b(i,j)$ is the path-loss from i to j, N_r represents the effect of the thermal noise, and γ_0 is a threshold. We assume that the transmitting power of the nodes are constants (e.g., every node uses maximum power when transmitting).

Typically, a multi-hop radio network is sparsely connected in our application context, as the nodes are military units spread out in some terrain. A sample network with 20 nodes is shown in Figure 1.

In some previous work [19, 20] on STDMA scheduling, the authors assume that $P_i = P, \forall i \in N$, and $L_b(i,j) = L_b(j,i), \forall (i,j) \in A$. As these conditions are not necessary for our models or solution methods,

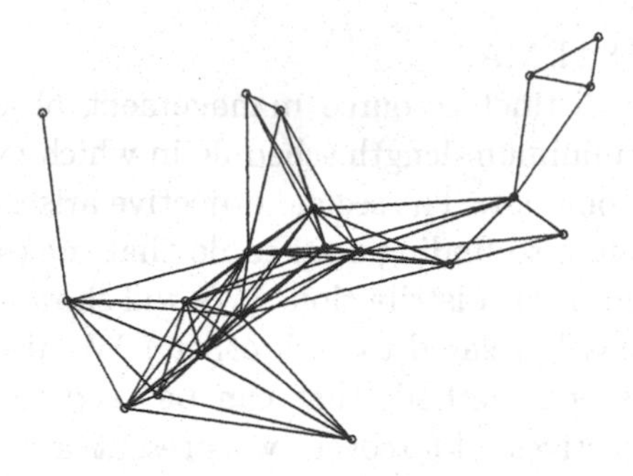

Figure 1: *A sample network.*

we will make no particular assumption about the parameters except that $0 < L_b(i,j) < 1, \forall(i,j) \in A$, $P_i > 0, \forall i \in N$, and $\gamma_0 \geq 0$. As a consequence of dropping the assumption of bi-directionality of the links, we will always treat the underlying network using a directed graph in our discussion of STDMA scheduling. A bi-directional link is thus modeled by two directed links in the set A.

STDMA scheduling must account for limitations in the physical networking environment. Typically, the following assumptions are made when generating a schedule.

- A node can transmit or receive, but not both simultaneously.
- A node cannot receive from more than one node at the same time.
- A link is free of error if and only if the signal-to-interference ratio (SIR) meets a threshold.

The first two assumptions are related to the physical limitation of the wireless interface. Sometimes, these two limitations can be overcome by improving the physical implementation. The third assumption is more general in the sense that it applies virtually in all physical implementations. Mathematically, the SIR condition for link (i,j) can be stated as follows.

$$SIR(i,j) = \frac{P_i}{L_b(i,j)(N_r + \sum_{k \in K, k \neq i} \frac{P_k}{L_b(k,j)})} \geq \gamma_1. \tag{2}$$

In (2), K is the set of nodes that are in simultaneous transmission as node i transmits to node j. The term $\sum_{k \in K, k \neq i} \frac{P_k}{L_b(k,j)}$ is, therefore, the total interference experienced on link (i, j). Note the similarity between (1) and (2). The former can be considered as a special case of the latter with zero interference. The threshold γ_1 should be less than or equal to γ_0 (if $\gamma_1 < \gamma_0$, some interference margin is made when identifying the link set A).

2.2 Assignment Strategies

The time slots in an STDMA schedule can be assigned to nodes or to links, giving rise to two different assignment strategies. In node-oriented assignment, a node that is scheduled in a slot may use any of its outgoing links to transmit data to another node in that slot. If the node needs to transmit the same data to several nodes, it may simultaneously activate a subset (possibly all) of its outgoing links (i.e., multicast and broadcast).

In link-oriented assignment, transmissions are organized with respect to links instead of nodes. In every slot, transmission may occur between a set of point-to-point connections. Note that an underlying assumption in link-oriented assignment is unicast traffic. Consequently, the outgoing links of a node are scheduled in different time slots in link-oriented assignment.

Clearly, for any set of nodes that share a time slot using node-oriented assignment, the same number of links (one outgoing link of every node in the set) can be active simultaneously. Therefore, the maximum possible spatial reuse of link-oriented and node-oriented assignment strategies, in terms of, respectively, the numbers of links and nodes that can share a time slot, is higher in the former case. The latter is, on the other hand, more suitable for broadcast traffic [20].

Note that the SIR constraint discussed in Section 2.1 has different implications for the two assignment strategies. In node-oriented assignment, the schedule must ensure that, in any time slot, all of the outgoing links of the nodes of this slot satisfy the SIR condition (as these links may be activated simultaneously). In link-oriented assignment, a set of links can be scheduled in a time slot if the SIR condition for these links are met.

2.3 Network Throughput

The throughput is defined as the largest admissible traffic load that yields a finite network delay. We assume that data packets are of equal length, and that one packet can be sent in one time slot. Following the notation in [19, 20], let λ denote the total traffic load entering the network (i.e., the average number of arriving packets per time slot), and ρ_i and ρ_{ij} denote the proportions of λ that arrive at node i and link (i, j), respectively. The values of ρ_i and ρ_{ij} are determined by a routing strategy. The average traffic load arriving at node i and link (i, j) are thus $\lambda_i = \rho_i \lambda$ and $\lambda_{ij} = \rho_{ij} \lambda$.

Let $T = \{1, \ldots, |T|\}$ denote the set of time slots in the STDMA schedule. (The length of a data frame is hence $|T|$, and the schedule repeats every $|T|$ slots.) Let h_i denote the number of slots in which node i is scheduled using node-oriented assignment. The average number of packets that node i can transmit per time slot is thus $h_i/|T|$. The delay in the network is finite only if the packet queue is finite at all nodes. To assure a finite delay at node i, inequality $\lambda_i \leq h_i/|T|$ must hold. Because $\lambda_i = \rho_i \lambda$, this inequality can be written as $\lambda \leq (h_i/|T|)/\rho_i$. The throughput is the largest λ that satisfies this inequality for all nodes, and is thus

$$\lambda_N^T = \min_{i \in N} \frac{\frac{h_i}{|T|}}{\rho_i}. \tag{3}$$

A similar derivation for the case of link-oriented assignment leads to the following expression of the throughput, where h_{ij} is the number of time slots assigned to link (i, j).

$$\lambda_L^T = \min_{(i,j) \in A} \frac{\frac{h_{ij}}{|T|}}{\rho_{ij}}, \tag{4}$$

Simulations performed in [22] show that (3) and (4) are very close to the true throughput values.

3 Problem Definitions

We consider two objectives in STDMA scheduling. If the schedule does not account for the traffic distribution, a natural measure of efficiency is the schedule length. Thus, the objective is to minimize the number of time slots, such that every node or link receives at least one slot. If, however, we wish the schedule to be optimized for a particular traffic distribution, then the corresponding objective is to maximize the throughput. In this case, the length of the schedule is typically given. We will use minimum-length scheduling and maximum-throughput scheduling, respectively, to refer to problems with these two objectives.

Combining the two objectives with the two assignment strategies gives arise to four scheduling problems. A summary of the input, objectives, and constraints of these four problems is shown in Table 1.

Note that for problems MNP and TNP, the first and second constraints effectively state, respectively, that a node does not transmit and receive simultaneously, and that two nodes do not transmit to a third node at the

Table 1: Problem definitions.

Problem MNP (minimum-length scheduling, node-oriented assignment)	
Input	A directed graph $G = (N, A)$.
	Path-loss between node pairs, $L_b(i,j), \forall i,j \in N$.
	Transmitting power of the nodes, $P_i, \forall i \in N$.
	Noise effect N_r.
	Two thresholds γ_0 and γ_1.
Objective	Minimize the number of time slots.
Constraints	The two end nodes of a link must be assigned different time slots.
	Two nodes, both having directed links to a third node, must be assigned different time slots.
	If a node is scheduled in a time slot, then the SIR of all outgoing links of the node must be at least γ_1.
	Every node is assigned at least one time slot.
Problem MLP (minimum-length scheduling, link-oriented assignment)	
Input	Same as MNP.
Objective	Same as MNP.
Constraints	Two links having a common node (irrespective of the link directions) must be assigned different time slots.
	If a link is scheduled in a time slot, then the SIR of the link must be at least γ_1.
	Every link is assigned at least one time slot.
Problem TNP (maximum-throughput scheduling, node-oriented assignment)	
Input	A directed graph $G = (N, A)$.
	Path-loss between node pairs, $L_b(i,j), \forall i,j \in N$.
	Transmitting power of the nodes, $P_i, \forall i \in N$.
	Noise effect N_r.
	Two thresholds γ_0 and γ_1.
	A schedule length $\|T\|$.
Objective	Maximize the throughput of node-oriented assignment.
Constraints	Same as MNP. (The last constraint is however redundant for TNP.)
Problem TLP (maximum-throughput scheduling, link-oriented assignment)	
Input	Same as TNP.
Objective	Maximize the throughput of link-oriented assignment.
Constraints	Same as MLP. (The last constraint is however redundant for TLP.)

same time. For link-oriented assignment, these conditions correspond to the first constraint in MLP and TLP.

The two objectives, to minimize the schedule length and to maximize the throughput, are very closely related to each other. In fact, MNP is equivalent to TNP, if the latter has equal traffic load for all nodes and the schedule length $|T|$ is equal to the minimum possible length (i.e., the optimum of MNP). To realize this equivalence, let T^* denote the optimal length of MNP. Consider, for the same graph of MNP, an instance of TNP with $|T| = T^*$ and $\rho_i = \rho, \forall i \in N$. The following observations can be made for this pair of problems. First, in any optimal schedule of MNP, there is at least one node that receives exactly one time slot, because otherwise the schedule length T^* cannot not optimal. Therefore, all optimal schedules of MNP have a throughput of $(1/T^*)/\rho$. Second, an optimal schedule of TNP is also optimal in MNP because, by construction, $|T| = T^*$. Moreover, an optimal schedule of TNP assigns at least one time slot to every node, because otherwise the throughput is zero, and the schedule cannot be optimal. Finally, as for MNP, there is at least one node that receives exactly one time slot in any optimal schedule of TNP (otherwise T^* is not optimal in MNP). Consequently, the optimal throughput of TNP is also $(1/T^*)/\rho$. It then follows immediately that the sets of optimal schedules coincide for this particular pair of problems. Solving one of the two problems is hence equivalent to solving the other. A similar equivalence can be derived for the problem pair MLP and TLP.

From a computational perspective, all four problem defined in this section are $\mathcal{NP}$-hard. For MNP and MLP, the $\mathcal{NP}$-hardness results are proved in [4] using reductions from graph coloring problems. Because MNP and MLP are special cases of TNP and TLP, respectively, we conclude that the latter two are also $\mathcal{NP}$-hard.

4 Prior Work

The principle of multi-hop radio networks appeared more than two decades ago. Some early references on (heuristic) algorithms for access control can be found, for example, in [3, 9, 10, 11, 28, 29].

A couple of remarks can be made on previous work on STDMA scheduling problems. First, the interference parameters are often not explicitly taken into account. A typical assumption is that nodes having a certain spatial separation (e.g., more than two hop away from each other) can transmit simultaneously without causing any interference. Second, most of earlier references did not address traffic-sensitive scheduling.

Among the problems defined in Section 3, minimum-length scheduling with node-oriented assignment has been addressed by a number of authors. Ephremides and Truong [15] used a heuristic for this problem (the SIR constraint was however not considered). The authors also provided the

$\mathcal{NP}$-hardness result for this problem. Later on, Sen and Huson [35] proved that the problem remains $\mathcal{NP}$-hard even for graphs with a special structure (planar point graphs). A distributed algorithm was discussed in [14, 15]. Assuming that every node knows its neighbor structure within two hops, the algorithm builds up a skeleton of the STDMA schedule. Additional nodes are then successively added to the schedule. Control information is exchanged in some reserved time slots. Alternative approaches for implementing distributed algorithms include information exchange through a separate control channel [7], and the use of a token that circulates through the nodes [33].

The authors of [17] used a neural network approach to solve the problem in [15]. A variation of the problem, in which a node can receive from a (limited and pre-defined) number of neighbors was addressed by Chou and Li [13].

Chlamtac and Faragó [8] exploited the theory of Galois fields to derive an algorithm for minimum-length scheduling with node-oriented assignment. The algorithm only needs some global network parameters (the number of nodes and the maximum node degree) when computing a schedule. Nice features of the algorithm include topology transparency (i.e., it is not dependent on any specific topology), fairness (all nodes receive the same amount of time slots), worst-case performance guarantee, and polynomial time complexity. A generalization of the algorithm was presented in [6].

Minimum-length scheduling with link-oriented assignment was solved using heuristics in [31]. The SIR constraint was not explicitly modeled. Interference was handled using the concept of interfering links, i.e., links for which the SNR does not allow communication but is strong enough for causing interference. A distributed algorithm for link-oriented assignment was presented in [26].

Hajek and Sasaki [23] studied the problem of finding a minimum-length schedule for a given demand pattern, subject to the first constraint discussed in Section 3. The authors showed that this problem is a generalization of finding the fractional chromatic index of a graph. As a result, the problem is solvable in polynomial time. Stevens and Ammar [37] compared the delay performance of node-oriented assignment to that of link-oriented assignment. Approximation algorithms for STDMA scheduling were presented in [24, 32]. Some recent advances in minimum-length scheduling (for both assignment strategies) were reported in [4, 5].

A few authors have addressed traffic-sensitive STDMA scheduling. Shor and Robertazzi [36] presented an (node-oriented) algorithm that uses the traffic load as priorities to allocate the time slots. The authors also described a distributed algorithm, which, however, does not explicitly make use of any traffic information. Extensive simulation results of traffic-sensitive scheduling were presented in [19, 20].

An STDMA schedule is fair if no node is discriminated in resource allocation. Note that an optimal schedule in terms of, for example, average network throughput, is not necessarily fair. The issue of fairness and some possible solutions were discussed in [12, 34].

Rather than looking at the scheduling algorithm alone, some authors have focused more on protocol design and implementation. We refer to [2, 30, 39, 40] for some examples of protocols. Among these, Young [39, 40] presented a component of the Soldier Phone system, the unified slot assignment protocol (USAP). Multiple types of time slots are used in this protocol. (For example, some slots are used for exchange of control information.) The protocol uses network information within a two-hop neighborhood to avoid conflicts and interference. The protocol supports both assignment strategies, as well as several heuristic scheduling algorithms.

Toumpis and Goldsmith [38] analyzed the capacity region of STDMA networks. The capacity region is the set of transmission rates (of links) that can be achieved in STDMA. Complexity issues related to verifying whether a set of transmission rates is realizable or not were studied in [1].

Finally, we would like to mention the work by Mehrotra and Trick [27] on the graph coloring problem. The two minimum-length scheduling problems, MNP and MLP, are extensions of the graph coloring problem. The column generation method discussed later on in Section 6 works similar to the method described in [27].

5 Mathematical Models

5.1 Node-slot and Link-slot Formulations

One possibility of formulating the four STDMA scheduling problems is to enumerate time slots explicitly, and associate the nodes (or the links) to the slots. We refer to this type of formulations as node-slot and link-slot formulations, respectively, for the two assignment strategies. The variables, objective functions, and constraints of these formulations are summarized in Table 2.

In formulation MNP-NS, the objective function (5) is used to minimize the total number of time slots. Constraints (6) ensure that every node is assigned at least one slot. Constraints (7) state that a slot is used (i.e., $y_t = 1$) if it is assigned to any node. Constraints (8) model the first two transmission

Table 2: Mathematical models: Node-slot and link-slot formulations.

MNP-NS (Node-Slot Formulation of MNP)	MLP-LS (Link-Slot Formulation of MLP)
$x_{it} = 1$ if time slot t is assigned to node i, otherwise $x_{it} = 0$. $y_t = 1$ if time slot t is used, otherwise $y_t = 0$.	$x_{ijt} = 1$ if time slot t is assigned to link (i, j), otherwise $x_{ijt} = 0$. $y_t = 1$ if time slot t is used, otherwise $y_t = 0$. $v_{it} = 1$ if node i is transmitting in time slot t, otherwise $v_{it} = 0$.
$\min \sum_{t \in T} y_t$ (5) $\sum_{t \in T} x_{it} \geq 1, \forall i \in N,$ (6) $x_{it} \leq y_t, \forall i \in N, \forall t \in T,$ (7) $x_{it} + \sum_{j:(j,i) \in A} x_{jt} \leq 1, \ \forall i \in N, \forall t \in T,$ (8) $\frac{P_i/N_r}{L_b(i,j)} x_{it} + \gamma_1(1 + M_{ij})(1 - x_{it}) \geq \gamma_1(1 + \sum_{k \in N: k \neq i,j} \frac{P_k/N_r}{L_b(k,j)} x_{kt}),$ $\forall (i,j) \in A, \forall t \in T,$ (9) $x_{it} \in \{0,1\}, \forall i \in N, \forall t \in T,$ (10) $y_t \in \{0,1\}, \forall t \in T.$ (11)	$\min \sum_{t \in T} y_t$ (12) $\sum_{t \in T} x_{ijt} \geq 1, \forall (i,j) \in A,$ (13) $x_{ijt} \leq y_t, \forall (i,j) \in A, \forall t \in T,$ (14) $\sum_{j:(i,j) \in A} x_{ijt} + \sum_{j:(j,i) \in A} x_{jit} \leq 1,$ $\forall i \in N, \forall t \in T,$ (15) $x_{ijt} \leq v_{it}, \forall (i,j) \in A, \forall t \in T,$ (16) $\frac{P_i/N_r}{L_b(i,j)} x_{ijt} + \gamma_1(1 + M_{ij})(1 - x_{ijt}) \geq \gamma_1(1 + \sum_{k \in N: k \neq i,j} \frac{P_k/N_r}{L_b(k,j)} v_{kt}),$ $\forall (i,j) \in A, \forall t \in T,$ (17) $x_{ijt} \in \{0,1\}, \forall (i,j) \in A, \forall t \in T,$ (18) $v_{it} \in \{0,1\}, \forall i \in N, \forall t \in T,$ (19) $y_t \in \{0,1\}, \forall t \in T.$ (20)
TNP-NS (Node-Slot Formulation of TNP)	**TLP-LS (Link-Slot Formulation of TLP)**
$x_{it} = 1$ if time slot t is assigned to node i, otherwise $x_{it} = 0$.	$x_{ijt} = 1$ if time slot t is assigned to link (i, j), otherwise $x_{ijt} = 0$. $v_{it} = 1$ if node i is transmitting in time slot t, otherwise $v_{it} = 0$.
$\max \ z$ (21) $z \leq \frac{\sum_{t \in T} x_{it}}{\rho_i \lvert T \rvert}, \forall i \in N,$ (22) (8), (9), (10).	$\max \ z$ (23) $z \leq \frac{\sum_{t \in T} x_{ijt}}{\rho_{ij} \lvert T \rvert}, \forall (i,j) \in A,$ (24) (15), (16), (17), (18), (19).

constraints in Section 3, i.e., different time slots must be assigned to two nodes if they are the two end nodes of a link, or if both have links to a third node. The SIR constraint is defined in (9). To see the equivalence between (9) and (2), note that if $x_{it} = 0$ (i.e., slot t is not assigned to node i), then (9) becomes redundant, provided that the value of the parameter M_{ij} is sufficiently large. If $x_{it} = 1$, then (9) reads $\frac{P_i/N_r}{L_b(i,j)} \geq \gamma_1(1+\sum_{k \in N:k \neq i,j} \frac{P_k/N_r}{L_b(k,j)} x_{kt})$, which is same as (2). To ensure that the SIR constraint is redundant when $x_{it} = 0$, the parameter M_{ij} can be set to the worst-case interference that may occur to link (i, j), i.e., $M_{ij} = \sum_{k \in N:k \neq i,j} (P_k/N_r)/L_b(k, j)$.

The variable definitions and constraints of MLP-LS are very similar to those of MNP-NS. Note, however, that two sets of variables (y and v) are needed to distinguish between slot utilization and node activation.

In formulation TNP-NS, the term $\sum_{t \in T} x_{it}$ in (22) is the number of time slots assigned to node i (among a total of $|T|$ slots). Constraints (22) ensure, together with the objective function (21), that z is the network throughput defined earlier in (3). A similar interpretation can be derived for (23) and (24) in TLP-LS.

Although being straightforward, the formulations in Table 2 are not efficient from a computational standpoint. A direct solution of these models using a linear integer solver (e.g., CPLEX [25]) works only for networks with few (less than 10) nodes. There are two main reasons for this. First, the models contain a lot of symmetry. There are, for example, many solutions that correspond to the same assignment but with different time slots allocated, and swapping the nodes (or links) of any two slots does not affect the objective function value. Moreover, the linear programming (LP) relaxations of these formulations tend to be very weak, making large networks out of reach of state-of-the-art integer solvers.

5.2 Formulations Based on Transmission Groups

Alternative formulations of our optimization problems can be obtained using transmission groups. A transmission group is simply a group of nodes, or a group of links, that can share a time slot.

Let L_N and L_A denote the sets of transmission groups of nodes and links, respectively. A schedule can then be represented using one variable for every group. Doing so gives us formulations based on transmission groups, which are summarized in Table 3. In MNP-TG and TNP-TG, s_{il} is an indication parameter that is one if group l contains node i, and zero otherwise. The corresponding indication parameter for the other two formulations is s_{ijl}.

In formulations MNP-TG and MLP-TG, (25) and (32) minimize the total number of groups (i.e., the total number of allocated slots). Constraints (26) and (33) make sure that every node and every link, respectively, receives at least one slot. For maximum-throughput scheduling, the throughput is defined using (28) and (29) in TNP-TG, and using (35) and (36) in TLP-TG. Constraints (30) and (37) define the schedule length.

Note that, in contrast to the formulations in Section 5.1, the formulations in Table 3 do not contain the SIR constraint (2), nor the other two transmission constraints discussed in Section 2.1. These constraints are hidden in the definitions of the sets of feasible transmission groups, L_N and L_A.

Table 3: Mathematical models: Transmission-group based formulations.

MNP-TG		MLP-TG	
$x_l = 1$ if transmission group l is assigned any time slot, otherwise $x_l = 0$.			
$\min \sum_{l \in L_N} x_l$	(25)	$\min \sum_{l \in L_A} x_l$	(32)
$\sum_{l \in L_N} s_{il} x_l \geq 1, \forall i \in N,$	(26)	$\sum_{l \in L_A} s_{ijl} x_l \geq 1, \forall (i,j) \in A,$	(33)
$x_l \in \{0,1\}, \forall l \in L_N.$	(27)	$x_l \in \{0,1\}, \forall l \in L_A.$	(34)
TNP-TG		**TLP-TG**	
$x_l =$ the number of time slots assigned to transmission group l.			
$\max z$	(28)	$\max z$	(35)
$z \leq \frac{\sum_{l \in L_N} s_{il} x_l}{\rho_i \lvert T \rvert}, \forall i \in N,$	(29)	$z \leq \frac{\sum_{l \in L_A} s_{ijl} x_l}{\rho_{ij} \lvert T \rvert}, \forall (i,j) \in A,$	(36)
$\sum_{l \in L_N} x_l = \lvert T \rvert,$	(30)	$\sum_{l \in L_A} x_l = \lvert T \rvert,$	(37)
$x_l \geq 0$, integer, $\forall l \in L_N.$	(31)	$x_l \geq 0$, integer, $\forall l \in L_A.$	(38)

The definitions of L_N and L_A effectively decompose our scheduling problems into two parts. The first part concerns the assignment of time slots to transmission groups. This part is formulated mathematically in Table 3. The second part involves the generation of (suitable) members in L_N and L_A. This decomposition is exploited in the column generation method in the next section.

6 Solution Methods

6.1 A Column Generation Method

The formulations presented in Table 3 have very simple constraint structures. The complexity of the formulations lies in the sizes of L_N and L_A. For networks of a realistic size, there are far more members in the two sets than what can be handled efficiently in a linear integer solver. However, we observe that, for these formulations, solving the LP-relaxations yields very sharp bounds to the integer optima. In addition, for the LP-relaxations, the sizes of L_N and L_A can be overcome using a column generation method, which handles the two sets implicitly rather than using an explicit description of all the members of the two sets.

Column generation decomposes a linear program into a master problem and a subproblem. The former contains only a subset of the columns (variables). The latter is a separation problem for the dual LP, and is used to check optimality, i.e., whether additional columns need to be added to the master problem or not. To apply the column generation method to the formulations in Table 3, the sets L_N and L_A are replaced by two subsets $L_N^0 \subseteq L_N$ and $L_A^0 \subseteq L_A$. This, together with the removal of integrality constraints, give four master problems. To ensure the initial feasibility of the master problems of MNP-TG and MLP-TG, we can set $L_N^0 = \{\{i\}, i \in N\}$ and $L_A^0 = \{\{(i,j)\}, (i,j) \in A\}$ (this, in fact, corresponds to TDMA). After solving a master problem, the method examines whether any new transmission group should be added to L_N^0 (or L_A^0). In LP terms, this is equivalent to examining whether there exists any element $l \in L_N$ for which the corresponding variable x_l has a negative reduced cost in MNP-TG and MLP-TG, or any element $l \in L_A$ for which the corresponding variable x_l has a positive reduced cost in TNP-TG and TLP-TG. To this end, the method solves a subproblem. The subproblem amounts to enumerating the members of L_N (or L_A), in order to find a group for which the reduced cost has the right sign (if any such group exists). It is important to note, however, that this enumeration is performed implicitly. Instead of using an explicit and complete description of L_N (or L_A), the subproblem contains variables and constraints that are necessary to characterize any feasible transmission group. The solution space of the subproblem is thus an equivalent, but implicit representation of L_N (or L_A). In other words, any feasible solution to the subproblem corresponds to a feasible transmission group. Moreover, the objective function of the subproblem expresses the reduced costs of transmission groups. The optimal subproblem solution is the group having the minimum reduced cost for MNP-TG and MLP-TG, or maximum reduced cost for TNP-TG and TLP-TG.

The subproblems of the four transmission-group based formulations are summarized in Table 4. Note the similarities between the subproblems and the formulations in Table 2. The difference is that the constraints of the subproblems are defined for one single time slot (instead of all time slots).

If solving a subproblem yields a negative reduced cost for MNP-TG and MLP-TG, or a positive reduced cost for TNP-TG and TLP-TG, the solution (i.e., a new transmission group) is added to the corresponding master problem. The master problem is re-optimized, after which the column generation procedure proceeds to the next iteration. If, on the other hand, no transmission group has a reduced cost with the right sign, then the corresponding LP-relaxation is solved to optimality.

The solutions of the LP-relaxations have a nice interpretation in TNP-TG and TLP-TG, namely that the value of $x_l/|T|$ is the proportion of time slots (transmission resource) assigned to group l.

Table 4: The subproblems.

Subproblem of MNP-TG	Subproblem of MLP-TG
Parameter β_i is the dual variable of (26). Variable $s_i = 1$ if node i is included in the transmission group, otherwise $s_i = 0$.	Parameter β_{ij} is the dual variable of (33). Variable $s_{ij} = 1$ if link (i, j) is included in the transmission group, otherwise $s_{ij} = 0$. Variable $v_i = 1$ if node i is transmitting, otherwise $v_i = 0$.
$\min 1 - \sum_{i \in N} \beta_i s_i$ (39) $s_i + \sum_{j:(j,i) \in A} s_j \leq 1, \forall i \in N,$ (40) $\frac{P_i/N_r}{L_b(i,j)} s_i + \gamma_1(1 + M_{ij})(1 - s_i) \geq \gamma_1(1 + \sum_{k \in N: k \neq i,j} \frac{P_k/N_r}{L_b(k,j)} s_k), \forall (i,j) \in A,$ (41) $s_i \in \{0,1\}, \forall i \in N.$ (42)	$\min 1 - \sum_{(i,j) \in A} \beta_{ij} s_{ij}$ (43) $\sum_{j:(i,j) \in A} s_{ij} + \sum_{j:(j,i) \in A} s_{ji} \leq 1, \forall i \in N,$ (44) $s_{ij} \leq v_i, \forall (i,j) \in A,$ (45) $\frac{P_i/N_r}{L_b(i,j)} s_{ij} + \gamma_1(1 + M_{ij})(1 - s_{ij}) \geq \gamma_1(1 + \sum_{k \in N: k \neq i,j} \frac{P_k/N_r}{L_b(k,j)} v_k), \forall (i,j) \in A,$ (46) $s_{ij} \in \{0,1\}, \forall (i,j) \in A,$ (47) $v_i \in \{0,1\}, \forall i \in N.$ (48)
Subproblem of TNP-TG	**Subproblem of TLP-TG**
Parameter β_i is the dual variable of (29). Parameter α is the dual variable of (30). Variable $s_i = 1$ if node i is included in the transmission group, otherwise $s_i = 0$.	Parameter β_{ij} is the dual variable of (36). Parameter α is the dual variable of (37). Variable $s_{ij} = 1$ if link (i, j) is included in the transmission group, otherwise $s_{ij} = 0$. Variable $v_i = 1$ if node i is transmitting, otherwise $v_i = 0$.
$\max \sum_{i \in N} \beta_i s_i - \alpha$ (49) (40), (41), (42).	$\max \sum_{(i,j) \in A} \beta_{ij} s_{ij} - \alpha$ (50) (44), (45), (46), (47), (48).

6.2 A Greedy Heuristic

Solving the LP-relaxations yields lower or upper bounds for the four optimization problems. Our numerical experiments show that these bounds are very sharp. However, as the LP solutions may contain fractional-valued variables, we need a procedure for obtaining integer feasible solutions. An iterative greedy algorithm can be used for this purpose. In one iteration, the algorithm constructs a transmission group, and assigns a time slot to the group. The algorithm is designed for the two maximum-throughput scheduling problems, but works nicely for the two minimum-length scheduling problems as well.

For problem TNP, one iteration of the algorithm is as follows. The algorithm sorts the nodes in an ascending order by their throughput values (for a node, this is the ratio between the proportion of time slots assigned to the node so far and the traffic load), and stores the result in a list. The algorithm scans through the list, and picks as many nodes as possible to form a transmission group. The throughput values of the nodes are then updated, and the algorithm proceeds to the next iteration. It can be easily shown that the algorithm has a polynomial time complexity for any of the four problems. Below is a formal description of the algorithm, where S_t is the group of nodes that are assigned time slot t, h_i is the number of time slots assigned to i, $\bar{\lambda}_i$ is the throughput of i, and Q is the list.

1. Set $h_i = 0, \bar{\lambda}_i = 0, \forall i \in N$, and $S_t = \emptyset, \forall t \in T$. Set $t = 1$.
2. Use $\bar{\lambda}_i$ to sort the nodes in ascending order. Store the result in Q.
3. Repeat until Q is empty:
   (a) Let i^* be the first element in Q.
   (b) Set $Q = Q \setminus \{i^*\}$.
   (c) If $S_t \cup \{i^*\}$ is a feasible transmission group: Set $S_t = S_t \cup \{i^*\}$, $h_{i^*} = h_{i^*} + 1$, and $\bar{\lambda}_{i^*} = \frac{h_{i^*}}{\rho_{i^*}|T|}$.
4. Let $t = t + 1$. If $t > |T|$, terminate. Otherwise go to Step 2.

It is straightforward to adapt the algorithm to problem TLP, by defining all the quantities for links instead of nodes. Once we have applied the algorithm for a schedule length of $|T| - 1$, increasing the schedule length to $|T|$ does not require a computation from scratch. In fact, only the transmission group for slot $|T|$ needs to be computed, because the groups for slots $t = 1, \ldots, |T| - 1$ will remain the same. It is therefore sufficient to run the algorithm once for length $|T|$ to obtain integer solutions for all schedule lengths between one and $|T|$.

The above algorithm can also be used to find feasible solutions to MNP and MLP. Because the throughput of a node (or a link) is zero before it is assigned any time slot, the algorithm effectively minimizes the schedule length. For problems MNP and MLP, the algorithm terminates as soon as the overall throughput becomes greater than zero, i.e., when every node (or every link) is assigned at least one time slot.

7 Numerical Results

Numerical results were obtained using three test sets provided by the Swedish Defense Research Agency. Each set contains five networks of the same size in the number of nodes. The traffic in the test networks are unicast, and its distribution is uniform (one unit of traffic demand between every pair of nodes). The load parameters, ρ_i and ρ_{ij}, were obtained using minimum-hop routing.

The column generation method was implemented using a modeling language AMPL [16] and a linear integer solver CPLEX [25]. The program was run on a Sun UltraSparc station with a 400 MHz CPU and 1 GB RAM. The greedy algorithm was implemented in MATLAB. For each network, we ran the algorithm for a schedule length $|T| = 1000$. Doing so gave solutions for all schedule lengths between one and 1000.

The numerical results are summarized in Tables 5 and 6. Table 5 shows the results for MNP and MLP. The format is as follows. The two columns denoted by 'TDMA' display the schedule lengths of TDMA (i.e., no spatial reuse). Note that the values of these two columns are, respectively, equal to the number of nodes $|N|$ and the number of links $|A|$. For each of the two problems, the results obtained using the column generation method are displayed in three columns: 'LP' that shows the optimum of the LP-relaxation, 'Iter.' that shows the number of iterations, and 'Time' that shows the solution time in seconds. Finally, the schedule length found by the greedy algorithm is shown in the two columns denoted by 'Greedy'.

Table 6 shows the results for the two maximum-throughput problems, TNP and TLP. The format of the table is similar to that of Table 5, but the entries of 'TDMA', 'LP', and 'Greedy' are throughput values instead of schedule lengths. Also, the throughput shown for the greedy algorithm is the best value among 1000 slots. Note that the table contains two additional columns: 'MNP' and 'MLP'. These two columns show the throughput of the LP solutions obtained for the two minimum-scheduling problems. By comparing the values in these two columns to others, we can examine how well minimum-length scheduling, which does not account for any particular traffic distribution, performs in terms of throughput. (The two tables do not include solution time of the greedy algorithm, because the algorithm needs only a few seconds for solving any of the test networks.)

We observe that STDMA significantly improves network performance when comparing to TDMA. For minimum-length scheduling, the number of time slots of TDMA can be reduced by a factor of up to four using STDMA. For maximum-throughput scheduling, the throughput of STDMA can be as much as 35 times higher than that of TDMA. Second, link-oriented assignment permits more spatial reuse as well as higher throughput for unicast traffic than what is possible in node-oriented assignment.

The formulations using transmission groups, and, in addition, the col-

Table 5: Numerical results for minimum-length scheduling.

	MNP					MLP				
	TDMA	Column generation			Greedy	TDMA	Column generation			Greedy
	$\|N\|$	LP	Iter.	Time		$\|A\|$	LP	Iter.	Time	
N20–1	20	9	24	4	10	44	13	85	84	18
N20–2	20	10	17	3	11	48	16	42	21	19
N20–3	20	9	27	5	11	58	21	73	55	24
N20–4	20	12	14	3	12	66	27	65	45	32
N20–5	20	12	18	3	13	76	40	44	28	44
N40–1	40	14	69	67	16	154	39	173	1281	47
N40–2	40	15	40	25	17	194	52	168	1424	62
N40–3	40	22	34	34	23	236	70	257	2675	89
N40–4	40	27	26	25	29	266	106	290	4076	128
N40–5	40	28	14	19	28	288	123	285	3164	139
N60–1	60	15	106	480	19	250	51	236	2311	67
N60–2	60	18	75	213	23	302	73	236	2314	87
N60–3	60	21	93	288	27	354	97	248	2539	105
N60–4	60	26	43	108	26	376	91	769	16667	119
N60–5	60	26	53	204	30	396	114	321	3814	131

Table 6: Numerical results for maximum-throughput scheduling.

	TNP						TLP					
	TDMA	MNP	Column generation			Greedy	TDMA	MLP	Column generation			Greedy
			LP	Iter.	Time				LP	Iter.	Time	
N20–1	0.087	0.193	0.299	14	2	0.299	0.087	0.295	0.447	53	19	0.444
N20–2	0.088	0.175	0.350	16	3	0.350	0.080	0.240	0.474	26	10	0.473
N20–3	0.091	0.202	0.374	14	3	0.373	0.072	0.199	0.597	65	36	0.589
N20–4	0.085	0.142	0.453	13	3	0.453	0.089	0.217	0.660	73	46	0.654
N20–5	0.103	0.171	0.444	20	4	0.444	0.067	0.127	0.657	102	70	0.652
N40–1	0.045	0.128	0.385	45	25	0.381	0.025	0.100	0.564	179	563	0.549
N40–2	0.047	0.125	0.397	49	27	0.394	0.020	0.076	0.646	260	1304	0.623
N40–3	0.068	0.123	0.436	20	17	0.433	0.038	0.130	0.787	261	1157	0.760
N40–4	0.065	0.096	0.520	21	20	0.515	0.077	0.194	0.807	248	1128	0.755
N40–5	0.090	0.128	0.456	25	35	0.451	0.027	0.062	0.931	559	12201	0.867
N60–1	0.042	0.170	0.447	53	75	0.444	0.020	0.100	0.790	522	24020	0.740
N60–2	0.039	0.129	0.435	63	115	0.428	0.020	0.085	0.724	402	4324	0.684
N60–3	0.043	0.121	0.451	54	113	0.446	0.023	0.083	0.756	428	3249	0.715
N60–4	0.054	0.126	0.433	55	132	0.428	0.019	0.079	0.753	521	4431	0.704
N60–5	0.051	0.118	0.457	64	185	0.450	0.020	0.069	0.782	601	5874	0.729

umn generation method, are effective for STDMA scheduling problems. The solution times of the column generation method look reasonable for most cases, although occasionally the computing time exceeds several hours. (We note that there are quite some space left for method refinements, such as a more sophisticated implementation and problem reduction techniques.) The optimal solutions of the LP-relaxations tell what at best can be achieved using STDMA. Note that the LP bounds are very sharp, that is, these bounds give very good estimations to the optimal performance of an STDMA system. (Results in [4] indicate that for minimum-length scheduling, the gap between the LP bound and the greedy solution is mainly due to the latter.)

Looking at the results of the greedy algorithm, we observe that for maximum-throughput scheduling, this algorithm constantly generates a near-optimal schedule within 1000 time slots. More detailed solution statistics (not shown in Table 6 due to the space limitation) reveal that the throughput became close to optimum in less than 200 slots for all test networks. For minimum-length scheduling, on the other hand, the performance of the greedy algorithm is not always consistent. This is not surprising, because in this case the algorithm does not utilize any network-specific information to determine in which order the nodes or links should be treated. Nevertheless, the algorithm gave satisfactory results for most test networks.

For small networks (of 20 nodes), the throughput of a minimum-length schedule is not too far away from what can be achieved in maximum-throughput scheduling. For large networks, the difference in throughput becomes quite significant. It should however be pointed out that it is considerably more complex to design a protocol for maximum-throughput scheduling than for minimum-length scheduling.

8 Conclusions and Perspectives

Some intelligence is required in STDMA to schedule the transmission activities. In this chapter, four optimization problems are identified in this context. The relations between them are clarified, and models and solution methods are presented.

One important conclusion from our study is that the mathematical formulations based on transmission groups and the column generation method provide us with an effective tool to assess the theoretical performance limits of STDMA, and thereby to evaluate heuristic, on-line algorithms as well as practical STDMA implementations. Having the maximum achievable performance at hand, it can be concluded that STDMA does, indeed, have a great potential of utilizing the network resource efficiently. As another conclusion, the greedy algorithm has a close-optimal performance in maximizing the network throughput. Due to its low computational complexity, the algorithm is an interesting candidate for distributed implementations.

There are many remaining research challenges in designing multi-hop radio networks. Having access to the theoretical limits in both the minimum length and the maximal throughput, a natural question to ask is to what extent these limits can be achieved in practice, when the two assumptions in Section 1 (off-line computation and full knowledge of interference) are replaced by more realistic ones. Thus, the next major step of research in this area is to design distributed, on-line algorithms that can approximate these theoretical limits. It will be very interesting (and challenging as well), for example, to adapt the greedy algorithm to distributed computations. To do this, a number of issues need to be resolved. Among these, the following, interrelated issues are of particular importance for future research.

- *Distributed computation.* In a distributed computing environment, full knowledge of the traffic distribution (which is, obviously, dynamic in nature) is very hard, if not impossible, to obtain. Hence, practical algorithms need to operate with uncertain and partial information. One possibility to deal with this difficulty is to use some static attributes to approximate the expected traffic load. In addition, a sensitivity analysis would provide insights into whether the solution of the greedy algorithm (or other heuristics) will deviate widely from optimum in presence of minor perturbation in the problem parameters.

- *Information sharing.* A second challenge is how to share the link quality information (i.e., the path-loss values) among the nodes. STDMA involves some signaling overhead to generate a schedule. The more information is shared, the more efficient the schedule becomes. At some point, however, the amount of signaling traffic becomes unreasonable with respect to network capacity. Thus some compromise has to be made between information sharing and solution optimality. One option is to limit information sharing to a local environment (e.g., a node exchanges information only with nodes within a certain hop limit). The amount of signaling traffic depends also on the network architecture, in particular whether clustering is used or not.

- *Mobility.* Mobility poses another challenge in the design of access control schemes. Mobile radio units lead to constant changes in the network topology. Benchmarking the network performance for a mobile scenario remains straightforward; the column generation method can be applied to a series of static scheduling problems, derived from a sequence of snapshots of the mobile scenario. A practical scheduling algorithm must, on the other hand, monitor topology changes and update the schedule in an on-line environment. The frequency of updating the schedule gives rise to a further dimension in designing STDMA protocols. Clearly, the issue of mobility should be tackled in conjunction with the two issues above.

Another research direction, which is worth mentioning, is the interplay between STDMA and other networking layers. One such example is the role of power control. Throughout the chapter, we have assumed that the power used in transmission is given. A more complex scenario arises if the units can adjust their transmitting power. Power control can be used for several purposes, such as to reduce the interference, to increase the link bandwidth, as well as to control the network topology. (The issue of power control is addressed in some other chapters of this book.) Another example is the routing protocol, which effectively determines the traffic load of individual nodes and links. One interesting topic for forthcoming research is whether and how these issues can be integrated with STDMA scheduling, and the impact of such an integration on system design and optimization.

Acknowledgment The authors wish to thank Jimmi Grönkvist at the Department of Communication Systems, Swedish Defense Research Agency (FOI), for the technical discussions and the test data. This work is partially financed by CENIIT (Center for Industrial Information Technology), Linköping Institute of Technology, Sweden.

References

[1] E. Arikan, Some complexity results about packet radio networks, *IEEE Transactions on Information Theory* Vol. IT-30 (1984) pp. 681-685.

[2] L.Bao and J.J. Garcia-Luna-Aceves, Collision-free topology-dependent channel access scheduling, *Proceedings of IEEE MILCOM '00* (2000) pp. 507-511.

[3] D.J. Baker and J.Wieselthier, A distributed algorithm for scheduling the activation of links in a self-organizing, mobile, radio networks, *Proceedings of IEEE ICC '82* (1982) pp. 2F.6.1-2F.6.5.

[4] P. Björklund, P. Värbrand and D. Yuan, Resource optimization of STDMA in ad hoc networks, *Ad Hoc Networks* (In press, 2004).

[5] P. Björklund, P. Värbrand and D. Yuan, Resource optimization of spatial TDMA in ad hoc radio networks: A column generation approach, *Proceedings of IEEE INFOCOM '03* (2003) pp. 818-824.

[6] S. Boztas, A robust multi-priority topology-independent transmission schedule for packet radio networks, *Information Processing Letters* Vol. 55 (1995) pp. 291-295.

[7] I. Cidon and M. Sidi, Distributed assignment algorithms for multi-hop packet radio networks, *IEEE Transactions on Computers* Vol. 38 (1989) pp. 1353-1361.

[8] I. Chlamtac and A. Faragó, Making transmission schedules immune to topology changes in multi-hop packet radio networks, *IEEE/ACM Transactions on Networking* Vol. 2 (1994) pp. 23-29.

[9] I. Chlamtac and A. Lerner, A link allocation protocol for mobile multihop networks, *Proceedings of IEEE GLOBECOM '85* (1985) pp. 238-242.

[10] I. Chlamtac and A. Lerner, Fair algorithms for maximal link activation in multihop radio networks, *IEEE Transactions on Communications* Vol. Com-35 (Jul. 1987) pp. 739-746.

[11] I. Chlamtac and S.S. Pinter, Distributed nodes organization algorithm for channel access in a multihop dynamic radio network, *IEEE Transactions on Computers* Vol. 36 (1987) pp. 728-737.

[12] A.-M. Chou and V.O.K. Li, Fair spatial TDMA channel access protocols for multihop radio networks, *Proceedings of IEEE INFOCOM '91* (1991) pp. 1064-1073.

[13] A.-M. Chou and V.O.K. Li, Slot allocation strategies for TDMA protocols in multihop packet radio networks, *Proceedings of IEEE INFOCOM '92* (1992) pp. 710-716.

[14] A. Ephremides and T. Truong, Distributed algorithm for efficient and interference-free broadcasting in radio networks, *Proceedings of IEEE INFOCOM '88* (1988) pp. 1119-1124.

[15] A. Ephremides and T. Truong, Scheduling broadcasts in multihop radio networks, *IEEE Transactions on Communications* Vol. 38 (1990) pp. 456-460.

[16] R. Fourer, D.M. Gay and W. Kernighan, *AMPL - A Modeling Language for Mathematical Programming*, (Boyd & Fraser, Danvers, MA, 1993).

[17] N. Funabiki and Y. Takefuji, A parallel algorithm for broadcast scheduling problems in packet radio networks, *IEEE Transactions on Communications* Vol. 41 (1993) pp. 828-831.

[18] A.J. Goldsmith and S.B. Wicker, Design challenges for energy-constrained ad hoc wireless networks, *IEEE Transactions on Wireless Communications* (August 2002) pp. 8-27.

[19] J. Grönkvist, Traffic controlled spatial reuse TDMA in multi-hop radio networks, *Proceedings of the 9th IEEE International Symposium on Personal, Indoor and Mobile Radio Communications* (1998) pp. 1203-1207.

[20] J. Grönkvist, Assignment methods for spatial reuse TDMA, *Proceedings of the First Annual Workshop on Mobile and Ad Hoc Networking and Computing* (2000) pp. 119-124.

[21] J. Grönkvist, A. Hansson and J. Nilsson, A comparison of access methods for multi-hop ad hoc radio networks, *Proceedings of IEEE VTC 2000-spring* (2000) pp. 1435-1439.

[22] J. Grönkvist, *Assignment Strategies for Spatial Reuse TDMA*, (Licentiate thesis, Royal Institute of Technology, Stockholm, Sweden, 2002).

[23] B. Hajek and G. Sasaki, Link scheduling in polynomial time, *IEEE Transactions on Information Theory* Vol. 34 (1988) pp. 910-917.

[24] S.O. Krumke, M.V. Marathe and S.S. Ravi, Models and approximation algorithms for channel assignment in radio networks, *Wireless Networks* Vol. 7 (2001) pp. 575-584.

[25] ILOG, *Ilog cplex, user's manual*, (Aug. 2000).

[26] R. Liu and E.L. Lloyd, A distributed protocol for adaptive scheduling in ad hoc networks, *Proceedings of the IASTED International Conference on Wireless and Optical Communications (WOC 2001)* (2001) pp. 43-48.

[27] A. Mehrotra and M.A. Trick, A column generation approach for graph coloring, *INFORMS Journal on Computing* Vol. 8 (1996) pp. 344-354.

[28] R. Nelson and L. Kleinrock, Spatial TDMA: A collision-free multihop channel access protocol, *IEEE Transactions on Communications* Vol. 33 (1985) pp. 934-944.

[29] M. Post, P. Sarachik and A. Kershenbaum, A biased greedy algorithm for scheduling multi-hop radio networks, *Proceedings of the 19th Annual Conference on Information Sciences and Systems* (1985) pp. 564-572.

[30] L.C. Pond and V.O.K. Li, A distributed time-slot assignment protocol for mobile multi-hop broadcast packet radio networks, *Proceedings of IEEE MILCOM '89* (1989) pp. 70-74.

[31] C.G. Prohazka, Decoupling link scheduling constraints in multihop packet radio networks, *IEEE Transactions on Computers* Vol. 38 (1989) pp. 455-458.

[32] S. Ramanathan and E.L. Lloyd, Scheduling algorithms for multihop radio networks, *IEEE Transactions on Networking* Vol. 1 (1993) pp. 166-177.

[33] R. Ramaswami and K.K. Parhi, Distributed scheduling for broadcasts in a radio network, *Proceedings of IEEE INFOCOM '89* (1989) pp. 497-504.

[34] A. Sen and J.M. Capone, Scheduling in packet radio networks - A new approach, *Proceedings of IEEE GLOBECOM '99* (1999) pp. 650-654.

[35] A. Sen and M.L. Huson, A new model for scheduling packet radio networks, *Proceedings of IEEE INFOCOM '96* (1996) pp. 1116-1124.

[36] J. Shor and T.G. Robertazzi, Traffic sensitive algorithms and performance measures for the generation of self-organizing radio network schedules, *IEEE Transactions on Communications* Vol. 41 (1993) pp. 16-21.

[37] D.S. Stevens and M.H. Ammar, Evaluation of slot allocation strategies for TDMA protocols in packet radio networks, *Proceedings of IEEE MILCOM '90* (1990) pp. 835-839.

[38] S. Toumpis and A.J. Goldsmith, Capacity regions for wireless ad hoc networks, *Proceedings of IEEE ICC '02* (2002) pp. 3168-3173.

[39] C.D. Young, USAP multiple access: Dynamic resource allocation for mobile multihop multichannel wireless networking, *Proceedings of IEEE MILCOM '99* (1999) pp. 271-275.

[40] C.D. Young, USAP multiple access: Transmitter- and receiver-directed dynamic resource allocation for mobile, multihop, multichannel, wireless networking, *Proceedings of IEEE MILCOM '00* (2000) pp. 549-553.

QoS Enhancement for the IEEE 802.11e Distributed Wireless LANs

Yang Xiao
Computer Science Division
The University of Memphis, Memphis, TN 38152
E-mail: `yangxiao@ieee.org`

Haizhon Li
Computer Science Division
The University of Memphis, Memphis, TN 38152
E-mail: `hli1@memphis.edu`

Contents

1 Introduction

The IEEE 802.11 standards have been very successful and popular in recent years. Both the medium access control (MAC) layer and the phsical (PHY) layer are specifed in the IEEE 802.11 standard family, i.e. IEEE 802.11 [1], IEEE 802.11e [2], IEEE 802.11b [3], and IEEE 802.11a [4], and IEEE 802.11g [12], etc.

The IEEE 802.11 MAC employs mandatory contention-based channel access function called Distributed Coordination Function (DCF), and an optional centrally controlled channel access function called Point Coordination Function (PCF) [1]. The DCF adopts a carrier sense multiple access with collision avoidance (CSMA/CA) and binary exponential backoff. It is treated as a wireless version of the most successful local area network (LAN), IEEE 802.3 (Ethernet), which adopts CSMA with collision detection (CSMA/CD) and binary exponential backoff. Both the IEEE 802.11 DCF and IEEE 802.3 enable fast installation with minimal management and maintenance costs, and are very robust protocols for the best-effort service. The popularity of the IEEE 802.11 market is largely due to the DCF, whereas the PCF is barely implemented in today's products due to its complexity and inefficiency for the normal data transmissions, even though it has some limited Quality of Service (QoS) support.

However, the current DCF is unsuitable for multimedia applications with QoS requirements [5, 6, 16]. Under the DCF, a station might have to wait an arbitrarily long time to send a frame so that real-time applications such as voice and video may suffer [5]. One possible solution is to provide a good priority scheme for the DCF. Simple DCF priority schemes can be easily designed with minor changes in the DCF, and they are quite effective [6, 16]. Prioritized QoS will be useful for those multimedia applications that can live without rigid QoS. One advantage of prioritized QoS is that it is simple to implement and looks like DiffServ model in the IP networks.

To support the MAC-level QoS, the IEEE 802.11 Working Group is currently working on the standardization of IEEE 802.11e [2], which is in the final stage. The emerging IEEE 802.11e standard provides QoS features and multimedia support to the existing 802.11b [3] and 802.11a/.11g [4, 12] WLAN standards, while maintaining full backward compatibility with these standards. The IEEE 802.11e MAC employs a channel access function, called Hybrid Coordination Function (HCF), which includes a contention-based channel access and a contention-free centrally controlled channel access mechanism. The contention-based channel is also referred to as Enhanced Distributed Coordination Function (EDCF). The EDCF provides a priority scheme by differentiating the inter-frame space, the initial window size, and the maximum window size.

The rest of the chapter is organized as follows. We briefly describe the IEEE 801.11/.11b/.11a/.11g PHY in Section 2. The original IEEE 802.11 MAC and the IEEE 802.11e EDCF are introduced in Section 3 and Section 4, respectively. We introduce two MAC enhancements of IEEE 802.11e in Section 5. Related work for the EDCF is summarized in Section 6. We evaluate the EDCF differentiated mechanism via extensive simulations in Section 7. We finally conclude this chapter in Section 8.

2 IEEE 802.11/.11b/.11a/.11g PHY

In this section, we briefly introduce the IEEE 802.11/.11b/.11a/.11g PHY as follows.

The IEEE 802.11 PHY [1] supports 1 and 2 Mbps raw data rates including three different physical layer implementations: frequency hopping spread spectrum (FHSS), direct sequence spread spectrum (DSSS), and infrared (IR). The FHSS utilizes the 2.4 GHz Industrial, Scientific, and Medical (ISM) band, and the basic access rate of 1 Mbps uses two-level Gaussian frequency shift keying (GFSK). The DSSS uses the 2.4 GHz ISM frequency band, where the 1 Mbps basic rate is encoded using differential binary phase shift keying (DBPSK), and a 2 Mbps enhanced rate uses differential quadrature phase shift keying (DQPSK). The IR specification is designed for indoor use only and operates with nondirected transmissions, and its encoding of the basic access rate of 1 Mbps is performed using 16-pulse position modulation (PPM) and the enhanced access rate of 2 Mbps is performed using 4-PPM modulation.

The IEEE 802.11b PHY [3] supports 1, 2, 5.5, and 11 Mbps raw data rates via Complementary Code Keying (CCK) and Direct Sequence Spread Spectrum (DSSS) modulation schemes at 2.4 GHz. The IEEE 802.11b PHY is the most widely used PHY.

The IEEE 802.11a PHY uses a convolutionally coded adaptation of Orthogonal Frequency Division Multiplexing (OFDM) for encoding and transmission called coded OFDM (COFDM), which is a frequency division multiplexed (FDM) multi-carrier communications scheme that includes the application of convolutional coding to achieve higher raw data rates. The IEEE 802.11a standard operates at the 5 GHz band, thereby avoiding the crowded 2.4 GHz band where IEEE 802.11b and Bluetooth operate. IEEE 802.11a uses COFDM to realize the full 6-54 Mbps range of data rates. OFDM is a multicarrier communications scheme in which a single high-rate data stream is split into lower-rate data streams that are subsequently transmitted in parallel over a number of subcarriers. The subcarriers overlap and the inter-carrier spacing are chosen such that all the subcarriers are orthogonal to each other. The IEEE 802.11 standard specifies a channel spacing of 20 MHz with a 16.56 MHz transmission bandwidth per channel. Each subcarrier is spaced 312.5 kHz from adjacent subcarriers, and each is modulated independently. The physical format of IEEE 802.11a [4] is illustrated in Figure 1. Data rates for IEEE 802.11a are 6, 9, 12, 18, 24, 36, 48, and 54 Mbps.

The IEEE 802.11g PHY, an extension of IEEE 802.11b and IEEE 802.11a, supports up to 54 Mbps at 2.4 GHz where IEEE 802.11b and Bluetooth operate, with the PHY technology of IEEE 802.11a. IEEE 802.11g is a superset of the 802.11b PHY, including the 802.11b modulation schemes and the OFDM schemes originally defined for 802.11a PHY at 5 GHz.

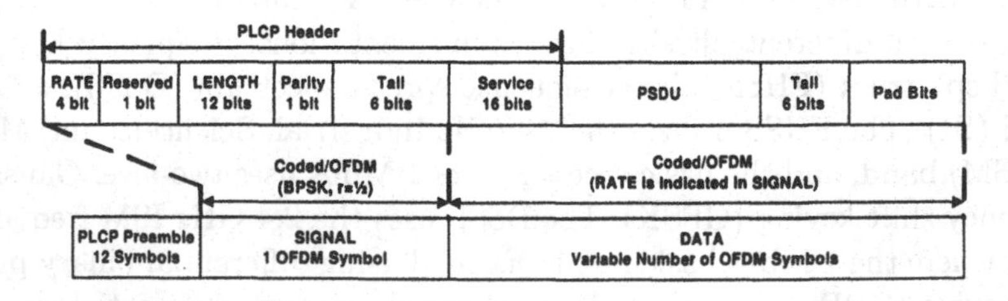

Figure 1: The PHY frame format for the IEEE 802.11a

3 IEEE 802.11 MAC

The IEEE 802.11 MAC employs a mandatory DCF and an optional PCF. These functions determine when a station, operating within a Basic Service Set (BSS) or Independent BSS (IBSS), is permitted to transmit. There are two types of 802.11 networks: Infrastructure Network, i.e. BSS, in which an access point (AP) is present and Ad hoc Network, i.e. IBSS, in which an AP is not present.

In the long run, time is always divided into repetition intervals called *superframes*, shown in Figure 2. Each *superframe* starts with a beacon frame, and the remaining time is further divided into an optional contention-free period (CFP) and a contention period (CP). The DCF works during the CP and the PCF works during the CFP. If the PCF is not active, a *superframe* will not include the CFP. However, the beacon frame is always sent no matter whether the PCF is active or not. The beacon frame is a management frame for synchronization, power management, and delivering parameters. In a BSS, an AP sends beacon frames. In an IBSS, any mobile station that is configured to start an IBSS will begin sending beacon frames. As other mobile stations join that IBSS, each station, a member of the IBSS, is randomly chosen for the task of sending beacon frames. Beacon frames are generated in regular intervals called target beacon transmission time.

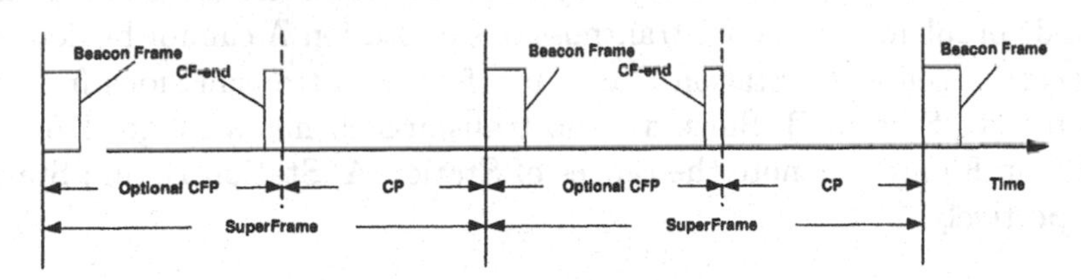

Figure 2: Superframes

3.1 The DCF

The DCF defines a basic access mechanism and an optional request-to-send/clear-to-send (RTS/CTS) mechanism. In the DCF, a station with a frame to transmit monitors the channel activities until an idle period equal to a distributed inter-frame space (DIFS) is detected. After sensing an idle DIFS, the station waits for a random backoff interval before transmitting. The backoff time counter is decremented in terms of slot time as long as the channel is sensed idle. The counter is stopped when a transmission is detected on the channel, and reactivated when the channel is sensed idle again for more than a DIFS. In this manner, stations, deferred from channel access because their backoff time was larger than the backoff time of other stations, are given a higher priority when they resume the transmission attempt. The station transmits its frame when the backoff time reaches zero. At each transmission, the backoff time is uniformly chosen in the range $[0, CW - 1]$ in terms of timeslots, where CW is the current backoff window size. At the very first transmission attempt, CW equals the initial backoff window size CW_{min}. After each unsuccessful transmission, CW is doubled until a maximum backoff window size value CW_{max} is reached. After the destination station successfully receives the frame, it transmits an acknowledgment frame (ACK) following a short inter-frame space (SIFS) time. If the transmitting station does not receive the ACK within a specified ACK Timeout, or it detects the transmission of a different frame on the channel, it reschedules the frame transmission according to the previous backoff rules.

The DCF provides a channel access mechanism with equal probabilities to all stations contending for the same wireless medium. If an AP is present, STAs are not allowed to transmit frames to other STAs that are not APs.

The above mechanism is called the basic access mechanism. In such a mechanism, hidden node problem may happen, shown in Figure 3. The hidden node problem exists since transmissions of Station A cannot be detected using carrier senses by Station C but interfere with transmissions from the Station C to Station B. Since a radio transmission has a range, RA, RB, and RC in Figure 3 denote the ranges of Station A, Station B, and Station C, respectively.

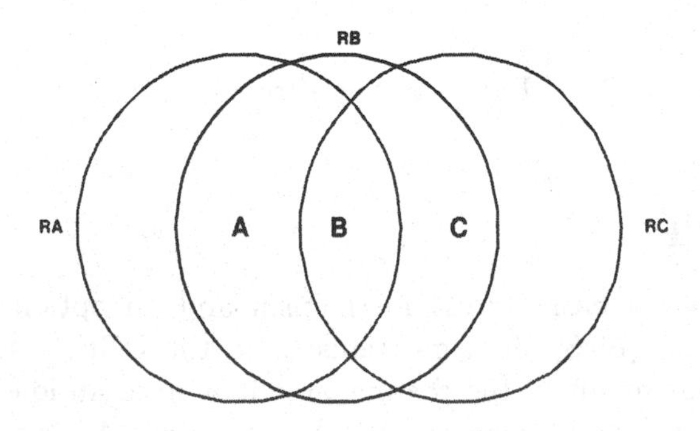

Figure 3: Hidden Node Problem

To reduce the hidden station problem, an optional four-way data transmission mechanism called RTS/CTS is also defined in DCF. In the RTS/CTS mechanism, before transmitting a data frame, a short RTS frame is transmitted. The RTS frame also follows the backoff rules introduced above. If the RTS frame succeeds, the receiver station responds with a short CTS frame. Then a data frame and an ACK frame will follow. All four frames (RTS, CTS, data, ACK) are separated by an SIFS time. In other words, the short RTS and CTS frames reserve the channel for the data frame transmission which follows. For example, in Figure 3, RA for Station A's RTS transmission and RB for Station B's CTS transmission are overlapped but not equal. Therefore, after the transmission of Station A's RTS and Station B's CTS, the channel is reserved for the data transmission followed, and any station in either RA or RB will not transmit. A hidden station such as Station C of Station A, who is in RB but not in RA, will not interfere with Station A's data transmission since it hears Station B's CTS transmission.

Data frames are transmitted at the data rate, and ACK frames are transmitted at the control rate. The data rate may be not the same as the control rate. The frame formats for data frames and ACK frames are shown in Figure 4 and Figure 5, respectively [1]. As illustrated in the figures, a data frame has 28 bytes of overhead including the MAC header and the FCS field, and an ACK frame is 14 bytes in length.

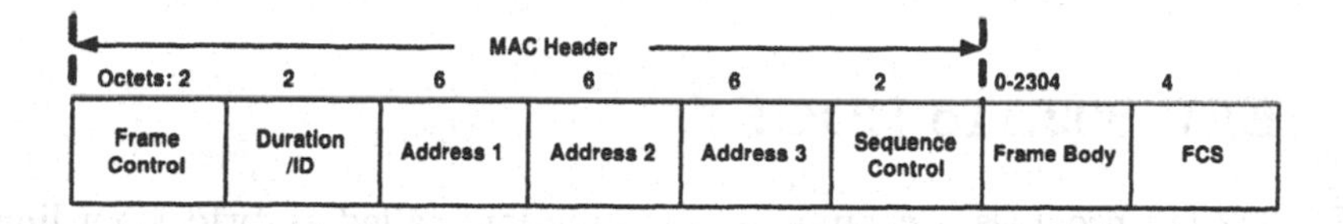

Figure 4: Data Frame Format

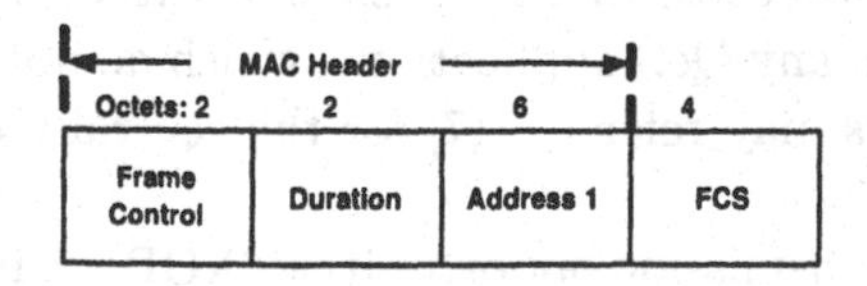

Figure 5: ACK format

3.2 The PCF

The PCF is an optional centrally controlled channel access function, which provides contention-free (CF) frame transfer. The PCF is designed for supporting time-bounded services, which can provides limited QoS. It logically sits on top of the DCF, and performs polling, enabling polled stations to transmit without contending for the channel. It has a higher priority than DCF by adopting a shorter Inter-frame space (IFS) called point inter-frame space (PIFS), i.e. SIFS<PIFS<DIFS.

If the PCF is implemented, a CFP under the PCF and a CP under the DCF alternate over time. A CFP and a CP forms a *superframe*. The AP, where the point coordinator (PC) is normally located, senses the medium idle for a PIFS interval, and then transmits a beacon frame to initiate a CFP (in other words, to initiate a *superframe*). After a SIFS time, the PC sends a poll frame to a station to ask for transmitting a frame. The poll frame may or may not include data to that station. After receiving the poll frame from the PC, the station with a frame to transmit may choose to transmit a frame after a SIFS time. When the destination station receives the frame, an ACK is returned to the source station after a SIFS time. The PC waits a PIFS interval following the ACK frame before polling another station or terminating the CFP by transmitting a CF-End frame. If the PC receives no response from the polled station for a PIFS interval, the PC can poll next station or terminate the CFP by transmitting a CF-End frame.

4 IEEE 802.11e EDCF

IEEE 802.11e provides a channel access function, called Hybrid Coordination Function (HCF) to support applications with QoS requirements. The HCF includes both a contention-based channel access and a centrally controlled channel access. The contention-based channel access of the HCF is also referred to as Enhanced Distributed Coordination Function (EDCF). In this chapter, we only consider EDCF since (1) it is simpler, and (2) it is expected to support many QoS applications, which do not require strict QoS provisioning. Readers may refer to [17] for the centrally controlled channel access.

A new concept, transmission opportunity (TXOP), is introduced in IEEE 802.11e. A TXOP is a time period when a station has the right to initiate transmissions onto the wireless medium. It is defined by a starting time and a maximum duration. A station cannot transmit a frame that extends beyond a TXOP. If a frame is too large to be transmitted in a TXOP, it should be fragmented into smaller frames.

The EDCF works with four Access Categories (ACs), which are virtual DCFs as shown in Figure 6, where each AC achieves a differentiated channel access. This differentiation is achieved through varying the amount of time a station would sense the channel to be idle and the length of the contention window during a backoff. The EDCF supports eight different priorities, which are further mapped into four ACs, shown in Table 1. Access Categories are achieved by differentiating the arbitration inter-frame space (AIFS), the initial window size, and the maximum window size. For the AC $i (i = 0, ...3)$, the initial backoff window size is $CW_{min}[i]$, the maximum backoff window size is $CW_{max}[i]$, and the arbitration inter-frame space is $AFIS[i]$. For $0 \leq i < j \leq 3$, we have $CW_{min}[i] \geq CW_{min}[j]$, $CW_{max}[i] \geq CW_{max}[j]$, and $AFIS[i] \geq AFIS[j]$, and at least one of above inequalities must be "not equal to". In other words, the EDCF employs $AFIS[i]$, $CW_{min}[i]$, and $CW_{max}[i]$ (all for $i = 0, ...3$) instead of DIFS, CW_{min}, and CW_{max}, respectively. If one AC has a smaller $AFIS$, CW_{min} or CW_{max}, the AC's traffic has a better chance to access the wireless medium earlier.

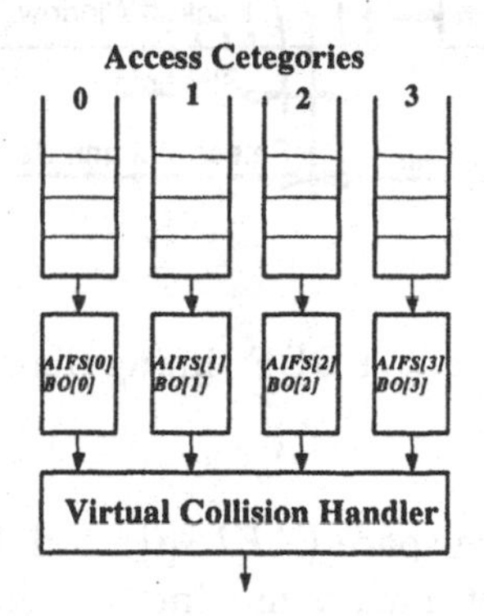

Figure 6: Virtual transmission queues, where $BO[i]$ stands for the backoff counter for AC i

Figure 7 shows the EDCF timing diagram, where 3 ACs are shown: i, j, and k. Figure 6 shows four transmission queues implemented in a station, and each queue supports one AC, behaving roughly as a single DCF entity in the original IEEE 802.11 MAC. It is assumed that a payload from a higher layer is labeled with a priority value, and it is enqueued into the corresponding queue according to the mapping in Table 1. Each queue acts as an independent MAC entity and performs the same DCF function,

Table 1: Priority to access category mapping

PRIORITY	AC	DESIGNATION
1	0	BEST EFFORT
2	0	BEST EFFORT
0	0	BEST EFFORT
3	1	VIDEO PROBE
4	2	VIDEO
5	2	VIDEO
6	3	VOICE
7	3	VOICE

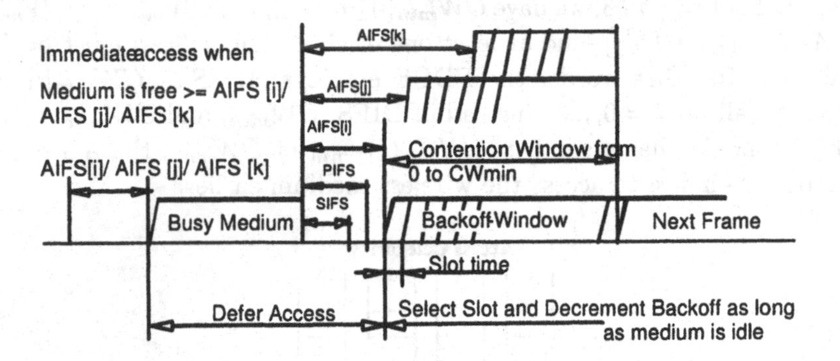

Figure 7: EDCF timing diagram

with a different inter-frame space ($AFIS[i]$), a different initial window size ($CW_{min}[i]$), and a different maximum window size ($CW_{max}[i]$). Each queue has its own backoff counter ($BO[i]$), which acts independently in the same way as the original DCF backoff counter. If there is more than one queue finishing the backoff at the same time, the highest AC frame is chosen to transmit by the virtual collision handler. Other lower AC frames whose backoff counters also reach zero will increase their backoff counters with $CW_{min}[i](i = 0, ...3)$, accordingly. Furthermore, we have $AFIS[i] \geq PIFS$, where $PIFS$ is point (coordination function) inter-frame space.

The values of $AIFS[i]$ $(i = 0, ..., 3)$, $CW_{min}[i]$ $(i = 0, ..., 3)$ and $CW_{max}[i]$ $(i = 0, ..., 3)$ are referred to as the EDCF parameters, which will be announced by the QoS Access Point (QAP) via periodically transmitted beacon frames.

5 IEEE 802.11e MAC Enhancements

IEEE 802.11e not only provides QoS enhancements introduced in the previous section, but also introduces some enhancements for MAC efficiency such as direct link protocol and block acknowledgement protocol.

5.1 Direct Link Protocol

The Direct Link Protocol (DLP) allows QoS stations (QSTAs) to transmit frames directly to another QSTA by setting up such data transfer when a QoS AP (QAP) is present. The need for this protocol is motivated not only by the fact that the intended recipient may be in Power Save Mode, in which case it can only be woken up by the QAP, but also by being efficient. The DLP allows the sender and the receiver to exchange rate set and other information. Furthermore, the DLP messages can be used to attach security information elements. This protocol prohibits the stations going into power-save for the active duration of the Direct Stream. The DLP does not apply in an ad hoc network, where frames are always sent directly from one QSTA to another. A direct link can be built by following sequences:

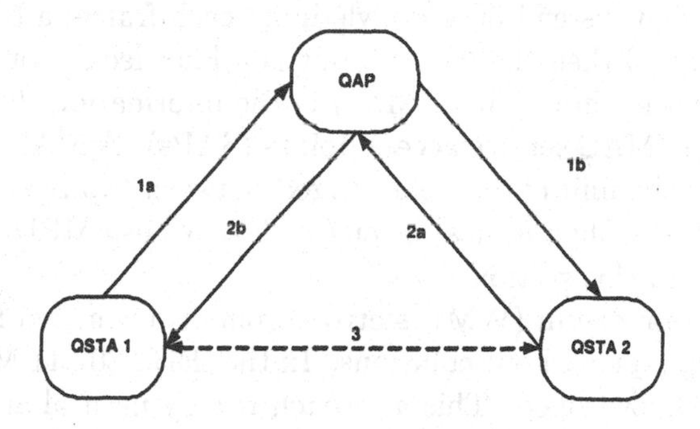

Figure 8: Direct Link Protocol

- QSTA-1 that has data to send invokes the DLP and sends a DLP-request frame to the QAP, shown in Figure 8 (1a). This request contains the rate set, and (extended) capabilities of QSTA-1, as well as the MAC addresses of QSTA-1 and QSTA-2.

- If QSTA-2 is associated in the BSS, the QAP shall forward the DLP-request to the recipient, QSTA-2, shown in Figure 8 (1b).

- If QSTA-2 accepts the direct stream, it shall send a DLP-response frame to the QAP, shown in Figure 8 (2a).
- The QAP shall forward the DLP-response to QSTA-1, shown in Figure 8 (2b), after which the direct link becomes active and frames can be sent from QSTA-1 to QSTA-2 and from QSTA-2 to QSTA-1, shown in Figure 8 (3).

When the direct link is active, QSTA-1 may use DLP-probes to gauge the quality of the link between QSTA-1 and QSTA-2. The direct link becomes inactive when no frames have been exchanged as part of the direct link for the duration of aDLPIdleTimeout defined in [2]. After the timeout, frames with destination QSTA-2 shall be sent via the QAP.

5.2 Block Acknowledgment Protocol

Block Acknowledgement Protocol (BAP) was proposed in IEEE 802.11e [2]. It is also called Burst Transmission and Acknowledgment (BTA) [21, 22], Contention-Free Burst (CFB) [14], Group Transmission and Acknowledgements (GTA) [17], and Delayed Group Acknowledgement (DGA), etc. The idea of BAP is that instead of acknowledging each frame, a burst of frames is received first, and then the whole burst is acknowledged one time.

A MAC Service Data Unit (MSDU) is the information that is delivered as a unit between MAC service access points (SAPs). A MAC protocol data unit (MPDU) is the unit of data exchanged between two peer MAC entities using the services of the physical layer (PHY). We use MPDUs and frames interchangeably in this section.

Since wireless medium (WM) is error-prone, transmitted frames can be corrupted easily, even without collisions. In the IEEE 802.11 MAC protocol, each frame is acknowledged. This approach is very natural and robust, but it introduces quite an amount of overhead. In order to reduce the overhead, BAP is currently being discussed in the IEEE 802.11e task group [2].

The BAP mechanism allows a burst of frames to be transmitted before any acknowledgement. After sending a burst of frames, the sender sends a burst acknowledgement request (BlockAckReq) frame, and the receiver must respond by sending the burst acknowledgement (BlockAck) frame, in which the correctly received frames' information is included. All the frames, including BlockAckReq frame and BlockAck frame, are separated by an SIFS period.

The sender should first win a transmission opportunity (TXOP) using a channel access mechanism before starting a burst. The burst length is limited, and the amount of state that must be kept by the receiver for receiving frames is bounded.

If the BlockAck indicates that an MPDU was not received correctly, the sender shall retry that MPDU subject to MPDU's appropriate retry limit. Retransmitted burst data MPDU's shall preserve their original relative order. The receiver shall maintain a burst acknowledgement record consisting of a transmitter address and a 32-octet bitmap of received MPDU sequence numbers. These hold the acknowledgement state of the burst data received from that sender.

In [21, 22], we show that BAP greatly improves throughput and delay.

6 Related work for the EDCF

In this section, we provide a short survey on the EDCF related work, especially for service differentiation.

6.1 Service Differentiation

Service differentiation has been studied by two approaches, i.e. simulation approaches and analytical approaches.

6.1.1 Simulation Approaches

Several priority studies have been reported in the literature for the DCF. Deng and Chang [5] proposed a priority scheme by differentiating the backoff window: the higher priority class uses the lower portion of the window and the lower priority class uses the high portion of the window. Aad and Castelluccia [7] proposed a priority scheme achieved by differentiating interframe spaces (IFS). Veres and Campbell et al. [8] proposed priority schemes by differentiating the initial backoff window size and the maximum window size. Pallot and Miller [9] proposed three priority schemes: static priority scheduling (SPS), prioritized DIFS time mechanism, and prioritized backoff time distribution mechanism (PBTDM). In PBTDM, the backoff time is chosen in the current window range with different distributions for different priorities. Note that Deng and Chang's scheme [5] is one special case of PBTDM.

For the EDCF, Mangold and Choi et al. [10] introduced the IEEE 802.11e EDCF and provided performance studies via simulations. Xiao [11] introduced the updated EDCF, and studied priority schemes of the EDCF via both simulation and an analytical model. Choi [14] provided an updated simulation analysis for the EDCF. Romdhani, Ni, and Turletti [28] consider internal collision issues within the same station. Xiao, Li, and Choi [19] proposed two-level protection and guarantee mechanisms for voice and video traffic in the EDCF.

6.1.2 Analytical Models

Bianchi [23, 24] proposed a simple and accurate analytical model to compute saturation throughput for the original DCF, and Ziouva and Antonakopoulos [25] improved Bianchi's model by deriving saturation delay. Xiao and Bianchi [20] provided an updated model with a comprehensive performance analysis, as well as a simple and accurate delay model. All these models are not for priorities, but for the original DCF.

Xiao [6, 13] proposed an analytical model to evaluate backoff-based priority schemes by differentiating the initial window size, the backoff window-increasing factor, and the maximum backoff stage. Ge and Hou [15] proposed an analytical model for p-persistent WLAN. Xiao [11] proposed a rough model to analyze AFIS differentiation. Bianchi and Tinnirello [26] proposed an analytical model for AFIS differentiation. Xiao [16, 18] proposed an updated analytical model for backoff-based priority schemes by strengthening underline assumptions for both IEEE 802.11 and IEEE 802.11e.

6.2 MAC Enhancements

Xiao [21, 22] provided an analysis for BAP. Choi studied CFB in [14].

7 Evaluation via Simulations

In this section, we evaluate the EDCF for voice streams (AC=3), video streams (AC=2), and data (AC=0) in terms of throughput per flow, delay per flow, total throughput, fairness index factor, failure transmission, as well as effects of buffer size on delay, queuing delay ratio, drop ratio, and buffer overflow ratio, where queuing delay ratio is defined as the ratio of queuing delay with total delay per flow; drop ratio is defined as the ratio of dropped frames/traffic (in terms of Mbps) with total arrival traffic per flow; buffer overflow ratio is defined as the ratio of dropped frames/traffic due to buffer overflow (in terms of Mbps) with total arrival traffic per flow. The reason to define the performance metric of queuing delay ratio is to see how much percentage of queuing delay is in delay.

7.1 Simulation Setup

In our simulations, we have the following parameters unless stated otherwise: $AIFS[3] = 25\mu s$; $CW_{min}[3] = 8$; $CW_{max}[3] = 64$; $AIFS[2] = 25\mu s$; $CW_{min}[2] = 16$; $CW_{max}[2] = 128$; $AIFS[0] = 34\mu s$; $CW_{min}[0] = 32$; $CW_{max}[0] = 1024$; the default buffer sizes per station for voice, video, and data are 0.02M bytes, 0.15M bytes, and 0.15M bytes, respectively; the total simulation time is 200 seconds. Each video stream rate is 4.86 Mbps, which is generated by a constant inter-arrival time 2.5 ms with a mean payload size 1464 bytes. Each voice stream rate is 0.0832 Mbps, which is generated by a constant inter-arrival time 20 ms with a mean payload size 208 bytes. Each data station has traffic of 1 Mbps, which is generated by an exponential inter-arrival time 12 ms with a mean payload size 1500 bytes. We adopt IEEE 802.11a, and the data rate and control rate are 54 Mbps and 24 Mbps, respectively. We assume that all the stations are within the transmission range. Initially, there are one voice stream, one video stream, and one data QSTA in the system. Beacon interval is 100ms, which is also used for the measurement interval. By default, we assume an ad hoc network or an infrastructure network in which all traffic is going out of the network. The latter case is equivalent to the DLP in some sense. In our simulations, each station only has one video stream, voice stream, or data. In other words, multiple streams/data within the same station are not simulated, i.e. internal virtual collisions, shown in Figure 6, are not considered in our simulations.

7.2 Throughput, Delay, Fairness Index Factor, and Failure Transmission under Small Traffic Load

In the small traffic load, one voice stream, one video stream, and one data station are added for each 10 seconds until there are total 10 voice streams, 3 video streams, and 3 data QSTAs in the system. The buffer sizes per station for voice, video, and data are 0.02M bytes, 0.15M bytes, and 0.15M bytes, respectively;

Figure 9 shows throughputs for voice, video and data. Figure 9 (a) shows that 10 voice streams are added totally, and Figure 9 (b) shows that 3 video streams are added totally. Figure 9 indicates that video streams need large bandwidth.

Figure 10 shows throughputs per flow for voice, video and data. We observe that voice and video streams obtain their throughput requirements, i.e. 4.86 Mbps for a video stream and 0.0832 Mbps for a voice stream.

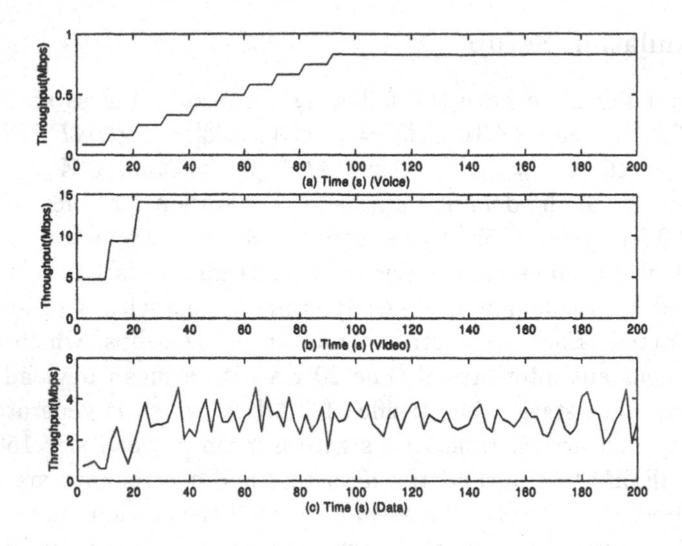

Figure 9: Throughputs (Mbps)

Figure 10 also indicates that the EDCF is effective in terms of throughput differentiation. As illustrated in the figure, with the DLP, throughput has been greatly improved for all ACs. We also observe that sometime a data station traffic is larger than 1 Mbps although the arrival rate of a data station is 1Mbps. There are two reasons as follows. The first reason is that 1 Mps of data traffic is the mean value so that there is a random factor with respect to the measurement interval. The second and the most important reason is that when a frame arrives, it unusually cannot be transmitted immediately, but waits in the buffer/queue that introduces much bursty. On average, data throughput is not larger than 1 Mbps.

Figure 11 shows delays of voice, video, and data. As illustrated in the figure, delays for all ACs are very small and similar. Delay Differentiation between ACs is not good due to the small traffic load. In fact, under a small traffic load condition, service differentiation is not important since all requests can be satisfied. As illustrated in the figure, with the DLP, delay has been greatly improved for all ACs.

Figure 12 shows the total throughput vs. the simulation time. The highest throughput is around 18 Mbps in this simulation. We observe that there are two throughput-jumps since a video stream and a data station are added per jump at the early time of the simulation. Later on, when voice streams are added, the total throughput changes very little since a voice

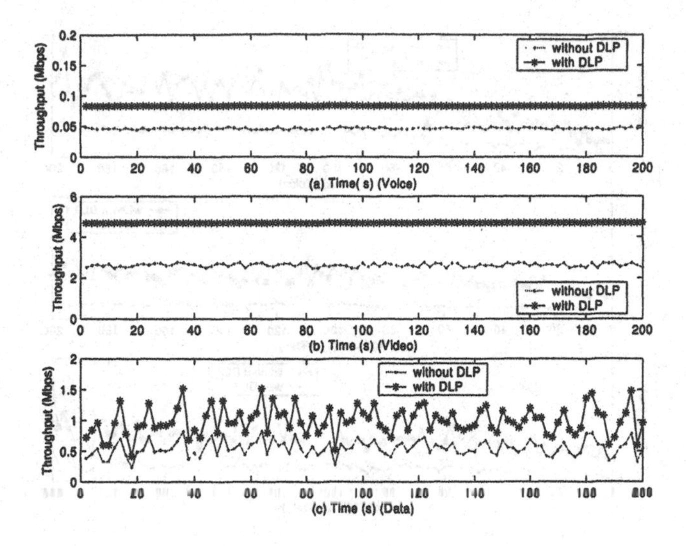

Figure 10: Throughputs (Mbps) per flow

stream's traffic is small.

Figure 13 shows that failure transmission in time (ms) increases as the traffic load increases, and remains relative constant when the traffic remains relatively constant. We observe that there are two big-jumps of failure transmission at the early time of the simulation since a video stream and a data station are added per jump.

Figure 14 shows fairness index factor defined in [27], where the nearer to 1 the value, the fairer it is. As illustrated in the figure, it is very fair among flows (data stations) within the same AC under such a traffic load.

7.3 Throughput, Delay, Fairness Index Factor, and Failure Transmission under Large Traffic Load

In the large traffic load, one voice stream, one video stream, and one data station are added for each 10 seconds until there are total 20 voice streams, 5 video streams, and 5 data QSTAs in the system. The buffer sizes per station for voice, video, and data are 1M bytes, 7.5M bytes, and 7.5M bytes, respectively.

Figure 15 shows throughputs for voice, video and data. Figure 15 (a) shows that 20 voice streams are added totally, and Figure 15 (b) shows that 5 video streams are added totally. Compared with the Figure 9 under small

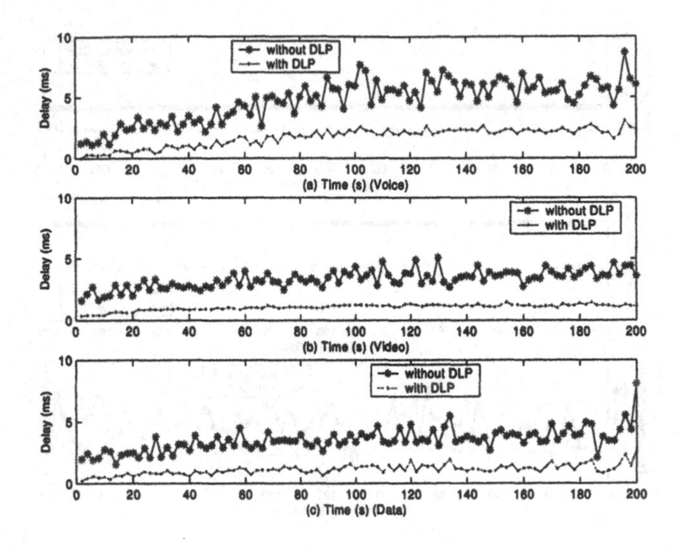

Figure 11: Delays (ms) of video, voice, and data

traffic, Figure 15 shows that video throughput decreases under a large traffic load condition.

Figure 16 shows throughputs per flow for voice, video and data. As illustrated in the figure, voice throughput is good throughout the simulation, whereas video throughout decreases as the traffic load increases and therefore is degraded. The reason that voice and video reacts differently for different traffic load conditions is that an AC with a high demanded throughput such as video is more likely to be affected by a high traffic load. Another reason is that voice is a higher class.

Figure 17 shows delays of voice, video, and data. As illustrated in the figure, voice delay is pretty good, i.e. at most 7 ms, whereas video delay and data delay are very high when the traffic load is high. Video delay reaches 15s at the simulation time 160s and reaches 17s at the end of simulation. This is intolerable for real-time video. Data delay reaches as high as 20s, but this is not important.

Figure 18 shows the total throughput vs. the simulation time. We observe that there are 4 throughput-jumps at the early time of the simulation since a video stream and a data station are added per jump. The highest throughput reaches about 28 Mbps. Later on, the total throughput decreases as the traffic load increases since a lot of collisions happen. However, compared with Figure 12, Figure 18 has a larger throughput at the

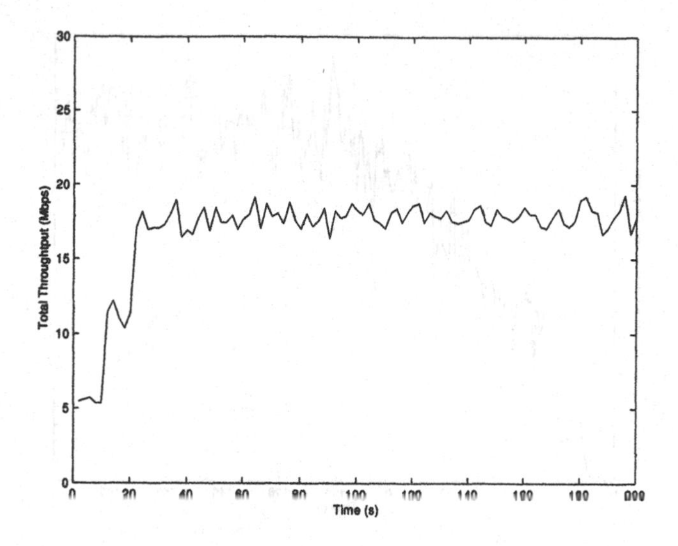

Figure 12: Total Throughput (Mbps)

end of simulation. The reason is that a relatively higher traffic load (not extremely high) is more likely to produce a higher throughput, but real-time traffic such as video and voice throughputs are more likely to be degraded.

Figure 19 shows that failure transmission in time (ms) increases as the traffic load increases. We observe that there are big-jumps of failure transmission at the early time of the simulation since a video stream and a data station are added per jump. Compared with Figure 13, Figure 19 has a much higher failure transmission time.

Figure 20 shows fairness index factor defined. As illustrated in the figure, it is very fair among flows within the same AC except data throughput. Compared with Figure 14, Figure 20 indicates that a high traffic load may cause a lower AC/class's throughput fairness factor degrade.

7.4 Effects of Buffer Size on Delay, Queuing Delay Ratio, Drop Ratio, and Buffer Overflow Ratio

In this section, we study effects of buffer size on delay, queuing delay ratio, drop ratio, and buffer overflow ratio. We have three buffer sizes in this subsection: small size, large size, and infinite size; the small buffer sizes per station for voice, video, and data are 0.02M bytes, 0.15M bytes, and 0.15M bytes, respectively; the large buffer sizes per station for voice, video, and

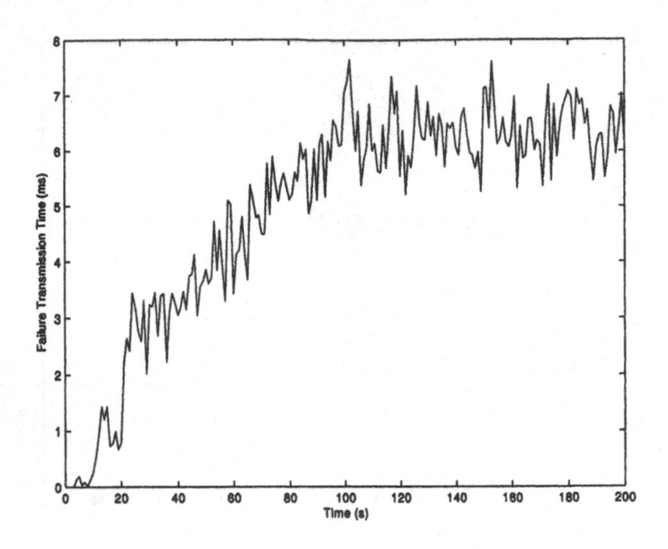

Figure 13: Failure transmissions in time (ms)

data are 1M bytes, 7.5M bytes, and 7.5M bytes, respectively.

Figure 21 shows delay with different buffer sizes. We observe that for both video transmission and data transmission, when traffic load is high, a smaller buffer size has a lower average delay. One reason is that with a smaller buffer size, longer delayed frames caused by too many frames in the buffer are more likely to be dropped, and therefore these dropped frames are not counted in calculating delay since only successfully transmitted frames are counted. Another reason is that if the queue is long when a frame arrives, it takes a longer time for the frame to move to the front of the queue, therefore causes a longer delay. Voice transmission has no such an effect in this simulation setting since voice belongs the highest class and has much smaller frame sizes. With a smaller buffer size, larger frames are more likely to be dropped. Furthermore, voice frames are small so that the buffer size for voice is relatively large under the current simulation setting.

Figure 22 shows queuing delay ratio, which is defined as the ratio of queuing delay with (total) delay. as illustrated in the figure, when traffic load is high, queuing delay for video is almost 100%. In other words, much of the delay time is caused by queuing! This observation has never been reported before according to our knowledge. When traffic load is high, much percentage of delay for data is also caused by queuing delay. Voice transmission has no such an effect in this simulation setting since voice

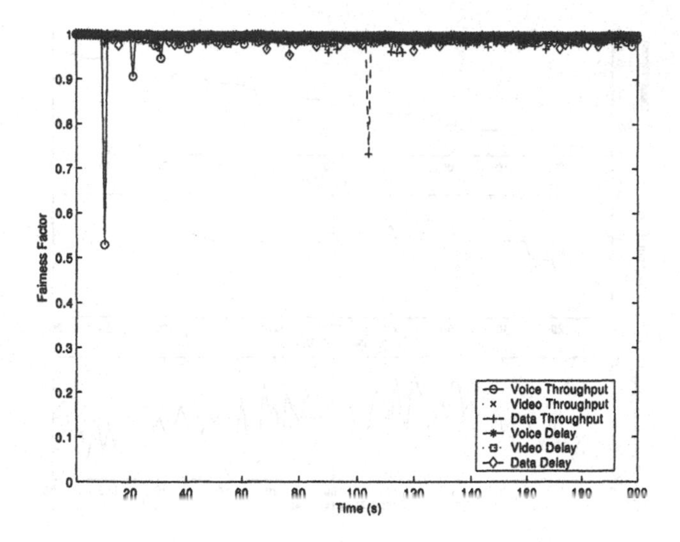

Figure 14: Fairness Index Factor

belongs the highest class. We also observe that a buffer size has no effects on queuing delay (ratio).

Figure 23 shows frame drop ratio with different buffer sizes. As illustrated in the figure, a smaller buffer size has a larger drop ratio. When traffic load is high, drop ratio for video reaches almost 28% in this simulation setting.

Figure 24 shows buffer overflow ratio with different buffer sizes, where buffer overflow ratio is defined as the ratio of dropped frames/traffic due to buffer overflow (in terms of Mbps) with total arrival traffic per flow. As illustrated in the figure, a smaller buffer size has a larger buffer overflow ratio. For video, buffer overflow ratio is very high (near 100%) when traffic load is high. In other words, when traffic load is high, many of dropped video frames are due to limited buffer size. We also observe that that buffer overflow ratio is zero when the buffer size is infinite.

8 Conclusions

In this chapter, we introduce the emerging IEEE 802.11e standard, as well as the related work for the EDCF. We give a performance analysis for the EDCF via extensive simulations for QoS enhancements to support multimedia applications such as voice and video over the wireless medium. The key

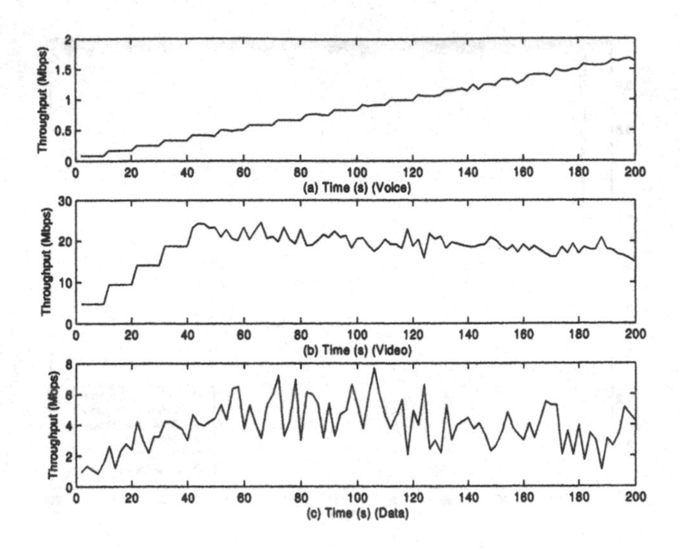

Figure 15: Throughputs (Mbps)

simulation results are summarized as follows.

- Service differentiation is important only when traffic is relatively large. Under a small traffic load condition, service differentiation is not important because all requests can be satisfied.

- Under a small traffic load, voice throughput and video throughput are good. However, under a large traffic load, video throughput is degraded with much more than voice throughput. The reason that voice and video reacts differently for different traffic load conditions is that an AC with a high demanded throughput such as video is more likely to be affected by a high traffic load. Another reason is that voice is a higher class.

- The highest throughput reaches about 28 Mbps in our simulations. A relatively higher (not extremely high) traffic load is more likely to produce a higher throughput, but real-time traffic such as video and voice throughputs are more likely to be degraded.

- Under a large traffic load, video delay is too high and intolerable although total throughput is a little higher.

- Fairness is pretty good among flows(data stations) within the same

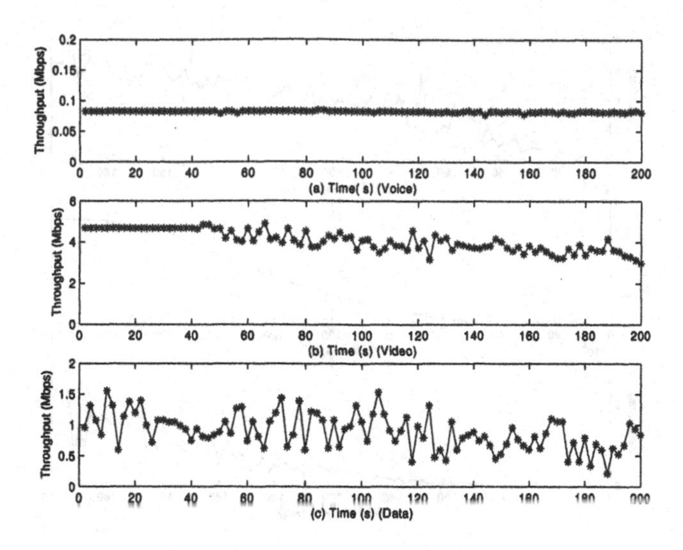

Figure 16: Throughputs (Mbps) per flow

AC. However, lower AC/class may have a better chance to have a degraded throughput fairness factor when traffic load is high.

- When traffic load is high, much percentage of both video delay and data delay is caused by queuing delay.
- When traffic load is high, many of dropped video frames are due to limited buffer size.

In order to guarantee QoS for voice and video, admission control is highly recommended. In [19], we proposed two-level protection and guarantee mechanisms, which well addressed these issues.

References

[1] IEEE 802.11 WG, Part 11: Wireless LAN Medium Access Control (MAC) and Physical Layer (PHY) specification, Standard, IEEE, (Aug. 1999).

[2] IEEE 802.11e WG, Draft Supplement to Part 11: Wireless Medium Access Control (MAC) and physical layer (PHY) specifications: Medium Access Control (MAC) Enhancements for Quality of Service (QoS), IEEE Std 802.11e/D3.3.2, (Nov. 2002).

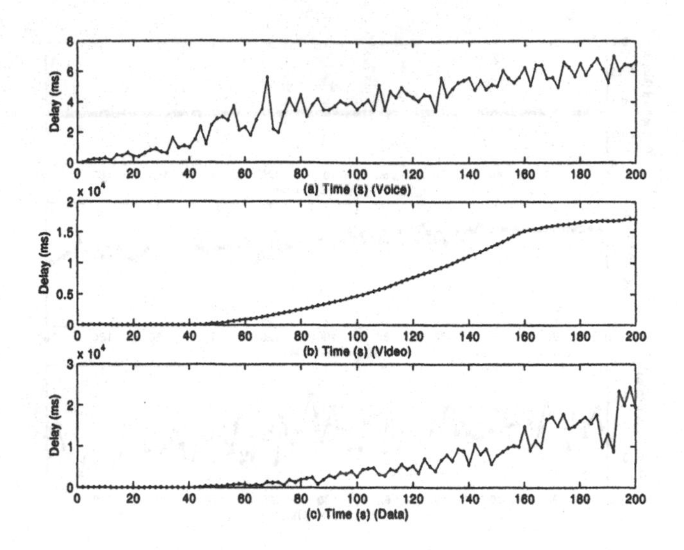

Figure 17: Delays (ms) of video, voice, and data

[3] IEEE 802.11b, Part 11: Wireless LAN Medium Access Control (MAC) and Physical Layer (PHY) specification: High-speed Physical Layer Extension in the 2.4 GHz Band, IEEE, (Sep. 1999).

[4] IEEE 802.11a WG, Part 11: Wireless LAN Medium Access Control (MAC) and Physical Layer (PHY) specification: High-speed Physical Layer in the 5GHz Band, (Sep. 1999).

[5] D.-J. Deng and R.-S. Chang, A priority Scheme for IEEE 802.11 DCF access method, *IEICE Transactions on Communications* Vol. E82-B No.1 (1999) pp. 96-102.

[6] Y. Xiao, A simple and effective priority scheme for IEEE 802.11, *IEEE Communications Letters* Vol. 7 No. 2 (2003) pp. 70-72.

[7] I. Aad and C. Castelluccia, Differentiation mechanisms for IEEE 802.11, *Proceedings of IEEE INFOCOM 2001*.

[8] A. Veres, A. T. Campbell, M. Barry, and L.-H. Sun, Supporting differentiation in wireless packet networks using distributed control, *IEEE Journal of Selected Area in Communications* Vol. 19 No. 10 (2001) pp. 2081-2093.

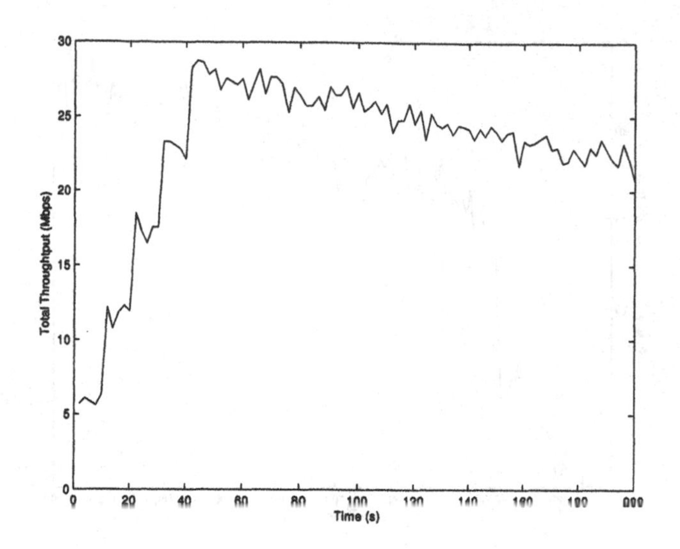

Figure 18: Total Throughput (Mbps)

[9] X. Pallot and L. E. Miller, Implementing message priority policies over an 802.11 based mobile ad hoc network, *Proceedings of IEEE MILCOM 2001.*

[10] S. Mangold, S. Choi, P. May, O. Kein, G. Hiertz, and L. Stibor, IEEE 802.11e wireless LAN for Quality of Service, *Proceedings of EW 2002.*

[11] Y. Xiao, Enhanced DCF of IEEE 802.11e to support QoS, *Proceedings of IEEE WCNC 2003.*

[12] IEEE 802.11g/D3.0, Draft Supplement to Part 11: Wireless LAN Medium Access Control (MAC) and Physical Layer (PHY) specifications: Further Higher-Speed Physical Layer Extension in the 2.4 GHz Band, (Jul. 2002).

[13] Y. Xiao, Backoff-based priority schemes for IEEE 802.11, *Proceedings of IEEE ICC 2003.*

[14] S. Choi, J. d. Prado, S. Shankar N, and S. Mangold, IEEE 802.11e contention-based channel access (EDCF) performance evaluation, *Proceedings of IEEE ICC 2003.*

[15] Y. Ge and C.-J. Hou, An analytical model for service differentiation in IEEE 802.11, *Proceedings of IEEE ICC 2003.*

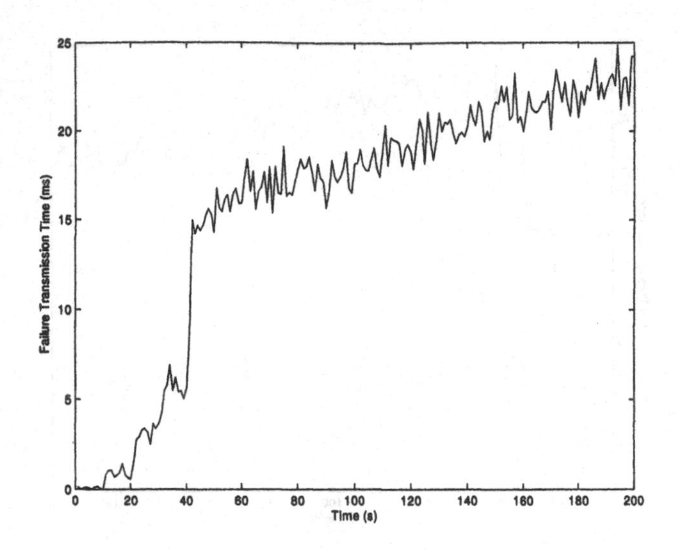

Figure 19: Failure transmission in time (ms)

[16] Y. Xiao, Performance analysis of priority schemes for IEEE 802.11 wireless LANs, *submitted to IEEE Transactions on Wireless Communications* (Nov. 5, 2002).

[17] Y. Xiao, IEEE 802.11e: a QoS provisioning at the MAC layer, *IEEE Wireless Communications*, (accepted, Jul. 2003).

[18] Y. Xiao, An analysis for differentiated services in IEEE 802.11 and IEEE 802.11e wireless LANs, *Proceeding of IEEE ICDCS 2004*.

[19] Y. Xiao, H. Li, and S. Choi, Protection and guarantee for voice and video traffic in IEEE 802.11e wireless LANs, *Proceedings of IEEE INFOCOM 2004*.

[20] Y. Xiao and G. Bianchi, Modeling saturation performance of the IEEE 802.11 MAC, *(Wiley) Journal of Wireless Communications and Mobile Computing*, (accepted Sep. 2003).

[21] Y. Xiao and J. Rosdahl, Performance analysis and enhancement for the current and future IEEE 802.11 MAC protocols, *ACM SIGMOBILE Mobile Computing and Communications Review (MC2R)*, special issue on Wireless Home Networks Vol. 7 No. 2 (Apr. 2003) pp. 6-19.

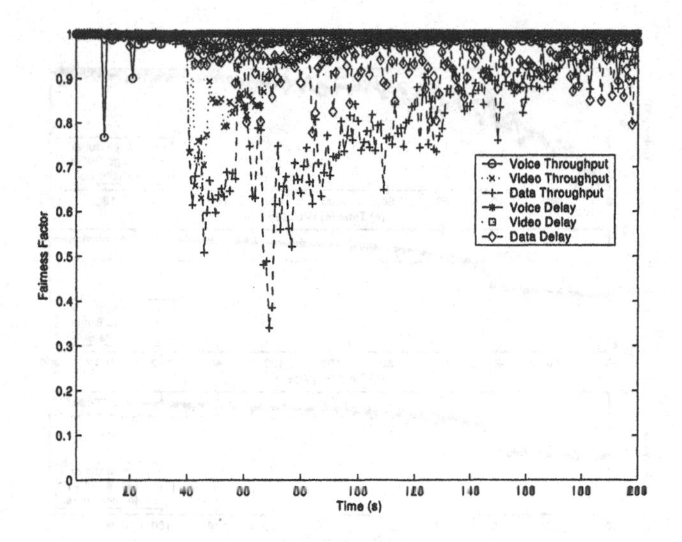

Figure 20: Fairness Index Factor

[22] Y. Xiao, MAC performance analysis and enhancement over 100 Mbps data rates for IEEE 802.11, *Proceesdings of IEEE VTC 2003 Fall.*

[23] G. Bianchi, IEEE 802.11-saturation throughput analysis, *IEEE Communications Letters* Vol. 2 No. 12 (Dec. 1998) pp. 318-320.

[24] G. Bianchi, Performance analysis of the IEEE 802.11 distributed coordination function, *IEEE J-SAC* Vol. 18 No. 3 (Mar. 2000) pp. 535-547.

[25] E. Ziouva and T. Antonakopoulos, CSMA/CA performance under high traffic conditions: throughput and delay analysis, *Computer Communications* Vol. 25 (2002) pp.313-321.

[26] G. Bianchi and I. Tinnirello, Analysis of priority mechanisms based on differentiated inter frame spacing in CSMA-CA, *Proceedings of IEEE VTC* (2003-Fall).

[27] R. Jain, The art of computer systems performance analysis: techniques for experimental design, measurement, simulation, and modeling, *Wiley- Interscience*, (New York, NY, Apr. 1991).

[28] L. Romdhani, Q. Ni, and T. Turletti, Adaptive EDCF: enhanced service differentiation for IEEE 802.11 wireless ad hoc networks, *Proceedings of IEEE WCNC 2003.*

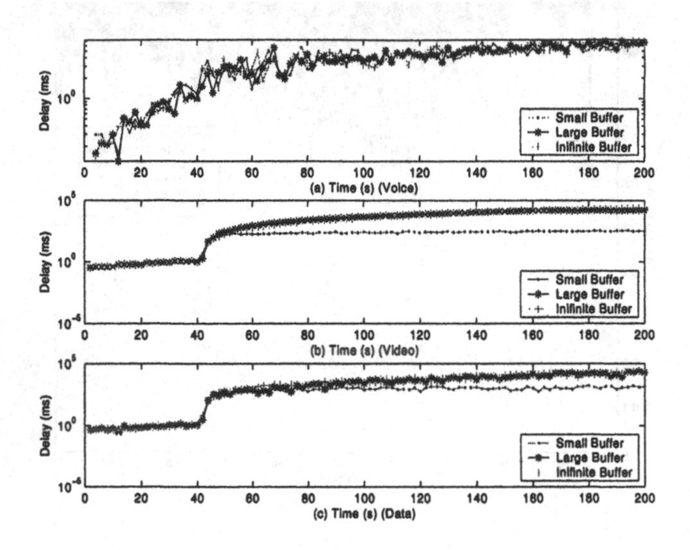

Figure 21: Delay (ms) with different buffer sizes

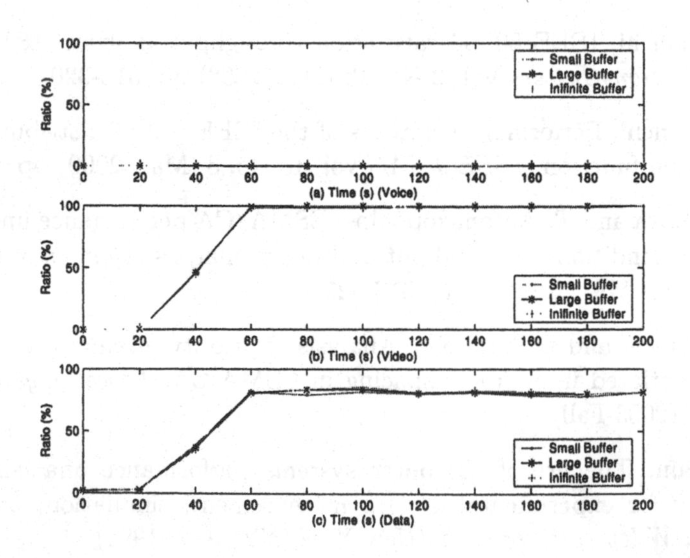

Figure 22: Queuing Delay Ratio (%) with different buffer sizes

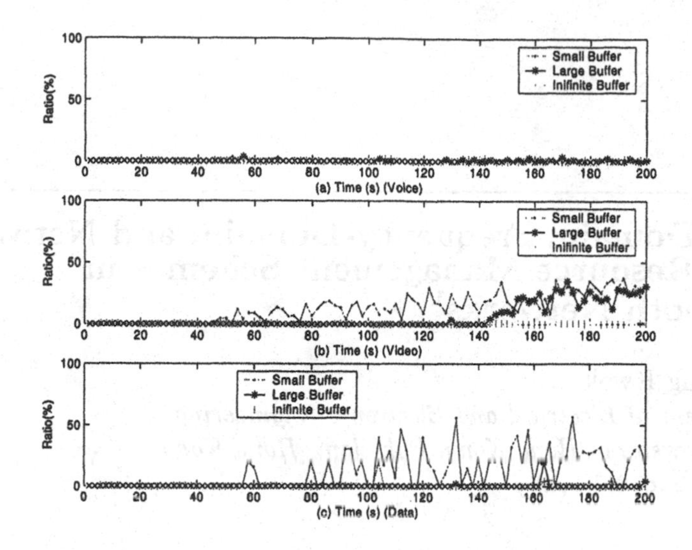

Figure 23: Frame Drop Ratio (%) with different buffer sizes

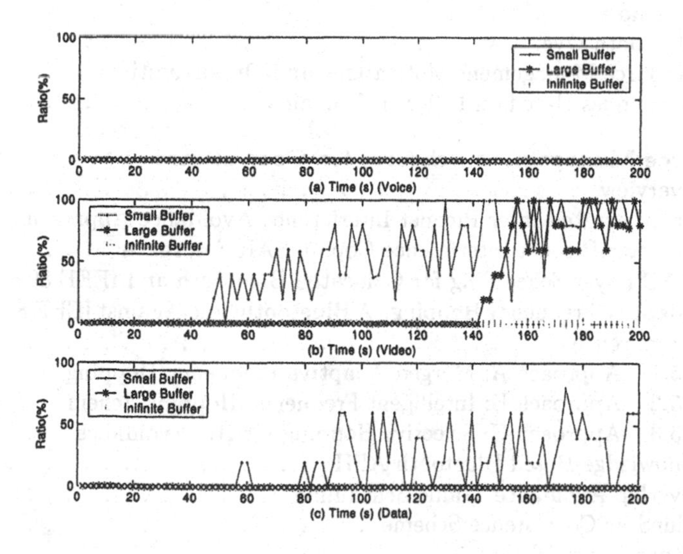

Figure 24: Buffer overflow ratio (%) with different buffer sizes

Time-Domain, Frequency-Domain, and Network Level Resource Management Schemes in Bluetooth Networks[1]

Yu-Kwong Kwok
Department of Electrical and Electronic Engineering
The University of Hong Kong, Pokfulam, Hong Kong
E-mail: ykwok@hku.hk

Contents

[1]This work was supported by a grant from the Research Grants Council of the HKSAR Government under project number HKU 7162/03E.

1 Introduction

1.1 Bluetooth Overview

Bluetooth[2] is a short-range wireless technology initially developed by Ericsson in 1994 [8, 7, 23, 29]. The technology is designed to be a low cost and low power "cable replacement" solution. Occupying the unlicensed 2.4 GHz ISM (Industrial, Scientific, and Medical) frequency band, Bluetooth uses a low power (three classes; Class 1: 100 mW (+20 dBm); Class 2: 2.4 mW (+4 dBm); Class 3: 1 mW) and its nominal transmission range is about 10 m. A total of 79 frequency channels, separated by 1 MHz, are defined. Frequency hopping spread spectrum (FHSS) using these 79 channels (in some countries, e.g., Japan, Spain, France, only 23 channels are used) is employed. Specifically, the frequency used for a particular transmission between two devices can be expressed as: $f = 2.402\text{GHz} + n\text{MHz}$, $n = 0, \ldots, 78$ (in Japan, Spain, and France, the starting frequencies are 2.473 GHz, 2.449 GHz, and 2.454 GHz, respectively, and the values of n range from 0 to 22 only). The frequency hopping rate is 1600 hops per second. A simple binary GFSK (Gaussian Frequency Shift Keying) modulation scheme is used and the symbol rate is 1 Mbps.

At the MAC (medium access control) layer, Bluetooth employs a TDD (time-division duplex) master-slave coordination. Specifically, multiple Bluetooth devices can form a network structure called a *piconet*, in which there is a single master and not more than seven active slaves. The Bluetooth MAC layer frame is divided into time-slots each of which is of 625 μs duration (corresponding to the hopping rate of 1600 hops/s). Thus, each time-slot is handled by a distinct carrier frequency, as illustrated in Figure 1. For each slave, the master first sends it a downlink packet, thereby also polling the slave to see if it has any uplink packet to send. If so, the slave is allowed to do so in the immediately next time-slot, using a different frequency as governed by the pre-defined frequency hopping pattern (different frequency hopping patterns can be used and a particular pattern is chosen depending on the master's clock and unique BD_ADDR [8, 7]). The master addresses each of its active slaves by using a three-bit identifier called AM_ADDR (that is why only seven active slaves are supported). Bluetooth provides several baseband packet types occupying different number of time-slots (1, 3, or 5), as shown in Table 1. The timing of the multi-slot packets are shown in Figure 2. Note that the frequency used is unchanged for a multi-slot packet. Specifically, the DHx (here, x indicates the number of time-slots occupied)

[2]Named after the King Harald Blaatand of Denmark (tenth century A.D.).

packet types are without FEC (forward error correction) bits incorporated. On the other hand, the DMx packet types incorporate a 2/3 rate FEC—a (15, 10) shortened Hamming code [8].

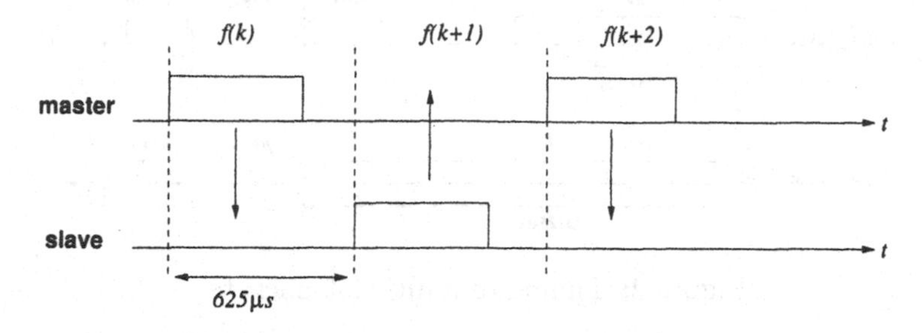

Figure 1: A snapshot of a Bluetooth frame.

Table 1: Bluetooth packet types.

Type	Max. Payload (bytes)	Symmetric rate (kbps)	Asymmetric rate (kbps)	
DM1	17	108.8	108.8	108.8
DH1	27	172.8	172.8	172.8
DM3	121	256.0	384.0	54.5
DH3	183	384.0	576.0	86.4
DM5	224	286.7	477.8	36.3
DH5	239	433.9	723.2	57.6

There are two types of links in Bluetooth: ACL (asynchronous connectionless) and SCO (synchronous connection oriented). ACL refers to a point-to-multipoint link between the master and all its slaves in a single piconet. The master can exchange packets with any slave in any order (although usually straightforward round robin is used), including a slave that is already engaged in an SCO link. Packet retransmissions are provided in ACL link. On the other hand, in an SCO link, a fixed bandwidth is allocated between a point-to-point connection involving the master and a single slave. The master is responsible for maintaining the SCO link by reserving time-slots for the slave at regular intervals. Thus, the slave may transmit in the reserved time-slots no matter if it detects a downlink packet is addressed to it from the master. SCO time-slots may be spaced to use every slot-pair, every second slot-pair, or every third slot-pair. Consequently, the maximum number of SCO links that can be supported is three. While DM1 is the packet type that can be employed in both ACL and SCO link, the other

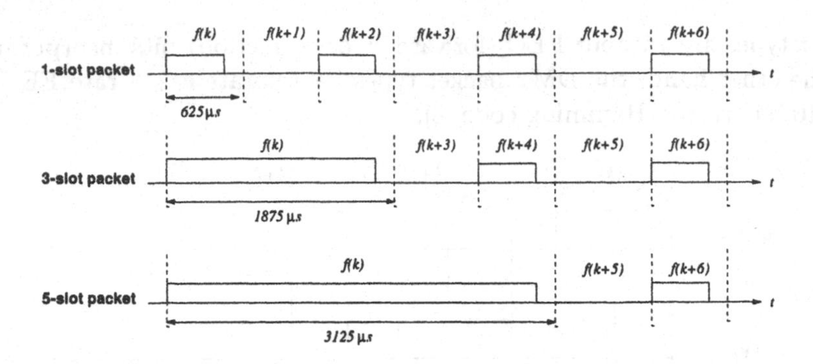

Figure 2: Timing of multi-slot packets.

DM and DH packet types are relevant to ACL link only. Furthermore, an SCO link uses HVx packet types with 1/3 FEC [8] (again, x is the number of time-slots used).

To set up an initial connection so that a device can join a piconet (or when the piconet is initially set up), an *inquiry* procedure is executed (the detailed state diagram can be found in [7, 8]). Specifically, a potential master, in the *inquiry* state, transmits an ID (identity) packet with an inquiry access code (IAC), which is common to all Bluetooth devices. Using 32 out of the 79 channels (i.e., the so-called 32 "wake-up" carriers), the master transmits the IAC over each channel in turn. On the other hand, a potential slave also periodically enters the *inquiry scan* state—to search for IAC packets in its vicinity. When the slave receives the IAC, it enters the *inquiry response* state such that it replies with an FHS (frequency hop synchronization) packet, containing its unique BD_ADDR and timing information. The slave then enters the *page scan* state to await the paging from the master to establish the connection. Here, paging means that the master actively calls the slave with a paging packet uniquely addressed to the slave. The master, however, may or may not respond with an FHS packet immediately because it may want to remain in the inquiry state to exhaustively find all potential slaves. After all slaves are found by the inquiry procedure, the master calculates a proper hopping sequence for its piconet using the slaves' BD_ADDR and timing information. The master then pages each slave to set up connections. The page message contains a device access code (DAC) of the specific slave (made up with the lower part of the device's unique BD_ADDR). A slave may be in any of the following four connected states:

Active. The slave gets an AM_ADDR code and actively participate in the piconet.

Sniff. The slave only participates in the piconet for a certain number of time-slots and then sleeps for a specific period of time (negotiated when connection is set up). The slave also gets an AM_ADDR code.

Hold. The slave does not support ACL packet exchange and is in a reduced power status. However, the slave also gets an AM_ADDR code. As such, the slave may also possibly participate actively in another piconet (i.e., becomes a *bridge* node), while keeping its connected status in the current piconet.

Park. The slave is still logically a member of the piconet but it does not have an AM_ADDR code. Slaves in this mode have the least level of activity.

A simplified Bluetooth protocol stack is shown in Figure 3. The details about the physical and baseband layer are described above. The LMP (link manager protocol) layer is responsible for link establishment between Bluetooth devices, as well as link management tasks such as authentication, encryption, and negotiation of baseband packet sizes. The L2CAP (logical link control and adaptation protocol) layer is responsible for adapting upper-layer protocols to the baseband layer, i.e., taking care of the segmentation and reassembly (SAR) process. The SDP (service discovery protocol) layer is responsible for providing the following data to the querying devices: device information, details of the available services, and the characteristics of the devices. Finally, RFCOMM (radio frequency communications) layer is a virtual serial port that is designed to provide a seamless support to upper layer protocols such as PPP, TCP/IP, etc.

An important aspect of Bluetooth is that multiple piconets can be connected together to form a larger scale network called *scatternet*, making use of some bridge nodes (e.g., slaves exploiting the hold mode), as illustrated in Figure 4. However, in the current Bluetooth specifications [7], the procedure for constructing a scatternet and the detailed operations of a bridge node (whether it is a master/slave bridge or a slave/slave bridge) are undefined as yet.

1.2 Resource Management Motivations and Opportunities

Although Bluetooth is seemingly a simple short-range communication technology, there is in fact a large space for research to improve its performance under different situations. Specifically, the performance of a Bluetooth network can be significantly enhanced by using judicious resource management

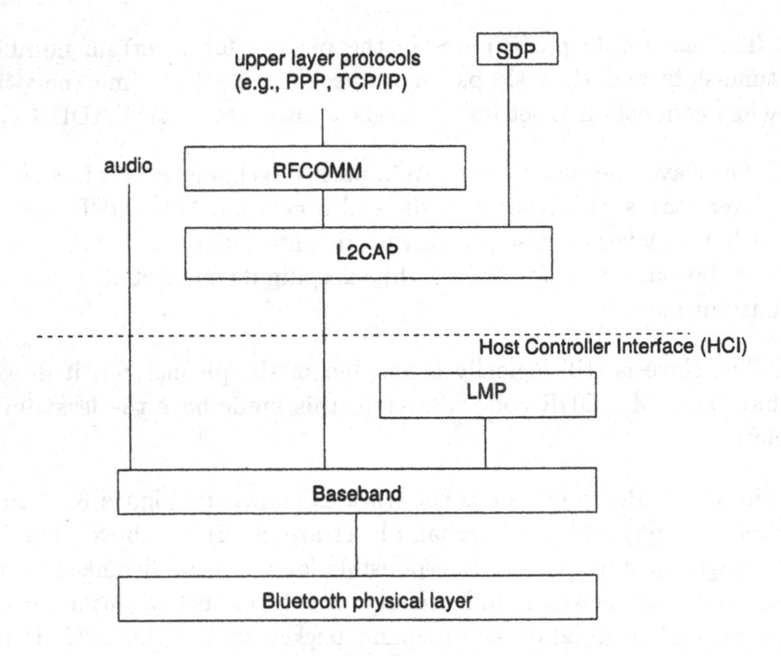

Figure 3: A simplified Bluetooth protocol stack.

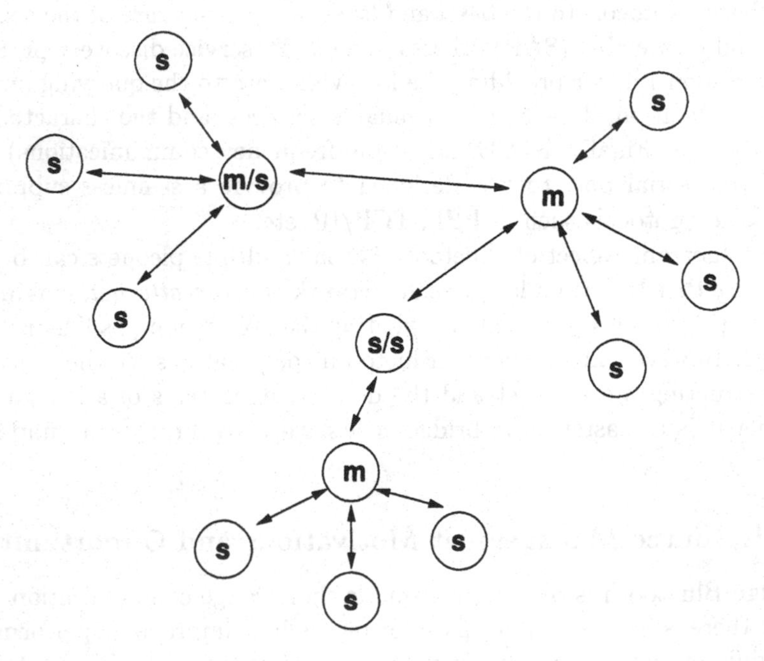

Figure 4: A Bluetooth scatternet.

schemes that can intelligently make use of the various aspects of resources such as time-slots, frequency channels, etc. There are broadly two groups of resource management schemes aiming at different functional objectives: *coexistence* and *connectivity.* This is illustrated in the taxonomy shown in Figure 5.

Coexistence Mechanisms. Because Bluetooth is inherently a short-range technology, we can expect that multiple independent piconets can exist even within a small area (e.g., office space). Under such a situation, the independent piconets, possibly using uncoordinated frequency hopping patterns, can interfere with other. Such coexistence of uncoordinated FHSS transmissions can easily lead to packet collisions, and in turn, to degraded performance for the different piconets involved.

On the other hand, the ISM band is free and as such, there are many different wireless devices (open or proprietary) that are using this spectrum. The most notable example is the IEEE 802.11b [27], which uses the DSSS technique occupying a 22 MHz passband in the spectrum. Again independent operations of a nearby Bluetooth piconet in close proximity of an IEEE 802.11b wireless LAN can lead to severe performance degradation for both.

In summary, the coexistence problem must be tackled with care and efficiency in order that the Bluetooth (and IEEE 802.11b) can be truly ubiquitous.

Connectivity Problems. Although the specifications define how a Bluetooth master and its slaves are connected together, there is no consideration about performance (in terms of time-slot utilization and/or access delay) and user quality of service (QoS). Specifically, at the piconet level, a straightforward round robin mechanism is commonly implemented in nowadays commercial devices. This is obviously too simple to provide optimized performance, leading to a limited scalability of the Bluetooth technology.

On the other hand, as described earlier, at the scatternet level, the construction process is as yet undefined, making scatternet formation still largely a research topic instead of a commercial scenario. Because of Bluetooth's limited resources in terms of power and time-slots, scatternet formation is a difficult problem in that practical devices may not be able to execute construction schemes that are of a high complexity (in both time and space). Furthermore, once a scatternet is formed, how a bridge node should operate (time-sharing itself to two or more piconets) is another void in the specifications. Similar to the scatternet formation process itself, the bridge node operation mechanism is also a difficult problem to tackle because the bridge node can easily become a scatternet-wide bottleneck.

Finally, even if after we set up a scatternet, we still need to deal with the "device discovery" problem—a device wants to communicate with a remote device that is a member of the scatternet but is out of its range. To achieve this, we need to have a routing policy, defining how to set up a multi-link (or multi-hop) communication path and a packet forwarding mechanism.

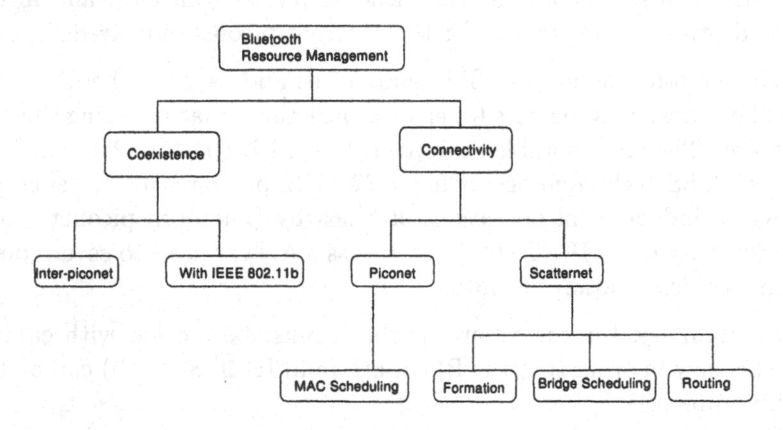

Figure 5: A taxonomy of Bluetooth resource management schemes.

1.3 Techniques Based on Different Domains

In this chapter, we provide an extensive survey of more than 20 recently proposed efficient techniques for the Bluetooth resource management problems under different situations. These techniques can be broadly classified into different domains:

Time-Domain. There are many techniques suggested for the coexistence and piconet/scatternet connectivity problems that make use of a time-domain perspective. Specifically, these techniques attempt to perform a judicious scheduling of time-slots or selection of packet types (i.e., DHx) in order to avoid interference or perform inter-piconet time-sharing. Some piconet scheduling algorithms are targeted to provide QoS to the slaves which may have different traffic demands.

Frequency-Domain. Because FHSS is used in Bluetooth, there is an opportunity for us to manipulate the hopping pattern in order to achieve, most notably, coexistence purposes. That is, by using a carefully designed hopping sequence, interference among piconets or other sources (e.g., IEEE 802.11b) can be completely eliminated.

Network-Level. At the scatternet level, the connectivity problem is largely tackled by network level message-exchange. Specifically, independent devices autonomously execute a common distributed algorithm to cluster with nearby devices, and eventually connected with every other device to form a single scatternet. Furthermore, in the routing aspect, various well known *ad hoc* routing techniques [14, 41] are applied.

2 Resource Management Schemes for Coexistence

2.1 Overview

As low cost and low power Bluetooth gadgets proliferate, we can foresee the situation where multiple independent piconets coexist in close vicinity. Thus, uncoordinated frequency hopping can easily result in frequency collisions, leading to transmission errors. This is one of the possible coexistence scenarios. On the other hand, because the unlicensed ISM band is free, there are already a crowd of wireless technologies making use of the spectrum. The most notable one is the IEEE 802.11b wireless LAN (WLAN) devices. Indeed, experimental results from the IEEE 802.15 Task Group 2 indicated that the interference between IEEE 802.11 and Bluetooth can severely degrade the performance of both systems when their separation is less than 2 m. As both Bluetooth and IEEE 802.11b are popular short-range wireless technologies, we cannot hope to eliminate any one of them. Thus, judicious mechanisms are needed to manage this coexistence problem.

Coexistence solutions can be broadly classified into two types: collaborative and non-collaborative [34]. In a collaborative solution, the two interfering sources (e.g., two coexisting independent piconets or one Bluetooth device coexists with one IEEE 802.11b device) need to exchange information so that they can synchronize their transmissions to eliminate collisions completely. In the case of two interfering Bluetooth piconets, the two masters can coordinate with each other to orthogonally multiplex their transmissions. In the case of Bluetooth device coexisting with an IEEE 802.11b device, the two devices (e.g., colocated within the same machine) can be controlled by a centralized scheduler to time-multiplex the transmissions. Thus, mainly time-domain coordination schemes are used. On the other hand, in a non-collaborative scheme, the two interfering sources do not communicate with each other. There are two different non-collaborative approaches. In the first approach, both interfering sources actively sense the medium to avoid transmission or adjust the power to capture the medium. In the second approach, one of the interfering source, usually Bluetooth, tries to adjust itself to accommodate the IEEE 802.11b device. Specifically, an adaptive frequency hopping mechanism can be implemented in Bluetooth so as to avoid using frequencies that fall within the spectrum occupied by IEEE 802.11b. Indeed, frequency domain techniques are more commonly used in this heterogeneous coexistence scenario.

In the following, we first describe techniques proposed to handle inter-piconet coexistence problem. We then discuss solutions for harmonizing the coexistence of Bluetooth and IEEE 802.11b WLAN. We also describe some of the performance studies about the coexistence problem.

2.2 An Integrated Inter-Piconet Interference Avoidance Approach

In [52], Sun et al. observed in their simulations that inter-piconet interference can degrade the system throughput by more than 30%. In simple terms, with N coexisting piconets and assuming that all 79 channels are used in the frequency hopping process, the probability of hopping into the same frequency in one slot is: $1-P_N^{79}/79^N$, where P_N^{79} is the number of permutations of N objects chosen from 79 distinct objects. However, for a five-slot packets (e.g., DM5 or DH5), the probability becomes: $1-(P_N^{79}/79^N)^5$. Motivated by these observations, they suggested an integrated inter-master time-slot scheduling and data segmentation/reassembling (SAR) policy, which is a collaborative scheme designed for a single logical scatternet. The bridge nodes in the scatternet play a critical role in that each bridge node needs to relay the BD_ADDR and clock of a master to another master, and vice versa. As such, each of its masters can compute the hopping sequences of other masters connected by the same bridge node. Thus, each master performs interference avoidance MAC scheduling as follows. Before a packet is sent to a slave (assume that this is an ACL link), a master, denoted by $\mathcal{M}_1$, computes the frequency hopping patterns of other masters for the next ten slots. Suppose that at the i-th slot ($10 \geq i \geq 1$), there is a collision involving another master $\mathcal{M}_2$. Further suppose that $\mathcal{M}_2$'s BD_ADDR is greater than that of $\mathcal{M}_1$. Then, according to the scheduling algorithm, $\mathcal{M}_1$ yields the transmission right to $\mathcal{M}_2$ at slot i. The master $\mathcal{M}_1$ can then use the first

$i - 1$ slots to communicate with the slave. Let the master's packet size be m (i.e., $m = 1, 3, 5$). The slave can send, on the uplink, the largest packet that can fit into the remaining $i - 1 - m$ slots. The master $\mathcal{M}_1$ conveys this information to the slave using two bits in the poll packet.

If there is an SCO link in a piconet, the corresponding master is always given a higher priority in the interference avoidance process, irrespective to its BD_ADDR (of course, if two masters are having SCO links, tie is broken by their BD_ADDR values as usual). Another implication of an SCO link is that a master cannot send a five-slot packet to a slave because according to the Bluetooth specification, SCO packet exchange has to be repeated every six slots. Indeed, the masters do not need to look-ahead for ten slots. Instead, looking ahead for four slots is enough. Furthermore, if there are three SCO links in a piconet, no ACL link can be supported. In [52], the simulations using BlueHoc [6] indicated that this frequency domain scheduling approach can significantly improve the overall scatternet throughput and end-to-end delay. A limitation of this approach is that if an independent piconet exists in close proximity of the scatternet, interference may still occur.

2.3 An Inter-Piconet Interference Coping SAR Approach

While the scheme described in Section 2.2 attempts to avoid inter-piconet interference, a non-collaborative technique, called the IBLUES (Interference Aware Bluetooth Segmentation) algorithm, proposed by Cordeiro and Agrawal [18], is more aggressive. Specifically, in IBLUES, instead of trying to avoid interference, each master computes a packet transmission successful probability before proceeding to exchange packets with a slave. The successful probability is determined by several important factors: the packet type (e.g., number of slots, DM or DH), the traffic load of the interference source (e.g., a nearby piconet master) and the distance between the master and the slave. The probability is derived from an analytical model suggested in [19]. Based on the probability for each packet type, the master then selects the type that gives the highest probability of successful transmission for its downlink and uplink communications with the slave. In this approach, the estimations of the traffic load and the distance are very important. To obtain an accurate estimation of traffic load, LOAD_UPDATE messages are periodically exchanged among different masters in the same scatternet through the bridge nodes. On the other hand, the distance is estimated by assuming that the distributions of the slaves are uniform. It is mentioned in [18] that IBLUES can also handle the coexistence problem involving an IEEE 802.11

WLAN by using a UBM (useful bandwidth monitor) which dynamically tracks the channel condition. However, it is unclear as to how this UBM can be implemented in practice. Furthermore, the frequent exchanging of LOAD_UPDATE messages can also be a significant overhead.

2.4 MAC Layer Scheduling for Colocated Bluetooth and IEEE 802.11

In this section we begin the discussion on coexistence mechanisms for coordinating Bluetooth and IEEE 802.11 WLAN. A collaborative coexistence mechanism, called MEHTA (MAC enhanced temporal algorithm) [33], is proposed within the IEEE 802.15 TG2. To make use of MEHTA, an IEEE 802.11 device and a Bluetooth node must be colocated in the same physical unit. MEHTA works by using a centralized controller that monitors the Bluetooth and the IEEE 802.11 traffic. Specifically, MEHTA is based on the exchange of information between the two colocated radio systems. Implemented at the MAC layer, the controller provides per-packet authorization of all transmissions from both IEEE 802.11 and Bluetooth. With perfect knowledge about the two channels, the controller can predict any possible collision. When a collision is foreseen, MEHTA schedules transmissions based on simple rules determined by the packet types. For instance, IEEE 802.11 acknowledgment packets are given a higher priority than any Bluetooth packets, while Bluetooth SCO traffic is assigned a higher priority than any IEEE 802.11 data packets.

Another collaborative coexistence scheme, called AWMA (alternatig wireless medium access) [47], is also proposed within the IEEE 802.11 TG2. The AWMA algorithm is a TDMA MAC scheduling scheme. Similar to MEHTA, AWMA also requires the two channels residing in the same physical unit. The IEEE 802.11 and Bluetooth tranceivers are activated alternately to completely avoid overlapped transmissions. Specifically, AWMA is based on the beaconing mechanism in IEEE 802.11 in that the IEEE 802.11 transceiver sends out a beacon at a periodic interval and each beacon-to-beacon interval is divided into two subintervals: one for IEEE 802.11 traffic and one for Bluetooth traffic. AWMA is not applicable if an SCO link exists because the periodic nature (recall that master-slave packet exchange in an SCO link has to repeat every six slots), the SCO link traffic will also jam into the beacon interval allocated for IEEE 802.11.

The two MAC layer scheduling techniques can also be enhanced by integrating them with a collaborative physical layer solution, which is called deterministic frequency nulling scheme [55]. The premise is that with respect to IEEE 802.11 DSSS signal, Bluetooth signal is a narrowband interference source. Thus, the IEEE 802.11 transceiver can be nulled at the particular frequency that Bluetooth is using. The crux of this scheme is that the IEEE 802.11 receiver must know clock and BD_ADDR of the Bluetooth master so that the frequency hopping pattern and timing can be computed.

2.5 Adaptive Frequency Hopping: A Bluetooth Solo Against IEEE 802.11 Interference

The original Bluetooth MAC layer is based on *pseudo-random frequency hopping*. Specifically, the frequency selection process is independent of the corresponding channel conditions. Generally, it performs well with non-static and narrowband interference. However, this is not the case when interferer like IEEE 802.11b WLAN exists nearby; where over 20 MHz of the spectrum is engaged for a certain period of time. It is neither non-static nor narrow. Thus, AFH (adaptive frequency hopping) is introduced in order to deal with this prevalent neighborhood.

AFH is a non-collaborative mechanism that differs from pseudo-random frequency hopping by taking into account the channel condition in the selection process. It enables the device modifying its hopping sequence dynamically, thereby minimizing interference and enabling coexistence with other devices in the 2.4 GHz ISM band.

Generally, AFH can further divide into 4 sub-problems: (i) device identification; (ii) channel classification; (iii) classification information exchange; and (iv) adaptive frequency hopping mechanism. Device identification is mainly used for backward compatibility and checking if a device can work in AFH mode. Channel classification is the process used for distinguishing the channel quality. There are a variety of implementation methods and the process is very important because the partition sequence generated by AFH mechanism is based on the result of this part. Classification information exchange is the process for master and slave to exchange the results of their measurement. AFH mechanism is the specific algorithm to choose the hop frequency. The goal of the algorithm is to avoid as many bad channels as possible. In this chapter, we will focus on the AFH mechanism.

Generally, any new AFH mechanism is placed between the original hop selection kernel and the frequency synthesizer. Firstly, the channels are partitioned into two sets—"good" or "bad"—by some other technologies. A new partition sequence is generated according to the channels' conditions. The newly generated partition sequence is then compared with the original hopping sequence. The adapted hop frequency is generated after the checking of a remapping function. Furthermore, AFH for SCO and ACL connections are treated differently due to their highly different characteristics. In this study, we focus on ACL connection only

AFH is considered to be a promising solution in non-collaborative coexistence schemes. A number of proposals have been submitted to 802.15 TG2 for consideration as the TG2 coexistence mechanism. Three major AFH mechanisms for ACL connection have been studied.

2.5.1 Approach A: Merged Adaptive Frequency Hopping

This method [54] has been selected and adopted as TG2 AFH mechanism. It introduces 2 new blocks: sequence generator and remapping functions. The sequence generated specifies when and which frequency to use, and the remapping function is used for maintaining the pseudo-random nature within the partition which behaves as the original hop selection kernel.

The system running this AFH mechanism examines the channel's quality and maintains a list of "good" channels and "bad" channels initially. The sequence generator then produces a binary partition sequence which depends on the number of "good" and "bad" channels being available. The system maintains these channel lists and refresh them periodically. The generated partition sequence is compared to the original hopping sequence, and remapping will take place whenever the two sequences are not referring to the same channel list.

The merged adaptive frequency hopping mechanism can operate in either Mode L or Mode H, depending on $N_{\min}$, the minimum number of channels that FCC (Federal Communications Commission) requires Bluetooth to hop over. When the number of available good channels is greater than $N_{\min}$, the device will operate in Mode L where only good channels are remapped to; otherwise, it will operate in Mode H so that some of the bad channels are still used in order to comply with the FCC regulations.

2.5.2 Approach B: Intelligent Frequency Hopping Scheme

This method [4] works in a similar manner as the adopted one in the sense that it also keeps track of good and bad channel lists and generates a binary partition sequence. However, instead of carrying out a remapping process, it uses a look-ahead algorithm which attempts to cull all the good channels in the original hopping sequence. This could be done by comparing the original hop sequence with the channel lists.

2.5.3 Approach C: Selective Hopping for Hit Avoidance

This approach [13] is also similar to the adopted one while it introduces a concept of rearrangement. Inspired by the duty cycle of WLAN traffic that causes interference, it tries to avoid it by temporarily rearranging the hopping sequence in the upcoming N channels. It can also incorporate scheduling for ACL which will induce transmission delay that does not exist in other schemes.

2.6 Knowledge Based Bluetooth AFH

The above mentioned AFH mechanisms heavily depend on the lists of good and bad channels, and their validity. After a scrutiny of these AFH mechanisms, two major deficiencies can be identified:

1. *Memory concern*: In order to make the adopted AFH mechanisms work, the master device must compile a list of good and bad channels, recording all channels' conditions and their associated information in real-time. Thus, it imposes a great pressure on the memory requirement on handheld devices. Indeed, as shown in [10, 11], reducing the memory size by half could double the collision rate.

2. *Power concern*: The adopted AFH mechanism heavily depends on the results of channel classification because the formation of the partition sequence depends on the number of "good" or "bad" channels, and the remapping function need to map "bad" channels to/from "good" channels. Thus, the accuracy of the classification results is crucial for these coexistence schemes. However, the channel conditions could keep on varying. For example, a channel just sensed to be "good" might turn bad right after the classification. As a result, the channel lists need to be updated frequent enough to reflect accurately the current situation. To be exact, the updating rate of the lists should be faster than or equal to the changing rate of the channel's condition (i.e., the coherence times of the fading processes). This basic requirement imposes stress on the computing power and hence, leads to rapid power drainage, especially in a fast changing environment (e.g., mobility is high). The simulation results in [10, 11] indicate that improper updating rate could result in three times higher collision rate compare to a proper one.

In view of these practical deficiencies, a new AFH approach, called *Knowledge Based AFH* (KBAFH), is suggested in [10, 11]. KBAFH has the following distinctive features:

- No need to maintain good and bad channel lists and record all information associated with the channels;
- Performance insensitive to memory and power constraints;
- Emerging interference sources can be added in the future; and
- Exploit potential performance gain based on more information.

Recall that the TG2 AFH scheme outperforms the original pseudo-random frequency hopping mechanism because it considers the channel quality also when choosing a hop frequency. KBAFH is a further improved approach in that the algorithm works by taking into account extra parameters in the network environment before generating a hop sequence. In particular, KBAFH takes into account "type of interference source" and its "transmission characteristics", instead of just the channels condition only. What makes this idea feasible is the fact that, unlike other stochastic phenomena, it is practically feasible for us to know what kind of potential interference source exists and their corresponding transmission characteristics. With these pieces of information known beforehand, we can estimate what the potential interference source is and their transmission bandwidth and transmission time, etc., and hence, an optimized choice of hop sequence can possibly be obtained.

KBAFH makes use of a *repository* where the extra information resides. Whenever there is any conflict in the ISM spectrum, the system will check the repository and estimate what kind of interference exists and thereby shift to an appropriate working mode accordingly. A repository for 2.4GHz ISM must be built before the proposed KBAFH can start. We use IEEE 802.11b non-overlapping channel to illustrate our ideas. In order to let the repository aware of this, we need to determine the physical layer specification of IEEE 802.11b [27], and then input the information into the repository, say, in the following format: {Type: IEEE 802.11b; carrier: 2412, 2437, 2462; span: 22}. During the operations, when collisions are detected around these ranges, the KBAFH based Bluetooth system will know which frequency range suffers from the same interference problem. Here, it should be noted that this simplified example is just for illustration purpose. In actual practice, more then one opportunity might exist and more sophisticated design for each operating mode should be considered.

A critical design issue is when the KBAFH scheme should start and terminate. In practice, it can start from the beginning or triggered by a threshold collision rate. Collision rate can be monitored by the master continuously. In the case where collision rate keeps increasing, the KBAFH system will infer that the interference type is out of the knowledge of the repository, which will then fall back to pseudo-random frequency hopping mode.

The general operation of the KBAFH is as follows: check the collision rate to see if it exceed threshold; trigger KBAFH and then go into the repository; then, based on the information provided by the repository, the system will estimate what the existing interference source is in the surrounding; finally appropriate operation mode will be chosen to cope with the interference. Meanwhile, the collision rate will be monitored to ensure the estimation accuracy. If the estimation turns out to be not accurate, other alternatives might be attempted. Specifically, if the collision rate is even higher than the theoretical maximum produced by the pseudo-random frequency hopping scheme, the system will fall back to pseudo-random frequency hopping so that the collision rate will be upper-bounded and the inference will not adversely affect other users.

We can see that in KBAFH, system performance depends on the intelligence built in the repository. Vigilant analysis and intelligent handling methods must be designed beforehand, especially in a heterogeneous environment.

2.7 Overlap Avoidance Traffic Scheduling

Unlike the frequency-domain techniques mentioned above, the schemes proposed by Chiasserini and Rao [15] are non-collaborative integrated time- and frequency-domain scheduling techniques, employing similar principles as in the integrated SAR approaches discussed earlier in Section 2.2. They proposed two schemes: V-OLA (voice overlap avoidance) scheduling and D-OLA (data overlap avoidance) scheduling. V-OLA is designed to be implemented in an IEEE 802.11 device to prevent its transmission from colliding with the voice packets (via an SCO link) transmission of a Bluetooth device, by carefully controlling the duration of the IEEE 802.11 packets. D-OLA is designed to be implemented in the Bluetooth device to prevent its data transmission from using those frequencies that fall within the spectrum used by the IEEE 802.11 device. Thus, V-OLA makes the transmissions of Bluetooth and IEEE 802.11 orthogonal in time, while D-OLA makes the transmissions orthogonal in frequency. Specifically, in V-OLA, the IEEE 802.11 device detects the time intervals that are occupied by interfering transmissions by using the clear channel assignment (CCA) procedure. The

IEEE 802.11 device then properly adjusts its transmissions so that each IEEE 802.11 packet falls within the inter-packet spacing of two SCO packets. On the other hand, in D-OLA, the Bluetooth device is responsible for determining which particular frequencies are within the spectrum used by the IEEE 802.11 device. The Bluetooth device can achieve this by using a learning approach or assessing the received signal strength indicator (RSSI) before starting operation. The Bluetooth device then keeps a list of such problematic frequencies that are subsequently avoided being used by SAR policies similar to those described in Section 2.2. In summary, using the integrated two-part approach, the Bluetooth device just goes ahead its transmission if the link is an SCO connection, relying on IEEE 802.11 to accommodate the transmission. If the transmission is an ACL link, the Bluetooth device adjusts it in order to accommodate the transmission from IEEE 802.11. Simulation results indicate that the integrated scheme exhibits significant performance improvements over an uncoordinated system in terms of goodput and packet delay, especially when the traffic load is high.

2.8 BlueStar Coexistence Scheme

The study by Cordeiro et al. [17] considers a two-tier short-range wireless network structure called BlueStar. In the network, there are two types of devices: pure Bluetooth devices and dual channel (Bluetooth plus IEEE 802.11) devices. The latter devices are called BWGs (Bluetooth wireless gateways). The major rationale of this structure is that the BWGs enable the short-range PAN devices to be connected to a longer range WLAN, and in turn, to the wired infrastructure. Obviously a coexistence mechanism is needed to make this network structure work. The coexistence mechanism used consists of two parts: BCS (Bluetooth carrier sensing) and AFH. Specifically, the devices in the network, no matter using a WLAN or Bluetooth transceiver, perform carrier sensing (CS) before initiating transmissions. CS is readily available in WLAN devices but Bluetooth currently does not include such a CS mechanism, and thus, needs to rely on learning or signal strength measurements. When a device senses that the medium is in use, it refrains its transmission and waits until the competing transmission finishes. AFH is required for Bluetooth transceivers because it is difficult for IEEE 802.11 WLAN devices to detect Bluetooth transmissions due to the much lower power of the latter.

2.9 Other Approaches

An integrated transmission power control and rate scaling scheme [48] is proposed within the IEEE 802.15 TG2. In practice, according to the perceived channel quality, IEEE 802.11 devices can adjust the transmission rate by using different modulation and coding schemes. The technique suggested in [48] extends this channel adaptive idea so that the IEEE 802.11 transceiver uses a lower power for a higher data rate.

In [12], a fuzzy based centralized controller is proposed for a two-tier system in which mobile devices can be assigned either Bluetooth or IEEE 802.11b WLAN channel for communication. The fuzzy controller takes into account two parameters, speed of device and the data rate required, in making the fuzzy logic based decisions. Specifically, incorporating the estimated values of these two parameters into a set of pre-defined fuzzy logic rules, the controller decides on whether a Bluetooth or IEEE 802.11b channel is allocated to the device. Thus, this fuzzy logic based technique is essentially a time-domain scheduling scheme at the service level. It is shown [12] that this fuzzy logic based approach can improve system performance in terms of blocking probability, system utilization, and end-to-end delay. A limitation of the system architecture considered is that each mobile device has to possess both Bluetooth and IEEE 802.11b capabilities.

2.10 Performance Studies

In [40], Pasolini describes an analytical performance study about the aggregate throughput of a set of nearby independent piconets. Assuming one-slot packet transmissions and that devices are uniformly distributed, the probabilities of packet loss due to collisions are derived. These probabilities critically depend on several parameters: the radius of the whole region considered, the maximum radius of a piconet, the slot efficiency factor (i.e., the ratio of packet and slot duration), maximum distance between two transmitting devices that a collision can be detected, and the number of channels used. Using a worst-case analysis, a lower bound of aggregate bandwidth is determined. Consequently, the lower bound can be used to deduce the optimal number of piconets that should be set up within a fixed size region so that the worst-case aggregate bandwidth is maximized. According to the numerical results shown in [40], when the ratio of region radius to the maximum piconet radius is 3, the optimal number of piconets is close to 300 and the corresponding lower bound aggregate bandwidth is about 31 Mbps.

For the coexistence of Bluetooth and IEEE 802.11, Howitt [24, 25] suggests rigorous mathematical modeling of the various performance measures including: packet error rate, number of packet retransmissions, and transmission latency. These performance measures are analytically related to one crucial quantity—the probability of packet collision, denoted by $\Pr[C]$. The probability $\Pr[C]$, in turn is derived by taking into account the expected number of interferers, the probability of time coincidence, and the probability of frequency coincidence. Howitt applies similar mathematical techniques in another analysis of inter-piconet interference [26] for multiple independent piconets. As indicated by the simulation results, the analytical model provides results within 95% confidence bounds of the empirical data. Low power environment (at 0 dBm) and high power environment were also compared. It was found that a 40% decrease in the number of scenarios in which the collision probability is less than 10%.

Conti et al. [16] also derives a close form expression for the packet error probability using bit-level error analysis. Their analysis takes into account many practical parameters such as thermal noise, propagation impairments, interference, modulation formats, channel coding techniques, frequency hopping, packet structures, and traffic load. The model is also applied in environments with and without line of sight propagation.

Kumar and Gupta [32] present a detailed analysis of the capacity of a single Bluetooth piconet. They then extend their results to the case where multiple independent piconets coexist. Their mathematical modeling also considers numerous practical parameters: hopping sequence, transmitted power, receiver sensitivity, modulation, forward error correction (FEC), coherence bandwidth, coherence time, and spatial characteristics. Performance is then analyzed in terms of aggregate throughput. Zhen et al. [58] also describe a simple analytical performance model of Bluetooth coexistence and other interference sources such as microwave oven are also considered. There are also a number of simulation performance studies about Bluetooth coexistence [1, 22, 50]. In particular, in [1], it was found that incorporated with STBC (space time block code), the performance of the Bluetooth system considered is enhanced significantly.

2.11 Discussion

While purely collaborative approaches may not be practicable (e.g., the independent piconets are owned by different people, or it is infeasible in some situations to colocate Bluetooth and IEEE 802.11b in the same machine), passively non-collaborative approaches may not be very effective either (e.g., it may not be possible to completely avoid frequency collisions using AFH).

A hybrid approach, in which the system gradually evolves from purely non-collaborative to semi-collaborative in the sense that the interfering devices learn the characteristics of the transmission environment so as to appropriately adjust the hopping sequence or power level. In such a scheme, an accurate learning component is needed. Furthermore, a cross layer design is also necessary in that the interference avoidance mechanism can take into account the traffic requirements of the applications. For example, in some cases, it may not hurt to miss some packets but the overall delay is important. The interference avoidance component can make better decisions with such information. The KBAFH approach [10, 11] is a very first step in this direction.

3 Resource Management Schemes for Connectivity: Piconet and Scatternet Time-Slot Scheduling

3.1 Overview

Within a piconet, the master completely controls the channel access of its slaves. In the default Bluetooth arrangement, a straightforward round robin polling algorithm is used—the master sends each slave its downlink packet, thereby polling the slave to see if it has any uplink packet to be transmitted; and this process is strictly repeated for each and every other slave. However, this simple round robin scheme can be very inefficient in that some slaves may not have any packet to send when they are polled. Furthermore, from a quality of service (QoS) point of view, some slaves may need more frequent service in order to satisfy its traffic demand (e.g., video), while some other slaves may just sparingly use its allocated slots. Thus, an efficient time-slot scheduling scheme should take into account the QoS of different slaves.

Time-slot scheduling is also an important problem for bridge nodes in a scatternet. Specifically, a bridge node needs to be time-shared by two or more piconets. Thus, we can expect that a bridge node is under an even higher pressure due to simultaneous traffic demands from different piconets. Consequently, a judicious time-sharing scheme is needed to satisfy the possibly heterogeneous traffic demands while maximizing the utilization of the links.

3.2 Piconet Queue-State-Dependent Scheduling

Kalia et al. [30] are among the first researchers to consider the inefficiency of straightforward round robin time-slot scheduling at the Bluetooth MAC layer. They classify master-slave communication pairs into three classes. The first class pairs are those that can completely utilize the downlink and uplink slots (e.g., 1-1, 1-3, and 3-1, for four consecutive slots). The second class pairs are those that lead to 25% wastage (e.g., 3-0 and 0-3 for four consecutive slots). The third class pairs are those that lead to 50% wastage (e.g., 1-0 and 0-1). Kalia et al. then propose three different scheduling policies:

1. HOL-PP (head-of-line priority policy): The master-slave pairs are served in a weighted round robin manner with first class having the highest weight while third class having the lowest weight. Specifically, when a first class slave is polled, it is given P slot-pairs (for uplink and downlink). On the other hand, for a second class slave, it is served only for $3P/2$ times. Finally for a third class slave it is given only one single slot-pair when it is polled. Thus, this HOL-PP scheme is designed to maximize utilization.

2. Because HOL-PP pays no attention to the fairness issue, Kalia et al. also consider another scheme called HOL-KFP (HOL K-fairness policy). In this scheme, the extra allocations to first class (or second class) with respect to a fair allocation without regard to utilization are accounted. Similarly, the deficit allocations to third class are also recorded. The maximum difference between these two extremes are continuously monitored and is restricted to be bounded by a constant K, which serves as a unfairness tolerance threshold. When the difference goes beyond this value, a third class slave is also given its due service without regard to utilization.

3. HOL-KFP assumes an error-free environment in that a master-slave pair can always proceed. However, the channel quality of a particular master-slave pair may not be good (even if the slave is a first class), and thus, should not be allocated time-slots which will be wasted due to channel errors. Thus, a modified HOL-KFP scheme is suggested and the scheme works by considering the channel state in the weighted round robin process.

Kalia et al. also propose two SAR policies. The first policy is called I-SAR (intelligent SAR) which works by selecting the right number of slots suitable for the upper layer traffic (e.g., video). Another policy is called PR-SAR (partial reordering SAR) which works by dispatching the second packet in the queue (at the master or slave) when the HOL packet cannot be accommodated in the allocated slots. Simulation results indicate that the fairness based policies (HOL-KFP and the modified version) work better.

3.3 Piconet SAR and Scheduling

Das et al. [20] investigate on efficient strategies for SAR and time-slot scheduling in a single piconet to enhance system performance of ACL link in terms of utilization and delay. They consider two different SAR policies:

SAR-BF. A *best-fit* (BF) mechanism, this strategy tries to segment upper layer packets using the baseband packet size that can achieve the highest utilization.

SAR-OSU. Although this scheme is called *optimal slot utilization* (OSU), in fact its objective is to reduce the overall delay by using the least number of baseband packets, with possibly larger than required sizes.

For example, suppose we have a 556 bytes L2CAP packet to be segmented. Assume that the capacities of a five-slot, three-slot, and one-slot packet are: 339 bytes, 183 bytes, and 27 bytes, respectively. Using SAR-BF, the L2CAP packet is first segmented into one five-slot packet and the remaining 217 bytes fit better into a three-slot packet than a five-slot one, in terms of utilization. Finally, the residual 34 bytes are broken into two one-slot packets. On the other hand, using SAR-OSU, the L2CAP packet is again first segmented into one five-slot packet. However, the remaining 217 bytes can also be fitted into one single five-slot packet, resulting in the minimum number of baseband packets used.

Das et al. also consider three different time-slot scheduling policies:

1. Adaptive flow-based polling (AFP): In this policy, the polling interval of each slave is dynamically adjusted according to their traffic intensity as observed by the master. Specifically, with every slave starting with the same nominal polling interval (e.g., a straightforward round-robin), the polling interval of a slave is doubled if the master finds that the slave does not have an uplink packet to send after it is polled. On the other hand, if the flow-bit of the uplink packet received from a slave is set to 1, indicating that the slave's queue is longer than a pre-defined threshold, the master reduces the slave's polling interval

to the minimum value. However, each slave is still allowed to send one packet at a time when it is polled.

2. Sticky: In this policy, whenever the master finds that a slave has its flow-bit set to 1, the master allows the slave to send *num_sticky* packets all at once (in the simulations, it is found that setting *num_sticky* to 16 gives the best performance).

3. Sticky with AFP: Combining the above two policies together.

In their simulations using a two-state Markov chain wireless channel model with TCP sources, they find that SAR-OSU gives better performance. Combining SAR-OSU with all the three time-slot scheduling policies, they find that the Sticky (alone) and AFP (alone) policies exhibit good performance, in particular in terms of end-to-end delay.

3.4 Piconet Link-State History Based Scheduling

Deb et al. [21] present a simple but analytically sound approach to channel adaptive time-slot scheduling in a Bluetooth piconet. Specifically, based on the ACK or NACK packets received (induced by downlink transmissions from the master) from a slave, the master infers the channel quality of the slave. A two-state channel model is assumed such that when the channel is in the bad state, the master simply does not allow the slave to make use of the link (i.e., time-slot is not assigned to it). Moreover, in contrast to most existing approaches, a frequency-specific channel model is used in that the channel state is frequency dependent. The master keeps a counter C_{ij} for slave i at frequency j to record the channel state at frequency j. Whenever a slave transmits an ACK packet on the uplink, the counter is reset to zero. On the other hand, if a NACK is transmitted, indicating a failure, the counter is incremented. When the counter reaches a threshold TH_1, the master stops scheduling any time-slot to the slave when its turn is due. Yet the counter is still incremented. When the counter eventually reaches another threshold TH_2, the master resets the counter to zero and in the next round, the slave gets scheduled a time-slot again.

In this scheme, essentially the master keeps on tolerating transmission failures of the slave until the threshold TH_1 is reached, in the hope that the errors are due to transient channel quality fluctuations. Indeed, if the slave transmits an ACK in time before TH_1 is reached, the counter is reset to zero such that the transient channel problem is "forgotten." However, beyond TH_1, the master considers that the channel is changed to the bad state, which is expected to persist for a period of time. Thus, the slave is forced to miss $(TH_2 - TH_1)$ rounds of transmissions, in the hope that after this period of time, the channel reverts to the good state. Obviously, the selection of the parameters TH_1 and TH_2 is critical in obtaining a good performance. Through their detailed analytical modeling and simulations, Deb et al. find that $TH_1 = 4$ and $TH_2 = 100$ are reasonable choices.

3.5 Piconet Sniff Scheduling

Lin and Tseng [36] tackle the sniff scheduling problem in a single Bluetooth piconet. Consider the sniffing arrangement of a slave k as shown in Figure 6, the master needs to appropriately assign values to the sniff window size $\mathcal{T}_k$ and the active window size $\mathcal{N}_k$. The objective is for power saving and maximum utilization of the active window while satisfying the slave's traffic demand. For each slave k, the master computes a parameter U_k which is the utilization fraction of the assigned active window (i.e., $0 \leq U_k \leq 1$). With B_k denoting the backlog size of slave k (i.e., the number of packets awaiting transmission), the following parameter is computed:

$$W_k = \alpha U_k + (1 - \alpha)\frac{B_k}{B_{max}} \tag{1}$$

where B_{max} is the maximum buffer space and α is a constant between 0 and 1 for adjusting the importance of U_k and B_k. The value of W_k is then compared against the pre-defined upper and lower bounds. If the value of W_k is beyond the range, the slave's sniff parameters need to be adjusted. Specifically, a new ratio of $\mathcal{N}_k$ to $\mathcal{T}_k$, denoted by S_k, is determined:

$$S_k = \left(\frac{\mathcal{N}_k}{\mathcal{T}_k}\right)\left(\frac{W_k}{\delta}\right) \tag{2}$$

Note that in Equation (2), the original values of $\mathcal{N}_k$ and $\mathcal{T}_k$ are used. This ratio S_k is then used to compute new values of $\mathcal{N}_k$ and $\mathcal{T}_k$.

Their scheduling model considers multiple slaves simultaneously. Thus, with Equation (2), we still need to determine the ordering of the sniff windows of different slaves. Two policies are suggested: longest sniff interval first (LSIF) and shortest sniff interval first (SSIF). These policies work by search through a two-dimensional resource matrix representing the time-slots availability in the piconet under different scheduling scenarios. Through simulations, they also find that the scheduling algorithms proposed are quite accurate and can dynamically adjust the scheduling decisions based on up-to-date sniff-related parameters governed by the slaves'

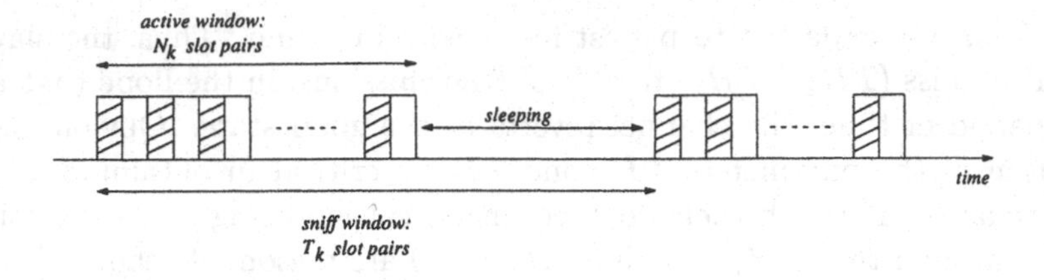

Figure 6: A sniff window.

traffic patterns. Most importantly, their schemes consider multiple slaves at the same time.

3.6 Scatternet Nodes Randomized Rendezvous Scheduling

Rácz et al. [43] formulate the time-slot scheduling problem of a bridge node participating in scatternet-wide communications as a *check-point* or rendezvous scheduling problem. In general, a bridge node has two or more links to distinct neighbors (from different piconets) in the scatternet. Thus, the bridge node has to perform rendezvous (or to "meet") with different neighbors at different times. Rácz et al. suggest a randomized algorithm for scheduling the rendezvous events. The algorithm is called PCSS (pseudo random coordinated scatternet scheduling) which works by randomly computing the rendezvous point based on the master's clock and the slave's device address. A rendezvous event will continue until one of the nodes leaves for attending another rendezvous. In order to adapt to various traffic conditions, the PCSS algorithm monitors the utilization of links in a coarse-grain manner and adjusts the inter-rendezvous time period accordingly.

3.7 Locally Coordinated Scheduling in Scatternets

Tan and Guttag [53] use a different approach in scheduling rendezvous events for bridge nodes in a scatternet. Their algorithm, called LCS (locally coordinated scheduling), works by computing the duration of the next rendezvous based on the queue size and past history of transmissions such that the duration is just large enough to exchange all the backlogged data. Furthermore, LCS computes the start time of the next meeting based on whether the data rate observed is increasing, decreasing or stable so that it responds to varying traffic conditions quickly without wasting resources. Moreover, LCS groups together meetings with the same traffic characteristics to reduce wasted bandwidth of nodes and end-to-end latency. LCS also aligns rendezvous events at various parts of the scatternet in a hierarchical fashion (assuming that the scatternet is a tree structure) so that the number of parallel communication is high, increasing system-wide throughput significantly. Finally, LCS is designed to reduce the amount of time a node spends transmitting packets while the receiver is not ready, thereby conserving energy.

3.8 Maximum Distance Rendezvous Scheduling in Scatternets

Johansson et al. [28] suggest a simple approach in determining a rendezvous schedule. The basic idea of their algorithm, called MDRP (maximum distance rendezvous point) scheduling, is that rendezvous points should be as far from each other as possible. To achieve this, the master in a piconet considers the rendezvous points of all the bridge nodes within its piconet. In determining a new rendezvous point with a bridge node, the master then computes the maximum allowable distance. Thus, the new rendezvous point is chosen as the middle slot of the largest interval between successive rendezvous points. These rendezvous points are then periodically repeated in the next *super-frame* which is implemented by using the sniff mode of the bridge node. It should be noted that the MDRP algorithm is radically different from the schemes described above in that it explicitly considers all bridge nodes simultaneously within a single piconet.

3.9 An Integrated Time-Slot Scheduling Approach

Shek and Kwok [45, 46] suggest an integrated scheme which contains four main mechanisms to address the different facets of the problem, namely Compensation-Based Time-Slot Assignment (CTSA), Traffic Differentiation Queueing (TDQ), Adaptive Master-Slave Switching (AMSS), and an Enhanced AODV algorithm for ad hoc routing.

Compensation-Based Time-Slot Assignment (CTSA) is a scheduling algorithm tailor-made for Bluetooth. With the formation of scatternet, some nodes have to take up the role of routers or bridges to interconnect different piconets together. As Bluetooth uses a frequency hopping scheme, a node cannot listen to two piconets simultaneously. As a result, it can only synchronize and participate in one piconet at a time. This implies that these bridge nodes have to multiplex their time among different piconets. However, as Bluetooth adopts a centralized polling scheme for time slots allocations, these nodes have to compete for time slots with all the other nodes when they switch back to another piconet. Consequently, the resources acquired by these nodes from the piconet are fewer. In a scatternet environment, traffic routing across the network can be abundant. The bridge nodes play an important role in routing packets across different piconets and they become the bottlenecks of the system. Most of the existing scheduling algorithms allocate time slots according to the loads of the queues, and if we apply them to Bluetooth directly, we may have ignored the bridging and uplink traffic problems. CTSA is an advanced scheduling scheme for Bluetooth. It exploits the properties of the absence of bridge nodes to calculate the time slots allocation for each nodes more effectively.

Traffic Differentiation Queueing (TDQ) is a technique to separate different traffic to different queues in Bluetooth's link layer. Traffic is differentiated as "self traffic" and "forward traffic" in this scheme and they are put into two different queues. Those being generated by the Bluetooth host itself are classified as "self traffic" while those being forwarded to it are classified as "forward traffic". The Bluetooth link layer serves these queues based on an adaptive ratio manner. This scheme helps routing packets across scatternet more efficiently and can improve the throughput and delay in a large extent.

Adaptive Master-Slave Switching (AMSS) is a hold time calculation scheme between bridge nodes, which decides the duration for a node to stay in a piconet. As bridge nodes in Bluetooth network need to multiplex their time to different piconets in a scatternet, the time spent on a piconet greatly affects the overall system performances. Hence, an adaptive calculation of the hold time is required. AMSS addresses this problem and uses an adaptive method based on queue lengths and utilization ratio to determine the hold time for a piconet. This scheme allocates time between piconets in a more effective way.

Enhanced AODV for Ad Hoc Routing is a routing protocol developed to facilitate ad hoc routing in Bluetooth network. AODV is used because it is one of the most representative ad hoc routing protocols and it performs well in most of the wireless network environment. In our enhanced AODV routing protocol, we have modified AODV for Bluetooth so that it becomes suitable for Bluetooth's narrow bandwidth and small packet size features. Our modifications are mainly about reducing the overheads in the AODV protocol, making the modified version more suitable for the Bluetooth environments.

3.10 Performance Studies

Capone et al. [9] present a simulation performance study on several piconet time-slot scheduling schemes:

- A benchmark scheme, denoted as B1, which is assumed to have complete knowledge about the queue lengths of all slaves; B1 does not switch to the next slave until the current slave's queue is exhausted;
- Pure Round Robin (PRR): The master polls the slaves in a fixed order; PRR is not exhaustive;
- Exhaustive Round Robin (ERR): Similar to PRR, the master polls the slaves in a fixed order but the server does not switch to the next slave until both the master and the slave queues are empty;
- Exhaustive Pseudo-Cyclic Master Queue Length (EPM): Dynamic cyclic order is defined at the beginning of each cycle and each master-slave pair is visited exactly once in a single cycle; the polling is in decreasing order of master to slave queue lengths;
- Limited and Weighted Round Robin (LWRR): The operation is similar to a normal weighted round robin but the weights are dynamically changed according to the observed queue status; Specifically, the weight of a slave is reduced by 1 if it does not have any packet to send when it is polled by the master.

Based on their simulation results using exponential traffic sources and TCP sources, the performance of ERR is better than PRR and EPM in terms of delay. However, an exhaustive serving scheme such as ERR can be subject to complete "capturing" by a single selfish slave. In this regard, the LWRR scheme performs the best among all schemes.

J. Mišić and V. B. Mišić [39] present a detailed and rigorous mathematical analysis of the performance of slave/slave bridges and master/slave bridges in a scatternet environment. They consider two simple scheduling policies, namely the limited service and exhaustive service, in their analytical modeling. The probability distributions of access delay and end-to-end delay for both intra-piconet and inter-piconet bursty traffic sources are considered. Specifically, the mean access delay is lower in the scatternet topology with an slave/slave bridge because both masters in the two piconets can spend more time serving their slaves. The same argument can be applied in the case for the lower mean end-to-end delay for local traffic. On the other hand, the average end-to-end delay for non-local traffic (i.e., inter-piconet) is lower for the master/slave configuration. This is because of the lower number of hops that the packets have to traverse. Finally, exhaustive scheduling is found to consistently outperform limited service scheduling. This is explicable by the fact that under a low load, limited service scheduling tends to waste slots but exhaustive service serves all the packets at once. Yet under a high load, this "exhaustive" effect diminishes and the two scheduling algorithms show similar performance.

3.11 Discussion

TDMA slots scheduling is in fact an old problem. Thus, it is a bit surprising to note that many previously proposed efficient techniques (e.g., fair queueing approaches) are not yet applied in the time-slot scheduling problem for a piconet or even a bridge node in a scatternet. The deficit round robin (DRR) algorithms are however studied in [45] and are found to exhibit quite good performance. On the other hand, the channel adaptive approach presented in [21] is interesting yet highly relevant in a practical situation. Indeed, it would be even more attractive if the scheduler also selects appropriate FEC coded packets (DMx types) in the time-slot scheduling problem, to further enhance the overall performance in terms of goodput. Bridge node rendezvous scheduling is a complicated problem and the currently available solutions are by and large not very mature. Perhaps this multi-dimensional scheduling problem needs to be tackled by an optimization approach.

4 Resource Management Schemes for Connectivity: Scatternet Formation and Routing

4.1 Overview

Unlike a wired network whose topology is set up manually by the planned connections among machines, in a wireless ad hoc network, the devices can spontaneously and autonomously connect with each other without human intervention. This is more so in a Bluetooth scatternet [2]. Simply put, scatternet formation can be done either in a centralized or distributed manner. In the former case, a leader is elected and then it instructs other devices within range to connect among each other to form the topology (e.g., a tree) according to the leader's desire. In the distributed case, each device acts autonomously and a scatternet can be formed after iterative merging of localized clusters.

The premise underlying scatternet formation as described above is that device discovery (i.e., routing) can be automatically done or at least very much eased. However, if the connectivity among devices is only useful for a short period of time, the overhead involved in forming a scatternet may not be worthwhile. In this case, on-demand routing is a better approach [41, 42].

In the following, we first describe several interesting algorithms (semi-centralized or distributed) for scatternet formation. We then discuss several routing strategies that can work with or without an underlying scatternet constructed a priori.

4.2 BTCP: Bluetooth Topology Construction Protocol

Salonidis et al. [44] pioneer in the quest for a distributed approach to scatternet formation. Their algorithm is called BTCP (Bluetooth topology construction protocol) and consists of three phases. The first phase is the most critical step and involves an exhaustive leader election. Specifically, all the n initially independent devices participate in the leader election process. Thus, a crucial requirement of BTCP is that all the devices have to be within each other's range. Indeed, as mentioned in [44], $n \leq 36$. Each device has a variable called VOTE. At the beginning of the leader election process, the value of VOTE is 1 for every device. Each device then performs inquiry and inquiry scan alternately in order to set up connections with nearby device. When a connection is set up, the two devices compare their VOTE variables. The device with a larger value of VOTE becomes a winner, while the other becomes a loser. Tie is broken by comparing the BD_ADDR—a larger BD_ADDR wins. The winner then adds the loser's VOTE to its own value and breaks the connection. Afterward, the winner continues the alternate switching and attempting to set up connections. On the other hand, the loser will just continually perform page scanning to wait for being scanned by potential masters in the final phase of the scatternet formation process. This VOTE comparison duel is repeatedly executed until there is only one single winner, signifying the end of the first phase.

The solitude winner is then the global centralized controller in the subsequent phases. In the second phase, the global winner gets any "special requests" from the other devices before making topology construction decisions. Having global and complete information (e.g., number of devices, each device's clock and BD_ADDR, etc.) the global winner instructs other devices to form the scatternet topology which satisfies any pre-defined constraints (e.g., node degree must be less than three). Specifically, some devices are assigned to be pure slaves, while other devices are assigned as masters or bridge nodes (slave/slave bridges or master/slave bridges). In the final phase, actual connections are made among devices, according to the global winner's instructions.

4.3 Bluetree Scatternet Formation

Zaruba et al. [57] suggest a scatternet formation protocol called bluetree. In this protocol, every node is required to perform a pre-processing step which aims to locate the neighbors. Specifically, each node, upon power up, actively performs inquiry and inquiry scanning to set up point-to-point connections with its neighbors. Then the connected nodes exchange information such as clock and BD_ADDR. After this pre-processing step, a node is designated as the blueroot, which has complete information about the geographical locations and properties of all nodes in the system. The blueroot then initiates a spanning tree formation process with itself as the root of the spanning tree. An interesting property is that some nodes may not be physically connected in the resulting spanning tree because some leaf nodes may be assigned to some master nodes that already have seven slaves. Zaruba et al. also propose an enhancement of the protocol by designating two or more nodes to be blueroots. These blueroots then independently form spanning trees and eventually these trees are merged together to form a single logical scatternet graph. A limitation of bluetree is that the root node may become a bottleneck of inter-piconet communications (e.g., two piconets that belong to different branches of the root).

4.4 Bluenet Scatternet Formation

Wang et al. [56] propose another three-phase algorithm for scatternet formation, called Bluenet. In contrast to bluetree described above, in Bluenet there is no designated root node. Thus, the resultant scatternet graph may or may not be a tree. In the first phase of the algorithm, similar to that in bluenet, the initially independent Bluetooth nodes perform inquiry and inquiry scanning to set up local connections with nodes in the neighborhood and in turn exchange information about each other. After this information acquiring step, nearby nodes cluster together and one of them becomes the master to form a piconet. Thus, at the end of phase one, there are numerous piconets distributed across the area. In phase two, any independent nodes that are still not connected to any piconet are required to perform page scanning, awaiting paging from nearby nodes that already connected to some piconets. The only constraint in this process is the maximum number of slaves a piconet is designated to handle (in their simulations, this value is five). Afterward, in phase three, the separate piconets in the system are connected to form a single scatternet by having the slaves in each piconet actively set up outgoing links to other slaves in other piconets. In their simulation study, it was found that the Bluenet structure can carry a higher volume of traffic then bluetree.

4.5 BlueConstellation Scatternet Formation

Basagni and Petrioli [3] also suggest a three-phase protocol for scatternet formation. In their algorithm, each piconet is called a BlueStar and the resultant scatternet is called a BlueConstellation. The initial phase of the algorithm is very similar to that in bluetree and Bluenet described above—a physical topology discovery process for obtaining mutual information about neighboring nodes via the exchanging of FHS (frequency hop synchronization) packets. In the second phase, BlueStars (i.e., piconets) are formed. In each BlueStar, the master is elected by comparing the weights of clustered nodes in the vicinity. Tie is broken by the unique IDs. In the final phase, the BlueStars are connected together to form the BlueConstellation. In the scatternet construction process, the masters of individual piconets that are within three hops distance from each other are qualified to make connections. Furthermore, the master that has a larger weight is the initiator of the connection. Specifically, such a master starts a paging process so as to recruit a slave of a neighboring piconet to join its piconet. Effectively such a slave becomes a bridge node.

4.6 Randomized Scatternet Formation

Law et al. [35] suggest a novel randomized algorithm for forming a scatternet from an initial set of totally independent Bluetooth devices in an iterative manner. The algorithm is elegant yet highly practical in that it has a time complexity of $O(\log n)$ (with probability of $1 - 1/n^{\theta(1)}$) and a message complexity of $O(n)$, where n is the number of Bluetooth devices. The key structures are the *components* and *leaders*. Initially, every device is a leader of a component consisting of only itself. During the scatternet formation process, an independent piconet's leader is the master and a scatternet's leader is one of the masters. The core mechanism is the iterative randomized *retiring* of leaders (i.e., changing a leader to a non-leader) and the maintenance of two invariants:

1. Each leader either has no slave, or has at least one unshared slave.
2. Each leader has no more than k slaves ($k = 7$ according to the Bluetooth specification).

The crux of the algorithm is the randomized expansion of components (from an independently lonely device, to a piconet, then to a scatternet, and so on). In each iteration, each leader randomly decides (governed by a probability p, $1/3 < p < 2/3$) whether to perform a SEEK operation (i.e., an inquiry) through which a new slave (that performs a SCAN, i.e., an inquiry scan) is connected; otherwise, the leader itself performs SCAN or instructs an unshared slave to perform SCAN in case the leader has such a slave. After the matching, two components are merged to form a larger component. This is followed by a restructuring of the resultant component in order to maintain the two invariants and the key feature that each component has a single leader (i.e., some leaders are retired after each iteration).

The algorithm has many salient features. For example, in the resulting scatternet, the node degree of each device is at most two, and as such, network bottleneck is avoided. Furthermore, the resultant scatternet has at most $\lfloor (n-2)/(k-1) \rfloor + 1$ piconets. This is very close to the lower bound: $\lceil (n-1)/k \rceil$ (see [35] for the proofs). Having a small number of piconets can help to reduce the inter-piconet interference. Finally, the network diameter is found, through simulations, to be $O(\log n)$ also.

4.7 Scatternet Formation Based on Search Trees

Sun et al. [51] propose a scatternet formation scheme that enables "self-routing"—nodes do not need to use routing tables to keep reachability information about other nodes in the scatternet. Specifically, their scheme aims to construct an seven-way search tree connecting all the nodes in the system. It is a seven-way search tree because each node is a master of a piconet having at the most seven slaves. Thus, the bridge nodes in the tree act as both master and slave. An essential requirement is that all nodes are within range of each other such that pairing of nodes to form links can be done arbitrarily independent of their geographic locations. The root of the tree is the one with the largest value of BD_ADDR. Having obtained all the characteristics of nodes within the system (e.g., BD_ADDR and clock), the root instructs other nodes to form the tree based on a tree-sort like procedure. In the process, an invariant for each node a in the tree is maintained: $\max(c_i) < \min(c_j)$ for all $i < j$, where c_i and c_j are children of a; $\max(b)$ and $\min(b)$ are respectively the largest and smallest value of BD_ADDR in the sub-tree rooted at node b. With such a sorted tree structure, routing becomes very easy in that if a node a wants to reach another node b, node a just needs to compare its BD_ADDR with that of b and then recursively traverses the tree until it locates b. The root also periodically goes into inquiry mode to see if any new node wants to join the scatternet. If the root finds such a node, it inserts the new node into appropriate place in the search tree. A practical limitation of this protocol is that the root can be heavily loaded with tree maintenance tasks and also become a bottleneck in inter-piconet communications.

4.8 Scatternet Formation Based on de Bruijn Graphs

Song et al. [49] propose another self-routing scatternet structure based on an elegant structure called *de Bruijn graph*. In a set of n initially independent Bluetooth nodes (again the nodes are all within range of each other), the one with the smallest BD_ADDR is selected as the leader, which also holds a token. The leader picks a group of 2^s nodes (including itself) into the master set for forming the scatternet backbone based on the de Bruijn graph structure. Note that s is chosen such that $2^{s-1} < \lceil n/6 \rceil \leq 2^s$. The leader then assigns itself with an address label 0^s (i.e., an s-bit number in which all digits are zero), and each node in the master set with a unique s-bit address label in the range from $0 \cdots 01$ to $1 \cdots 11$. The following procedure is then executed.

1. The token holding node (initially the leader, with address label 0^s) selects two nodes from the remaining nodes (outside of the master set) to be slave/slave bridge nodes. These bridge nodes get address labels $(x_1, \ldots, x_s, 010)$ and $(x_1, \ldots, x_s, 101)$, respectively. The three-bit portions (i.e., 010 and 101) are also used as the Bluetooth medium access code within the piconet. The token holding node then uses the bridge node with label $(x_1, \ldots, x_s, 010)$ to connect to a neighbor master with address label $x_2 x_3 \cdots x_s 0$. Similar, the bridge node with label $(x_1, \ldots, x_s, 101)$ is connected to another neighbor master with address label $x_2 x_3 \cdots x_s 1$.

2. The token holding node then selects from the remaining nodes another three (or less) nodes to be *pure* slaves.

3. The token is then passed to the successor master (i.e., the master with next value of address label) and the whole process is repeated again.

Using the above procedure (see [49] for more details), a scatternet with a de Bruijn graph backbone is formed. For example, a 48-node scatternet is shown in Figure 7. With the de Bruijn backbone structure, self-routing is again an easy process. Specifically, if a master with address label $x_1x_2\cdots x_s$ wants to reach another master with address label $y_1y_2\cdots y_s$, the message can be routed through the following chain of master nodes: $x_1x_2\cdots x_s$, $x_2x_3\cdots x_sy_1$, $x_3x_4\cdots x_sy_1y_2$, ..., $x_sy_1\cdots y_{s-1}$, $y_1y_2\cdots y_s$. If a pure slave node wants to reach another pure slave node in another piconet, the former just passes the message to its master which then follows the above routing rule to reach the corresponding master of the latter. Finally, if a slave/slave bridge wants to reach a node in another remote piconet, it just randomly selects one of its two masters and passes the message to it, which then again follows the above routing rule.

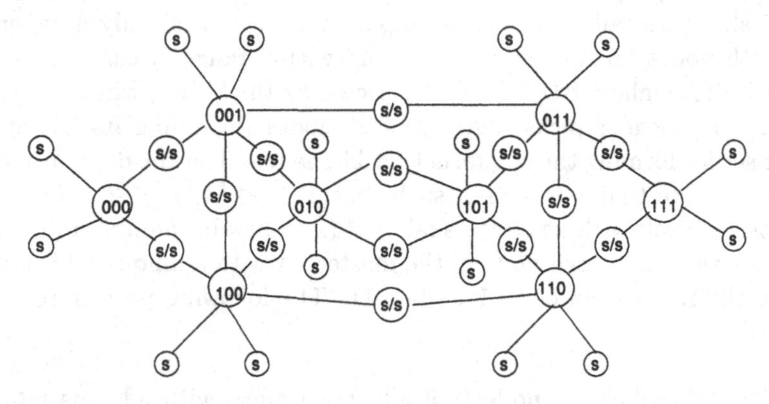

Figure 7: An example de Bruijn graph based scatternet core.

Compared with a tree-like structure discussed earlier, using the de Bruijn graph structure leads to many salient features: each bridge node only participates in at most two piconets; self-routing is easy; the diameter of the graph is $O(\log n)$; the congestion of every node is at most $O(\frac{\log n}{n})$ assuming that the traffic demand is evenly distributed. The structure can also be incrementally changed at a low time complexity when nodes join or leave the network.

4.9 BlueRing Scatternet Formation

Lin et al. [37] suggest a ring topology for constructing Bluetooth scatternets. Similar to many approaches described above, their algorithm assumes that all nodes within the system are within range of each other. In the initial step, nodes exchange information about their characteristics (e.g., BD_ADDR, clock, etc.). A leader, with the largest BD_ADDR, is elected among all nodes. The leader then designates several other nodes in the system to be masters as well, and this group of master nodes form a ring structure. Afterward, each master recruits nodes from the remaining nodes to form its own piconet. One such slave is designated as a slave/slave bridge for connecting two neighboring masters in the ring. With this ring structure, again self-routing is easy because a message is just circulated around the ring until it finally reaches the destination's master. However, compared with the de Bruijn graph structure discussed in Section 4.8, the diameter of a ring is $O(n)$ instead of $O(\log n)$, and as such, the delay may be quite significant for a large ring.

4.10 RVM Routing

In this section we begin our survey of several routing protocols for Bluetooth scatternets. Bhagwat and Segall [5] pioneer in the quest for a practical routing scheme for Bluetooth scatternets. Designed from a practical perspective, their scheme, called RVM (routing vector method) works by using the source routing algorithm. Specifically, when a node wants to reach a remote node in another piconet, the former first searches a route to reach the latter. The source node then embeds the whole route discovered in the packets which are forwarded incrementally through connected piconets. In the route searching process, similar to many *ad hoc* routing protocols [41], a flooding of route search packets are sent across the entire scatternet. During the flooding process, the intermediate nodes receiving the search request append their own addresses into the search packet before forwarding it onward. The destination, upon receiving the search packet, extracts the route embedded in it, and then unicasts a reply message (of course, also containing the entire route) to the source. The source can then extract the route and make use of it in the subsequent data transmissions. Thus, the RVM algorithm is basically a direct application of traditional *ad hoc* routing techniques, without much exploitation of the distinctive characteristics of Bluetooth and the structure of the underlying scatternet.

4.11 On-Demand Scatternet Routing

Based on the observation that static scatternet formation may not be suitable for networks with low traffic demand and severely power-limited devices, Liu et al. [38] propose an on-demand approach in setting up scatternet routes. Essentially, the nodes in the system are neither logically nor physically connected until there is a need for packet transmissions. This on-demand approach is in sharp contrast to the rationale behind the various scatternet formation strategies described above—to maintain connectivity at all time. Of course, the demerit of on-demand approach is that the route set-up delay can be very large. Indeed, this is the major problem tackled by Liu et al. in their study [38]. Their on-demand routing approach is again very similar to those common *ad hoc* routing protocols. Specifically, the source node floods a route request packet across the entire system, intermediate nodes record their addresses in the route request packet, and the destination eventually extracts the route which is then used for unicasting the route reply to the source.

The crux of their scheme is an elaborate strategy to reduce the set-up delay, in particular the delay introduced by the route request incremental forwarding process. The difficulty involved in a totally on-demand setting, compared to the situation where an underlying scatternet exists, is that the neighboring nodes do not have connection set up a priori that can be used for relaying the route request packet. Thus, in the most straightforward approach, the route request packet is relayed after a connection is set up between two neighboring nodes. Specifically, the route request packet can be forwarded through the connection either at the L2CAP layer or at the LMP layer. However, in both cases, the delay is dominated by the connection set-up process involving inquiry and inquiry scanning. Indeed, the inquiry and inquiry scanning process lead to a delay of 10.24 seconds for one hop. Thus, for a multihop relaying of route request packets, the delay can be unacceptably large. In view of this obstacle to the practicality of the routing process, Liu et al. [38] make an intelligent suggestion in that the inquiry packet itself can be used for carrying the route request information. Using this technique, the per hop delay is reduced to 2.56 seconds. The resultant scatternet route is composed of alternating master and slave nodes, as illustrated in Figure 8. Based on their simulations using BlueHoc [6], it is found that the network utilization is the highest (at 36.16%) for DH5 packets.

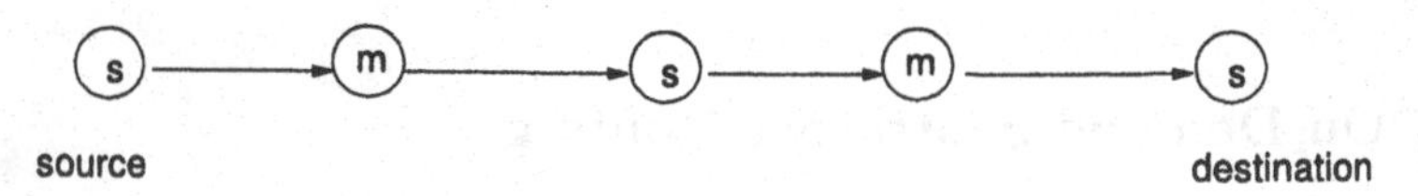

Figure 8: An example scatternet route with alternating masters and slaves.

4.12 Zone Routing Protocol

Kapoor and Gerla [31] also propose a new scheme to address the route set-up overhead problem in Bluetooth scatternet. However, in their scheme, it is assumed that an underlying scatternet structure already exists. Specifically, the entire system is partitioned into disjoint zones and the size of each zone is governed by the MAXHOPS parameter. Within a zone, each and every node maintains the complete routing information of other nodes in the same zone. Such complete information is determined by complete exchange within the zone using LMP broadcasts. The routing across zones is accomplished by using an on-demand routing protocol such as the AODV (ad hoc on-demand distance vector) algorithm, which is also based on flooding route request packets. Thus, whenever a node has information to send to a remote node (outside its piconet), it first checks its zone routing hash table to see if the destination is within the same zone. If the destination is not in the same zone, it triggers an AODV algorithm across the zones. Simulation results indicate that the value of MAXHOPS is better set to be two to four in order not to incur a large overhead in terms of route information storage space.

4.13 Discussion

The various distributed algorithms for scatternet formation are all interesting yet practicable given their moderate or low complextiy. However, an even more interesting problem is to construct the scatternet by taking into account the traffic requirements of the participating devices. Such traffic information can also be exchanged during the first phase of the scatternet formation process. The scatternet so constructed can then provide more support, in the form of less traffic relaying burden, to those heavily loaded devices. On the other hand, routing approaches applied in Bluetooth are largely those suggested for general ad hoc networks. Such ad hoc routing protocols may not be the most appropriate in a Bluetooth scatternet because the devices in a Bluetooth environment are already clustered into at least independent piconets. Such a feature should be exploited in order to enhance the performance of ad hoc routing.

5 Further Research Opportunities

While many salient approaches to various resource management problems in Bluetooth networks have been proposed and shown to be effective, there are still a large number of interesting problems to be tackled. Indeed, we have described a number of such problems in the "Discussion" sections earlier. Here, we highlight three more interesting problems:

Cross-Layer Channel Adaptive SAR. Due to various inevitable signal propagation impairments such as multipath fading and shadowing, the quality (e.g., in terms of error rate or signal-to-noise ratio) of a wireless channel is always time-varying. That is, the channel quality may be very good in certain moments so that no FEC is required. In some other moments, the channel may be very noisy so that even a high level of FEC may not guarantee correct reception of packets. Of course, in other cases, the channel quality may be somewhere in between. The channel quality of a Bluetooth link is more so time-varying due to the crowded ISM band. Thus, overall throughput can be enhanced by using a channel adaptive approach—when the channel is good, DHx packets are used; when the channel is not good, DMx packets are used. Research issues include: how we can infer channel quality accurately; which packet type (DM1, DM3, or DM5) we should use; how we can take into account application level traffic requirements.

Integrated Traffic Engineering. While piconet scheduling, scatternet formation, and bridge node scheduling are independently tackled with success, it is good to design a total solution that takes into account the possibly heterogeneous traffic requirements of participating nodes. Furthermore, such a scatternet-wide integrated traffic engineering mechanism should be dynamic in that when the traffic pattern changes (determined by some traffic inference algorithms), the structure of the scatternet and/or the scheduling strategies of each piconet/bridge-node should be changed as well.

Theoretical Capacity of Scatternets. In some scatternet formation algorithms, the number of piconets, and hence the number of Bluetooth devices, can be very large. Given the modest data rates supported by Bluetooth, a large number of piconets may not make much sense in practice. Thus, it is interesting to analytically study the capacity of a scatternet possibly in terms of number of distinct source-destination pairs. Certainly the capacity is also limited by the inter-piconet interference and power consumption constraints.

6 Summary

Bluetooth is becoming a ubiquitous short-range wireless technologies in that many low cost commercial devices are already very popular. However, there are still many outstanding issues not defined in the specifications or not agreed upon by the research/industrial community. In this chapter, we focus on various resource management issues. Specifically, we consider resource management problems related to coexistence and connectivity. We present a detailed survey of more than 20 recently proposed techniques for tackling the resource management problems considered.

Acknowledgments

The author would like to thank Mr. Tyrone Kwok, Mr. Mark Yeung, and the anonymous referees for their careful reading and useful comments.

References

[1] A. K. Arumugam, A. Doufexi, A. R. Nix, and P. N. Fletcher, Investigation of the Coexistence of 802.11g WLAN and High Data Rate Bluetooth Enabled Consumer Electronic Devices in Indoor Home and Office Environments, *IEEE Transactions on Consumer Electronics* Vol. 49 No. 3 (Aug. 2003) pp. 587-598.

[2] S. Baatz, M. Frank, C. Kühl, P. Martini, and C. Scholz, Bluetooth Scatternets: An Enhanced Adaptive Scheduling Scheme, *Proceedings of INFOCOM 2002*, Vol. 2 (Jun. 2002) pp. 782-790.

[3] S. Basagni and C. Petrioli, Multihop Scatternet Formation for Bluetooth Networks, *Proceedings of VTC Spring 2002*, Vol. 1 (May 2002) pp. 424-428.

[4] A. Batra, An Intelligent Frequency Hopping Scheme for Improved Bluetooth Throughput in an Interference-Limited Environment, IEEE 802.15-01/082r1, `http://www.ieee802.org/15/pub/TG2-Draft.html` (Jan. 2001).

[5] P. Bhagwat and A. Segall, A Routing Vector Method (RVM) for Routing in Bluetooth Scatternets, *Proceedings of the IEEE International Workshop on Mobile Multimedia Communications* (Nov. 1999) pp. 375-379.

[6] BlueHoc Simulator, `http://oss.software.ibm.com/bluehoc/`, (2003).

[7] Bluetooth Special Interest Group, *Specifications of the Bluetooth Systems* Vol. 1 (v.1.0B Core, Dec. 1999).

[8] J. Bray and C.F. Sturman, *Bluetooth: Connect without Cables*, (Upper Saddle River, N.J., Prentice Hall, 2002).

[9] A. Capone, M. Gerla, and R. Kapoor, Efficient Polling Schemes for Bluetooth Picocells, *Proceedings of ICC 2001* Vol. 7 (Jun. 2001) pp. 1990-1994.

[10] M. C. H. Chek, *Coexistence Techniques for Heterogeneous Mobile Devices Operating in Uncoordinated Radio Spectrum*, M.Phil. Thesis, Department of Electrical and Electronic Engineering, The University of Hong Kong, (Dec. 2003). (available: http://www.eee.hku.hk/~ykwok/students/thesis-mphil-ChekCH.pdf)

[11] M. C. H. Chek and Y.-K. Kwok On Adaptive Frequency Hopping with Resource Constraints in an *Ad Hoc* Uncoordinated Wireless Environment: Performance Analysis and A New Algorithm Based on a Cross-Layer Design, *Proceedings of the 7th International Symposium on Parallel Architectures, Algorithms, and Networks*, (Hong Kong, China, May 2004).

[12] J.-L. Chen, H.-S. Wu, and H.-W. Tzeng, Bluetooth and IEEE 802.11 Coexistence Service Architecture with Fuzzy-Based Switching Scheme, *Proceedings of the 22nd International Conference of the North American Fuzzy Information Processing Society* (Jun. 2003) pp. 506-512.

[13] K. C. Chen, H. K. Chen and C. C. Chao, Selective Hopping for Hit Avoidance, IEEE 802.15-01/057r2, `http://www.ieee802.org/15/pub/TG2-Draft.html`, (Mar. 2001).

[14] X.-Z. Cheng, X. Huang, and D.-Z. Du (eds.), *Ad Hoc Wireless Networking*, (Kluwer Academic Publishers, 2003).

[15] C.-F. Chiasserini and R. R. Rao, Coexistence Mechanisms for Interference Mitigation in the 2.4 GHz ISM Band, *IEEE Transactions on Wireless Communications* Vol. 2 No. 5 (Sep. 2003) pp. 964-975.

[16] A. Conti, D. Dardari, G. Pasolini, and O. Andrisano, Bluetooth and IEEE 802.11b Coexistence: Analytical Performance Evaluation in Fading Channels, *IEEE Journal on Selected Areas in Communications* Vol. 21 No. 2 (Feb. 2003) pp. 259-269.

[17] C. de M. Cordeiro, S. Abhyankar, R. Toshiwal, and D. P. Agrawal, A Novel Architecture and Coexistence Method to Provide Global Access to/from Bluetooth WPANs by IEEE 802.11 WLANs, *Proceedings of the IEEE International Conference on Performance, Computing, and Communications* (Apr. 2003) pp. 23-30.

[18] C. de M. Cordeiro and D. P. Agrawal, Employing Dynamic Segmentation for Effective Co-Located Coexistenc between Bluetooth and IEEE 802.11 WLANs, *Proceedings of GLOBECOM 2002*, Vol. 1 (Nov. 2002) pp. 195-200.

[19] C. de M. Cordeiro, D. Sadok, and D. P. Agrawal, Piconet Interference Modeling and Performance Evaluation of Bluetooth MAC Protocol, *Proceedings of GLOBECOM 2001* Vol. 5 (Nov. 2001) pp. 25-29.

[20] A. Das, A. Ghose, A. Razdan, H. Saran, and R. Shorey, Enhancing Performance of Asynchronous Traffic over the Bluetooth Wireless Ad Hoc Network, *Proceedings of INFOCOM 2001* Vol. 1 (Apr. 2001) pp. 591-600.

[21] S. Deb, M. Kapoor, and A. Sarkar, Error Avoidance in Wireless Networks Using Link State History, *Proceedings of INFOCOM 2001*, Vol. 2 (Apr. 2001) pp. 786-795.

[22] N. Golmie, R. E. van Dyck, and A. Soltanian, Interference of Bluetooth and IEEE 802.11: Simulation Modeling and Performance Evaluation, *Proceedings of the 4th ACM International Workshop on Modeling, Analysis and Simulation of Wireless and Mobile Systems* (Jul. 2001) pp. 11-18.

[23] J. C. Haartsen and S. Mattisson, Bluetooth—A New Low-Power Radio Interface Providing Short-Range Connectivity, *Proceedings of the IEEE*, Vol. 88 No. 10 (Oct. 2000) pp. 1651-1661.

[24] I. Howitt, WLAN and WPAN Coexistence in UL Band, *IEEE Transactions on Vehicular Technology* Vol. 50 No. 4 Jul. 2001) pp. 1114-1124.

[25] I. Howitt, Bluetooth Performance in the Presence of 802.11b WLAN, *IEEE Transactions on Vehicular Technology* Vol. 51 No. 6 (Nov. 2002) pp. 1640-1651.

[26] I. Howitt, Mutual Interference between Independent Bluetooth Piconets, *IEEE Transactions on Vehicular Technology*, Vol. 52 No. 3 (May 2003) pp. 708-718.

[27] IEEE Std. 802.11b, *Part 11: Wireless LAN Medium Access Control (MAC) and Physical Layer (PHY) Specifications: Higher-Speed Physical Layer Extension in the 2.4GHz Band* (Sep. 1999).

[28] P. Johansson, R. Kapoor, M. Kazantzidis, and M. Gerla, Rendezvous Scheduling in Bluetooth Scatternets, *Proceedings of ICC 2002*, Vol. 1 (May 2002) pp. 318-324.

[29] P. Johansson, M. Kazantzidis, R. Kapoor, and M. Gerla, Bluetooth: An Enabler for Personal Area Networking, *IEEE Network*, Vol. 15 No. 5 (Sept./Oct. 2001) pp. 28-37.

[30] M. Kalia, D. Bansal, and R. Shorey, MAC Scheduling and SAR Policies for Bluetooth: A Master Driven TDD Pico-Cellular Wireless System, *Proceedings of the IEEE Workshop on Mobile Multimedia Communications* (Nov. 1999) pp. 384-388.

[31] R. Kapoor and M. Gerla, A Zone Routing Protocol for Bluetooth Scatternets, *Proceedings of WCNC 2003* Vol. 3 (Mar. 2003) pp. 1459-1464.

[32] A. Kumar and R. Gupta, Capacity Evaluation of Frequency Hopping Based Ad Hoc Systems, *Proceedings of the ACM SIGMETRICS International Conference on Measurement and Modeling of Computer Systems* Vol. 29 No. 1 (Jun. 2001) pp. 133-142.

[33] J. Lansford, MEHTA: A Method of Coexistence between Colocated 802.11b and Bluetooth Systems, IEEE 802.15-00/360r0, `http://www.ieee802.org/15/pub/TG2-Draft.html`, (Nov. 2000).

[34] J. Lansford, A. Stephens, and R. Nevo, Wi-Fi (802.11b) and Bluetooth: Enabling Coexistence, *IEEE Network*, Vol. 15 No. 5 (Sep./Oct. 2001) pp. 20-27.

[35] C. Law, A. K. Mehta, and K.-Y. Siu, A New Bluetooth Scatternet Formation Protocol, *Mobile Networks and Applications* Vol. 8 No. 5 (Oct. 2003) pp. 485-498.

[36] T.-Y. Lin and Y.-C. Tseng, An Adaptive Sniff Scheduling Scheme for Power Saving in Bluetooth, *IEEE Wireless Communications Magazine*, Vol. 9 No. 6 (Dec. 2002) pp. 92-103.

[37] T.-Y. Lin, Y.-C. Tseng, K.-M. Chang, and C.-L. Tu, Formation, Routing, and Maintenance Protocols for BlueRing Scatternet of Bluetooth, *Proceedings of the 36th Hawaii International Conference on System Sciences*, (Jan. 2003) pp. 313-322.

[38] Y. Liu, M. J. Lee, and T. N. Saadawi, A Bluetooth Scatternet-Route Structure for Multihop Ad Hoc Networks, *IEEE Journal on Selected Areas in Communications* Vol. 21 No. 2 (Feb. 2003) pp. 229-239.

[39] J. Mišić and V. B. Mišić, Bridges of Bluetooth County: Topologies, Scheduling, and Performance, *IEEE Journal on Selected Areas in Communications* Vol. 21 No. 2 (Feb. 2003) pp. 240-258.

[40] G. Pasolini, Bluetooth Piconets Coexistence: Analytical Investigation on the Optimal Operating Conditions, *Proceedings of ICC 2003*, Vol. 1 (May 2003) pp. 198-202.

[41] C. E. Perkins (Ed.), *Ad Hoc Networking*, (Addison-Wesley, 2000).

[42] B. J. Prabhu and A. Chockalingam, A Routing Protocol and Energy Efficient Techniques in Bluetooth Scatternet, *Proceedings of ICC 2002* Vol. 5 (May 2002) pp. 3336-3340.

[43] A. Rácz, G. Miklós, F. Kubinszky, and A. Valkó, A Pseudo Random Coordinated Scheduling Algorithm for Bluetooth Scatternets, *Proceedings of MobiHOC 2001*, (Oct. 2001) pp. 193-203.

[44] T. Salonidis, P. Bhagwat, L. Tassiulas, and R. LaMaire, Distributed Topology Construction of Bluetooth Personal Area Networks, *Proceedings of INFOCOM 2001* Vol. 3 (Apr. 2001) pp. 1577-1586.

[45] L. L.-Y. Shek, *Ad Hoc Routing and Time-Slot Scheduling in Bluetooth Networks*, M.Phil. Thesis, Department of Electrical and Electronic Engineering, The University of Hong Kong, Dec. 2002. (available: http://www.eee.hku.hk/~ykwok/students/thesis-mphil-ShekLiza.pdf)

[46] L. L.-Y. Shek and Y.-K. Kwok, An Integrated Approach to Scatternet Traffic Management in Bluetooth Ad Hoc Networks, *Computer Networks*, (to appear in 2004).

[47] S. Shellhammer, "Colocated Collaborative Coexistence Mechanism: TDMA of 802.11 and Bluetooth," IEEE 802.15-01/025r0, `http://www.ieee802.org/15/pub/TG2-Draft.html`, (Jan. 2001).

[48] M. B. Shoemake, "Proposal for Power Control for Enhanced Coexistence," IEEE 802.15-00/081r0, `http://www.ieee802.org/15/pub/TG2-Coexistence-Mechanisms.html`, (Jan. 2001).

[49] W.-Z. Song, X.-Y. Li, Y. Wang, and W. Wang, dBBlue: Low Diameter and Self-Routing Bluetooth Scatternet, *Proceedings of the 2003 MOBICOM Joint Workshop on Foundations of Mobile Computing* (Sep. 2003) pp. 22-31.

[50] L. Sydanheimo, M. Keskilammi, and M. Kivikoski, Performance Issues on the Wireless 2.4 GHz Band in a Multi-System Environment, *IEEE Transactions on Consumer Electronics* Vol. 48 No. 3 (Aug. 2002) pp. 638-643.

[51] M.-T. Sun, C.-K. Chang, and T.-H. Lai, A Self-Routing Topology for Bluetooth Scatternets, *Proceedings of the International Symposium on Parallel Architectures, Algorithms, and Networks* (May 2002) pp. 13-18.

[52] M.-T. Sun, S. Wang, C.-K. Chang, T.-H. Lai, H. Sawatari, and H. Okada, Interference-Aware MAC Scheduling and SAR Policies for Bluetooth Scatternets, *Proceedings of GLOBECOM 2002*, Vol. 1 (Nov. 2002) pp. 11-15.

[53] G. Tan and J. Guttag, A Locally Coordinated Scatternet Scheduling Algorithm, *Proceedings of IEEE LCN 2002* (Nov. 2002) pp. 293-303.

[54] B. Treister, H. B. Gan, K. C. Chen, H. K. Chen, A. Batra, and O. Eliezer, Adaptive Frequency Hopping: A Non-Collaborative Coexistence Mechanism, IEEE 802.15-01/252r0, `http://www.ieee802.org/15/pub/TG2-Coexistence-Mechanisms.html`, (May 2001).

[55] R. E. van Dyck and A. Soltanian, IEEE 802.15 Clause 14.1 - Collaborative Colocated Coexistence Mechanism, IEEE 802.15-01/364r0, `http://www.ieee802.org/15/pub/TG2-Draft.html`, (Jul. 2001).

[56] Z. Wang, R. J. Thomas, and Z. Haas, Bluenet - A New Scatternet Formation Scheme, *Proceedings of the 35th Hawaii International Conference on System Sciences*, (Jan. 2002).

[57] G. V. Zaruba, S. Basagni, and I. Chlamtac, Bluetrees - Scatternet Formation to Enable Bluetooth Based Ac Hoc Networks, *Proceedings ICC 2001* Vol. 1 (Jun. 2001) pp. 273-277.

[58] B. Zhen, Y. Kim, and K. Jang, The Analysis of Coexistence Mechanisms of Bluetooth, *Proceedings of VTC Spring 2002* Vol. 1 (May 2002) pp. 419-423.

Energy-Efficient MAC Layer Protocols in Ad Hoc Networks

Fang Liu, Kai Xing, Xiuzhen Cheng, and Shmuel Rotenstreich
Department of Computer Science
The George Washington University
Washington, DC 20052
E-mail: {fliu,kaix,cheng,shmuel}@gwu.edu

Contents

1 Introduction

A mobile ad hoc network (MANET) is defined as an autonomous system of mobile routers (and their associated hosts) connected by wireless links - the union of which forms an arbitrary graph. It is characterized by fast deployment, dynamic multi-hop topology, self-organization without typical infrastructure support, etc. These properties are desirable in situations such as battlefields, where network connectivity is temporarily needed, or fixed infrastructures are unavailable, expensive, or infeasible to deploy. However, wide deployment of MANET has not come yet due to many technical challenges, among which energy issue is a fundamental one. Typical wireless devices are powered by small-sized batteries, whose replacement is very difficult or even impossible in some applications (e.g. disaster relief operation). Therefore, power conservation is one of the most important design considerations for MANET. It has attracted a large number of researchers in recent years [3].

Power conservation in an ad hoc network is the procedure of determining the transmit power of each communication terminal such that a design objective (e.g. network lifetime, throughput, etc.) can be satisfied. There are two major reasons for transmit power control. First, transmitting at a high power may increase the interference to co-existing users and therefore degrade network throughput. Power saving mechanisms have been shown to be able to decrease multi-user interference, and hence increase spatial channel reuse and the number of simultaneous single-hop transmissions. One direct benefit of this increase is the enlarged overall traffic carrying capacity of the network [27]. Second, energy-efficient schemes can impact battery life, consequently prolonging the lifetime of the network. Current power control mechanisms include low-power wireless access protocols, power-aware routing for ad hoc and sensor networks [60] [51], and node-level energy-efficient information processing [16]. In this paper, we will focus on energy aware MAC (Media Access Control) layer protocols for ad hoc networks.

MAC layer is the sublayer of the data link layer that is responsible for coordinating and scheduling of transmissions among competing nodes. As claimed by [63], MAC protocols could significantly reduce the power consumption of mobile terminals in MANETs. The energy-aware MAC protocols in a multi-hop self-organizing mobile ad hoc network must simultaneously satisfy the following three objectives. First, MAC protocols should facilitate the creation of the network infrastructure. Second, MAC protocols are in charge of fairly and efficiently sharing the wireless channels among a number of mobile terminals. In MAC layer channel scheduling, packet collision among different users should be reduced or even completely avoided, and the bandwidth should be fully-utilized. These two goals are conflicting with each other. Therefore MAC protocols should be carefully designed to balance them based on network requirements. Third, MAC protocols should be energy-aware for extending battery lifetime. Supporting power management to save energy is required for battery-powered mobile nodes in MANETs. Actually, this is the motivation of our paper. The power conservation mechanisms for MANETs also must support multi-hop forwarding. They should be distributed, as there is no centralized control to rely on.

The remainder of this paper is organized as follows. In Section 2, we identify major sources of energy waste in MANETs and discuss low-power MAC design principles. From Section 3 to section 8, we survey and analyze different energy-aware mechanisms together with their applications in energy-aware MAC protocols. The techniques covered include channel reuse, power controlled scheme, power off mechanism, dual channels, and antenna-based power efficient schemes. Detailed analysis and comparison will also be given. We conclude this paper in Section 9.

2 Sources of Power Waste and Low-Power MAC Design Principles

2.1 Major Sources of Energy Waste

The major sources of power waste in mobile computing devices include radio communication and data processing, with radio communication often being the dominate source of energy consumption. Data processing involves the usage of CPU, memory, hard drive, etc. Its energy consumption is relatively negligible compared with that of the radios. The energy expenditure in radio communication includes the power consumed by transmitting and receiving devices of all nodes along the path from source to destination, together with their neighbors that can overhear the transmission. Actually, there is a tradeoff on energy consumption between data processing and radio communication, the two energy consuming factors [33]. For example, data compression techniques are introduced in [13] to reduce packet length and therefore achieve energy saving in radio communication, but the cost of computation is increased.

Let's first take a look at the characteristics of energy consumption in a radio interface so that we can easily understand the motivations of the energy efficient mechanisms discussed in the following sections. In mobile ad hoc networks, communication related energy consumption includes the power consumed by the radios at the sender, receiver and intermediate nodes in the route from the source to the destination. Actually, at any time a mobile node in MANETs must be in one of the following four modes: transmit, receive, idle listening, and sleep. When a node is in transmit or receive mode, it is transmitting or receiving a packet. Idle listening mode means the node is neither transmitting nor receiving a packet, but is doing channel monitoring. This mode consumes power because the node has to listen to the wireless medium continuously in order to detect the arrival of the packet that

Table 1: LUCENT IEEE 802.11 WAVELAN PC CARD (2Mbps) CHARACTERISTICS

Modes	Energy Consumption
Sleep Mode	14 mA
Idle Mode	178 mA
Receive Mode	204 mA
Transmit Mode	280 mA

it should receive, so that the node can switch to receive mode [45]. When in the sleep mode, nodes do not communicate at all. Receive and idle mode consume similar amount of power, while transmit mode requires slightly larger amount. Nodes in sleep mode consume extremely low power. As an example, we illustrate the energy consumption of different modes for the 2.4GHz DSSS Lucent IEEE 802.11 WaveLAN PC "Bronze" (2Mbps) wireless network interface card [21] in Table 1. Note that mobile ad hoc nodes must keep on monitoring the media for possible data transmission. Thus most of the time nodes must be in idle listening mode instead of sleeping. Actually, a network interface operating in ad hoc status has a constant idle power consumption, which reflects the cost of listening to the wireless channel. Many measured results have shown that the energy spent by idle listening is 50~100% of that by receiving. In other words, idle listening consumes only slightly less energy than actually receiving traffic. Thus, significant energy is consumed even when there is no traffic in the MANETs. Further, the energy expenditure for the radio interface to transit from one mode to anther is not negligible because the transition time can not be infinitesimally short. For example, the transition between transmit and receive modes typically takes 6 to 30 μs, while the transition from sleep to transmit or receive generally takes even more time (250 μs) [29]. Mode transitions have significant impact on energy consumption of wireless nodes.

Besides the power consumption in transmit, receive and idle listening, there exists other significant energy expenditure in packet retransmission, node overhearing and protocol overhead [62]. Retransmission is caused by collision. When a packet is corrupted, it must be discarded and transmitted again. Retransmission increases energy consumption. In fact, due to the lack of a centralized authority in mobile ad hoc networks, transmissions of packets from distinct mobile terminals are more prone to overlap, resulting in more serious packet collisions and energy loss. Overhearing means a node picks up packets that are destined for other nodes. Wireless nodes will consume power unnecessarily due to overhearing transmissions of their neighboring nodes. Protocol overhead is generated by packets dedicated for network control and header bits of data packets. It should be reduced as much as possible because transmitting data packet headers or control packets also consumes energy, which results in the transmission of less amount of useful data packets.

2.2 Low-Power MAC Design Guidelines

As stated in Subsection 2.1, the major energy waste comes from idle listening, retransmission, overhearing and protocol overhead [62]. Thus there is no wonder why all power-aware MAC protocols try hard to reduce energy waste from one or all of the above sources. To make a MAC protocol energy efficient, at least one of the following design guidelines must be obeyed:

- **Minimize random access collision and the consequent retransmission**

Collisions should be avoided as far as possible since otherwise the followed retransmission will lead to unnecessary energy consumption and longer time delay. Actually, one of the fundamental tasks of any MAC protocol is to avoid collisions so that two interfering nodes do not transmit at the same time. The simplest ways for collision avoidance in a general network include code division multiple access (CDMA), time division multiple access (TDMA), and frequency division multiple access (FDMA). However, for mobile ad hoc networks there exist many special issues that need to be addressed for a MAC protocol design. For example, because of the nonexistence of fixed base stations in MANETs, mechanisms to avoid collision among mobile nodes must be distributed. Since collision avoidance may result in substantial overhead, which will burn more energy, tradeoffs must be explored to achieve reasonable solution. Further analysis on channel reuse mechanisms based on scheduling will be given in Section 3. Such schemes are designed to increase the channel utility and at the same time to avoid collisions.

- **Minimize idle listening**

In typical MANET systems, receivers have to be powered on all the time. This results in serious energy waste. Since the power consumed in idle listening is significant, as indicated in Subsection 2.1, we should pay attention to the energy conservation in nodes other than the source and destination. Ideally the radio should be powered on only when it needs to transmit or receive packets, thus remove the unnecessary monitoring of the media. Recently, energy-aware MAC protocols that require nodes be in sleep mode periodically for energy conservation have been proposed. When in sleep mode, nodes neither transmit nor receive packets; but they must be woken up to idle mode first for attending traffic relay. Sleep mode requires more than an order of magnitude less power than idle mode. Hence, intelligently switching to sleep mode whenever possible will generally lead to significant energy saving. This topic will be addressed in detail in Section 5.

- **Minimize overhearing**

Wireless nodes consume power unnecessarily due to overhearing the transmissions of their neighbors. This is often the case in a typical broadcast environment. For example, as the IEEE 802.11 wireless protocol defines, receivers remain on and monitor the common channel all the time. Thus the mobile nodes receive all packets that hit their receiver antennae. Such scheme results in significant power consumption because only a small number of the received packets are destined to the receiver or needed to be forwarded by the receiver. One solution to this problem is the introduction of a control channel for the transmission of control signals that will wake up the nodes only when needed. Such dual channel mechanism will be discussed in Section 6. Another solution for overhearing avoidance is to power off interfering nodes after they hear an RTS or CTS packet [62]. [54] proposes to broadcast a schedule that contains the data transmission starting times for each mobile nodes.

- **Minimize control overhead**

Protocol overhead should be reduced as much as possible, especially for transmitting short packets [62]. Due to the large channel acquisition overhead, small packets have disproportionately high energy costs. Header compression can be used to reduce packet length, thus achieving energy savings. Since significant energy is consumed by the mobile radio when switching between transmit and receive modes, packet aggregation for header overhead reduction will be useful. When mobile nodes request multiple transmission slots with a single reservation packet, the control overhead for reservation can be reduced. Allocating contiguous slots for transmission or reception to reduce the turnaround also helps to achieve low power consumption [11].

- **Explore the tradeoff between bandwidth utilization and energy consumption**

As stated by [21], energy consumption and bandwidth utilization are substantively different metrics. But they are strongly related to each other. To conserve power, its radio must be turned off if the node does not participate in the traffic dissemination, as the energy spent in receiving and discarding packets (this happens when overhearing and idle listening) is significant. Further, to shun the energy consumption resulted from packet retransmission, a node need to be powered off if the media is busy. This greatly decreases the channel utilization, thus decreases network throughput. Therefore, the tradeoff between the bandwidth utilization and energy consumption must be exploited for throughput improvement. Scheduling the channel efficiently among neighboring nodes is a challenging problem.

To design a good MAC protocol for MANETs, designers must take into account energy, bandwidth, delay, channel quality, etc.. Other factors that play important roles in power-aware MAC protocol design include network-wide traffic pattern (broadcast versus point to point traffic, short packet vs. long packet, etc) and per node operation mode (promiscuous mode vs. non-promiscuous mode). Different tradeoffs must be explored based on application requirement (operation time, availability of infinite power supply, etc.) such that multiple factors can be considered together. We can simplify the design goal of power-controlled MAC protocols as follows: to increase the overall network throughput while maintain low energy consumption for packet processing and radio communication. We will discuss different mechanisms for power-controlled MAC protocols in detail in the following sections. All of them are trying to achieve the above goal.

3 Scheduling-based Mechanism

From the above analysis, we know that collisions occurred in MAC layer is one of the major sources of energy waste in mobile ad hoc networks. Therefore, reducing or avoiding the data link layer collision is the first issue we should consider in power-controlled MAC protocol design. This is the motivation of the scheduling-based mechanism for low power MAC. The simplest techniques in this category include frequency-division multiple access (FDMA), time-division multiple access (TDMA), and code-division multiple access (CDMA). Such classification is based on the domain in which the channel resources are shared by multiple simultaneous transmissions, as shown in Figure 1. FDMA divides the available frequency band into multiple non-overlapping channels. Nodes access the media at the same time but operate on different frequency channel. On the other hand, TDMA allocates

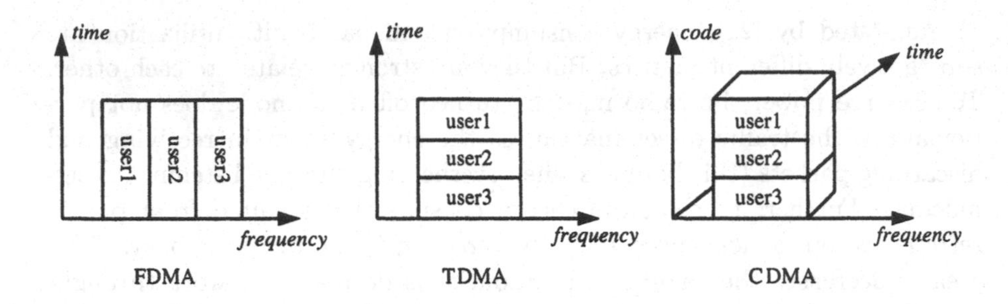

Figure 1: FDMA/TDMA/CDMA

different time slot to each user. Nodes share the same frequency band but access the media at different time slot. CDMA relies on orthogonal codes. Simultaneous transmissions can coexist on the same frequency channel at the same time if each is assigned a unique code. These three techniques have been well studied in literature. FDMA is applied to the first-generation mobile system, while TDMA and CDMA are widely used as the multi-access scheme for the second-generation system. Now, CDMA seems to attract the most attention in research. Actually, it has been shown that the capacity provided by CDMA is up to six times of that by TDMA/FDMA-based solutions [23]. Thus in this paper, we will put more effort on CDMA ad hoc networks. We will talk about the motivation of scheduling-based MAC protocols in Subsection 3.1. Then we will discuss the three techniques together with their applications in mobile ad hoc networks in Subsections 3.2– 3.4. Example scheduling-based low power MAC protocols for ad hoc networks will be examined in Subsection 3.5.

3.1 Motivation

With the introduction of FDMA, TDMA, and CDMA, wireless links occupied by different transmissions can be maintained active at the same time without collision. The scheduling in frequency, time, or code domain, is essential to coordinate the transmissions of independent users. This mechanism assigns each user a predetermined and fixed portion of the wireless bandwidth resource. Such scheduling-based assignment can eliminate the interference among different users. With scheduled access, each node can only transmit with the pre-assigned frequency or code, or at the pre-assigned set of slots. Therefore no collision in the MAC layer can occur. In this way, energy waste due to packet retransmission can be avoided. Another direct benefit of collision avoidance is the improvement of network throughput and effective channel utilization. Further, in TDMA, since nodes know their schedules ahead of time, they can simply turn off their radios to save energy if it is not the time for them to grep the channel. Thus the chance of overhearing and idle listening can greatly decreased, and the corresponding energy expenditure can be reduced too.

Compared with random access MAC protocols, in which nodes contend for the channel resource whenever they have packets to send, scheduling-based MAC protocols can achieve better energy conservation due to collision avoidance. However, the complicated control involved in the setup and maintenance of a schedule may compensate the saved energy obtained from collision avoidance, if it does not exacerbate the energy consumption. On the other hand, scheduling-based MAC protocols may not work well for bursty traffic, as the schedules are fixed, regardless of the user's real need. Another restriction on the application of these techniques is the topology dynamism. Scheduling-based protocols require that the network topology remains stable or changes slowly such that a schedule can effect for longer time to compensate the high setup overhead. Thus, they cannot be applied to highly mobile or other dynamic environments directly [25].

3.2 CDMA Ad Hoc Networks

3.2.1 CDMA

CDMA is based on direct sequence spread spectrum (DSSS) encoding technique. In a typical CDMA system, all users share the same frequency, but each sender-receiver pair is assigned a unique pseudo-random noise (PN) code [48]. Data packets are *xor*-ed with the PN code before transmission and then *xor*-ed again with the same PN code at the intended receiver to ensure the correct decoding. When each signal is assigned a distinct PN code, several coded messages can simultaneously occupy the same channel.

CDMA improves the overall system capacity by making it possible for several independently coded signals to occupy the same channel. Thus multiple users can "coexist" and transmit simultaneously with minimal interference. This feature is crucial for networks with bursty traffic, high node density, and asymmetric transmission ranges. CDMA can effectively average the interference across transmissions at different spatial scales, hence enhance capacity and relax power control requirements. For example, by using CDMA, the channel efficiency has been improved by a factor of 4 with respect to TDMA in the Qualcomm cell phone. And as mentioned earlier,

CDMA has been shown to provide up to six times as large as the capacity of TDMA or FDMA based solutions [23]. CDMA can also protect users from outside interference and jamming. Some other desirable features of CDMA include signal degradation, multi-path fading resistance, and frequency diversity [41].

However, there are also some deficiencies. First, CDMA degrades network throughput for bursty traffic. Orthogonal codes may be wasted when users have no data to send. Second, CDMA can not avoid idle listening. Nodes must keep on sensing the channel to receive possible traffic. As mentioned earlier, such idle listening consumes a large amount of energy. Third, data transmission rate is limited.

3.2.2 CDMA Ad Hoc Networks

In a CDMA ad hoc network, mobile nodes serve as CDMA transceivers. This topic has been extensively studied in recent years. As claimed by [56], analysis of a CDMA ad hoc system is quite difficult due to the uncoordinated behavior of nodes. In [6], the performance comparison between centrally-controlled and ad hoc CDMA wireless LAN networks is presented; while in [7], the performance comparison between CDMA ad hoc networks and cellular networks is given. The results show that for the same spreading gain and error control coding, CDMA ad hoc systems have higher throughput and shorter packet delay than centrally-controlled CDMA wireless LAN and cellular networks under light traffic load; however, cellular networks and centrally-controlled ad hoc systems outperform CDMA ad hoc networks when the traffic is heavy.

The advantages of CDMA in ad hoc networks are summarized in the following:

- Capacity improvement. By increasing the number of successful receptions at the link-layer, the capacity of ad hoc networks is enhanced.
- Energy saving. CDMA improves parallelism, thus decreases the possible delay a node must go through when waiting for grabbing the channel. This helps to decrease idle listening. On the other hand, predetermined schedules result in collision avoidance, which helps in reducing energy consumption.
- Routing overhead reduction. It is easy to implement multi-path routing in CDMA ad hoc networks, which reduces the overhead significantly.

However, CDMA ad hoc networks still face quite a few challenges:

- Near-Far effect [46]. When cross-correlations between different CDMA codes are non-zero, the induced multi-access interference results in collisions at the receiver. This problem can cause a significant reduction in network throughput, thus it must be carefully considered. When combined with power control issues, things become even more complex. Since each terminal can communicate with several other nodes simultaneously in ad hoc networks, the transmit power must be controlled to avoid overhearing in near-far situations.

- Code design. Distinct codes are required for different communication parties. Thus a spreading-code protocol is needed to decide which code to use for packet transmission and which to use for monitoring the channel in anticipation of packet reception [56]. Such problem is usually formulated as Graph Coloring, a well-known NP-complete problem.

- Recoding. Due to the topology dynamism of ad hoc networks, recoding is necessary. Otherwise collisions may happen. However, recoding brings about additional overhead on channel acquisition at the receiver side and synchronization between the sender and the receiver. Such process is expensive and must be minimized as well [26].

3.3 TDMA

In TDMA systems, the time axis is divided into a number of frames with fixed length. Each frame is further divided into a number of time slots. A user can only transmit in the pre-assigned time slots. By this way, interference from neighboring nodes is reduced. Compared with random access MAC protocols, TDMA protocols is advantageous in energy conservation. This is because dedicated time slots for each user have been pre-determined. Therefore collision and the corresponding overhead can be avoided.

However, the scalability of TDMA is not as good as that of random access scheme [62]. Applying TDMA protocols in wireless systems usually require all nodes to form real communication clusters, such as in Bluetooth and LEACH [30]. The control of inter-cluster communication and interference is complex, especially when nodes are mobile as in MANETs. Further, TDMA requires the sender and receiver to be perfectly time-synchronized, which is not trivial. Designers need to consider factors such as the timing difference, clock shift, propagation delay, etc..

3.4 FDMA

FDMA is a technology that transmits multiple signals simultaneously in time domain. Each signal only occupies a portion of the entire available bandwidth. The basic idea is stated as follows. The spectrum is divided into many channels using discrete frequencies, called *subcarriers*. Different terminals are allocated a different number of subcarriers, depending on their data rate requirement. By this way, all terminals can transmit simultaneously without collision.

Orthogonal Frequency Division Multiplexing (OFDM) is an emerging modulation scheme. OFDM distributes the data over a large number of carriers that are spaced apart at precise frequencies. This spacing prevents the demodulators from seeing frequencies other than their own. Such a design results in high spectral efficiency, which is important to the limited wireless resources. The IEEE 802.11a standard uses OFDM in the 5.8-GHz band [70]. OFDM has also been included in the standardization for 3G systems [71].

3.5 Related Protocols

3.5.1 CA-CDMA

In [41], a novel *Controlled Access* CDMA (CA-CDMA) protocol is proposed to solve the notorious near-far problem for CDMA ad hoc networks. In CA-CDMA, all nodes use a common spreading code in the control channel to contend for the channel resource with a modified RTS/CTS mechanism. Data packets are transmitted in the data channel encrypted with different terminal-specific codes. By using two non-overlapping frequency bands, it is easy to get a code in the control channel orthogonally with each code in the data channel. Hence, nodes can transmit and receive at the same time over the control and data channels no matter how much the signal power is. This makes it possible to allow interfering nodes to transmit simultaneously. CA-CDMA adjusts the transmitting power according to the channel-gain information so that concurrent transmissions are possible. Figure 2 shows the code assignment scheme used in CA-CDMA [41]. The results show that CA-CDMA can improve the network throughput by 280% and save 50% energy consumption when compared with IEEE 802.11.

Energy savings are achieved in CA-CDMA by using orthogonal codes to achieve the goal of channel reuse. Based on the orthogonality between the control channel and the data channel, concurrent transmissions are made possible such that the throughput is improved with the same or less energy

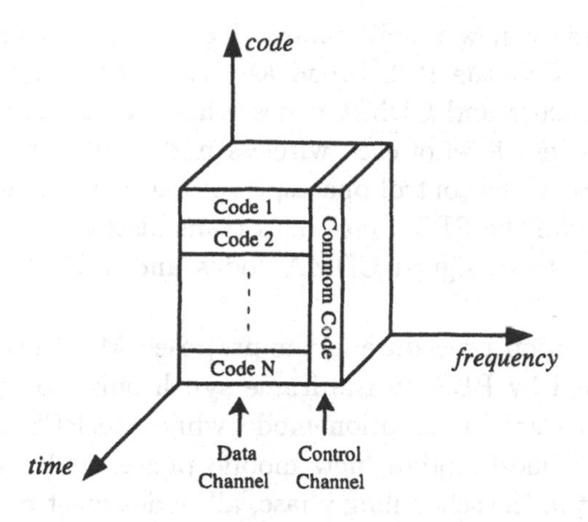

Figure 2. Code Assignment in CA-CDMA

consumption. In fact, CA-CDMA utilizes power control and dual channel mechanisms, which will be discussed in detail in the following Sections 4 and 6.

3.5.2 Energy-Efficient MAC Based on CDMA/TDMA

In [32], a MAC protocol based on CDMA/TDMA for energy-limited single-hop ad hoc networks is proposed. The main idea is based on the observation that radio module consumes more power in transmission mode than in reception mode, and the least power in sleep/idle mode. The motivation is to inform each terminal through MAC layer control when to wake up from sleep mode and when to sleep from transmission/reception mode in order to save battery power. This motivation is similar to that of the power control mechanism to be discussed in the following section. In this protocol, such a design objective is achieved by a scheduling and reservation method based on CDMA/TDMA.

There are two different kinds of mobile terminals defined in this MAC protocol: a pseudo base stations (PBS) and normal mobile nodes. The PBS is introduced to emulate a base station (BS) in infrastructure networks. It is responsible for the centralized control of collecting requests from mobile terminals and allocating CDMA codes and TDMA slots. The protocol is sketched below. The PBS first broadcasts synchronization message to all the other terminals at the start of each frame. Terminals with packets to deliver must send request and power level update messages to the PBS in the request/update/new mobile phase. New terminals register themselves with the PBS. Next the PBS broadcasts the scheduling information containing TDMA slots and CDMA codes whose assignment is based on the priority and battery level of each wireless node. After exchanging information during these three control phases, terminals communicate directly with no mediation from the PBS. Since data transmission in the communication phase uses only pre-assigned CDMA codes and TDMA slots, no collision will occur.

The transmission procedure in the proposed MAC protocol is based on frames controlled by PBS. In the frame synchronization phase, all normal mobile terminals are in reception mode while the PBS is in transmission mode. In the request/update/new mobile phase, nodes with no traffic to send sleep. But in the scheduling phase, all nodes must be waken-up for the reception of the schedule. A node can sleep again if it is not the source and target of any traffic. This strategy avoids the idle listening and overhearing for a traffic-free mobile. But the overhead incurred at the three control phases consumes non-negligible power. On the other hand, the schedule for data transmission ensures collision free communication, which reduces the energy for packet retransmissions. The direct data traffic between mobile nodes is independent of the PBS, further saving energy.

PBS plays an important role in power conservation. First, The PBS selection/reselection is based on battery power. Only high power node can serve as the PBS. Second, the scheduling is based on the power level of each normal terminal. Nodes with lower power have higher priority in allocating radio resources. This can effectively reduce the possibility of early termination of low power nodes. By this way, the battery energy of high power terminals is shared within all nodes and the network lifetime is extended. Simulation results show that this protocol can effectively protect mobile terminals from collision and retransmission when traffic is heavy. Low power terminals work better than high power nodes, especially in energy efficiency. However, with the centralized control by the PBS, this protocol can only support single-hop wireless ad hoc networks.

4 Power Control Techniques

Besides scheduling-based schemes, all remaining MAC mechanisms are contention-based (also named random access MAC), since mobile nodes must contend for the channel resources before transmission. Such contention schemes provide a simple but robust method for mobile nodes to efficiently share wire-

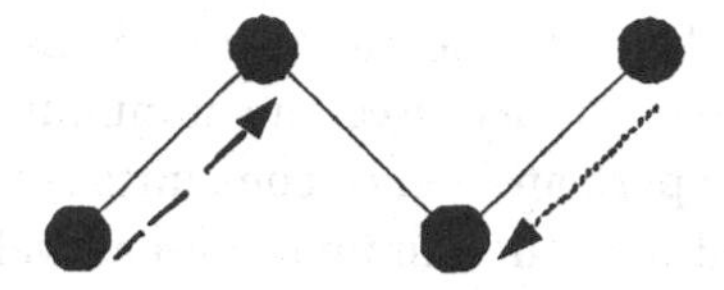

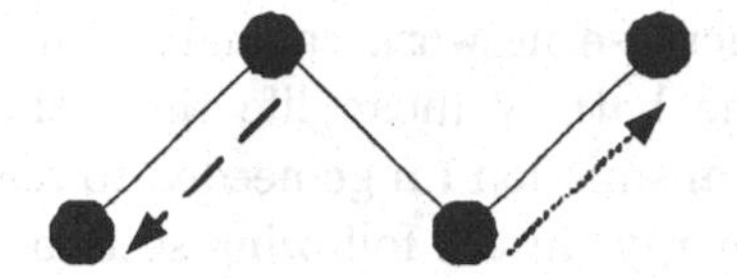

Figure 3: 802.11: Spatial reuse with CSMA/CA

less links and are especially suitable for the dynamic topology of MANETs. The introduction of contention, However, also means additional power for channel sensing, ACK schemes, retransmissions etc. Thus, conservations mechanisms are especially important to the design of contention-based MAC protocols. From this section on, we will focus on some mechanisms for energy conservation in contention-based MAC protocols. First, we will discuss the power control mechanism which allows nodes to modify the transmission power levels for network capacity improvement.

4.1 Mechanism

The most widely used MAC protocol in MANETs is the IEEE 802.11 DCF (CSMA/CA+ RTS/CTS) mechanism. In 802.11, mobile nodes try to avoid collisions with carrier sensing before transmission. If the channel is busy, the node will defer transmission and enter into backoff state. Otherwise, the nodes will begin the RTS/CTS dialog process to capture the channel and then transmit the packets.

The CSMA/CA scheme effectively reduces the amount of possible collisions. The RTS/CTS is also helpful since it will reserve the channel spatially and temporarily. RTS/CTS exchange is helpful in avoiding hidden terminal problem, since any node overhearing a CTS message cannot transmit for the duration of the transfer. However, this process severely limits available bandwidth. Measurements show that the flow can only obtain about 2% of available bandwidth. Actually, such artificial restrictions may bring about a waste of the wireless link resources. For example, in the following examples (Figure 3), the two data flows are compatible, but this is not allowed by the CSMA/CA scheme [72].

Hence, a new kind of energy conservation scheme is proposed to reduce the active link power levels to minimum values so that more than one link can access the channel simultaneously thereby excluding the prevention from parallel transmission by CSMA. The low power transmission scheme can help increase network capacity. Furthermore, such a scheme can also increase the battery mean life time, since each node only acquires the minimum transmission range needed to reach its counterpart and hence consumes less energy. In the following section, we will detail how the minimal power level can be found and applied.

The first benefit is the energy conservation when every node tries to transmit at the minimal level. This brings another benefit of spatial reuse improvement of the wireless channel, by letting more users transmit at the same time. Thus the network capacity is improved. And thirdly, by reducing the power to a minimal level, the transmission range needed is also reduced to the minimal one, thus helping in avoiding some unnecessary overhearing and reducing co-channel interference with neighboring nodes. All the above benefits help achieve the goal of energy savings in the MAC layer.

However, such a scheme will result in an inherent asymmetry. Poojary et al. [47] investigated the system performance using the IEEE 802.11 MAC protocol in which different nodes may transmit at different power levels. Results show that such scheme can lead to unfairness, since low power nodes may be overwhelmed by the higher power nodes in accessing and using channel resources. Collisions may also happen and in some cases may degrade the performance. This is even worse when the mobile nodes are of high density. A simple idea to overcome this deficiency is to broadcast CTS or its variation multiple times such that more nodes clear the channel for the initiating node, as proposed by [47]. However, simulation study shows that the potential gain of this strategy is outweighed by the corresponding overhead.

Another problem is that such multiple power levels scheme may cause additional collisions (Figure 4, [20]). When node A transmits with a low power level, node A may not detect the transmission and hence possible collision may occur because of the interference between the transmission from a to b and from A to B. This may be solved by using adaptive power control loop scheme [1], which adjusts transmission power adaptively according to a weighted history data. Another solution is to use an optimal common transmit power level for the whole network instead of per-node transmission power levels. COMPOW is one such example [43].

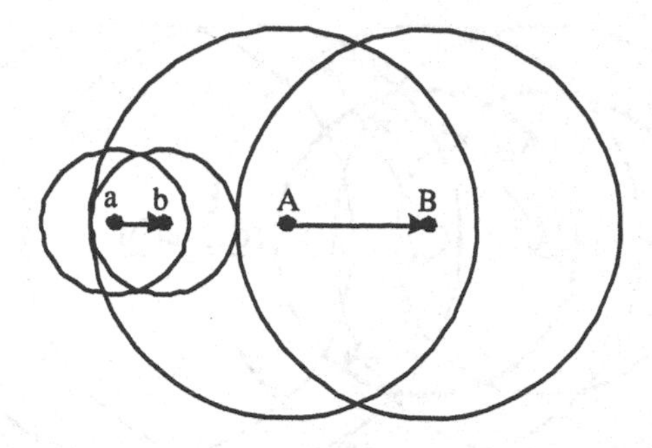

Figure 4: Problem with different power levels

4.2 Related Protocols

4.2.1 PCM

A Power Control MAC (PCM) protocol is proposed in [34] to allow per-packet selection of transmit power. In PCM, RTS/CTS packets are transmitted with a max power level, P_{max}. But for data packets, they are transmitted with a lower power level. In order to avoid a potential collision caused by the reduced carrier sensing zone, during the DATA packet transmission PCM periodically increases the transmission power to p_{max}. ACK packets are transmitted with the minimum required power to reach the source node. Figure 5 shows the power level used in PCM.

By periodically increasing the power level for data transmission, PCM effectively reduces the amount of possible collisions. This way, retransmission is avoided as much as possible, and correspondingly, the goal of energy savings is achieved. Results show that PCM can achieve a throughput comparable to the IEEE 802.11 but with less energy consumption. However, PCM requires a frequent increase and decrease in transmission power levels, hence the implementation is not easy.

4.2.2 PCMA

The Power Controlled Multiple Access (PCMA) Protocol [40] proposes a flexible "variable bounded power" collision suppression model and allows variable transmit power levels on a per-packet basis. Similar with IEEE 802.11, PCMA uses RPTS/APTS handshake to determine the minimal transmission power required for successful packet reception. The difference lies in that PCMA introduces a second channel, the busy tone channel, to implement the noise tolerance advertisement. During data transmission pe-

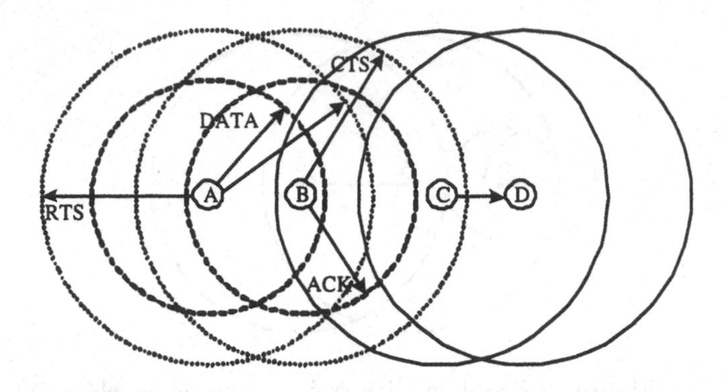

Figure 5: PCM Scheme: Data packets are transmitted with a periodically increased power level

riods, each active receiver will periodically send a busy tone to advertise the maximum additional noise power it can tolerate. Any potential transmitter must first sense the channel for busy tones to determine the upper bound of its transmit power for a minimum time period (determined by the frequency with which the busy tones are transmitted). Actually, PCMA uses the signal strength of a received busy tone message to bound the transmission power of neighboring nodes. This way, power control mechanism is realized and spatial reuse is achieved.

PCMA works effectively in energy conservation since it allows more concurrent data transmission compared with IEEE 802.11 standard by adapting the transmission ranges to be the minimum value required for successful reception on the receiver side. Results show that PCMA can improve the throughput performance by more than a factor of 2 compared to the IEEE 802.11 for highly dense networks. The throughput gain over 802.11 will continue to increase as the connectivity range is reduced. What's more, the power controlled transmission in PCMA helps increase channel efficiency at the same time preserving the collision avoidance property of multiple access protocols.

Energy efficient mechanisms using busy tone is discussed in Section 6. Similar to PCMA, [67] also uses the power control schemes with busy tone.

4.2.3 DCA-PC

A Dynamic Channel Assignment with Power Control (DCA-PC) is proposed in [66]. Similar with PCMA, this power control protocol uses one control channel to transmit all the control packets (RTS, CTS, RES etc). The difference is that multiple data channels are assigned on demand. In DCA-PC, the pair of source and destination nodes uses a RTS/CTS dialogue to decide which channel to grab and which power level to use for data transmission. A RES message is used to reserve the data channel. Then data packets and ACKs are transmitted on the reserved data channel using the assigned power level.

In DCA-PC, all the control packets are transmitted with a maximal power level in order to warn the neighboring nodes of the communication. The data packets are transmitted with proper power levels for channel reuse. Results show DCA-PC can achieve a higher throughput with the same energy consumption compared to DCA [65] protocol, which includes no power control mechanisms.

DCA-PC is the first protocol to realize the mechanisms of power control and multi- channel medium access together in MAC protocols of MANETs. By using multiple channels, it is easier to increase the throughput, reduce normalized propagation delay per channel, and support quality of service. We discuss the multi-channel scheme for energy efficient MAC in the following Section 6.

4.2.4 PCDC

The Power Controlled Dual Channel (PCDC) Medium Access Protocol [42] also uses two channels like PCMA, one control channel and one data channel. However, PCDC is the first to utilize the inter-layer dependence between the MAC and network layers to provide an efficient and comprehensive power control scheme. The idea is based on the observation that the transmission power has direct impact not only on the floor reserved for the next transmission but also on the selection of the next hop node. Hence, the interaction between the MAC and network layers can help for an effective power control scheme. In order to select the lowest possible power level while maintaining the network connectivity and proper MAC function, PCDC uses a distributed algorithm to compute a minimal connectivity set(CS) for each node. By controlling the transmission power of a route request (RREQ) packet, PCDC broadcasts the RREQ packets to the connectivity set only, hence the MAC can effectively control the set of candidate next-hop nodes.

Since RREQ packets are only transmitted to the nodes in the connectivity set, it is easy to control the potential contention. Hence, the process to find the destination in PCDC has low overhead, less contention and less power consumption. Compared with IEEE 802.11, PCDC achieves improvements of up to 240% in channel utilization and over 60% in throughput, and a reduction of over 50% in energy consumption [42]. However, the adaptive computing of the connectivity set may impose a lot of computing workload for each node.

4.2.5 Other Power Control Techniques

We have discussed the power control mechanism and some corresponding protocols used in MAC layer in the above section. Actually, power control can also be used in network layer, for example, COMPOW [43] and clustering scheme [36].

As the analysis in Section 4.1 shows, different transmission power may cause additional collisions. Hence, it is proposed to use an optimal common transmit power level by the nodes in the networks. According to [43], a per node throughput of $O(\frac{1}{\sqrt{nlogn}})$ can be obtained using a common power level, while with per node optimal levels the upper bound of the throughput is $O(\frac{1}{\sqrt{n}})$. Thus, common power for all nodes is near optimal.

The Common Power (COMPOW) Protocol [43] proposes to use a common power level to ensure bidirectional links. Every COMPOW node runs several instances of a proactive ad hoc routing protocol (eg. DSDV), with each at a different transmit power level. A set of routing tables are maintained, each contains the corresponding connectivity information. The optimum power is defined to be the minimum power whose routing table has the same number of entries as that of the routing table at maximum power level. which achieves the same level of network connectivity as the

COMPOW is designed to maximize the traffic capacity of the network, provide power aware routes and reduce the contention at the MAC layer. Thus, energy consumption is reduced at the same time that network capacity is increased. However, COMPOW completely relies on routing layer agents to converge to a common power level. This usually incurs significant overhead, especially for constantly moving mobile nodes. In clustering ad hoc networks, such a common power level may be more difficult or even infeasible to achieve, because of the hierarchical architecture.

Clustering by power control is used for non-homogenous ad hoc networks [36]. Clustering is used for non-homogenous scenarios. It classifies nodes hierarchically into clusters, dominated by "cluster-heads" and connected by "gateways". Cluster-heads are used as base stations to emulate power control as in an infrastructure network. The main idea of such a kind of power control is based on the observation of the extremely high energy cost of an idle interface. Thus by emulating the infrastructure (BSS) operating mode through clustering, energy consumption can be reduced by letting the cluster-heads control the intra-cluster data transmission. Different power levels are used in the intra-cluster and inter-cluster communication with a high power level among the cluster-heads and a low common power level among most of the intra-cluster communication. On the other hand, clustering can also help in reducing the route discovery overhead. They both help reduce energy consumption. However, the dynamic topology of MANETs brings about quick connectivity changes, which means a high overhead for cluster maintenance. Thus, protocol designers must consider latency and packet loss issues. Some new problems also arise. For example, how to form routing backbones to reduce network diameter, how to abstract network state information to reduce its quantity and variability, etc [17].

5 Power Off Mechanism

5.1 Mechanism

In typical wireless systems, receivers have to be powered on at all time to detect any possible signals that target them. However, this "always-on" results in significant unwanted energy consumption. As mentioned in Section 2, much more energy is consumed in idle state than in sleep state. For example, measurements have shown that idle listening consumes 50-100% of the energy required for receiving [62]. Therefore, idle listening should be avoided as much as possible. Ideally, the wireless radio is powered on only when there are packets waiting to be transmitted or received. Otherwise, the radio is powered off and the node is in sleep/doze state. By this way, the wasted monitoring effort is minimized.

The motivation can be further explained as the following. If the radio is powered on at all time, the ongoing transmission is overheard by all the neighbors of the transmitter, which frequently happens in ad hoc networks. In this scene, the neighboring nodes have to monitor the channel and consume power even though the packets are not directed to them. A large amount of energy is consumed unnecessarily in this case. Take a simple example: If a transmitter T has n neighbors, then the transmit energy needed for a m-packet transmission is $m * (E_{trans}$(transmitting energy per packet)$+E_{recv}$(receiving energy per packet). However, since the transmission may be heard by all its neighbors, the actual power will be $m*(E_{trans}+n*E_{recv})$. Obviously, it produces energy waste of $m*(n-1)*E_{recv}$.

Based on this observation, a new kind of power conservation mechanism is proposed in which some nodes are allowed to stay in doze/sleep state when they are not actively transmitting, receiving, or waiting for a channel

([11], [73], [31]). Obviously, this power off mechanism can save battery power. Thus, it helps prolong the lifetime of hosts and the whole network system. With the current hardware technology, this scheme requires the CPU, transceiver and other hardware to be switched off.

Several different methods have been proposed to realize this mechanism. One method is to exploit an additional control channel to notify the wireless terminals when to power off or power on, such as PAMAS [53, 52]. PAMAS will be studied in this section and the section covering the dual channel mechanism (Section 6). Another method is to periodically power down the terminal ([74], [75]). As proposed in [24], periodically shutting down a host results in great power saving during the sleep period. Based on this, considerably longer beacon intervals may be acceptable to realize the scheduling. Both the sender and the receiver can shut down the radio according to its schedule. Scheduling helps to avoid the effort of determining when to power the radio down. But this technique requires time synchronization, which makes practical scheduling hard. One solution is to use a master host to store the data while the destination is powered down and forwarding the data when it powers up. Another solution is to provide a framing structure for hosts to synchronize their active periods [74], [75]. However, scheduling locally or globally is expensive and is not so easy to implement in ad hoc networks.

Other techniques to realize the power off mechanism include multi-sleep mode [12] and power mode scheduling method [45]. Waking up terminals from sleep mode can be done with synchronization [55] or without synchronization [25]. We will discuss the protocols in the following subsection.

However, there exist problems to be considered for this power conservation mechanism, including when to switch to sleep mode, how long to sleep, and how a host can send or receive packets when in sleep mode. Actually, it is difficult to design a protocol that allows hosts to spend most of their times with the receivers switched off.

5.2 Related Protocols

5.2.1 PAMAS

The Power-aware Multi-access Protocol with Signaling (PAMAS) [53, 52] is proposed to conserve battery power by powering off nodes that are not transmitting or receiving. This is a combination of the original MACA protocol [35], and the use of a separate signal channel - the "busy tone" channel [64]. By using busy tone, the terminals are enabled to determine when and how long they should power off the radio. The determination must obey the following rules: If a host has no packets to transmit, then it should power the radio off if one of its neighboring nodes begins transmitting. Similarly, if at least one neighboring node is transmitting and another is receiving, the host should also power itself off because it cannot transmit or receive packets (even if its transmit queue is nonempty).

In the proposed protocol, each host makes the decision whether and when to power off the radio independently. As proposed in [53, 52], a host knows whether a neighboring node is transmitting because it can hear the transmission over the channel. Similarly, a host (with a nonempty transmit queue) knows if one or more of its neighbors is receiving because the receivers should transmit a busy tone when they begin to receive packets (and in response to the RTS transmissions). Thus, a host can easily decide when to switch to the sleep mode. And, PAMAS also gives several factors to determine the length of time for which nodes can be in sleep mode: empty transmit queue and t_probe(l) control packet.

The results show that PAMAS works effectively in power conservation. It achieves power saving from 10% (when the network is sparsely connected) to almost 70% (in fully connected networks) without affecting the delay and throughput behavior of the basic protocol.

5.2.2 S-MAC

An energy-efficient MAC protocol for wireless sensor networks, called S-MAC is proposed in [62]. Different from PAMAS, S-MAC uses the scheme of periodic listen and sleep to reduce the energy consumption by avoiding idle listening. However, this requires synchronization among neighboring hosts. And the latency is increased since a sender must wait for the receiver to wake up before transmission. But S-MAC uses synchronization to form virtual clusters of nodes on the same sleep schedule. This technique coordinates nodes to minimize additional latency.

Another difference from PAMAS is that S-MAC uses the in-channel signaling to put the nodes in sleep mode when its neighboring node is in transmission. The in-channel signaling helps reduce the overhearing problem and avoids the use of additional channel resource.

Compared with 802.11, S-MAC reduces the energy consumption by up to 50% for heavy traffic; and much more energy is saved for light traffic. It surely has very good energy conserving properties comparing with 802.11. Additionally, it can make trade-offs between energy and latency according to traffic conditions. This is a useful feature for sensor networks.

5.2.3 PicoNode's Multi-Channel MAC

A low power distributed MAC is proposed in UC Berkeley's PicoNode project [25]. A power saving mode based on waking up radio without synchronization is used in this MAC protocol. With a separate wake up radio, the normal data radio can be powered down when it is in idle listening state. Multiple channels and CSMA/CA are combined in the MAC for efficient energy usage. The multi-channel spread spectrum helps reduce collisions and retransmissions. It also helps reduce delay and increase throughput (see Section 6). The exploit of random access results in the avoidance of synchronization since it does not require any topology knowledge. Therefore there is no overhead in exchanging schedules and reservation information. All these measures help in energy consumption reduction. Simulation results show the proposed protocol can reduce the power consumption by 10-100 times compared with existing MAC protocols with traditional radio.

As defined by the MAC protocol, each terminal in the network is either in "mobile mode", or "static mode". Mobile nodes periodically broadcast a beacon through the wake-up channel to keep the neighboring nodes awake, thus maintaining a dynamic active zone within two hops. Channel assignment is conducted as the problem of two hop coloring in graph theory. Static hosts in the active zone remain awake. They go back to sleep mode again when no beacon has been received for a predefined period. Under two instances a node can be waken up: it has packets to send out, or it will receive packets from a neighbor. A node can be woken up by itself, or by a beacon from a neighboring node through the wake-up radio channel.

5.2.4 Power Management Using Multi Sleep States

A distributed power management policy is introduced in [12] for ad hoc networks. This policy aims to maximize energy conservation in battery-powered devices while satisfying the required traffic quality of service.

In this paper, L different states are used: the first $L-1$ states are sleep states, while the L-th state corresponds to the active state in which nodes can transmit or receive packets. Each sleep state is characterized by a certain amount of power consumption and a delay overhead. The deeper the sleep state, the less the power consumption and the longer the time to wake up. According to the desired QoS and its own battery status, each terminal must choose an appropriate sleep state when it has been idle for a certain time period equal to or greater than the corresponding timeout value. The sleeping node will switch back to the active state when it receives the signal from the Remote Activated Switch (RAS).

Simulation is conducted to study the power gain in a simple network scenario which assumes L = 4. Results show that power gain as high as 24% can be obtained even at high traffic load. However, this gain is achieved at the expense of a limited additional delay.

Compared with periodic sleep mechanisms, the biggest advantage of this power management policy is perhaps the remote signaling scheme. By using the Radio Frequency (RF) tags technology, this policy avoids the needs of clock synchronization. In this way, nodes are woken up only when necessary, rather than switched back to an active state periodically to check for potential traffic.

6 Multi Channel Mechanism

6.1 Mechanism

As mentioned before, the main function of MAC layer protocols is to control and coordinate the multiple access of wireless terminals to share the communication medium, while at the same time maintain high network utilization. Most MAC protocols assume that there is only one channel shared among different mobile nodes in ad hoc networks. Thus the essential design goal is to increase the channel utilization while avoid hidden terminal and exposed terminal problems. As shown in Figure 6, hidden terminals and exposed terminals cause collisions if no measures are taken. This problem is more serious if transmission delay is longer. In MAC layer, unnecessary collisions should be avoided, since retransmissions cause additional power consumption and further increase packet delay. MAC protocols based on RTS/CTS, such as [35], [8], [22] have been proposed to alleviate these problems. However, as the number of mobile terminals increase, more energy will be consumed for channel contention and the network performance will degrade quickly. On the other hand, as explained in the following, RTS/CTS-based protocols do not completely solve the hidden terminal and exposed terminal problems.

As shown in Figure 6, exposed terminals are allowed to send their RTS packets. However, they could not receive any CTS replies if another node is transmitting on the same channel. Similar scenario happens on the hidden terminals: they are forbidden to access the channel because they can not reply to RTS packets. Since the in-band transmission of RTS/CTS packets inhibits the data transmission of the exposed terminals and the data reception of the hidden terminals, the introduction of an additional control channel may be a proper solution to relieve the hidden terminal and

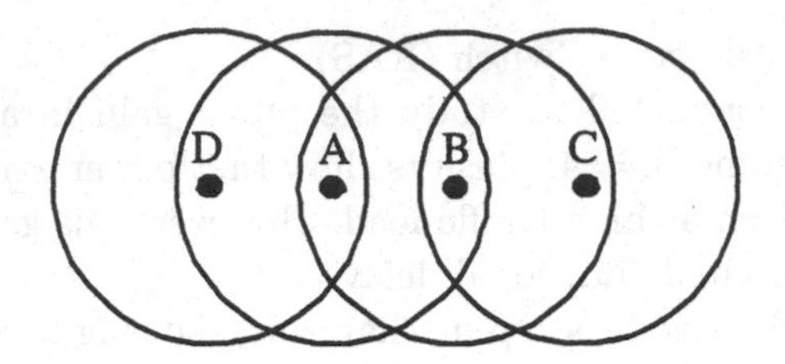

Figure 6: (a) Hidden terminal: A is sending packets to B. Since C is in the range of B but not A, it cannot sense the on-going transmission. Therefore collisions may happen at B if C transmits before the termination of the traffic from A to B. In this case, C is referred as the hidden terminal that needs to be notified explicitly. (b) Exposed terminal: A is sending packets to D. B is in the range of A and hence delays its transmission to C. B is called the exposed terminal. The postpone of B's transmission is unnecessary: these two transmissions can happen at the same time without collision.

exposed terminal problems. Motivated by this observation, multi channel mechanism has been proposed. With an additional control channel, the hidden/exposed problem is avoided. Therefore the corresponding unnecessary energy consumption is conserved.

One notice should be mentioned here: "channel" is used upon a logical level [65]. From the physical level, transmission can be based on CDMA or FDMA, and hence the channel can be a frequency band or an orthogonal code. In this section, we disregard the transmission technology and discuss those mobile hosts that can access multiple channels at the same time.

6.1.1 Basic Multi Channel Scheme

An ad hoc network built on multi channel scheme can be considered as a system composed of one control channel, together with one or more data channels. In other words, the overall bandwidth is divided into one control channel and n data channels. The introduction of the control channel is to resolve the contention that may happen on data channels and to distribute data channels among mobile nodes. Data channels are used to transmit data packets and acknowledgements. A special case is the "Dual Channel" scheme in which only one data channel is used along with a control channel. Examples include [18] and [67].

6.1.2 Busy Tone Scheme

Another scheme to resolve the contention on data channels is to use a busy tone channel. Busy tone scheme is first introduced in [59] that uses a busy tone to solve the hidden terminal problem. However, this protocol, named as Busy Tone Multiple Access (BTMA), relies on a centralized network operation.

Later, Wu and Li [64] propose a Receiver-Initiated Busy-Tone Multiple Access scheme (RI-BTMA). In this protocol, the busy tone provides two functions: to acknowledge the request for channel access and to prevent transmissions from other nodes. However, the performance of RI-BTMA is dependent on the synchronization, which is usually not easy to achieve in a distributed system. Synchronization is especially hard when considering the mobile behavior of the ad hoc networking environment.

Another scheme is proposed in [67], which combines the busy tone scheme with power control technique. In this protocol, the sender transmits the data and the busy tone at the minimum power level, while the receiver transmits its busy tone at the maximum power level. Each neighboring node estimates the channel gain from the busy tone and is allowed to transmit if its transmission is expected to add no more than a fixed "noise" value to the ongoing reception. This combination of busy tone and power control prevents channel interference and hence helps to reduce the power consumption.

6.2 Related Protocols

6.2.1 DCA

Wu et al. [65] propose a Dynamic Channel Assignment(DCA) protocol that assigns channels dynamically in an on-demand style. This protocol exploits one control channel to resolve contentions on data channels and assign data channels to mobile hosts. Multiple data channels are available for data transmission. In this protocol, all data channels are equivalent with the same bandwidth. Each host has two half-duplex transceivers, thus it can listen on the control channel and its data channel simultaneously.

This protocol is sketched below. For a mobile node A to communicate with B, A sends a RTS to B carrying its free channel list (FCL). Such list includes all information about the data channel condition around A. Then B matches this FCL with its channel usage list(CUL) to select a data channel (if any) for subsequent communication and replies A with a CTS. After receiving B's CTS, A sends a RES (reservation) packet to inhibit its neighborhood from using the same channel. Similarly, the CTS inhibits B's neighbors from using that channel. All these message are transmitted on the control channel. After this handshake protocol is done, DATA packets and their ACK messages are exchanged on the selected data channel.

Channels are assigned on demand in this protocol. There is no need for clock synchronization. Thus channels are used with little control message overhead. Results show that DCA suffers less collision and corruption compared with a simple 802.11-like multi channel protocol. The introduction of the control channel and multi data channel helps to reduce unwanted power consumption.

6.2.2 DBTMA

In Dual Busy Tone Multiple Access protocol (DBTMA) [18] [28], two busy tones, namely transmit busy tone and receive busy tone, are placed on the available spectrum at different frequencies with enough separation.

The receive busy tone provides two functions: 1) Acknowledge the sender that the channel has been successfully acquired, and 2) Notify its neighboring nodes of the following transmission and provides continuous protection for the on-going traffic. The transmit busy tone is used to protect the RTS packets. With these two busy tones, exposed terminals can establish their own transmission, since there is no need for them to monitor the channels to receive the acknowledgment from their intended receivers. Instead, the acknowledgment of the successful channel request will be sent by means of the receive busy tone. Furthermore, the hidden terminals can reply to the RTS requests by simply setting up its receive busy tone. Power control technique is also exploited in DBTMA. Simulation results show that DBTMA protocol is superior to RTS/CTS-based protocols, such as MACA [35], MACAW [8], and FAMA-NCS protocols, which works on a single channel. DBTMA achieves the performance gain as high as 140% over MACA and FAMA-NCS, 20% over RI-BTMA. It also reduces the number of possible collisions and corruptions. However, this scheme requires hardware support. Additional busy tone transmitters and sensing circuits need to be incorporated into each wireless terminal.

7 Antenna-based Mechanism

7.1 Mechanism

Antenna is an effective way used in ad hoc networks for energy conservation. There are a lot of factors resulting in energy consumption such as transmission/receiving, collision, interference and so on. To save energy for transmission and receiption is the main goal of antenna design. Different antenna designs lead to different reduction in energy consumption. Three kinds of antennas are discussed in this section, namely omni-antenna, directional antenna, and smart antenna. Omni-antenna is the pervasive way used in ad hoc networks due to its simplicity, although it consumes the most energy among these three kinds of antennas. Compared with omni-antenna, directional antenna achieves more energy saving by distributing the energy directionally and purposely. The most energy conserving method may be smart antenna that has more "intelligence" than directional antenna has.

Significant improvements and benefits can be achieved by using directional antennas or smart antennas. However, it is important to note that the deployment of directional or smart antennas involves additional complexity. Despite the complexity, typically there are many gains in throughput and power consumption reduction.

7.1.1 Omni-Antenna

Omni-antennas radiate or receive energy equally well in all directions. That is, all hosts have a 360 degree coverage angle and do not need to aim at each other for communication. The advantage of this approach is its simplicity. However, a lot of energy is wasted this way, since the power is broadcasted towards all directions and therefore attenuates rapidly with distance. Furthermore, it causes unwanted interference. It may be therefore advantageous to use directional antennas or smart antenna instead.

7.1.2 Directional Antenna

Directional antennas, with the advantage of reducing unwanted interference, can provide higher gains than that of omni-directional antennas. Also, directional antennas have the capability to receive/transmit more energy in one direction as compared to other antennas. In addition, the use of directional antennas allows hosts to communicate using less power than omni-directional ones, because the power savings of a directional antenna over an omni-directional one depend on the primary beam/lobe and the suppressed

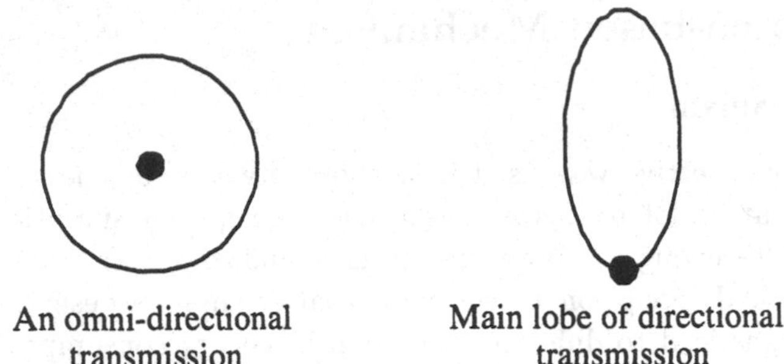

Figure 7: Antenna-based transmission

secondary lobes [2]. These characteristics can potentially provide larger frequency reuse, reduce collisions, interference. So directional antennas can be useful in power saving and increasing host and network lifetime in wireless ad hoc networks.

MAC design with directional antennas in ad hoc network is a relatively new area. Some researchers have proposed the use of directional antennas in packet radio networks to achieve better performance. In [68], directional transmissions were proposed in a slotted ALOHA packet radio network. Thereafter, more attention is paid to the MAC layer protocol design using sectorized and beamforming directional antennas for mobile ad hoc networks ([4], [50], [37], [49], [15], [57]). The usage of directional antennas offers benefits to mobile ad hoc networks since it helps isolate independent transmissions in the network and thereby increase the network performance.

7.1.3 Smart Antenna

Smart Antennas can null out interference from other hosts. A smart antenna contains an array of antennas elements, and each element has an associated "eight" that determines the weight. The antenna has complex signal processing techniques to decide which elements to receive signals (or transmission) from and how much power to use on each element.

7.2 Related Protocols

7.2.1 Directional MAC

In [37], directional antennas are applied to the IEEE 802.11a MAC protocol. RTS, data and ACK packets are sent directionally and a better performance is achieved than current MAC protocols since it allows simultaneous transmissions that are not allowed by the current MAC protocols.

The Directional MAC protocol works under the following assumptions: All the terminals in a region share a wireless channel and communicate on the shared channel. Each node is equipped with multiple directional antennas. Transmissions from two different nodes will interfere at some node X, even if at X, different directional antennas are used to receive the two transmissions. Simultaneous transmissions to different directions are not allowed at any node. Under these assumptions, several possible cases are considered and two different schemes are proposed in [37]: Directional MAC scheme 1 for using only directional RTS (DRTS) packets, and Directional MAC scheme 2 for using both DRTS and omni-directional RTS (ORTS) packets. The use of ORTS and CTS packets allows the corresponding recipients to determine the direction of the transmitters. These directions are then used for directional transmission and reception of the data packets. [37] also discussed an optimization using directional Wait-to-Send (DWTS) packets to prevent unnecessary retransmissions of RTS packets. However, it relies on an accurate tracking and locating technology, such as GPS or periodic location beaconing, which may be impossible in some cases.

7.2.2 DMACP

In [44], a new directional antenna based MAC protocol with power control (DMACP) is proposed which uses directional antennas along with power control technique. DMACP focuses on the adaptation of IEEE 802.11 so as to find practical solutions for: (a) Finding the directions of transmission/reception at mobile nodes, (b) Designing appropriate transmission and reception strategies for the MAC control packets to minimize interference amongst distinct pairs of communicating hosts, (c) Implementing the power control strategy for data transmission to reduce power consumption. In addition, [44] discuss how to take advantage of directional antennas and present some practical schemes for implementing directional RTS and CTS transmissions.

Results show that the use of directional antennas offers many benefits, such as significant power savings, network throughput improvement, and much less interference. However, all these benefits do not come for free. Different from the omni-directional antennas based scheme, the antennas of transmitters/receivers have to be aimed at each other before the communication starts. The implementation is complex, and the hidden terminals, deafness problems may also exist [14].

7.2.3 A MAC Protocol Using Directional Antennas

[49] avoids the requirements for hardware support by exchanging omni-directional RTS/CTS frames. The direction of a transmitting host is determined through measuring the signal strength, and hence the need for GPS receivers is eliminated. The scheme is based on the IEEE 802.11 protocol, except for some modifications for adapting the use of directional transmission. The key modification is a mechanism for the pair of communicating nodes to recognize the direction of each other through the RTS/CTS exchange. No location information is needed, which helps reduce the interference.

According to [49], the radio transceiver in each mobile node is assumed to be equipped with M directional antennas. Each of the antennas has a conical radiation pattern, spanning an angle of $2\pi/M$ radians. The M antennas in each host are fixed with non-overlapping beam directions, so as to collectively span the entire plane.

Simulation experiments indicate that by using the proposed protocol with 4 directional antennas per node, the average throughput in the network can be improved up to 2 to 3 times over those obtained by using CSMA/CA schemes based on RTS/CTS exchange with traditional omni-directional antennas. However, since the proposed protocol uses omni-directional RTS/CTS exchange, it loses the benefits of reduced transmission area.

7.2.4 DPC with Smart Antenna

A distributed power control(DPC) protocol is proposed for ad hoc network stations with smart antennas in [19]. In the proposed protocol, the receivers gather local interference information and send it back to the transmitters. Then the transmitter can use this feedback to estimate the power reduction factors for each activated link. The feedback information consists of the corresponding minimum SINR (signal to interference plus noise) during RTS, CTS, DATA and ACK transmission. Here DATA and ACK transmissions are in (beamformed) array-mode since smart antennas are used at both ends of the link.

In DPC protocol, the interference information is collected during both omni-directional RTS/CTS transmission and the beamformed DATA/ACK transmission. RTS /CTS packets are always transmitted with full power in omni-directional mode, and the power level of DATA/ACK transmission is determined by a power reduction factor which is determined by the maximum interference. According to the simulation results, significant performance improvement has been achieved compared with a system using conventional IEEE 802.11 protocol. And the results indicate that the DPC protocol enables the network to dynamically achieve capacities close to the optimal levels which is achieved by a system where the power control has been statically optimized.

8 Others

Besides the protocols discussed above, there exist other MAC layer techniques proposed to reduce energy consumption for wireless ad hoc networks.

The Multiple Access with Collision Avoidance protocol(MACA) [35] solves the hidden terminal and exposed terminal problems. Therefore MACA outperforms CSMA in wireless multi-hop networks. MACA uses three-way handshake, RTS-CTS-DATA, to reduce the number of collisions. MACAW [8] expands MACA to five-way handshake (RTS/CTS/DS/DATA/ACK). However, each additional handshake brings about a longer turn-around time, and more control (e.g. source-destination information) and checksum bits. This enlarged overhead clearly reduces the channel utilization and causes more power consumption. Therefore, decreasing the number of handshakes may be an effective way for reducing the control overhead. MARCH [61] and MACA-BT [58] explores this technique. MARCH reduces the control overhead by reducing the number of RTS's along the multi-hop path. MACA-BI uses a two-way handshake, RTR(Ready To Receive)-DATA, which reduces transmit/receive turn-around time (up to 25 microseconds) and control packet collisions. Simulation results confirm the superiority of MACA-BT to existing MACA type schemes in CBR traffic. However, its performance degrades in bursty traffic compared to MACA.

As we have discussed in Section 3, scheduling-based mechanisms are established by either dynamically exchanging and resolving channel requests, or prearranging a set of channels for individual nodes or links before hand. In this way, transmissions from these nodes or on these links are collision-free. Contentions are avoided, and power consumption is reduced. However, it is difficult to realize scheduling in distributed mobile ad hoc networks. A novel solution is to use GPS neighborhood information to realize scheduling([69] [5] [9]). For example, in SNDR (Sequenced Neighbor Double Reservation) [9], the neighbor sequenced method is used to realize scheduling, and the double reservation method is adopted to improve the throughput. No handshake or carrier sensing is needed in SNDR, but it avoids contentions, hidden terminal and exposed terminal problems. However, in dense ad hoc networks, or when the network topology changes very often, the control overhead will be very high. In this case, it consumes significant power and bandwidth.

9 Conclusion

As the dynamic, fast deployable ad hoc networks have many promising applications such as e-conference, emergency services, home networking, etc., more and more attention are focused on ad hoc network research, especially on energy-aware mechanisms. It is important to study how to reduce the power consumption while at the same time fully-utilize the bandwidth resource. Moreover, third generation wireless networks are supposed to carry diverse multimedia traffic that will consume more power than a normal data device. Thus energy-aware mechanisms play an important rule in future wireless networks. In this paper, we study the low-power MAC layer mechanisms. We not only analyze the design motivation of each mechanism but also study multiple example protocols in each category. However, MAC layer is not the only layer for power conservation. As claimed by [33], enhancing power efficiency can be achieved in the entire network protocol stack of wireless ad hoc networks. For example, it is common to use power conservation schemes in the hardware design of wireless systems. On the other hand, attention should also be paid on higher levels of the protocol stack.

References

[1] S. Agarwal, S.V. Krishnamurthy, R.H. Katz, and S.K. Dao, Distributed power control in ad-hoc wireless networks, *Proceedings of IEEE International Symposium on Personal, Indoor and Mobile Radio Communications(PIMRC)* (San Diego, 2001).

[2] C.A. Balanis, *Antenna Theory: analysis and design*, 2nd ed., (Wiley, 1997).

[3] N. Bambos, Towards power-sensitive network architectures in wireless communications: concepts, issues and design aspects, *IEEE Personal Communications* Vol.5 No.3 (1998) pp. 50-59.

[4] S. Bandyopadhyay, K. Hausike, S. Horisawa, and S. Tawara, An adaptive MAC and directional routing protocol for ad hoc wireless networks

using ESPAR antenna, *Proceedings of the ACM/SIGMOBILE MobiHoc* (October 2001).

[5] L. Bao and J.J. GarciaLunaAceves, Channel access scheduling in ad hoc networks with unidirectional links, *Proceedings of Workshop on Discrete Algorithms and Methods for Mobile Computing and Communications(DIALM)* (Rome, Italy, Jul. 2001).

[6] J.Q. Bao and L. Tong, A performance comparison between ad hoc and centrally controlled CDMA wireless LANs, *IEEE Transactions on Wireless Communications* Vol.1 (Oct. 2002) pp. 829-841.

[7] J.Q. Bao and L. Tong, A performance comparison of CDMA ad-hoc and cellular networks, *Proceeding of IEEE Global Telecommunications Conference (GLOBECOM'00)* (San Francisco, 2000) pp. 208-212.

[8] V. Bharghavan, A. Demers, S. Shenker, and L. Zhang, MACAW: a media access protocol for wireless LAN's, *Proceedings of the ACM SIGCOMM'94* (London, UK, 1994) pp. 212-25.

[9] Z. Cai, M. Lu, SNDR: a new medium access control for mult-channel ad hoc networks, *Proceedings of IEEE Vehicular Technology conference(VTC) 2000-Spring*, Vol.2 (May 2000) pp.966 - 971.

[10] J.C. Chen, K.M. Sivalingam, and P. Agrawal, Performance comparison of battery power consumption in wireless multiple access protocols, *Wireless Networks* Vol.5 No.6 (1999) pp. 445-460.

[11] J.C. Chen, K.M. Sivalingam, P. Agrawal, and R. Acharya, Scheduling multimedia services in a low-power MAC for wireless and mobile ATM networks, *IEEE Transactions on Multimedia* Vol.1 No.2 (1999) pp. 187-201.

[12] C.F. Chiasserini and R.R. Rao, A distributed power management policy for wireless ad hoc networks, *IEEE Wireless Communication and Networking Conference* (2000) pp. 1209-1213.

[13] J. Chou, D. Petrovic, and K. Ramchandran, A distributed and adaptive signal processing approach to reduce energy consumption in sensor networks, *INFOCOM 2003*.

[14] R.R. Choudhury and N.H. Vaidya, Impact of directional antennas on ad hoc routing, *Proceedings of IFIP Personal and Wireless Communications* (2003).

[15] R.R. Choudhury, X. Yang, R. Ramanathan, and N.H. Vaidya, Using directional antennas for medium access control in ad hoc networks, *Proceedings of the 8th annual international conference on Mobile computing and networking* (Atlanta, USA, March 2002).

[16] M. Chu, H. Haussecker, and F. Zhao, Scalable information-driven sensor querying and routing for ad hoc heterogeneous sensor networks, *IEEE Journal of High Performance Computing Applications* Vol.16 No.3 (2002) pp. 90-110.

[17] L.A. DaSilva, J.H. Reed, W. Newhall, Mobile ad hoc networks and automotive applications, *Available at: http://www.mprg.org/Tech_xfer/ppt/GMTutorialonAdhocNetworksfromVATech.pdf*

[18] J. Deng and Z.J. Haas, Dual busy tone multiple access(DBTMA): a new medium access control for packet radio networks, *Proceedings of the International Conference on Universal Personal Communication* (Oct 1998).

[19] N.S. Fahmy, T.D. Todd, V. Kezys, Distributed power control for ad hoc networks with smart antennas, *Proceedings of IEEE Vehicular Technology conference(VTC) 2002-Fall* Vol.4 (Sep. 2002) pp. 2141-2144.

[20] L.M. Feeney, Energy efficient communication in ad hoc wireless networks, *http://www.sics.se/ lmfeeney/wip.html.*

[21] L.M. Feeney and M. Nilsson, Investigating the energy consumption of a wireless network interface in an ad hoc networking environment, *INFOCOM 2001.*

[22] C.L. Fullmer and J.J. Garcia-Luna-Aceves, Floor acquisition multiple access (FAMA) for packet-radio networks, *Proceedings of the ACM SIGCOMM'95* (Cambridge, MA, 1995) pp. 262-273.

[23] K.S. Gilhousen, I.M. Jacobs, R. Padovani, A.J. Viterbi, Jr.L.A. Weaver, and C.E.Wheatley III, On the capacity of a cellular CDMA system, *IEEE Transactions on Vehicular Technology* Vol.40 No.2 (1991) pp. 303-312.

[24] G. Girling, J.L.K. Wa, P. Osborn, and R. Stefanova, The design and implementation of a low power ad hoc protocol stack, Presented at IEEE Wireless Communications and Networking Conference, (Chicago, September 2000). Available at www-lce.eng.cam.ac.uk/publications/files/tr.2000.13.pdf

[25] C. Guo, L.C. Zhong, and J.M. Rabaey, Low power distributed MAC for ad hoc sensor radio networks, *Proceedings of IEEE GlobeCom 2001* (San Antonio, Nov. 2001).

[26] I. Gupta, Minimal CDMA recoding strategies in power-controlled ad-hoc wireless networks, *Proceeding of 1st International Workshop on Parallel and Distributed Computing Issues in Wireless Networks and Mobile Computing* (San Francisco, Apr. 2001).

[27] P. Gupta and P.R. Kumar, The capacity of wireless networks, *IEEE Transactions on Information Theory*, Vol.46 No.2 (2000) pp. 338-404.

[28] Z.J. Haas and J. Deng, Dual busy tone multiple access(DBTMA) - a multiple access control scheme for ad hoc networks, *IEEE Transactions On Communications* Vol.50 No.6 (2002) pp. 975-985.

[29] P.J.M. Havinga, G.J.M. Smit, and M. Bos, Energy efficient wireless ATM design, *Proceedings of the second IEEE international workshop on wireless mobile ATM implementations (wmATM'99)* (Jun. 1999) pp. 11-22.

[30] W.R. Heinzelman, A. Chandrakasan, and H. Balakrishnan, Energy efficient communication protocol for wireless microsensor networks, *In 33rd Annual Hawaii International Conference on System Sciences*, (2000) pp. 3005-3014.

[31] C.-H. Hwang, A.C.-H. Wu, A predictive system shutdown method for energy saving of event-driven computation, *IEEE/ACM International Conference on Computer-Aided Design* (San Jose, CA, Nov. 1997), pp. 28-32.

[32] K.T. Jin and D.H. Cho, A new MAC algorithm based on reservation and scheduling for energy-limited ad hoc networks, *IEEE Transactions on Consumer Electronics* Vol.49 (Feb. 2003) pp. 135-141.

[33] C.E. Jones, K.M. Sivalingam, P. Agrawal, and J.C. Chen, Survey of energy efficient network protocols for wireless networks, *Wireless Networks* Vol.7 No.4 (2001) pp. 343-358.

[34] E.-S Jung and N.H. Vaidya, A power control MAC protocol for ad hoc networks, *Proceedings of ACM MOBICOM'02* (Sep. 2002).

[35] P. Karn, MACA - a new channel access method for packet radio, *Proceedings of the 9th ARRL/CRRL Amateur Radio Computer Networking Conference* (Sep. 1990) pp. 134-140.

[36] V. Kawadia and P.R. Kumar, Power control and clustering in ad hoc networks, *IEEE INFOCOM 2003*.

[37] Y.-B. Ko, V. Shankarkumar, and N.H. Vaidya, Medium access control protocols using directional antennas in ad hoc networks, *Proceedings of IEEE INFOCOM* (2000) pp. 13-21.

[38] G. Kulkarni, V. Raghunathan, M.B. Srivastava, and M. Gerla, Channel allocation in OFDMA based wireless ad hoc networks, *SPIE International Conference on Advanced Signal Processing Algorithms, Architectures, and Implementations*, (Jul. 2002).

[39] B. Mcfarland, G. Chesson, C. Temme, and T.M eng, The 5-UP protocol for unified multiservice wireless networks, *IEEE Communications Magazine* Vol.39 No.11 (2001) pp. 74-80.

[40] J. Monks, V. Bharghavan, and W. Hwu, A power controlled multiple access protocol for wireless packet networks, *Proceedings of IEEE Infocom* (Apr. 2001) pp. 219-228.

[41] A. Muqattash and M. Krunz, CDMA-based MAC protocol for wireless ad hoc networks, *Proceedings of the International Symposium on Mobile Ad Hoc Networking and Computing (MOBIHOC'03)* (Annapolis, Jun. 2003).

[42] A. Muqattash and M. Krunz, Power controlled dual channel (PCDC) medium access protocol for wireless ad hoc networks, *IEEE INFOCOM 2003*.

[43] S. Narayanaswamy, V. Kawadia, R.S. Sreenivas, and P.R. Kumar, Power control in ad-hoc networks: Theory, architecture, algorithm and implementation of the COMPOW protocol, *Proceedings of European Wireless Conference* (Feb. 2002) pp. 156-162.

[44] A. Nasipuri, K. Li, and U.R. Sappidi, Power consumption and throughput in mobile ad hoc networks using directional antennas, *Proceedings of the IEEE International Conference on Computer Communication and Networks(ICCCN2002)*, (Miami, Florida, Oct. 2002).

[45] S. PalChaudhuri and D.B. Johnson, Power mode scheduling for ad hoc networks, *Proceedings of the 10th IEEE International Conference on Network Protocols(ICNP'02)* (2002).

[46] R.L. Pickholtz, D.L. Schilling, and L.B. Milstein, Theory of spread spectrum communications - a tutorial, *IEEE Transactions on Communications* Vol. 30 (May 1982) pp. 855-884.

[47] N. Poojary, S.V. Krishnamurthy and S. Dao, Medium access control in a network of ad hoc nodes with heterogeneous transmit power capabilities, *Proceedings of ICC 2001* (Helisinki, 2001).

[48] J.G. Proakis, *Digital Communications*, (McGraw-Hill Inc., 2001).

[49] A. Nasipuri, S. Ye, J. You, and R.E. Hiromoto, A MAC protocol for mobile ad hoc networks using directional antennas, *Proceedings of IEEE Wireless Communications and Networking Conference(WCNC)*, (Chicago, IL, Sep. 2000).

[50] R. Ramanathan, On the performance of ad hoc networks with beamforming antennas, *Proceedings of ACM MobiHoc* (Oct. 2001) pp. 95-105.

[51] R. Shah and J. Rabaey, Energy aware routing for low energy ad hoc sensor networks, *Proceedings of IEEE Wireless Communications and Networking Conference(WCNC'02)* (Florida, Mar. 2002).

[52] S. Singh and C.S. Raghavendra, Power aware multi-access protocol with signaling for ad hoc networks, *ACM Computer Communication Review* Vol. 28 No. 3 (Jul. 1998) pp. 5-26.

[53] S. Singh and C.S. Raghavendra, Power efficient MAC protocol for multihop radio networks, *in The Ninth IEEE International Symposium on Personal, Indoor and Mobile Radio Communications* (Sep. 1998) pp. 153-157.

[54] K.M. Sivalingam, J.C. Chen, P. Agrawal, and M. Strivastava, Design and analysis of low-power access protocols for wireless and mobile ATM networks, *Wireless Networks* Vol.6 No.1 (2000) pp. 73-87.

[55] K. Sohrabi and G. Pottie, Performance of a novel self-organization protocol for wireless ad-hoc sensor networks, *Proceedings of the IEEE 50th Vehicular Technology Conference* (1999) pp. 1222-1226.

[56] E.S. Sousa and J.A. Silvester, Spreading code protocols for distributed spread-spectrum packet radio networks, *IEEE Transactions on Communications* Vol.36 No.3 (Mar. 1988) pp. 272-281.

[57] M. Takai, J. Martin, and R. Bagrodia, Directional virtual carrier sensing for directional antennas in mobile ad hoc networks, *Proceedings of ACM International Symposium on Mobile Ad Hoc Networking and Computing(MOBIHOC)* (Lausanne, Switzerland, Jun. 2002).

[58] F. Talucci and M. Gerla, MACA-BI(MACA By Invitation): A wireless MAC protocol for high speed ad hoc networking, *Proceedings of IEEE ICUPC'97* (San Diego, Oct. 1997).

[59] F.A. Tobagi and L. Kleinrock, Packet switching in radio channels: Part-II the hidden terminal problem in carrier sense multiple access models and the busy tone solution, *IEEE Transactions on Communications* Vol.23 No.12 (1975) pp. 1417-1433.

[60] C. Toh, Maximum battery life routing to support ubiquitous mobile computing in wireless ad hoc networks, *IEEE Communications Magazine* (Jun. 2001) pp. 138-147.

[61] C.K. Toh, V. Vassiliou, G. Guichal, and C.H. Shih, MARCH: a medium access control protocol for multihop wireless ad hoc networks, *Proceedings of IEEE Military Communications Conference(MILCOM)* Vol.1 (Los Angeles, 2000) pp. 512-516.

[62] Y. Wei, J. Heidemann, and D. Estrin, An energy-efficient MAC protocol for wireless sensor networks, *INFOCOM 2002*.

[63] H. Woesner, J-P. Ebert, M. Schlager, and A. Wolisz, Power-saving mechanism in emerging standards for wireless LANs: The MAC level perspective, *IEEE Personal Communication* Vol.5 No.3 (1998) pp. 40-48.

[64] C. Wu and V.O.K. Li, Receiver-initiated busy-tone multiple access in packet radio networks, *Proceedings ACM SIGCOMM'87 Workshop* Vol.17 No.5 (Stowe, Vermont, 1987) pp. 336-342.

[65] S.L. Wu, C.Y. Lin, Y.C. Tseng, and J.P. Sheu, A new multi-channel MAC protocol with on-demand channel assignment for mobile ad hoc networks, *Proceedings of International Symposium on Parallel Architectures, Algorithms and Networks(ISPAN'00)* (Dallas/Richardson, Texas, USA, 2000) pp. 232-237.

[66] S.L. Wu, Y.C. Tseng, C.Y. Lin, and J.P. Sheu, A multi-channel MAC protocol with power control for multi-hop mobile ad hoc networks, *The Computer Journal (SCI)* Vol.45 No.1 (2002) pp. 101-110.

[67] S.L. Wu, Y.C. Tseng, and J.P. Sheu, Intelligent medium access for mobile ad hoc networks with busy tones and power control, *IEEE Journal on Selected Areas in Communications* Vol.18 No.9 (2000) pp. 1647-1657.

[68] J. Zander, Slotted ALOHA multihop packet radio networks with directional antennas, *Electronics Letters* Vol.26 No.25 (2000) pp. 2098-2099.

[69] C. Zhu and M.S. Corson, A five-phase reservation protocol (FPRP) for mobile ad hoc networks, *Proceedings of IEEE Conference on Computer Communications(INFOCOM'98)* Vol.1 (San Francisco, 1998) pp. 322-331.

[70] IEEE Standard 802.11a-1999, Wireless LAN medium access control (MAC) and physical layer (PHY) specifications, *Supplement to IEEE Standard for Information Technology*, (Sep. 1999).

[71] Flarion Technologies, http://www.flarion.com/

[72] http://userver.ftw.at/ toumpis/ad_hoc_course/

[73] ETSI TC-RES, Radio equipment and systems(RES); high performance radio local area network(HIPERLAN) type 1; functional specification, *European Telecommunication Standard ETS 300 652* (Oct. 1996).

[74] Bluetooth specification volume 1, *Bluetooth Consortium*, (Jul. 1999).

[75] ISO/IEC and IEEE draft international standards, part 11: wireless LAN medium access control (MAC) and physical layer (PHY) specifications, *ISO/IEC 8802-11*, IEEE P802.1 l/D10, (Jan. 1999).

QoS-Based Routing in Wireless Mobile Networks

Eylem Ekici
Department of Electrical Engineering
Ohio State University, Columbus, OH 43210
E-mail: `ekici.2@osu.edu`

Contents

1 Introduction

Wireless ad hoc networks are formed by mobile devices that autonomously form a communication network over their wireless communication interfaces. These *mobile nodes (MN)* have the capability of moving arbitrarily, resulting in a dynamic network infrastructure. The concept of self forming networks without a wired infrastructure has applications in various domains ranging from disaster relief operations to digital battlefield. With the wide variety of application fields, it is not surprising that ad hoc networks have become the center of attention of the research community.

Ad hoc networking concept brings a multitude of problems to be addressed. Wireless channel access, routing, topology management, transport layer and many more problems cannot simply be solved completely based on the solutions developed for the wired networks. Mobile nodes in an ad hoc network are usually distributed over a given geographical region that prevents direct communication between all source-destination pairs. Therefore, mobile nodes in ad hoc networks must be able to act as relay nodes to enable end-to-end communication. In wired networks, packets may follow either static pre-configured paths or paths created via a routing protocol. Since the network topology of an ad hoc network is not known in advance, routes between source-destination pairs must be calculated dynamically, which is often used in wired networks, as well. However, the similarities end at this point. In ad hoc networks, calculated routes only have limited life times since the connection structure of the network changes due to nodal mobility. Augmented with limited resources on the wireless links, computational capacity, and power available at mobile nodes, routing in ad hoc networks becomes an important and non-trivial problem to be addressed.

Basic ad hoc network routing protocols can be classified in two groups. The *pro-active routing protocols* are based on the determination of the paths before there is any demand for communication. Routing tables in mobile nodes are configured ahead of time by the algorithm and maintained through periodic and trigger-based topology exchange messages. This way, all packets can be routed toward their destinations without delay as soon as they are generated. Destination-Sequenced Distance Vector Routing (DSDV) [1], Fisheye State Routing (FSR) [2], and Optimized Link State Routing (OLSR) [3] are examples of pro-active routing protocols for ad hoc networks. Since pro-active routing protocols rely on the dissemination of topology information in the network even though there may be no demand for data communication, they potentially incur high protocol overhead. The second group of the routing protocols for ad hoc networks are *reactive (or on-demand) protocols*. Protocols belonging to this class do not calculate the paths ahead of time. Paths are discovered at the time of the communication request. Hence, topological changes do not incur control communication overhead regardless of the network dynamics until a source-destination pair needs to be connected. Ad-Hoc On Demand Distance Vector (AODV) Routing Protocol [4], Temporarily Ordered Routing Algorithm (TORA) [5], and Dynamic Source Routing (DSR) [6] can be classified as reactive routing protocols. Although reactive routing protocols can potentially initiate redundant path discovery and establishment procedures, these negative effects can be reduced by listening to other path discovery procedures and caching already known paths. Despite these improvements, the path discovery delay, which is the time between generation of connection request and sending the first packet, can be be prohibitively large under reactive schemes. For a discussion of basic routing protocols in ad hoc networks, the reader is referred to [7].

In the last decade, the need for networks providing *Quality of Service (QoS)* guarantees has increased tremendously primarily due to increasing popularity of mission-critical real-time and multimedia applications. Ad hoc networks are considered for deployment in environments where multimedia and real-time applications with various QoS demands are essential to their basic operation. The quality of service guarantees delivered to applications include bandwidth, throughput, delay, delay jitter, error rate guarantees, etc., or a combination thereof. In this article, QoS provisioning in ad hoc networks is discussed from a network layer perspective. In the following sections, principles of QoS-based routing are reviewed and example QoS-based routing proposals for ad hoc networks are presented along with possible research directions for future work.

2 Quality of Service Based Routing

In its broadest form, Quality of Service refers to the contract between the service provider, i.e., the network, and customers, i.e., applications [8]. For any communication network to provide QoS guarantees to applications, following steps are essential:

1. **Determination of Requirements:** Many applications require by their nature service guarantees that are already quantified. However, QoS can also be defined based on the human perception, which, more

often than not, is based on a set of subjective measures. An example subjective requirement may be that the streaming video should not be "choppy." In order for networks to provide QoS guarantees, subjective and qualitative requirements must be translated to quantified ones.

2. **Network State Collection:** To provide QoS guarantees in the network, the available resources in the network must be determined. Depending on what particular metrics are utilized in QoS provisioning, most up-to-date information about resource availability in and between network elements must either be computed or measured.

3. **Route Calculation and Resource Allocation:** The routing protocol must calculate a route between the source and destination that can satisfy the QoS metrics required by the application. Once the path is calculated, necessary resources must be allocated on the calculated path. It is also the responsibility of the routing algorithm and resource management scheme to compute alternative paths and allocate new resources if the network conditions change.

4. **Session Termination and Resource Deallocation:** Once the communication session is over, all allocated resources must be released. Accordingly, the availability of resources in the network must be updated such that they are utilized by other data flows.

All these steps are very important to QoS provisioning in any network. However, in order to focus on the QoS-based routing techniques, we will make some simplifying assumptions. First, we assume that applications provide a clear, quantified set of QoS metrics that must be satisfied. The second and more important assumption is that we assume that network nodes are equipped with methods to determine availability of resources and are capable of allocating and deallocating them. Routing protocols may include routines to signal allocation and deallocation of resources, or they may rely on the availability of auxiliary signaling protocols. In any case, the details of the resource discovery and allocation methods are shielded from the routing protocol. Note that some of the protocols reviewed in this article include explicit methods to deal with the above mentioned functions, which will be very briefly described where necessary.

2.1 Network State

Network state refers to the collection of the status information about all resources in the network. Nodal connectivity, capacity of the links, bandwidth usage, computational capacity of nodes, bandwidth utilization, link delay, and other metrics can be used as descriptors of the network state. A network may choose to utilize one or more of these metrics to represent the network state. The selection of these descriptors affects the collection and maintenance of the network state. This point can be analyzed with the following example:
(i) Consider a wired network that only utilizes the nodal connectivity as the network state descriptor and ignores the resource availability on the links. In this case, the network state is stationary for long periods because the state changes only if there is a link failure or addition of a new link. Hence, the network state can be updated on an event-triggered basis.
(ii) Let us expand the state definition of the previous example by adding the available bandwidth on each link. Although nodal connectivity does not change during most of the lifetime of the network, available amount of bandwidth changes with addition and removal of flows. If the flows in the network are monitored and admitted by a central authority, then the state maintenance may be reduced to resource tracking in a central location. If flows are admitted to the network in a distributed fashion, then more frequent updates are necessary to maintain the state information. Updates can be obtained periodically or whenever there is a change in the resource availability exceeding a certain threshold value.

The state of the network can be maintained in several ways at varying levels [8]. The so-called *local state* is the state of a node and its adjacent links that is tracked by the node itself. Each node maintains the local state by monitoring the availability of resources (bandwidth, buffer space, etc.) or the effects of the combination of available resources and other factors (delay, error rate, etc.). The *global state* of the network is created by exchanging the local state information. If all nodes utilize the global state, then local states can be exchanged using a link state algorithm, which will be pieced together to represent the global state. The link state algorithms broadcast the information to all nodes in the network and all nodes make path calculations on their own behalf. Another possibility is to use a distance vector algorithm, which exchanges "cost" information to all possible destinations periodically. The cost of a path passing through a particular node is changed according to the methods used to calculate the cost of a path before it is

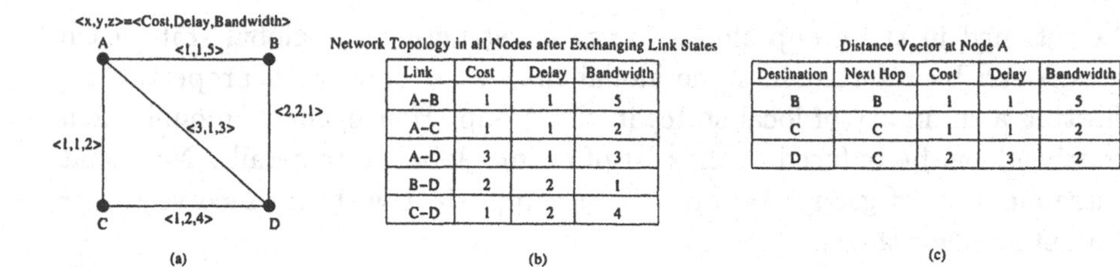

Link	Cost	Delay	Bandwidth
A–B	1	1	5
A–C	1	1	2
A–D	3	1	3
B–D	2	2	1
C–D	1	2	4

Destination	Next Hop	Cost	Delay	Bandwidth
B	B	1	1	5
C	C	1	1	2
D	C	2	3	2

Figure 1: A simple network and its representation in nodes with link state and distance vector algorithms

advertised to other nodes. If the global state information is to be used by a central authority for flow admission, then the local state exchanges can be converted to local state reports that are sent to only one node rather than broadcast to all nodes.

A simple network consisting of four nodes and five links is depicted in Figure 1.a. Each link is associated with three metrics, representing the cost of the link, delay, and the available bandwidth. Assuming that the links share all metrics in both directions, the topology table constructed using link state algorithm is shown in Figure 1.b. The distance vector created at node A using the distance vector algorithm is shown in Figure 1.c. It is assumed that the distance vector algorithm selects and forwards the least cost paths to all destinations. Note that the distance vector algorithm processes the path selection criteria during vector exchange, whereas the link state algorithm only collects local states to be processed later in the nodes.

Construction of global state of the network involves several communication events and may take considerable amount of time. Hence, one cannot assume that the constructed global state reflects the most up-to-date condition of the network. This effect is more pronounced in distance vector algorithms since the propagation of the cost information is performed incrementally at every time period, whereas the link state routing algorithms broadcast and relay local state as soon as they are generated/received.

Another important point is the amount of data to be collected to construct global state. In large networks, broadcasting individual local states to construct global state can cause very high protocol overhead and, due to larger delays experienced, can become invalid by the time they are received by nodes potentially using them. Instead, it is possible to organize nodes in groups and communicate an aggregate state of the group. If necessary, node groups can recursively form other groups and details of the group state can be retained in the group alone. When constructing the global state, each group can be inserted into the global map as a node with properties reflecting a summary of local states in the group. Hence, the communication overhead can be reduced at the cost of losing global state details. Note that dissemination of group state can be accomplished with distance vector or link state algorithms.

2.2 Route Calculation and Session Establishment

A path that satisfies all QoS requirements between the source and the destination is called *feasible*. A feasible path is called *optimal* if it has the least cost among all feasible paths. The global network state, either in full detail or in a summarized form, is used to calculate a path connecting the source-destination pair which satisfies the required QoS parameters. When all local states are collected in a node, the network can be represented as a graph with links that have several metrics associated with them. However, even though the global state is reconstructed, calculation of QoS-constrained paths is a non-trivial task. In fact, the computation of paths that satisfy two or more additive constraints (such as cost and delay) is shown to be NP-complete [9]. If a constraint is bottleneck based, such as bandwidth, multiple constraint problems can be simplified by eliminating all links (or nodes) that do not satisfy the bottleneck constraint. Provided that there is only one additive constraint left, the graph that contains only eligible links and nodes can be used to find the optimal path with a shortest path algorithm such as Dijkstra's [10]. Otherwise, remaining problems are addressed with suboptimal algorithms that rely on sequential elimination of possible routes [11]. The path can be computed either by the source or by a central authority, in which case the computed path is delivered to the source to be processed. Once the end-to-end path is determined, a signaling protocol such as RSVP [12] used to allocate resources along the path. At this point, it is important to emphasize the fact that these solutions are based on the assumption that precise global state information is available at the time of the computation. QoS-based route calculation problem with imprecise state information is addressed in [13].

Distance vector algorithms tend to incorporate the local states into routing decisions as the state information is propagated in the network. This approach enables the system to determine the availability of resources in a parallel way for all possible destinations. The local state information is incorporated into potential routing possibilities when distance vectors are exchanged. When there is a demand to reach a destination subject to QoS constraints, the source node compares the QoS requirements of the flow with the QoS levels that can be supported by known paths and makes a selection. Note that since it is impossible to forward all possible distance vectors from a node to all possible destinations, a selection among available alternatives is made during distance vector exchange. Therefore, resulting paths may not be optimal or feasible paths may have been eliminated during the distance vector exchange, causing the source not to find a path to destination. In case a feasible path is found, the resource reservation signal travels the path hop by hop toward destination, both validating and reserving resources along the route. After a path is selected and the resources are allocated, the packets are forwarded on the path where reservations have been made.

At this point, it is worth noting that, regardless of the network state collection and QoS path calculation method, there is a possibility that the resources known to be available may become unavailable between the time of the path calculation and the actual allocation of resources. It is possible for two almost simultaneous requests to contend for the same set of resources. Since there is a non-zero time between the selection of the QoS path and the actual allocation of resources, two independent path calculations may see resources available for them during calculation, but only one of them may obtain these resources whereas the other path setup fails because the remaining after the first allocation are not sufficient to sustain the second flow. In an extreme case, if two calculated QoS paths use the same set of links in opposite directions, both path setup attempts may fail when these requests meet. A good QoS routing protocol should avoid resolve such race conditions, and if they happen, resolve them such that a minimum number of path setup requests fail.

2.3 Session Termination

Allocation of resources introduce another dimension to the system state in the network. The resource allocation state affects the global state by changing the availability of resources, which in turn changes the state descriptors of the network. After the initial resource allocation, the resource allocation state must be maintained so that the flows utilize them while packet forwarding. The maintenance of allocated resources directly influences the way sessions are terminated, as well. The first way of resource allocation maintenance is to retain the allocation unless the connection is explicitly terminated. This *hard allocation* method requires that, once the source no longer has data to send, it signals the release of resources along the path, which returns the allocated resources to the available pool. Alternatively, resources may be allocated as *soft states* where, following the initial resource allocation, the resource allocation must be refreshed regularly to ensure availability. In case the allocation is not refreshed within a time period, resources are deallocated. A resource allocation scheme based on the soft state paradigm does not require the use of explicit disconnect messages. Depending on the choice of resource maintenance, explicit session termination messages may or may not be used.

3 QoS-Based Routing in Ad Hoc Networks

As presented in the previous section, QoS-based routing has its own unique challenges, especially regarding the variability of network state and its effects on routing and resource management. These challenges are augmented with additional constraints imposed by the nature of ad hoc networks. The special properties of ad hoc networks affect QoS-based routing schemes as described in the following section.

3.1 Effect of Ad Hoc Network Properties on QoS-Based Routing

Network Dynamics: QoS-based routing schemes need to adapt themselves to changing network state continuously. Even in wired networks where the topology changes are very seldom, the network state presents its dynamic nature due to changing availability of network resources. The network state dynamics is more pronounced in ad hoc networks since the nodes are free to move in arbitrary directions, causing the topology of the network to change constantly. As the topological changes affect the resource availability directly, the network dynamics cause more frequent local state propagation and exchange operations in ad hoc networks to reach the same level of global state representation in a static network. As an example, even though there may be no traffic carried in the network, the global state changes in an ad hoc network, whereas the global state would not change in a wired network under the same conditions. Consequently, it can be concluded that ad hoc networks require a higher frequency of local state exchanges if the QoS routes are calculated centrally.
Resource Dynamics: In wired networks, all resources available for a node are known by the node at all times. By inspecting the allocation of resource

for all data flows, a node can determine quite easily how much residual resources are available for other potential data flows. In ad hoc networks, it is possible to keep track of resources allocated within the node, such as buffer space and processing power. However, available wireless resources may not be determined as easily. In general, wireless resources are not permanently and exclusively allocated to individual mobile nodes as to increase the channel utilization. Although the maximum available bandwidth in the wireless channel can be assumed to be known, it is shared among mobile nodes that are within each other's communication range. Similarly, hidden terminals in a region affect the transmission schedules, and hence the available bandwidth, of nodes not directly in their communication ranges. Hence, the residual available bandwidth is not only a function of maximum available capacity and characteristics of flows carried by that node, but also of the flows traversing other nodes in the near vicinity. Hence, generation of local states by individual nodes may result in approximate rather than exact results.

Autonomy: By definition, ad hoc networks are autonomous entities decoupled from wired infrastructures in functioning and management. An ad hoc network is generally composed of nodes of similar properties, i.e., none of the nodes is designed to carry out specialized jobs. Consequently, the network must be able to manage itself in all aspects. Since there is no entity that can collect and control all network operations, the load is shared among mobile nodes. Load sharing can be accomplished by distributed calculations, by making each node responsible for a certain set of calculations (e.g., paths are calculated by source nodes), or, in case centralized calculations are required, by employing multiple central nodes (e.g., mobile nodes form clusters and cluster heads make calculations). Note that single hop wireless networks such as cellular networks and wireless local area networks do not face similar problems since the centralized computations can easily be performed by a node in the wired segment of the network.

3.2 Maintenance of QoS Paths

In addition to the steps outlined in Section 2, another important operation is imposed by the dynamic topology of ad hoc networks to enable QoS-based routing. In wired networks, the topology does not change unless there is a link or node failure in the network. Therefore, as long as nodes ensure that resources are allocated for flows as requested, it can be assumed that the network will deliver negotiated QoS guarantees. In case of a failure, affected paths can be recalculated based on the new topology information.

In ad hoc networks, topological changes are rules rather than exceptions. Maintenance of paths is essential to provide the negotiated QoS guarantees. There are two main mechanisms for path maintenance and recovery in ad hoc networks. The first one is based on the recalculation of paths. When a link on a path becomes unavailable due to mobility or failure, the node where the original path is calculated or path calculation is initiated is notified by intermediate nodes on the path that discover the broken link in the chain. Once the notification is received, a new path is calculated and resource allocation is performed. Once the new path is ready, the flow is shifted from the old path to the new path. This approach creates high quality paths at the cost of response time. During the path recalculation period, the packets in the network can be buffered, dropped, or routed on alternative best effort paths. If immediate recovery is required, multiple paths can be maintained between the source and the destination. However, this would require redundant resource allocations in a resource-limited environment.

The second method is based on the utilization of some portions of the original path. When a link becomes unavailable, a neighboring node on the path initiates a path discovery from that particular location to reach the source, the destination, or another intermediate node on the path to bridge the unavailable link. The resources in the remaining portions of the original path is retained and the old path is augmented with a new segment. This second method usually provides faster recovery than the first method, but the resulting paths may have higher cost than the alternatives. Redundant protection of segments of the path can also be utilized if vulnerable segments can be determined at the initial path setup.

It is also possible to combine the two methods and create temporary and fast solutions with the path augmentation and switch over the flow to a new path that is computed end-to-end once it is available. However, this hybrid scheme would incur higher protocol overhead than each method alone. Note that it is possible that there are no feasible paths in the network after the topology change or that the routing protocol is unable to find one. Both cases result in the forced termination of the communication on the path in question.

3.3 Example QoS-Based Routing Protocols

In literature, several proposals for QoS provisioning in ad hoc networks exist. In the following sections, proposals that mainly investigate methods for discovering paths that can provide QoS guarantees are overviewed. Although the protocols presented in this article are not exhaustive, they represent the state-of-the art of the subject. For all proposals presented below, it is assumed that mobile nodes discover their neighbors through packet exchanges and that all nodes are associated with unique IDs. Furthermore, unless otherwise stated, it is also assumed that the MAC layer handles resource reservations and emulates point-to-point links where necessary.

3.3.1 DSDV QoS Routing [14]

The DSDV QoS Routing protocol is one of the earliest QoS-based routing protocols proposed for ad hoc networks. In this proposal, the nodes are grouped into clusters and each cluster is assigned a distinct CDMA code. Within the cluster, medium access is accomplished with TDMA using the same CDMA code. By using different CDMA codes in every cluster, the computation of the available bandwidths on links is simplified. The propagation of bandwidth information is piggybacked on the distance vector packets of the DSDV [1] protocol, which normally carries a sequence number, previous hop, and a distance metric to the destination in question. Hence, the source of a flow can initiate a path setup if the available bandwidth on the path to the destination is greater than the one needed for the flow. Broken paths are recovered with the methods inherited from the DSDV protocol. The use of soft state paradigm is proposed for resource reservation, where the resources are released if a reserved slot is not used for several consecutive frames.

3.3.2 Core Extraction Distributed Ad Hoc Routing (CEDAR) [15]

The CEDAR protocol creates and maintains bandwidth-constraint paths in ad hoc networks. CEDAR is based on creation and maintenance of mobile node groups under the group heads referred to as dominators. Each dominator is responsible for a group of nodes in its n-hop neighborhood, and knows the full group state. The groups are formed such that the entire network is covered. The dominators communicate with each other over virtual links that may be composed of several hops in the ad hoc network. The dominators and the virtual links connecting them form the *core* of the network. The topology and local state information also travels past the group boundaries over the virtual links. When the nodes realize that the available bandwidth in one of their adjacent links changes more than a threshold value, they inform their dominators about the change. These core nodes inform other core nodes by broadcasting the information to other dominators in the core over virtual links. The information about increase and decrease in bandwidth travels in the core at different speeds. The increase waves travel in the core slower than the decrease waves such that decrease waves catch up with increase waves and cancel each other. Effectively, knowledge about unstable links with frequently changing bandwidth amounts is confined to a small neighborhood whereas stable links are advertisedin a larger range among the cores.

The bandwidth-constrained path calculation is performed collectively in the core of the network. As the first step, the location of the destination must be discovered if not already known. This is accomplished by broadcasting a discovery packet in the core, which records the core nodes it has traversed. The dominator of the destination replies with a packet containing the reversed core node list recorded in the discovery packet, which is called the *core path*. As the next step, the dominator of the source calculates a segment of the path using the full topology information of its group and the partial topology information of other groups known to it. During this phase, the dominator of the source tries to calculate a path to a node that belongs to the furthest away core node in the core path subject to bandwidth constraints. If there are multiple feasible path to reach the furthest group, the core node chooses the shortest one.[1] If the destination is not reached, the furthest core node in the core path is notified with the destination, the core path, the segment that is already computed, and bandwidth requirement. The intermediate core nodes repeat the calculation until the destination is reached or connection request fails due to insufficient resources.

An example path path calculation is depicted in Figure 2. To calculate the path from source S to destination D, the core path {I,II,III,IV} is discovered. Then core node I calculates the shortest-widest-furthest path from S toward D using its partial topology information. Assume that the path segment calculated by core node I is {S,I,2,3,4,6}. Then core node I notifies III about the segment it has calculated. III calculates the remaining part of the path that connects 6 with D over segment {6,III,8,D}.

For path maintenance, a two-phase method is used. Assume that a link fails on the path as a result of mobility. In the first phase, the node upstream

[1]The path segments calculated by core nodes using the described method are called "shortest-widest-furthest path."

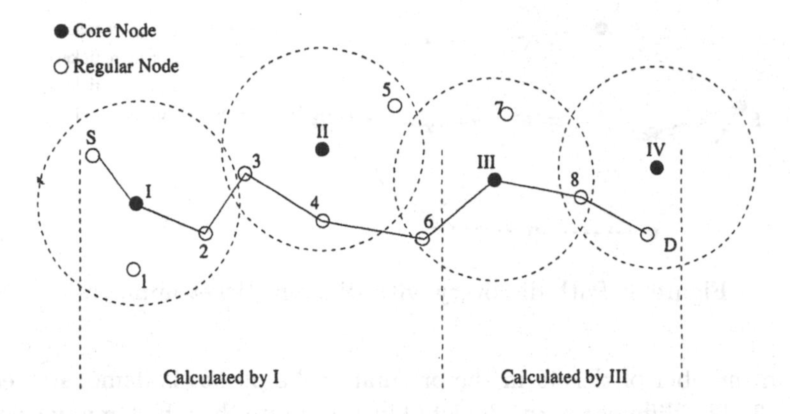

Figure 2: An example of path calculation under CEDAR

to the point of failure initiates a local route computation similar to the initial path calculation to create a path from itself to destination. Meanwhile, the source is also notified, which triggers another route calculation to find a better path to the destination as the second phase of path maintenance. The first phase is used to deliver packets already in transit and provides a fast solution while second phase create a long-term alternative path. QoS state maintenance for individual flows and connection termination details are not discussed in [15].

3.3.3 Ticket Based Probing (TBP) [16]

The Ticket Based Probing protocol is a distance vector based routing protocol that discovers paths subject to QoS constraints. It is also assumed that the topology changes do not occur too frequently, which would render QoS calculations meaningless. The paths are calculated on a per-connection basis using precise cost and possibly imprecise bandwidth and delay informations. The main idea is to search multiple paths from source to destination in parallel. Instead of flooding the network with probe packets to discover paths, TBP limits the overhead of searching by assigning a limited number of tickets for each probe packet, where each ticket corresponds to a possible path to be searched. Intermediate nodes are not allowed to increase the number of tickets, and consequently, the maximum number of paths to be searched is limited to the number of tickets generated by the source. Intermediate nodes receiving probe packets may split the tickets between multiple outgoing probe packets and search alternative paths in parallel. The effect of

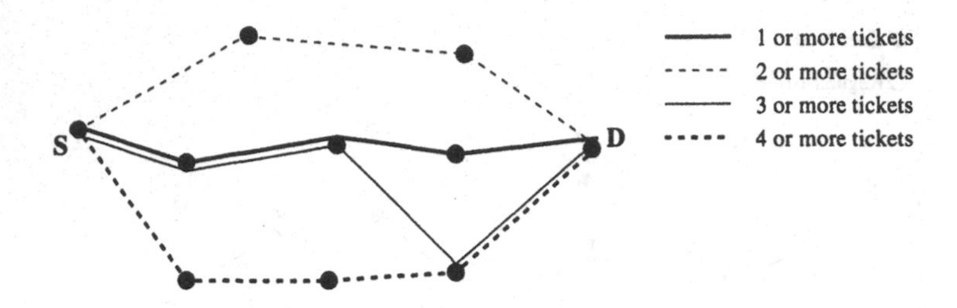

Figure 3: Path discovery with different ticket numbers

different number of tickets in the original probe packet is demonstrated in Figure 3. The different paths depicted in this figure show how many original tickets are needed such that the protocol probes a particular path between S and D. As can be seen, as the number of tickets is increased, the number of probed paths is increased, as well. In other words, the ticket number determines the search space of the path discovery algorithm.

TBP is used to discover delay-constrained minimum cost paths and bandwidth-constrained minimum cost paths. In both versions, two different type of tickets, marked as yellow and green, are utilized. In the delay-constrained minimum cost path version, yellow tickets are used to maximize the probability of finding a feasible path by searching low delay paths. When TBP is used to discover bandwidth-constrained minimum cost paths, the yellow tickets are used again to maximize the probability of finding a feasible path, but this time by following wider paths toward destination. On the other hand, the green tickets are used to find low cost paths. The paths discovered by probes with green tickets may or may not satisfy the delay/bandwidth constraint. In case of violation of a constraint, the probes are dropped and a notification is sent to the destination. Once the destination accounts for all tickets (either through receiving the probe packets or by receiving drop notifications), it selects the path with minimum cost among the paths corresponding to the received probe packets. (These paths are known to be feasible because the probe packets have been received by the destination.) A confirmation packet is sent back on the selected path toward the source, which also allocates necessary resources on the path. If the resource allocation fails, an alternative path is selected at destination and the same procedure is repeated.

TBP proposes different levels of path redundancy to deal with the dynamic network topology. Flows are assigned different redundancy levels depending on their reliability requirements. The first level of redundancy calls for establishment of parallel link-disjoint paths between the source and the destination. All paths carry copies of the same packet to the destination such that the communication is not disrupted when a path becomes unavailable. The second level of redundancy is similar to the first one in that resources on multiple parallel paths are allocated, but data is sent on the primary path. The resources of other paths are used by best-effort traffic unless needed after the primary path fails. The third level of redundancy is identical to the second level except that resources are allocated only on the primary path. Apart from the redundancy, TBP also supports two-hop path repairs where a failing or unavailable link is bridged between the upstream and downstream nodes around the node of failure. However, this recovery method does not work for bridging nodes more than two hops away from each other. In all cases, explicit resource deallocation is not required since the allocations are maintained as soft states refreshed by explicit messages as long as the resources are utilized.

3.3.4 AODV-Based QoS Routing Protocol [17]

The objective of the protocol presented in this paper is to address the bandwidth-constrained routing problem in flat ad hoc network architectures using TDMA.[2] This protocol is developed for small networks and low to medium mobility rates, and calculates bandwidth-constrained paths for flows using the principles of AODV [4]. In this paper, a method to calculate the bandwidth on a given path is also provided, which is a greedy scheme seeking local maximal bandwidth from the source to the next hop, given the sets of time slots used along the path to reach the current node. (The reader is referred to [17] for the details of bandwidth calculation procedure.) When a node wants to establish a bandwidth-constrained path, it broadcasts a route request packet (RREQ) that contains the flow information as shown in Figure 4.a. As the RREQ packets are propagated in the network, each node receiving the request calculates the bandwidth according to bandwidth and time slot allocation of the previous hops. If, at a certain point, it is discovered that a particular hop cannot maintain required bandwidth, RREQ is no longer forwarded. Nodes receiving multiple RREQ packets only forward the first RREQ packet that corresponds to a feasible partial path and discard subsequent RREQ packets for the same flow as shown in Figure 4.b.

[2]A similar routing algorithm based on bandwidth calculation in a system using CDMA over TDMA is presented in [18]. The routing protocol presented below [17] is chosen for presentation as the description was more detailed.

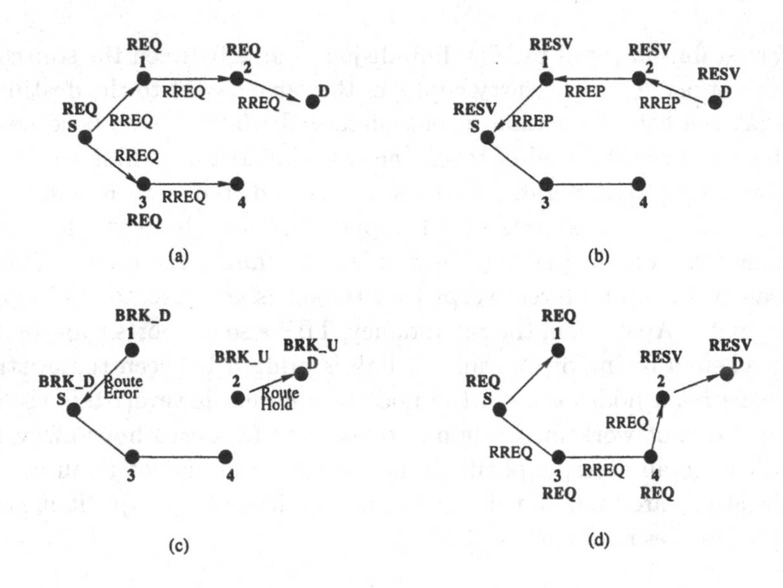

Figure 4: Path setup and rerouting with AODV-QoS protocol

As the RREQ packets are forwarded, the state of the node is set to REQ stating that resources are temporarily allocated. As the intermediate nodes, the destination node also reacts to the first feasible RREQ packet and sends back on the reverse direction a route reply packet (RREP) which converts the REQ state of the nodes on the selected path to RESV, indicating that resources are allocated for packet forwarding.

The proposed protocol deals with resource allocation using soft states as the underlying protocol AODV does. When the RREQ packet is forwarded in the network, many nodes will be set into REQ state, which goes back to neutral if they are not converted to RESV by a RREP packet. Similarly, the allocated resources, i.e. time slots, are released if RESV state is not refreshed by incoming packets of the flow. Note that, instead of using special control packets, data packets are used to refresh the reservations.

The route maintenance procedures proposed in [17] also utilize soft states. When a link becomes unavailable due to mobility, the node downstream to the point of failure sends a route hold packet toward the destination, indicating that an upstream link has failed and that they should enter BRK_U state and wait. Such a case is depicted in Figure 4.c, where the link between nodes 1 and 2 is broken because node 2 moves out of the communication range of node 1. Similarly, the node upstream sends a route error packet toward source, causing the upstream nodes to enter BRK_D state, indicating that a link downstream has failed. When the source receives the route error packet, it initiates a new path discovery procedure (Figure 4.d). The main idea behind this recovery scheme is to reuse already allocated resources if portions of the old and new paths overlap. As the RREQ packet is forwarded, if bandwidth is available, it converts neutral and BRK_D states to REQ, and converts BRK_U states to RESV directly. Intermediate nodes in BRK_U state are allowed to send RREP packets to the source to finalize the recovery process. In case of the source receiving multiple RREP packets, only the first one is processed and subsequent ones are discarded. With the use of soft states, this protocol achieves reuse of already allocated resources and solves the double resource booking problem.

3.3.5 INORA [19]

The INORA proposal for QoS-based routing is based on two other proposals: It uses TORA [5] for route computation in ad hoc networks, and INSIGNIA [20] for resource allocation and signaling.[3] TORA is an on-demand routing protocol that discover paths from source to destination by link reversal method. When a path to a destination is needed, a QUERY packet is flooded in the network. When the destination or an intermediate node that knows a path to destination receives the QUERY packet, it broadcasts an UPDATE packet that contains its relative height with respect to the destination. A node receiving the UPDATE packet sets its own height to a value more than the value contained in the UPDATE packet, creating a gradient toward the destination. The directed acyclic graph (DAG) generated by TORA maintains multiple paths to destination, which is used by the INORA proposal. INSIGNIA, on the other hand, is a signaling protocol which allocates, maintains, and releases resources for end-to-end flows. It is designed on the principles of in-band signaling and maintaining a soft state in the highly dynamic network topology. When allocating resources on a given path, the request is forwarded in a RES packet which indicates the requested resource amount. If a node in the path cannot allocate requested resources, the source is notified and packets are delivered in best effort mode. Note that INSIGNIA does not include mechanisms for path discovery and requires that paths are supplied by a routing protocol.

INORA protocol uses hop-by-hop feedback about bandwidth allocation

[3]The name of the protocol INORA is derived from these two protocols, INSIGNIA+TORA.

to search for alternative paths that can provide QoS guarantees. The paths calculated by TORA are used to forward INSIGNIA RES packets. Two feedback schemes, coarse and fine feedback, are supported under INORA. Under the coarse feedback scheme, the call admission feedback obtained from INSIGNIA is binary, i.e., the resources can either be allocated or not. When resources cannot be allocated at an intermediate node, an admission control failure (ACF) packet is sent to previous hop, which is not necessarily the source. This previous node selects the next best hop determined by TORA and forwards the request. ACF packets are generated and sent to the previous hop only when all possible paths starting from an intermediate node results in failure. Hence, the DAG generated by TORA is searched in a depth-first fashion. When using fine feedback scheme, the minimum and maximum bandwidth amounts requested by the flow are specified and the minimum-maximum bandwidth interval is divided into N classes. When reserving bandwidth on links, the call admission procedure generates ACF messages only if the minimum bandwidth level cannot be sustained. Otherwise, the maximum discrete amount of bandwidth from the minimum-maximum bandwidth interval is allocated, and a modified RES packet that contains a new maximum bandwidth is sent toward destination. The previous hop is notified about the amount of resources not allocated, which in turn searches alternative paths that can sustain the remaining bandwidth difference. Effectively, a large flow is split into smaller flows and delivered to the destination on alternative paths. The operation of INORA in a small network is depicted in Figure 5. The DAG from source S to destination D generated by TORA is shown in Figure 5.a. The numbers next to the links represent the available bandwidth on the corresponding links. Let S have a flow that has a minimum bandwidth requirement of 2 units and a maximum bandwidth requirement of 4 units. When the coarse feedback scheme is used, the flow follows the path marked with the dashed line in Figure 5.b since the entire flow must take the same path. However, when the fine feedback scheme is used, the flow can be split as shown in Figure 5.c, where two units are carried on the right path after bifurcation and 2 units carried on the left. Using the fine feedback scheme, available network resources can be utilized better since the bandwidth allocation need not be made as a whole.

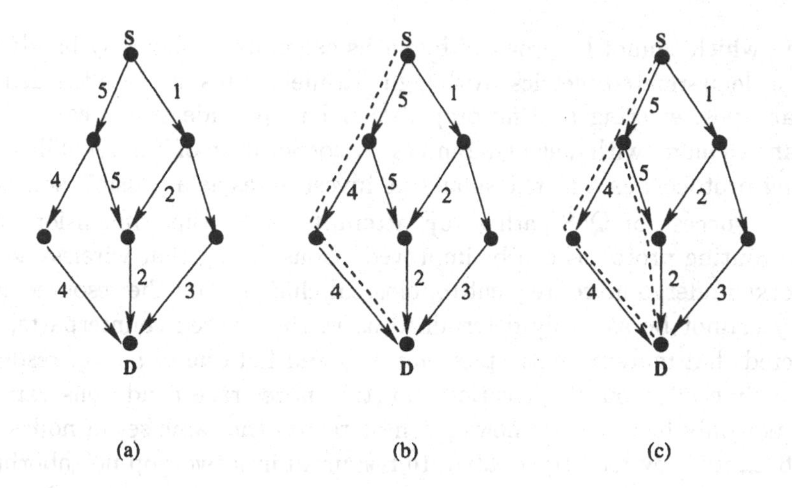

Figure 5: Example DAG generated by TORA and resource allocation options with INORA

4 Conclusions

QoS-based routing in ad hoc network is one of the most important topics that will enable the use of many applications in self-forming ad hoc network environments. These applications are expected to significantly improve effectiveness of operations conducted in locations without a wired infrastructure. Majority of the existing work on QoS-based routing for ad hoc networks aims to discover paths in this dynamic network environment that satisfy bandwidth or delay constraints. Although these efforts provide answers to many questions, there are other important issues that would increase the range of deployment scenarios and applications running on ad hoc networks.

One of the more important topics to be investigated is the scalability of QoS-based routing protocols for ad hoc networks. As the number of nodes in the network increases, the protocol overhead, especially regarding local state exchanges, increases. In literature, several solutions regarding the routing protocol scalability have been proposed [21]. The application of these methods to the QoS-based routing may result in higher performance in large scale networks.

Another important aspect of QoS-based routing is the selection of metrics considered for route calculation. Bandwidth, delay, and delay jitter are important metrics for multimedia applications. Mission-critical real-time applications, on the other hand, require delivery as well as timeliness guarantees, which cannot be provided by paths calculated using only bandwidth and/or delay-related metrics. Although parallel routes increase the delivery probabilities, existing routing proposals fail to provide statistical delivery guarantees along with delay guarantees. Incorporation of link reliability into routing protocols as a metric should be investigates as a possible solution.

The success of QoS-path setup attempts is another dimension along which routing protocols can be improved. Considering that wireless ad hoc networks undergo more frequent topological changes and the resource availability cannot be as easily determined as in their wired counterparts, it is expected that more path setup attempts would fail due to lack of resources at the time of resource allocation. Furthermore, race conditions can also arise not only because two flows attempt to use the same set of nodes, but also because they may try to allocate resources in a two-hop neighborhood, where simultaneous resource reservations may not be sustained. In either case, a failed attempt may require the system state to be generated anew, lengthening the path setup time and possibly resulting in unnecessary use of valuable wireless resources. Hence, routing protocols and applications requiring QoS guarantees may benefit from more efficient methods for determination of available network resources, their propagation in the network, and for resolving race conditions.

Although ad hoc networks operate independently from a wired infrastructure, it is very important to develop mechanisms that enable integration of QoS-based routing solutions with the QoS-support mechanisms of wired networks. Seamless integration of QoS-based paths in the ad hoc and wired domains would enable various new applications. Regarding the integration issue, the extension of the paths from wired to ad hoc domain can be investigated with an eye toward supplementing existing QoS support mechanisms for wired networks to operate in ad hoc networks.

References

[1] C.E. Perkins and P. Bhagwat, Highly Dynamic Destination-Sequenced Distance Vector Routing (DSDV) for Mobile Computers, *Proceedings of ACM SIGCOMM 94* (1994) pp. 234-244.

[2] M. Gerla, X. Hong, and G. Pei, Fisheye State Routing Protocol (FSR) for Ad Hoc Networks, *Internet Draft, draft-ietf-manet-fsr-03.txt,* (2002).

[3] P. Jacquet, P. Muhlethaler, A. Qayyum, A. Laouiti, L. Viennot, and T. Clausen, Optimized Link State Routing Protocol, *Internet Draft, draft-ietf-manet-olsr-07.txt,* (2002).

[4] C.E. Perkins and E.M. Royer, Ad-Hoc On-Demand Distance Vector Routing, *Proceedings of IEEE WMCSA 99* (1999) pp. 90-100.

[5] V.D. Park and M.S. Corson, A Highly Adaptive Distributed Routing Algorithm for Mobile Wireless Networks, *Proceedings of IEEE INFO-COM 97* (1997) pp. 1405-1413.

[6] D.B. Johnson, D.A. Maltz, Y. Hu, J.G. Jetcheva, The Dynamic Source Routing Protocol for Mobile Ad Hoc Networks (DSR), *Internet Draft, draft-ietf-manet-dsr-07.txt,* (2002).

[7] E.M. Royer and C.K. Toh, A Review of Current Routing Protocols for Ad Hoc Mobile Wireless Networks, *IEEE Personal Communications* (Apr. 1999), pp. 46-55.

[8] S. Chen and K. Nahrstedt, An Overview of Quality of Service Routing for Next-Generation High-Speed Networks: Problems and Solutions, *IEEE Network,* Vol.12 No.6 (Nov./Dec. 1998) pp.64-79.

[9] J.M. Jaffe, Algorithms for Finding Paths with Multiple Constraints, *IEEE Network* Vol.14 (1984) pp. 95-116.

[10] A. Kershenbaum, *Telecommunications Network Design Algorithms,* (McGraw Hill, 1993).

[11] S. Chakrabarti and A. Mishra, QoS Issues in Ad Hoc Wireless Networks, *IEEE Communications Magazine* Vol.39 No.2 (Feb. 2001) pp. 142-148.

[12] R. Braden, L. Zhang, S. Berson, S. Herzog, and S. Jamin, Resource ReSerVation Protocol (RSVP) - Version 1 Functional Specification, *RFC 2205,* (Sep. 1997).

[13] R. Guerin and A. Orda, QoS-based Routing in Networks with Inaccurate Information: Theory and Algorithms, *Proceedings of IEEE INFO-COM 97* (1997) pp. 75-83.

[14] T. Chen, J.T. Tsai, and M. Gerla, QoS Routing Performance in Multi-hop Multimedia Wireless Networks, *Proceedings of IEEE Sixth International Conference of Universal Personal Communications* Vol.2 (1997) pp. 557-561.

[15] R. Sivakumar, P. Sinha, and V. Bharghavan, CEDAR: Core Extraction Distributed Ad Hoc Routing Algorithm, *IEEE Journal on Selected Areas of Communications* Vol.17 No.8 (Aug. 1999) pp. 1454-1465.

[16] S. Chen and K. Nahrstedt, Distributed Quality-of-Service Routing in Ad Hoc Networks, *IEEE Journal on Selected Areas of Communications* Vol.17 No.8 (Aug. 1999) pp. 1488-1505.

[17] C. Zhu and M.S. Corson, QoS Routing for Mobile Ad Hoc Networks, *IEEE INFOCOM 2002* Vol.2 (2002) pp. 958-967.

[18] C.R. Lin, On-Demand QoS Routing in Multihop Mobile Networks, *IEEE INFOCOM 2001* Vol. 3 (2001) pp. 1735-1744.

[19] D. Dharmaraju, A. Roy-Chowdhury, P. Hovaresti, and J.S. Baras, INORA - A Unified Signaling and Routing Mechanism for QoS Support in Mobile Ad Hoc Networks, *Proceedings of International Conference on Parallel Processing Workshops* (2002) pp. 86-93.

[20] S. Kee, G. Ahn, X, Zhang, and A.T. Campbell, INSIGNIA: An IP-Based Quality of Service Framework for Mobile Ad Hoc Networks, *Journal of Parallel and Distributed Computing* Vol. 60 (2000) pp. 374-406.

[21] X. Hong, K. Xu, and M. Gerla, Scalable Routing Protocols for Mobile Ad Hoc Networks, *IEEE Network* (Jul./Aug. 2002) pp. 11-21.

Quality of Service Routing in Mobile Ad Hoc Networks

Imad Jawhar and Jie Wu
Department of Computer Science and Engineering
Florida Atlantic University, Boca Raton, FL 33431
E-mail: {imadj, jie}@cse.fau.edu

Contents

1 Introduction

As technology advances, wireless and portable computers and devices are becoming more powerful and capable. These advances are marked by an increase in CPU speed, memory size, disk space, and a decrease in size and power consumptions. The need for these devices to continuously communicate with each other and with wired networks is becoming increasingly essential. Mobile ad hoc networks (MANETs) open the door for these devices to establish networks on the fly, i.e., formally, a MANET is a collection of mobile devices which form a communication network with no pre-existing wiring or infrastructure. They allow the applications running on these wireless devices to share data of different types and characteristics. There are many applications of MANETs, each with different characteristics of network size (geographic range and number of nodes), node mobility, rate of topological change, communication requirements, and data characteristics. Such applications are conferences, classroom, campus, military, and disaster recovery. Each node is directly connected to all nodes within its own effective transmission range. Nodes in the network are allowed to move in and out of range of each other. Communication between nodes that are not within range of each other is accomplished by establishing and using multi-hop routes that involve other nodes which act as routers. New nodes can join the network at any time and existing nodes can leave the network as well.

Due to the dynamic nature of MANETs, designing communications and networking protocols for these networks is a challenging process. One of the most important aspects of the communications process is the design of the routing protocols used to establish and maintain multi-hop routes to allow the communication of data between nodes. A considerable amount of research has been done in this area, and multi-hop routing protocols have been developed. Most of these protocols such as the Dynamic Source Routing protocol (DSR) [29], Ad Hoc on Demand Distance Vector protocol (AODV) [31], Temporally Ordered Routing Protocol (TORA) [28], and others establish and maintain routes on a best-effort basis. While this might be sufficient for a certain class of MANET applications, it is not adequate for the support of more demanding applications such as multimedia audio and video. Such applications require the network to provide guarantees on the Quality of Service (QoS).

Some researchers have been active in the area of QoS support in MANETs, and have proposed numerous QoS routing protocols for this environment. Most of these protocols provide QoS support for the available bandwidth requirement for a given path. This is because bandwidth is the most critical parameter in most MANET applications due to the scarcity of this resource in the wireless environment. The protocols that are discussed in this chapter support quality of service to varying degrees, in different ways, and using various network and and communication models.

In this chapter, these different approaches are presented, discussed and classified according to which of the main existing best-effort routing algorithms (DSR, AODV, DSDV, TORA, etc) they extend or are most closely related. In addition, some protocols are based on new algorithms. The QoS routing protocols that are discussed operate in both the network layer and the medium access control (MAC) layer which is equivalent to the data link layer in the OSI model. There are also design approaches, such as the IP-Based quality of service framework for MANETs (INSIGNIA) [1][15][16] and the integrated mobile ad hoc QoS framework (iMAQ) [3], which are designed to support multimedia traffic and achieve better efficiency in terms of bandwidth and energy consumption through the implementation of inter-layer QoS frameworks. As these protocols are classified and presented, the networking layer, or layers in some cases, within which they operate, as well as their assumed QoS model will be identified and discussed.

The remainder of this chapter is organized in the following manner. First, an overview of the topic is presented which includes a brief look at the best-effort routing algorithms which are both most popular and most used. Then the typical models used for QoS routing protocols are discussed along with the different layers within which the QoS support mechanisms are included. Existing QoS routing protocols are then classified according to which best effort routing protocol they extend or to which they are closest in design. As these protocols are discussed, the QoS model and networking environment they assume is identified along with the communication layer within which they are designed to operate. The chapter is completed with a conclusions and future research section.

2 Overview

Most QoS routing algorithms represent an extension of existing classic (or major) best-effort routing algorithms. Many routing protocols have been developed which support establishing and maintaining multi-hop routes between nodes in MANETs. These algorithms can be classified into two different categories: on-demand (reactive) such as DSR, AODV, and TORA, and table-driven (proactive) such as Destination Sequenced Distance Vector protocol (DSDV). In the on-demand protocols, routes are discovered between a source and a destination only when the need arises to send data. This provides a reduced overhead of communication and scalability. In the table-driven protocols, routing tables which contain routing information between all nodes are generated and maintained continuously regardless of the need of any given node to communicate at that time. With this approach, the latency for route acquisition is relatively small, which might be necessary for certain applications, but the cost of communications overhead incurred in the continued update of information for routes which might not be used for a long time if at all is too high. Furthermore, this approach requires more memory due to significant increase in the size of the routing table. These requirements put limits on the size and density of the network. A third hybrid approach, the Zone Routing Protocol (ZRP), has also been proposed and attempts to reap the benefits of both methods. In ZRP, the network is divided into zones. A proactive table driven strategy is used for establishment and maintenance of routes between nodes of the same zone, and a reactive on-demand strategy is used for communication between nodes of different zones. This approach can be effective in larger networks with applications that exhibit a relatively high degree of locality of communication, where communication between nodes with close proximity to one another is much more frequent than that between nodes which are further apart.

Before presenting the current approaches for design and implementation of QoS routing protocols, it is important to briefly discuss the existing best-effort routing protocols which exist for MANETs. Many routing protocols have been designed to discover and maintain routes between source and destination nodes.

Among the most important and classic routing algorithms for MANETs that have evolved are three basic types. Each of these three basic types has its own advantages, disadvantages, and appropriateness of use in certain types of ad hoc networks depending on the mobility, number of nodes involved, node density, underlying link layer technology, and general characteristics of the environment and applications being supported. These three routing algorithms are: (1) reactive (on-demand) such as DSR (Dynamic Source Routing) protocol, AODV (Ad hoc On Demand Distance Vector) routing protocol, and TORA (Temporally Ordered Routing Alogrithm) protocol, and (2) proactive (table-driven) such as DSDV (Destination Sequenced Distance Vector) protocol. There are also other types of routing protocols designed for more scalability such as (3) the ZRP (Zone Routing Protocol), which is a hybrid framework for routing in ad hoc networks (proactive within the zone and reactive between zones), in addition to others, which will be mentioned subsequently.

2.1 DSR - the Dynamic Source Routing Protocol

DSR is one of the most well-known routing algorithms for ad hoc wireless networks. It was originally developed by Johnson, Maltz, and Broch [29]. DSR uses source routing, which allows packet routing to be loop free. It increases its efficiency by allowing nodes that are either forwarding route discovery requests or overhearing packets through promiscuous listening mode to cache the routing information for future use. DSR is also on demand, which reduces the bandwidth use especially in situations where the mobility is low. It is a simple and efficient routing protocol for use in ad hoc networks. It has two important phases, route discovery and route maintenance. The main algorithm works in the following manner. A node that desires communication with another node first searches its route cache to see if it already has a route to the destination. If it does not, it then initiates a route discovery mechanism. This is done by sending a Route Request message. When the node gets this route request message, it searches its own cache to see if it has a route to the destination. If it does not, it then appends its id to the packet and forwards the packet to the next node; this continues until either a node with a route to the destination is encountered (i.e. has a route in its own cache) or the destination receives the packet. In that case, the node sends a route reply packet which has a list of all of the nodes that forwarded the packet to reach the destination. This constitutes the routing information needed by the source, which can then send its data packets to the destination using this newly discovered route. Although DSR can support relatively rapid rates of mobility, it is assumed that the mobility is not so high as to make flooding the only possible way to exchange packets between nodes.

2.2 AODV - The Ad Hoc On-demand Distance-Vector Protocol

AODV [31] is another routing algorithm used in ad hoc networks. Unlike DSR (Dynamic Source Routing Algorithm) it does not use source routing, but like DSR it is on-demand. In AODV, each node maintains a routing table which is used to store destination and next hop IP addresses as well as destination sequence numbers. Each entry in the routing table has a destination address, next hop, precursor nodes list, lifetime, and distance to destination.

To initiate a route discovery process a node creates a route request (RREQ) packet. The packet contains the source node's IP address as well as the destination's IP address. The RREQ contains a broadcast ID, which is incremented each time the source node initiates a RREQ. The broadcast ID and the IP address of the source node form a unique identifier for the RREQ. The source node then broadcasts the packet and waits for a reply. When an intermediate node receives a RREQ, it checks to see if it has seen it before using the source and broadcast ID's of the packet. If it has seen the packet previously, it discards it. Otherwise it processes the RREQ packet. To process the packet the node sets up a reverse route entry for the source node in its route table which contains the ID of the neighbor through which it received the RREQ packet. In this way, the node knows how to forward a route reply packet (RREP) to the source if it receives one later. When a node receives the RREQ, it determines if indeed it is the indicated destination and, if not, if it has a route to respond to the RREQ. If either of those conditions is true, then it unicasts a route replay (RREP) message back to the source. If both conditions are false,i.e. if it does not have a route and it is not the indicated destination, it then broadcasts the packet to its neighbors. Ultimately, the destination node will always be able to respond to the RREQ message. When an intermediate node receives the RREP, it sets up a forward path entry to the destination in its routing table. This entry contains the IP address of the destination, the IP address of the neighbor from which the RREP arrived, and the hop count or distance to the destination. After processing the RREP packet, the node forwards it toward the source. The node can later update its routing information if it discovers a better route. This could be used for QoS routing support to choose between routes based on different criteria such as reliability and delay. To provide such support additional QoS attributes would need to be created, maintained, and stored for each route in the routing table to allow the selection of the appropriate route among multiple routes to the destination.

2.3 TORA - The Temporally Ordered Routing Algorithm

TORA [28] is the most well known LRR (Link Reversal Routing) algorithm [29] which provides a very adaptive type of routing. It is intended to be used in networks with rapidly changing topologies. It uses a strategy of de-coupling of far-reaching control message propagation from the dynamics of the network's topology. It is efficient to use TORA in networks where the rate of topology changes is not so fast as to make flooding the only form of transmitting messages and not so slow as to make the use of algorithms supporting shortest path calculations applicable. Therefore, the algorithm's applicability is a function of the network's size, rate of topological changes, and available bandwidth. TORA minimizes the network messages in reaction to changes in topology, which are caused by link activation and failure. The algorithm localizes the reaction to these topological changes. TORA does not maintain information sufficient to support shortest path calculation, and maintains only state information sufficient to form a DAG (directed acyclic graph) routed at the destination. The destination is therefore the only node with no outgoing links (a sink). The maintenance of the DAG provides loop free communication to the destination. It also allows the existence of multiple paths to the destination. This provides good reliability, which is desirable in ad hoc networks, and possible QoS extension support, by selecting paths with particular characteristics and that can support pre-specified QoS constraints.

TORA is source initiated and demand driven. Therefore, due to its nature, it forgoes optimal routing. It does not make sure to select the shortest possible path, even though it can be shown that due to the nature of RPY message propagation, shorter paths are more likely to form. However, it provides routing which is very adaptive and scalable with relatively small overhead bandwidth usage for control messages. In addition, lower delivery latency can be achieved.

In contrast with other earlier LLR (Link Reversal Routing) algorithms, TORA's key feature is its reaction to link failures. This reaction is structured as a temporally ordered sequence of diffusing computations with each computation consisting of a sequence of directed link reversals. Each link reversal sequence effectively conducts a search for alternative routes to the destination. The search mechanism in TORA often involves only a single pass of the distributed algorithm because it simultaneously modifies the routing tables during the outward phase of the search procedure itself. This is not the case in other approaches such as DSR and AODV which take three-pass procedures (i.e. route-error/route-request/route-reply) to discover new routes when a node loses its last route [29]. The algorithm uses a "physical or logical clock" to provide a temporal order of topological change events, which is used to structure the protocol's reaction to changes. More information on TORA is available in [29] and in [28].

2.4 DSDV - The Destination Sequenced Distance Vector Protocol

DSDV [29] is one of the most well known table-driven routing algorithms for MANETs. It is a distance vector protocol. In distance vector protocols, every node i maintains for each destination x a set of distances $\{d_{ij}(x)\}$ for each node j that is a neighbor of i. Node i treats neighbor k as a next hop for a packet destined to x if $d_{ik}(x)$ equals $min_j\{d_{ij}(x)\}$. The succession of next hops chosen in this manner leads to x along the shortest path. In order to keep the distance estimates up to date, each node monitors the cost of its outgoing links and periodically broadcasts to all of its neighbors its current estimate of the shortest distance to every other node in the network. The distance vector which is periodically broadcasted contains one entry for each node in the network which includes the distance from the advertising node to the destination. The distance vector algorithm described above is a classical Distributed Bellman-Ford (DBF) algorithm.

DSDV is a distance vector algorithm which uses sequence numbers originated and updated by the destination, to avoid the looping problem caused by stale routing information. In DSDV, each node maintains a routing table which is constantly and periodically updated (not on-demand) and advertised to each of the node's current neighbors. Each entry in the routing table has the last known destination sequence number. Each node periodically transmits updates, and it does so immediately when significant new information is available. The data broadcasted by each node will contain its new sequence number and the following information for each new route: the destination's address, the number of hops to reach the destination and the sequence number of the information received regarding that destination, as originally stamped by the destination. No assumptions about mobile hosts maintaining any sort of time synchronization or about the phase relationship of the update periods between the mobile nodes are made. Following the traditional distance-vector routing algorithms, these update packets contain information about which nodes are accessible from each node and the number of hops necessary to reach them. Routes with more recent sequence numbers are always the preferred basis for forwarding decisions. Of the paths with the same sequence number, those with the smallest metric (number of hops to the destination) will be used. The addresses stored in the route tables will correspond to the layer at which the DSDV protocol is operated. Operation at layer 3 will use network layer addresses for the next hop and destination addresses, and operation at layer 2 will use layer-2 MAC addresses.

2.5 Other Approaches

In addition to the standard routing protocols discussed above, there exist other protocols which use different approaches. The following are some of these protocols [29].

Location-Assisted Routing: This approach improves route discovery and maintenance with the use of localization information which is accomplished by keeping track of the position and velocity of the mobile node. Nodes not in that general direction can be excluded from the route discovery and maintenance process to reduce bandwidth consumption and control message communication overhead.

Fisheye Routing: This is a form of routing which has nodes keeping track of more topology data for closer nodes. It is similar to the Zone Routing Protocol strategy, but with blurred boundaries between zones. This strategy can be very useful to increase the scalability of routing protocols.

Cedar (Core Extraction Distributed Ad Hoc Routing): The strategy in this type of routing algorithm is to increase scalability by creating and maintaining a backbone for communication of route requests to avoid broadcasting such information on a network-wide basis. The difficulty is in managing these backbone nodes, which can move relative to each other. Wu and Li in [36] provide a good algorithm for constructing a core which is a connected dominating set. Further research in this area is needed.

3 QoS Routing Protocols: Models and Classification

In this section, the different QoS models used in literature are presented. This is followed by a classification of the current QoS routing protocols according to the best effort routing protocol they extend as well as the model and environment they assume, and the communication layer within which they operate.

3.1 QoS Models in MANETs

Depending on the application involved, the QoS constraints could be available bandwidth, end-to-end delay, delay variation (jitter), probability of packet loss, and so on. This kind of demand puts more pressure on the network and the routing protocols which are used to support the communications. Establishing multi-hop routes between nodes is not sufficient in this case. The discovered routes can only be considered if they provide guarantees of the QoS parameters, such as bandwidth required by the application. Let $m(u, v)$ be the performance metric for the link (u, v) connecting node u to node v, and path $(u, u_1, u_2, ..., u_k, v)$ a sequence of links for the path from u to v. Three types of constraints on the path can be identified [6][33]:

1. *Additive constraints*: A constraint is additive if $m(u, v) = m(u, u_1) + m(u_1, u_2) + ... + m(u_k, v)$. For example, the end-to-end delay (u, v) is an additive constraint because it consists of the summation of delays for each link along the path.

2. *Multiplicative contraint*: A constraint is multiplicative if $m(u, v) = m(u, u_1) \times m(u_1, u_2) \times ... \times m(u_k, v)$. The probability of a packet $prob(u, v)$, sent from a node u to reach a node v, is multiplicative, because it is the product of individual probabilities along the path.

3. *Concave constraint*: A constraint is concave if $m(u, v) = min\{m(u, u_1), m(u_1, u_2), ..., m(u_k, v)\}$. The bandwidth $bw(u, v)$ requirement for a path between node u and v is concave. This is due to the fact that it consists of the minimum bandwidth between the links along the path.

Wang and Hou [33] provide a list of twelve combinations with multiple constraints. It has been proven in [35] that any multiple constraints with two or more type 1 and/or type 2 constraints are NP-complete; otherwise, they are tractable. Approximation methods exist for QoS constraints that are NP-complete. *Sequential filtering* is a commonly used approach, where multiple paths between two nodes u and v that satisfy a single metric first (like bw) are found, then a subset of these paths is eliminated by optimizing over a second metric (like end-to-end delay), and so on.

In MANETs, node mobility often results in frequent topology changes, which presents a significant challenge when designing QoS routing protocols. High node mobility can make satisfying QoS requirements unreachable. Consequently, it is required that the network be *combinatorically stable* in order to achieve QoS support [2]. This means that the changes in network topology must be slow enough within a particular time window to allow the topology updates to propagate successfully as required in the network.

QoS support of MANETs requires availability of network state. However, due to mobility and constant topology changes, the cost of maintenance of the network state is expensive especially in large networks. In [4] the *imprecise network state model* is introduced. It provides a cost-effective method for providing QoS support based on imprecise network information. The majority of QoS routing protocols are reservation-based. Probe messages are sent through the network from the source to the destination in order to discover and reserve paths which satisfy a given QoS requirement. Due to the dynamic nature of the network, reserved QoS paths must be reaffirmed periodically by sending special control packets, called *refreshers*, along the path. Another approach, called *soft state*, relies on periodic time out at each node for path maintenance.

In addition, due to the difficulty of QoS support in the inherently dynamic environment of MANETs, some more "compromising principles" have been presented; *Soft QoS* and *QoS adaptation*. Soft QoS [9] indicates that there may be transient periods of time during which the QoS specifications are not honored. However, the QoS satisfaction is quantified by the total disruption time over the total connection time. This ratio must be above a specified threshold in order to fulfill the QoS requirements. In the *fixed−level QoS* approach, the reservation is defined in an n-dimensional space where the coordinates define the characteristics of the service [27]. On the other hand, QoS adaptation introduces the concept of *dynamic QoS*, where a range of QoS values, rather than a single point, is allowed to be specified by the application. This must be done through appropriate, flexible, and simple user interface which effectively maps the perceptual parameters into QoS constraints. The use of dynamic QoS provides more flexibility to the system and gives the network the ability to adjust the allocation according to the current availability of the required resources. The higher networking layers can then adapt to these changes. This adaptation can be achieved in different ways at the different layers of the architecture. The physical layer, for example, can adjust the transmission power to react to more frequent bit errors. The link layer can incorporate more error control (detection and correction) codes as well as automatic repeat requests (ARQ) in reaction to changes in link error rates.

At the other end of the OSI stack, namely the application layer (multimedia video conferencing for example), different compression techniques with varying compression ratios can be employed to adapt the application to the changes in bandwidth, delay, and error rates without drastically compromising the perceived audio and video quality. As more resources become available, the quality of the presentation can then be adjusted to take advantage of the added resources. In addition to compression algorithms, other techniques are being investigated at this level including layered encoding, rate shaping, adaptive error control, and bandwidth smoothing.

It is important at this point to state that the QoS model defines the general approach, goals, and framework for providing QoS support in a network. It does not specify a particular protocol, design, or implementation details. Providing QoS support is done at each of the layers of the OSI model starting from the application layer and ending with the physical layer. Various protocols and specifications such as QoS user interface, routing, signalling, resource reservation, and error checking, measuring, and correcting must work and coordinate together in a collaborative and complementary fashion in order to satisfy the QoS requirements of the underlying applications. In this chapter, we focus on QoS routing, which is one of the most critical components in providing QoS support in MANETs.

3.2 QoS Routing Protocols

In order to support applications with quality of service requirements, such as multimedia, different QoS routing protocols have been developed. Of the quality of service parameters that are required by these applications that were mentioned earlier, minimum bandwidth is the most common. To support such requirement, the application layer of the source node sends a request to the lower layers (in the OSI model) with a specific destination and an amount of bandwidth which is required in order to satisfy the communication needs of that particular session. Depending on the communication requirements used, this desired amount of bandwidth is represented in different ways. In the synchronous Time Division Multiple Access (TDMA) environment, which is described later in this chapter, the bandwidth is represented by the number of slots needed to be reserved in the TDMA frame. In order for the route to satisfy this requirement, each of its links must reserve that number of slots for this particular session. When the session is ended, the reserved slots are freed and allowed to be reserved for other sessions. On the other hand, in the asynchronous environment, the slot size is variable and the amount of bandwidth reserved on particular links of the path is represented by a duration and a start time within the super frame [19][21][26][32]. The location of the slots within the frames of the different links of a path have direct affect over the total end-to-end transmission delay of data for a particular source-destination pair during a session. Different algorithms which have been proposed which choose the start times and durations of the slots according to different policies. The policies can be minimize end-to-end delay, maximize the probability of success in applying the reservation algorithm, or compromise between the two objectives.

QoS routing protocols for MANETs can be classified into different categories depending on the best effort routing algorithm which they extend or most closely resemble (DSR, AODV, DSDV, TORA, etc.) Though most of the the QoS routing protocols are designed to operate within the network layer, some of the implementations go below the network layer into the MAC layer. In the following sections, different QoS routing protocols are presented.

4 Sample QoS Routing Protocols

The following samples of QoS routing protocols are presented grouped according to which best effort routing protocol they extend or to which they are most closely related. As each of the protocols is discussed, the following characteristics will be identified: (1) the network layer within which it resides. (2) The QoS model it assumes (TDMA, CDMA-over-TDMA, etc.) (3) The networking environment within which it is designed to operate (synchronous or asynchronous).

4.1 Extensions of DSR

Many of the QoS routing algorithms represent extensions of DSR, which is a popular on-demand protocol for MANETs. Since this category has a larger share of QoS protocols than its counterparts, both it and its associated protocols are presented with more detail than the others. Some of these protocols are shown in table 1 and are discussed below.

The QoS routing algorithm in [17] by Liao et al. is an extension of the DSR protocol. It is on-demand and can operate in a single channel/code or multiple channel/code environment. In the single channel case, 1-hop neighbor nodes (nodes that are within range of each other) transmit and receive on the same channel frequency in frequency division multiplexing (FDM) networks, or code in the code division multiplexing (CDM) networks. The implementation of the protocol in [17], assumes a TDMA synchronous networking environment (single channel mode). In this network, communication between nodes is done using a synchronous TDMA frame. The TDMA frame is composed of a control phase and a data phase

[24]. Figure 3 shows the TDMA frame structure for a TDMA network (or a TDMA cluster) of N nodes. Each node in the network has a designated control time slot (control slots 1 through N in this example), which it uses to transmit its control information, but, the nodes in the network must compete for use of the data

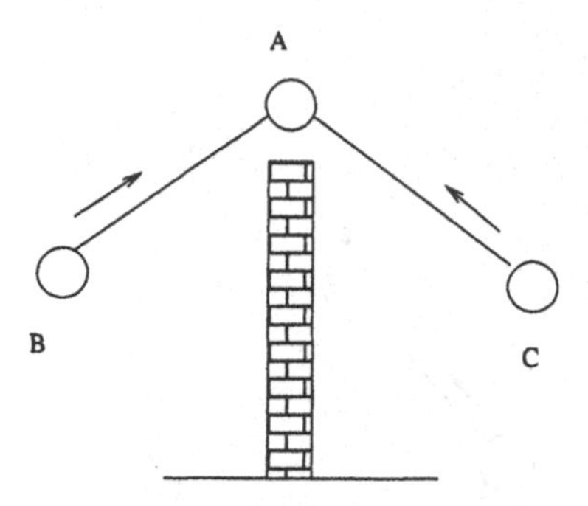

Figure 1: The hidden terminal problem creating a collision at node A.

time slots (data slots 1 through M in this example) in the data phase of the frame.

As mentioned earlier, the TDMA environment is a single channel model. This model is generally practical and less expensive because only a relatively simple transmission mechanism and antenna design is needed. However, this model imposes on the designer the constraint of the hidden terminal and exposed terminal problems. The routing protocol must account for these problems and have appropriate mechanisms to avoid hidden terminal interference on one hand, and maximize channel re-use by taking advantage of the exposed terminal transmissions on the other hand. Consider the example in figure 1. A *hidden terminal* problem in a wireless environment is created when two nodes which are out of range of each other, B and C for example, transmit to a third node A, which can hear them both. This creates a collision of the two transmissions at the "middle node" A. An *exposed terminal* is illustrated in figure 2. It is created when a node A, is within range of and between two other nodes B and C, which are out of range of each other. When A wants to transmit to one of them, node B for example, the other node, C in this case, is still able to transmit to a fourth node, D which is in C's range (but out of the range of node A). Here A is an exposed terminal to C but can still transmit to B.

Liao and Tseng [17] specify three rules which must be satisfied for proper slot allocation at a particular node. These rules are in place in order to prevent transmission collisions that are due to the nature of the wireless medium, and to avoid the hidden terminal problem. They state that a time slot t is considered free to be allocated to send data from a node x to a node y if the following conditions are true:

1. Slot t is not scheduled for receiving or transmitting in neither node x nor y.

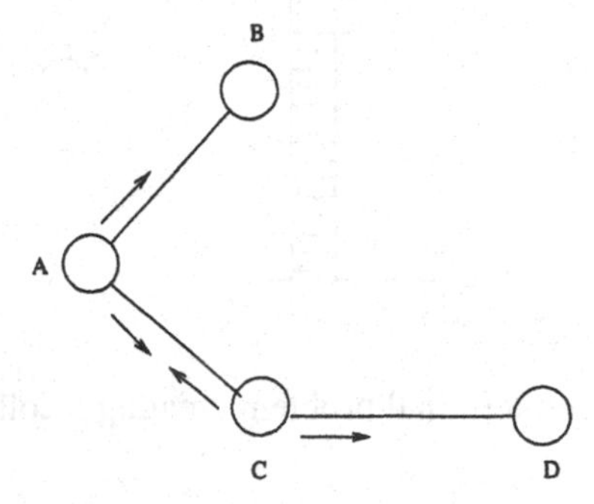

Figure 2: An exposed terminal is created at node A. Node A is exposed to node C's transmission to node D, but can still transmit to node B.

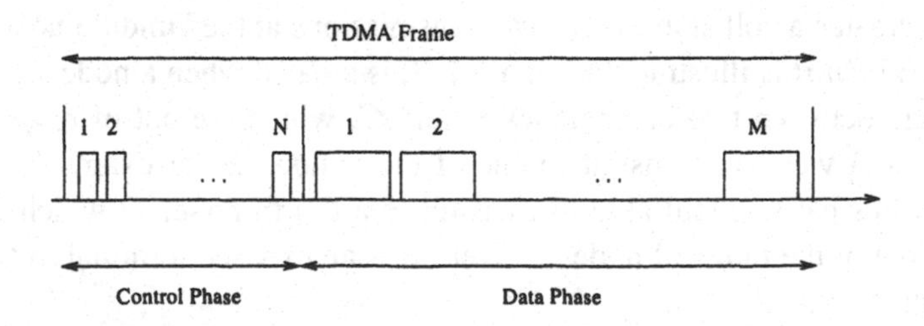

Figure 3: The structure of a TDMA frame for a network of N nodes and M data slots per frame. Each node has a fixed control slot. Nodes compete over the use of data slots.

2. Slot t is not scheduled for receiving in any node z that is a 1-hop neighbor of x.

3. Slot t is not scheduled for sending in any node z that is a 1-hop neighbor of y.

Race condition due to multiple reservations at intermediate nodes. The protocol in [17], like many of the other QoS routing protocols [10][11][18][17], do not account for the race conditions which can become more significant with increased node mobility, network density and higher traffic loads. These race conditions can arise when multiple path reservations are being processed at intermediate nodes at the same time. A solution for such race conditions is presented in [14]. The following is a brief description of the race conditions and the solution provided. Consider the example shown in Figure 4. Here, two QoS path reservations are taking place simultaneously. A path is being reserved from node A to C (source node A sends the QoS request message QREQ1 to propagate to destination node C) and another is sent from node D to E (source node D sends the QoS request message QREQ2 to propagate to destination node E). The two requests pass through a common intermediate node B. When node B receives QREQ1 (with say b slots required) from node A to node C, it allocates b slots and forwards the request. Let slot t be among these allocated slots. Before B receives the reply message, QREP1, which would confirm the QoS path reservation from node C to A and reserve the allocated slots, it is possible the another request, QREQ2, can arrive at node B. QREQ2 from node D requests to reserve a path from node D to node E passing through node B. In the algorithm in [17], node B would potentially go ahead and allocate one or more of the same slots, including slot t in this example, for the request, QREQ2, for the path from D to E. This constitutes a double allocation of the same slots to two different QoS paths. When the reply message, QREP1, arrives at B to confirm the QoS path reservation from C to A, node B will go ahead and confirm these slots, including slot t, and mark them as reserved in its ST and RT tables. Later, when the other reply message, QREP2, arrives at node B to confirm the QoS path from D to E, node B will potentially again reserve the same slots, including slot t in this example, for the second QoS path. Therefore, due to this race condition, the same slot t was reserved for two different QoS paths. This would create a conflict when the source nodes start using these reserved QoS paths to send data.

The conflict arises when the packets are transmitted from A to C and D to E simultaneously, and two data packets from two different paths arrive at node B. In this case, node B must decide which data packet it will actually send. The other data packet will be dropped. In this case, node B can, if the protocol requires, inform the other source of this error condition, or the source would simply time out the request. The corresponding source must then start the process of trying to reserve a new QoS path all over again. This leads to a decline in the throughput. In [14] an algorithm is proposed to fix this problem which is called the "race condition due to multiple reservations at an intermediate node".

Parallel race condition. The paper in [14] presents another possibility of a race condition which is due to a *parallel reservation problem*. An example of this race condition is shown in Figure 5. In this case, we have two parallel paths, ABCD and EFGH, being reserved. Two or more of the intermediate nodes belonging to the two parallel paths are 1-hop neighbors. In this case, node B, which belongs to the first path, and node F, which belongs to the other path are 1-hop neighbors. This is indicated in the figure using the dashed lines. The same relationship exists between nodes C and G. When QREQ1 is propagating from node A to D, the slots are allocated at the intermediate nodes. However, if the slot allocation information is not maintained by the nodes, say node B here, but only placed in the QREQ1 message, then no memory of this allocation is kept by the node, as is the case in [17]. This can cause another type of race condition, which we call the parallel reservation problem. This problem arises if, before QREQ1 propagates and is confirmed, the same process occurs with QREQ2 and node F allocates slots for the other QoS path and does not take into consideration the allocation of slots for QREQ1 at node B.

If both QREQ messages are successful in reserving their corresponding paths, a potential problem exists because the slot allocations at nodes B and F can be violating the slot allocation conditions mentioned earlier in this chapter. Nodes B and F each did the allocation based on information which did not consider the other 1-hop neighbor node's slot allocation for the corresponding parallel path being reserved. Again, if the two parallel paths are reserved successfully and data transmission is started along these paths, collisions will occur at the 1-hop neighbors belonging to the different parallel paths. In this example, nodes B and F would experience this collision in their transmissions. A similar situation can occur between any 1-hop neighbors belonging to the two parallel paths, for example, between nodes C and G of the same figure.

It is important to note that this parallel reservation problem can occur in any situation where the two paths have 1-hop neighbors, with each belonging to the other path. This would also be the case in the example presented in Figure 6, where nodes B and E are 1-hop neighbors who belong to two different QoS paths. In [14], an algorithm which fixes this *parallel reservation problem* is proposed.

A more detailed example of the multiple QoS path reservation competition is shown in Figure 7. Node A wants to request a QoS path to node C with $b = 3$ (i.e. the required bandwidth is 3 slots). Node A sends a QoS request, QREQ1, to reserve the path. The QREQ message travels through the nodes on its way to C

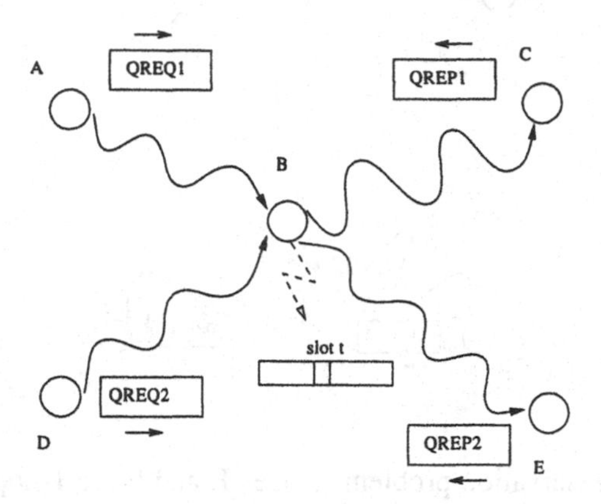

Figure 4: Multiple QoS path reservation competition. Two different QoS paths A..B..C and D..B..E are being reserved simultaneously, and they both pass through node B.

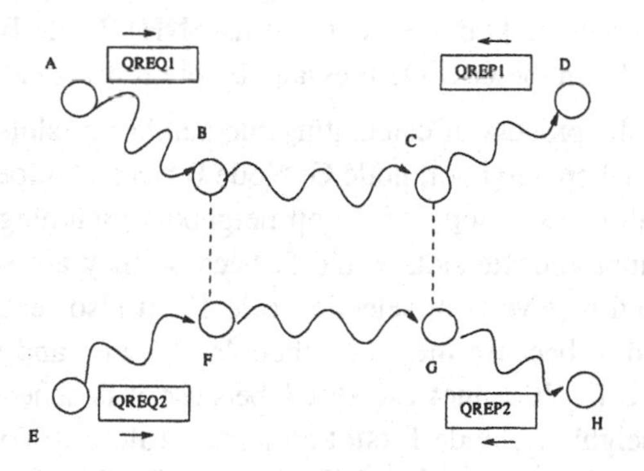

Figure 5: Parallel reservation problem. Nodes B and F (similarly, nodes C and G) are 1-hop neighbors which belong to two different QoS paths ,A..B..C..D and E..F..G..F, that are being reserved simultaneously.

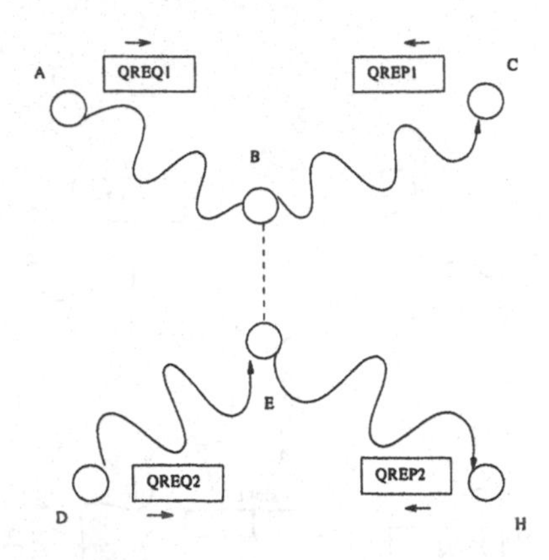

Figure 6: Parallel reservation problem. Nodes B and E are 1-hop neighbors which belong to two different QoS paths, A..B..C and D..E..F, that are being reserved simultaneously.

and arrives at node B. We see that node B has nodes F and G as 1-hop neighbors, and node G has node B and H as 1-hop neighbors. Node B will now try to allocate slots for this arriving QREQ1 message to send to each of its 1-hop neighbors, if there are b slots available to send from itself to this neighbor. It will calculate the number of slots available to each of those neighbors and will place those neighbors along with the allocated slots in the next hop list (NH). Node B will then include the next hop list (NH) in the QREQ1 message before it broadcasts (forwards) it.

Let's consider the process of calculating the number of slots available to send from node B to its 1-hop neighbor, node G. Node B has slot allocation information for itself and for all of its 1-hop and 2-hop neighbors including node G. Node B realizes that it cannot allocate slots 2 and 5, because they are scheduled by node B itself, to send and receive (slot selection rule 1). It also realizes that it cannot allocate slots 3 and 4, because they are scheduled to send and receive in node G (slot allocation rule 1). It cannot use slot 1 because it is scheduled to receive in one of its 1-hop neighbors, node F (slot allocation rule 2). Note that this is due to the hidden terminal problem; if node B sends to G using slot 1, this will cause a collision at node F which is using slot 1 to receive as well. Furthermore, node B cannot use slot 6, because it is scheduled to send in node H, which is a 1-hop neighbor of the node it intends to send to, node G (slot allocation rule 3). Note that this is another example of the hidden terminal problem, because if node B sends to node G using slot 6, it will cause a collision at node G. However, node B can use slot 7 to send to node G even though it is scheduled to send in node F. This is the exposed terminal problem. In fact, it would be more desirable for node B to allocate this slot to send to node G; this would increase channel reuse, a desired goal in wireless communications. Therefore, this leaves slots 7 through 12 which are free to send from node B to node G in this example.

After the calculation above, node B allocates slots 7, 8, and 9 to send from itself to G. It includes G in its next hop list NH along with the list of the slots 7, 8, and 9. It then broadcasts the QREQ1 message. In [17], node B does not keep track of this allocation which is only remembered in the forwarded QREQ1 message. So, until node B receives the corresponding QREP1 message which will be propagated from the destination C, slots 7, 8, and 9 in node B will remain $free$. They will only change status from $free$ to $reserved$ when and if the corresponding QREP1 message arrives from node C on its way to node A to confirm the QoS path A..FBG..C slot reservations. This poses no problem so long as no other requests arrive at node B during the period between forwarding QREQ1 and receiving the reply message QREP1. However, consider a situation where another request, QREQ2, arrives at node B from a source node D trying to reserve a QoS path from itself to node E with b=3 (i.e. the required bandwidth is 3 slots). Node B in this case will look at its slot status tables and will see no allocation for slots 7 through 12. It will then proceed to allocate some of these slots for this newly requested path. If the corresponding slot allocation procedure allocates slots 7, 8 and 9 for this new path and includes them in the next hop list, NH, then Node B will broadcast (forward) QREQ2 to node I which is on the path to node E. When QREP1 arrives at node B, it will change the status of slots 7, 8 and 9 to reserved. Afterwards, QREP2 will arrive at node B from node E on its way to node D. Node B will then have the problem of double allocation of slots 7, 8 and 9. In [17] the slots are reserved again (double reservation) for the second path. This will lead to a conflict at node B when data transmission using the two different paths starts. This is a multiple reservation problem due to a race condition at node B.

A similar example can be shown for the parallel reservation problem. This was described in Figure 6 where node B would select the slots to forward QREQ1 by considering only the status of the slots in node E prior to the allocation done by node E for the slots for QREQ2. When QREP1 returns to node B and QREP2 returns to node E, they both reserve the allocated slots. These slot selections can be in violation of the slot allocation rules and result in collisions when data transfer using the two different QoS paths begin.

In [14] a protocol is proposed to solve the race conditions described earlier and enhance network performance, especially in situations of increased node mobility, increased node density and higher traffic loads. The protocol uses a more conservative strategy. This strategy is implemented using the following features: (1) Three states for each slot: $reserved$, $allocated$, $free$. The three states are defined in the

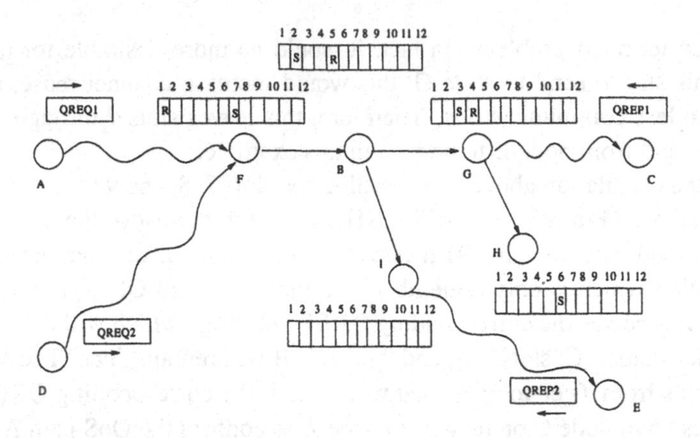

Figure 7: Multiple QoS path reservation competition. Two different QoS paths A..FBG..C and D..BI..E are being reserved simultaneously and they both pass through node B. The figure shows the slot reservation status before allocation to send from node B to G for QoS path requested by QREQ1. R: scheduled to receive. S: scheduled to send. empty: not scheduled to receive or send.

following manner: *Free*: not yet allocated or reserved. *Allocated*: in process of being reserved, but not yet confirmed. This means that the slot is allocated by a QREQ message but the corresponding QREP message has not yet arrived to confirm the reservation. *Reserved*: reservation is confirmed and the slot can be used for data transmission. (2) Discrimination between allocated (not yet reserved) and free slot status to alleviate the racing condition. (3) Wait-before-reject at an intermediate node with three conditions to alleviate the multiple reservation at intermediate node problem. (*conditon* 1: all required slots are available, *condition* 2: not-now-but-wait, and *condition* 3: immediate drop or reject of QREQ). (4) TTL timer for allocated and reserved slots. (5) TTL timers for maximum total QREQ propagation delay allowed, and for maximum total QREQ/QREP delay allowed (i.e. maximum QoS path acquisition time). More discussions and details about the proposed protocol is presented in [14]

In [18], Liao et al. present a multi-path QoS routing protocol which is also an extension of DSR. The protocol is designed to work in a CDMA-over-TDMA environment, but can be easily extended to other types of networks. The protocol enables a source node to search for a multi-path QoS route to a particular destination satisfying certain bandwidth requirements. A number of tickets are distributed from the source. The tickets can be further partitioned in to sub-tickets to search for a satisfactory multi-path. This protocol provides a higher success rate for finding a QoS path satisfying the required bandwidth requirements when the bandwidth is very limited. If a link along the path does not have the entire required bandwidth, the path does not immediately fail to be reserved; the protocol still searches for the possibility that a multi-link path exists from that node to intermediate nodes or to the destination whose aggregate bandwidth is equal to the required one. This increases the success rate of reserving the required path between the source and the destination especially in situations where the bandwidth is scarce and the network traffic is high. The number of tickets issued by the source can affect the performance and can be empirically adjusted. When the network bandwidth is sufficient, this protocol provides a performance similar to protocols which find a uni-path between the source and the destination.

In [37], Zhu et al. present a Five-Phase Reservation Protocol (FPRP) for QoS support in synchronous TDMA-based MANETs. FPRP perfoms the tasks of channel access and node broadcast scheduling simultaneously. It uses a contention-based mechanism for nodes to reserve TDMA slots. The protocol takes into consideration the hidden terminal interference in the reservation process. Reservation is made through a localized conversation between nodes in a 2-hop neighborhood. Due to its fully-distributed nature, it is scalable.

4.2 Extensions of AODV

The protocol presented in [10] and in [11] by Gerasimov et al., which is named QoS-AODV, is an extension of the well known AODV protocol. It is on-demand and designed to work in a TDMA network. This protocol combines information from both the network and data link layer. Unlike other protocols which make path bandwidth calculations only after paths to the destination have been discovered [5][10][13][23][24], QoS-AODV incorporates path finding with the bandwidth reservation mechanism. QoS-AODV is fully aware of the bandwidth resource availability by coupling together routing and MAC TDMA layers. As described earlier, the nodes compete for the slots contained in the data phase of the TDMA frame. In order for the source node to send data to a destination node, it must establish a virtual circuit (VC) connection with that destination. The VC establishment process includes route discovery, path bandwidth calculation and bandwidth reservation (data-phase-slot reservation) components. Each node keeps a *schedule* which contains information about both its own and its neighbor's time slots that are used for sending and receiving. A schedule is defined as a sequence of 1's and 0's where a number is the order of the corresponding slot in the data phase of the TDMA frame. The paper in [10] includes the algorithm used by each node to determine which slots are available to send to and receive from its neighbor, and to calculate link bandwidth scheduling from itself to each of its neighbors.

The link bandwidth information is used in the calculation of the path bandwidth schedules to source and destination nodes. Modified AODV HELLO messages are used which include slot scheduling information. The HELLO messages are sent either periodically or when link bandwidth information is changed.

In QoS-AODV, path discovery is done in the following manner. A source node that wants to send data to a particular destination determines if it has enough link bandwidth available to any of its neighbors. If it does not, it then denies the request initiated by its application layer. Otherwise, it creates a routing table entry for the requested application call ID and the destination address. Note that, in QoS-AODV, there is an entry in the routing table for each application call ID/source/destination triple instead of one per source/destination tuple as in the original AODV. The source node then sends the reservation request message, RREQ, which contains call ID and number of slots required for reservation, in addition to the standard AODV information.

When an intermediate node receives a RREQ message, it checks whether it already has an entry in its routing table corresponding to the received application call ID. The node then calculates the path bandwidth schedules using algorithms similar to ones presented in [24]. If the calculated path bandwidth to the source is insufficient, then the node does not forward the RREQ message. Otherwise, the intermediate node augments the RREQ message with path and link bandwidth parameters and broadcasts it further. The link bandwidth between two nodes is calculated as the intersection of their free slot schedules. The send link bandwidth (say of a link AB at a node A) is defined as the intersection of the free send slot schedule of the sender node (A) and the free receive slot schedule of the receiver node (B). The receive link bandwidth (say of a link AB at a node A) is defined as the intersection of the free receive slot schedule of the receiver node (A) and the free send slot schedule of the sender node (B). In addition to information corresponding to the original AODV protocol, the route table entry contains the addresses of three nodes along the path to the source, and link and path bandwidth schedules between those nodes. This information is needed in order to allocate slots that do not cause interference according to the slot allocation rules discussed before including the hidden terminal problem considerations.

An example of QoS-AODV route discovery is shown in Figure 8. A source node A needs to create a new VC to send data to a destination node E. Node A broadcasts an RREQ message which contains the call ID and the number of slots required for the QoS path. Upon receiving the RREQ message, node B, knowing that node A is one of its neighbors, determines that the available path bandwidth from A to B is equal to the receive link bandwidth from A to B. Path bandwidth AB is calculated as a portion of the receive link bandwidth AB. Node B then augments the RREQ message with the calculated link bandwidth AB and the address

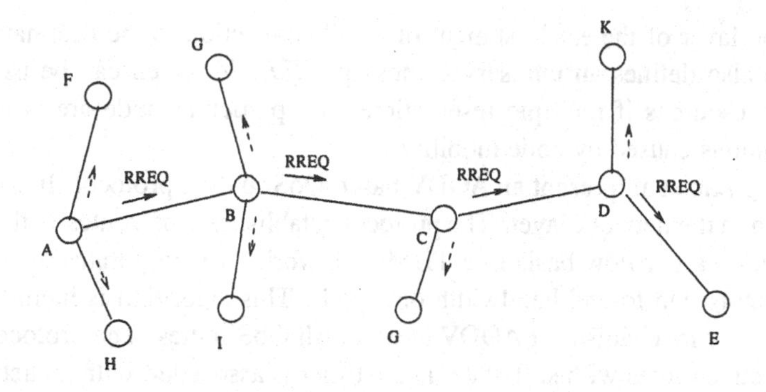

Figure 8: QoS-AODV - Propagation of RREQ message from source node A to destination node E. The RREQ message is forwarded only if the required bandwidth is available at each link. At each intermediate node, it is augmented with QoS bandwidth information

of A and rebroadcasts the RREQ message to all of its neighbors. When node C receives the propagated RREQ message from B, it knows that A is B's neighbor. Therefore it calculates the available path bandwidth using the AB and BC receive link bandwidth to avoid any interference conflicts including the hidden terminal problem. If the path bandwidths AB and BC contain the required number of slots, then C augments the RREQ message with the address of A and B, the receive link bandwidth BC, and the path bandwidth AB before it rebroadcasts it. Node D then receives the RREQ message; it calculates the path bandwidth to A using the link bandwidths AB, BC, and CD. If the calculated path bandwidth is sufficient, then D rebroadcasts the RREQ message after adding the address of C and B along with the receive link bandwidth CD and path bandwidth BC. When the destination node E receives the message, it uses the same algorithms to determine the path bandwidth scheduling CD and DE. Once the destination node E determines that there is enough bandwidth to the source node, it starts the reservation process by creating a reservation instance. The reservation parameters stored at each node along the VC for that VC ID include: (1) source and destination ID's, (2) application call ID, (3) next hop, previous hop and next hop bandwidth scheduling, and (4) reservation status. The destination node (E in this example) reserves MAC receive slots corresponding to previous hop bandwidth scheduling and composes a reservation message, RSV, which is a new message added to the AODV protocol. The RSV message is propagated back to the source, A in this example, by the intermediate nodes (B, C, and D in this example) which reserve corresponding MAC receive and send slots. When the source node receives the RSV message, it informs its application layer of the establishment of a VC connection to the destination. The algorithm also defines an unreserved message, $URSV$ which can be used to release slot resources if multiple reservations at a particular node are done due to race conditions caused by node mobility.

In [38], Zhu et al. present an AODV-based QoS routing protocol. It is designed to function in the network layer. The protocol establishes QoS routes with reserved bandwidth on a per flow basis in a TDMA network. It incorporates an algorithm for calculating end-to-end bandwidth on a path. This algorithm is included in the path discovery mechanism of AODV to establish QoS routes. The protocol in [38] protects active routes with soft-state, i.e., a timer is associated with an active route at a node and is refreshed every time the route is used. If the route is not used within a certain amount of time and the timer expires, the corresponding entry in the routing table is deleted. The protocol defines the five possible states of a QoS route, which indicate whether the route exists, and if so, if it is processed but not established, set up and used to forward packets, broken at upstream of the node and is being repaired, or broken at downstream of the node and is being repaired. Transitions among these states is done by either receiving or transmitting a packet, or expiration of the timer associated with the state. The paper defines eleven conditions and operations associated with transitions among these states. The QoS routing protocol builds different QoS routes for individual flows even between the same source and destination. The protocol is also capable of restoring a route when it breaks due to some topological change, which allows it to cope more robustly with some degree of node mobility. The simulation in the paper shows that the protocol produces higher throughput and lower delay than the best-effort AODV protocol. It performs best in smaller networks with low node mobility.

4.3 Extensions of TORA

The protocol in [10] by Gerasimov et al., named QoS-TORA, is based on the link reversal best effort protocol TORA. It is designed to work in a TDMA network where the bandwidth of a link is measured in terms of slot reservations in the data phase of the TDMA frame. This protocol makes use of information in the network and MAC layers.

QoS-TORA operates in the following manner which is illustrated in Figure 9. When a source desires to communicate with a particular destination node, it checks whether it has a best-effort path to that destination. If there is no path, it tries to establish one by sending the original TORA QRY packet as indicated in [28]. When there is at least one best-effort path to the destination, the source node sends a QoS specific BQRY message, which contains the number of slots needed by the application along with the application ID. This BQRY message is propagated to the destination along the best-effort path. When the BQRY message reaches the destination, it checks whether it has enough slots available to receive. If it does, it then broadcasts a QoS-TORA specific UBW message, which contains the application ID, number of slots required and the source ID. Upon receiving the UBW message, each intermediate node checks whether it has already received a UBW message with the same application ID from the same neighbor and whether there is an existing path to that destination node with the required bandwidth. If the node does not have a QoS path available or the new path contains a smaller number of hops, the new path bandwidth is saved, which corresponds to the path that is going through the neighbor from which the UBW message was received. The intermediate node calculates the path bandwidth based on the information for three nodes along the path to the destination. This information is necessary to make sure that the slot allocation is done according the the rules stated earlier which provide interference-free operation. The source node waits for the reception of several UBW messages from its neighbors before it starts the reservation process. This allows the source node to choose which neighbor it wants to use for the establishment of the QoS path. This is in contrast to AODV-based QoS protocols which have single table entries for each destination. This gives QoS-TORA more flexibility to respond to link breakage due to node mobility. The simulation experiments presented in [10] show considerable improvements in the probability of being able to find an end-to-end QoS path. Simulation also shows that QoS-TORA provides higher throughput under higher mobility circumstances. This is due to the fact that when a VC breaks, unlike the case in AODV-based QoS protocols, the source node might have another neighbor to start reservations, so the path discovery procedure can be skipped.

In [7] Dharmaraju et al. present another TORA-based QoS routing protocol for MANETs called INORA (INSIGNIA [1][15][16] + TORA). INORA is a network layer QoS support mechanism that makes use of the INSIGNIA in-band signalling mechanism and the TORA routing protocol for MANETs. In INORA, QoS signalling is used to reserve and release resources, and set up, tear down and renegotiate flows in the network. These reservations can be either hard state or soft state. The latter is more desirable in MANETs due to their dynamic nature. The INORA protocol operates the signalling mechanism independently from the TORA routing protocol. This provides decoupling of the two mechanisms and there is no interaction between. TORA provides the route between the source and the destination of a flow. Then the signalling mechanism (INSIGNIA) establishes resources for the route provided by TORA. INORA tries to find paths in the network that can satisfy the desired QoS requirements. In INORA, INSIGNIA asks TORA for alternative routes when the current route is not able to meet the QoS requirements. The INORA scheme provides load-balancing in the network which aids in the per-

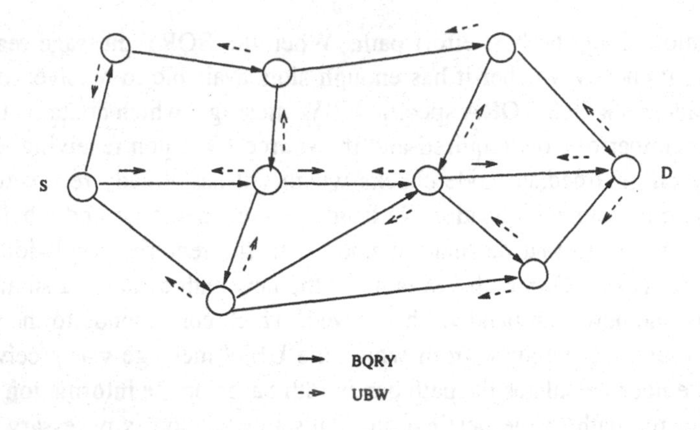

Figure 9: The DAG in the QoS-TORA protocol. The figure shows the propagation of the BQRY message from the source node, S, to the destination. The destination node, D, responds with the UBW message which contains the application ID, No. of slots, and source node ID.

formance of non-QoS flows. Future work will try to alleviate congestion in the wireless network by establishing QoS flows which avoid congested neighborhoods. The decoupling between the signalling and routing protocols allows for more flexibility in the design to incorporate load-balancing, congestion control, class-based admission control, and so on. This added flexibility comes at the price of more overhead when compared with other TORA-based QoS routing protocols which do not have the decoupling mechanism stated above.

In [12], Gupta et al. propose a framework for providing quality of service guarantees given limits on the rate of change in the topology. The protocol specifies changes which need to be incorporated in buffer requirements and play out times. The QoS support mechanisms reside in the network layer and constitute an extension of the INSIGNIA protocol mentioned earlier. The protocol also uses a soft state model (flows time out according to a soft state timer if not used to transmit data before timer expiration) which allows it to withstand link failures and route changes. The protocol relies on local action taken by a node detecting link failure to repair the path locally instead of generating an error message back to the source. The node is expected to buffer the received packets in the meantime before the path is repaired. The underlying protocol is assumed to be able to provide link failure information to the network layer. The IEEE 802.11 link layer protocol supports this operation. Route repair is done using a Route Repair Request (RRReq) packet which is sent to a limited area specified by a TTL (Time to Live in number of hops) field. Other nodes send a Route Repair Reply (RRRep) packet listing a path. The best path is used. The route restoration process and its ability to repair is limited by the rate of change of the topology. Otherwise, quality of service guarantees would not be possible and in such circumstances the application must be adaptive and able to live with the best-effort service.

4.4 Extensions of DSDV

In [26], Manoj et al. propose a MAC layer protocol named Real-time MAC (RT-MAC) for MANETs, which provides a bandwidth reservation mechanism. The protocol is designed to work in an asynchronous environment. The protocol relies on the flexibility of placement of reservation slots (of variable start and finish times) in the super-frame. The protocol makes use of holes (short free slots in the super-frame which otherwise cannot be utilized). The authors provide simulation which compares the protocol performance with the MACA/PR protocol, proposed by Lin and Gerla [22], and show that RTMAC outperforms MACA/PR in call blocking ratio, average end-to-end delay, packet delivery ratio, and provides less effect of the presence of best-effort traffic on real time traffic. RTMAC is an extension of DSDV [30]. It is responsible for finding an end-to-end path that satisfies the QoS bandwidth requirements. Bandwidth reservation for Constant Bit Rate (CBR) traffic is provided by dividing the transmission time into successive super-frames. This scheme can also be extended to support Variable Bit Rate (VBR) traffic as well.

The main concept in this approach is the flexibility of slot placement in the super-frame. Each super-frame consists of a sequence of reservation-slots (resv-slot). The time duration of a resv-slot is twice the maximum propagation delay. Each session between source and destination nodes requires the reservation of a block of consecutive resv-slots. A node must first reserve a set of resv-slots and a guard band to cushion the propagation delay (henceforth it is referred to as a connection-slot) on a super-frame and uses the same connection-slot to transmit in successive super-frames. A reservation table must be maintained by each node. The table contains information such as sender, receiver, and starting and ending time of the reservations that are active within its transmission range. This scheme is different from that of the TDMA environment because it requires no time synchronization (no need to maintain a global clock with the associated communication overhead) and uses a relative time for all reservation purposes [26]. Each node transmits its reservation table along with the route update packet of DSDV. The protocol includes the capability of a node to designate a specific connection-slot to be reserved for a particular connection, which gives the routing protocol the flexibility to position the connection slot. The protocol applies different schemes to reserve connection-slot (conn-slot). Different schemes can be used to allocate connection-

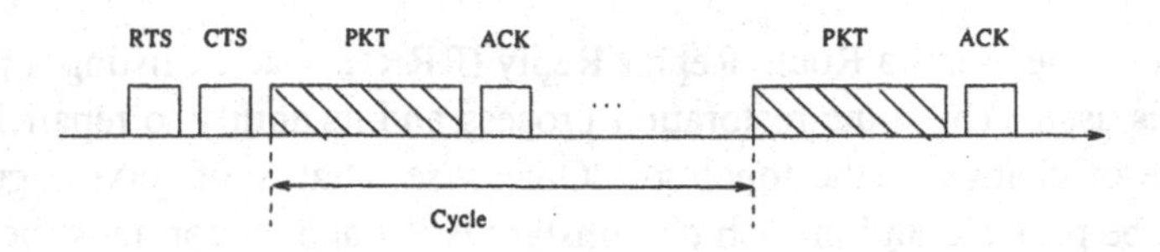

Figure 10: The MACA/PR protocol: RTS-CTS-PKT-ACK...PKT-ACK sequence. A $CYCLE$ is the maximum interval allowed between two real-time packets.

slots such as first fit (reserve slot in the immediate freely available connection-slot), best fit (place connection-slots at a place that succeeds the connection-slot on which the node receives the real-time packets) and fair fit (reserve connection-slots in a way which creates free slots that can be used for best-effort traffic). A source node desiring to transmit data to a certain destination node checks the reservation information of its neighbors, finds free slots that can be reserved, and initiates the reservation process for those free slots.

In [19], Lin introduces the MACA/PR (Multiple Access Collision Avoidance with Piggy-back Reservation) protocol. It is an asynchronous network based on the collision avoidance MAC scheme used in the IEEE 802.11 standard. MACA/PR avoids collisions due to the hidden terminal problem by establishing an RTS-CTS (request to send - clear to send) dialogue, which can be used as a building block to eliminate hidden terminal interference. The key components of the MACA/PR architecture are: a MAC protocol for transmission of data packets, the reservation protocol for setting up real-time connections, and the QoS routing algorithm which is an extension of the best-effort table-driven DSDV protocol. Figure 10 shows the MACA/PR protocol's RTS-CTS-PKT-ACK...PKT-ACK sequence. A $CYCLE$ is defined to be the maximum interval allowed between two real-time packets. The sender schedules its next transmission after a time CYCLE following the current data packet. The intended receiver enters the reservation in its reservation table and confirms it in the ACK returned to the sender.

The protocol has two features: 'flexible' reservations within a CYCLE (as opposed to slotted reservation of TDMA schemes), which is defined below, and QoS loop-free routing. The first data packet in the multimedia stream makes the reservations along the path. Once the first data packet is accepted on a link, a transmission window is reserved on that link at appropriate time intervals for all the subsequent packets in the connection. The window is released if it is idle for a specified number of cycles. The RTS-CTS exchange is used in the first packet transmission to set up the reservation. The subsequent data packets do not require the RTS-CTS exchange and are done using only packet transmission followed by corresponding acknowledgements.

The QoS routing protocol must first find a path which satisfies the QoS resource requirements desired by the application. The resource setup protocol starts hop-by-hop negotiation and setup along the path. Each node periodically broadcasts to its neighbors the {bandwidth, hop distance} pairs for the preferred paths (one per bandwidth value) to each destination. The number of preferred paths is the maximum number of slots (or packets) in a cycle. In the implementation in [19] each node keeps track of the bandwidth on the shortest path, and the maximum bandwidth (over all possible paths). A node drops real time packets with bandwidth requests which it cannot satisfy by the currently available path to the destination. If a link has no bandwidth or the bandwidth is below a predefined threshold, its weight is set to ∞. This will prevent any new connections from being established until an old connection releases some bandwidth. MACA/PR strikes a good compromise between the totally asynchronous, unstructured PRNET and the highly organized cluster TDMA.

5 Other QoS Routing Protocols and Related Issues

In addition to the categories of protocols mentioned above, there exist other protocols which are not direct extensions of DSR, AODV, TORA, and DSDV. In this section, some of these protocols are presented.

In [21], Lin and Gerla present a network architecture for multimedia. The architecture assumes a code division access scheme. Specifically, direct-sequence spread-spectrum for CDMA is used. The protocol operates at the level of the MAC layer. The nodes in the MANET are organized into *clusters*. All links in a cluster are assigned the same spread spectrum code. Any two nodes in a cluster are only one hop away from each other. A round robin scheme is used to provide channel access to the nodes in a cluster with bandwidth reservation for real time traffic. A virtual circuit is allocated at call setup time between a source and a destination node. The protocol is designed to work in an asynchronous environment. The protocol uses piggy-back reservations with packet transmissions to reserve time slots dynamically for packets from active voice sources without conflict. The protocol uses CSMA-PR (Code Division Multiple Access with Piggy-back Reservation) to resolve the conflict situation which arises when some traffic sources are trying to access the channel at the same time. Because CDMA is used, all clusters can operate simultaneously.

In [32] by Sheu et al., a distributed bandwidth allocation/sharing/extension (DBASE) protocol is proposed. It supports multimedia traffic with CBR (constant bit rate) and VBR (variable bit rate) traffic over ad hoc WLAN. This protocol functions at the MAC layer. It is asynchronous and uses RTS-CTS-asynchronous data-ack exchanges for channel access. In DBASE, $rt(realtime) - stations$ can reserve and free channel resources dynamically. The bandwidth allocation procedure is based on a contention (and back-off) process that only occurs before the first successful access and a reservation process after the successful contention. When the rt-station leaves, the bandwidth is immediately released by DBASE. The proposed protocol is compliant with the IEEE 802.11 standard. The simulation presented in the paper shows an improvement over the conventional 802.11 standard in terms of high channel utilization, low access delay and small delay variation for real-time services.

In [34] by Wang et al., a QoS Routing protocol with Mobility Prediction (QRMP) is presented. QRMP uses mobility prediction and QoS requirements on bandwidth and delay to select the most stable path. The route setup phase consists of request and reply stages. In addition to the usual routing information such as node_id and sequence_number, all QoS requirements information and $node\_info$ including related information of the node, e.g. link delay, link bandwidth and interface velocity, are considered. QRMP reduces route setup time and control overhead as well as increases packet delivery ratio by selecting the most stable route based on mobility prediction. The latter is also used to reduce the update message frequency.

In [8], Dong et al. propose an on-demand Supernode-based Reverse Labeling (SRL) algorithm for QoS provisioning, specifically bandwidth and delay, in MANETs. The algorithm utilizes a hierarchical structure, which is formed by dynamically electing $super\ nodes$. The other nodes are $slave\ nodes$ and are always one hop away from their corresponding super node. Slave nodes regularly communicate with their super node through periodic HELLO messages. The authors provide algorithms to perform effective route discovery and local route information management: virtual route discovery, reverse link labeling and dynamic route repairing. A node sends its QREQ message to its super node. A sequence of super nodes is considered a virtual route (VR). Delay requirement and accumulated delay fields are supported and can be used by delay sensitive applications. The simulation in [8] shows that SRL is efficient in terms of packet delivery ratio and average end-to-end delay. It also has reduced packet loss ratio and route request overhead caused by node mobility. A source node which needs to transmit information to a destination sends a route request message to its super-node. The request includes QoS requirements, which can be bandwidth and/or delay constraints depending on the needs of the application involved. The request propagates through intermediate super nodes to find the destination and its corresponding super node, which will then process the request in a manner similar to DSR [8]. Transmitted messages include Delay Requirement and Accumulated Delay fields for applications that have such requirements.

A summary of the classification of the protocols discussed in this chapter is presented in table 1. The table contains the following columns. First the QoS routing protocol is listed. Then, "Net. Layer" parameter indicates the networking layer within which the protocol is designed to operate. The "Syn./Asyn." parameter indicates whether the protocol operates within a synchronous or asynchronous environment. The "Comm. Mode" parameter indicates the communication network assumed such as TDMA, CDMA-over-TDMA, and so on. The "BE Routing Prot." parameter indicates the best effort routing protocol that is extended by or is most closely related to the corresponding QoS protocol. The "Proact./React." parameter indicates whether this QoS protocol is reactive (on-demand) or proactive (table-driven). Then the "Comments" field contains additional information about the QoS protocol. There are other parameters which can also be considered such as a protocol being location assisted or not, which were not included in the table.

6 Conclusions and Future Research

In this chapter, we discussed the existing QoS routing protocols. The different approaches taken by researchers who are active in this area were discussed. The most popular best effort routing protocols in MANETs, such as DSR, AODV, DSDV and TORA, and the different QoS parameters were presented with a brief description of each. Classification of the existing QoS routing protocols was done according to different criteria such the related best effort routing protocol, the OSI layer, communication model, and synchronization mechanisms used.

Significant advances are constantly taking place to increase the capabilities and use of wireless devices. Communication between these devices will become an essential part of their growth. As applications including audio and video multimedia are developed to support the growing services that these networked devices provide, the need for QoS guarantees to be given by lower layers of the network to the application layer will become an indispensable part of supporting communication. Many areas of research in this field provide considerable challenge and potential to enhance the growth and proliferation of MANETs and their applications. These areas include power consumption, resource availability, location management, inter-layer integration of QoS services, support for heterogenous MANETs, as well as robustness and security. Continued growth is expected in this area of research in order to develop, test and implement the critical building blocks to provide efficient and seamless communications in mobile ad hoc networks. QoS routing protocols will play an essential role in providing the required support mechanisms.

QoS Routing	Net. Layer	Syn./ Asyn	Comm. Mode	BE Routing	React./ Proact.	Comments
Gerasimov et al. [10]	net./ MAC	syn.	TDMA	AODV	react.	QoS-AODV.
Gerasimov et al. [11]	net./ MAC	syn.	TDMA	TORA	react.	QoS-TORA.
Ho et al. [13]	net.	syn.	TDMA	ODQoS	react.	ODQoS (On-demand QoS-based routing prot.)
Liao et al. [17]	net.	syn.	TDMA	DSR	react.	QREQ from source to dest. allocating slots. QREP from dest. to source reserves slots.
Liao et al. [18]	net.	syn.	C-o-T or FDMA	DSR	react.	Multi-path QoS (ticket-base) routing.
Manoj et al. [26]	MAC	asyn.	N/A	DSDV	proact.	Ext. of 802.11 DCF function.
Lin [19]	MAC	asyn.	N/A	DSDV	proact.	Flexible reservations within a CYCLE.
Lin et al. [21]	MAC	asyn.	C-o-TDM	N/A	react.	CDMA-over-TDMA. Each cluster has different code.
Lin et al. [25]	net.	syn.	C-o-T	DSDV	proact.	Destination does calc. of the path BW.
Lin [20]	net.	syn.	C-o-T	DSR	react.	RREQ packets to find paths and calc. BW.
Sheu et al. [32]	MAC	asyn.	N/A	Lower level	react.	Compliant with 802.11. RTS-CTS-Asyn. Data-ACK chan. access.
Wang et al. [34]	net.	syn.	N/A	QRMP (source r.)	react.	QoS routing with mobility prediction.
Dong et al. [8]	net.	gen.	gen.	SRL (DSR-like)	react.	Supernode-based Reverse Labeling Algorithm
Zhu et al. [38]	net.	syn.	TDMA	AODV	react.	BW calc. integrated with AODV prot.
Dharmaraju et al. [7]	net.	gen.	gen.	TORA	react.	INORA: Uses signalling done at higher level than routing
Gupta et al. [12]	net.	gen.	gen.	TORA	react.	Extension of INSIGNIA.
Zhu et al. [37]	net.	syn.	TDMA	FPRP (DSR-like)	react.	Five-phase reservation protocol.

Table 1. QoS Routing Algorithm Classification. Abbriviations: gen.: general which also indicates applicability to all cases of that classification (higher level); C-o-T: CDMA-over-TDMA; C-o-TDM: CDMA-over-TDM.

References

[1] G.-S. Ahn, A. T. Campbell, S.-B. Lee, and X. Zhang, Insignia, *Internet Draft, draft-ietf-manet-insignia-01.txt*, (Oct. 1999).

[2] S. Chakrabarti and A. Mishra, QoS issues in ad hoc wireless networks, *Communications Magazine, IEEE* Vol.39 No.2 (Feb. 2001) pp. 142-148.

[3] K. Chen, S. H. Shah, and K. Nahrstedt, Cross layer deisgn for data accessibility in mobile ad hoc networks, *J. Wireless Commun.* Vol.21 (2002) pp. 49-75.

[4] S. Chen. Routing support for providing guaranteed end-to-end quality-of-service, *http://www.cs.uiuc.edu/Dienst/UI/2.0/Describe/ncstrl.uiuc_cs/ UIUCDCS-R-99-2090, UIUCDCS-R-99-2090, University of Illinois at Urbana-Champain*, (Jul. 1999).

[5] T.-W. Chen, J. T. Tsai, and M. Gerla, QoS routing performance in mulithop, multimedia, wireless networks, *IEEE 6th International Conference on Universal Personal Communications Record* Vol.2 (Oct. 1997) pp. 557-561.

[6] X. Chen and J. Wu, Multicasting techniques in mobile ad hoc networks, *Chapter 2, The Handbook of Ad Hoc Wireless Networks, edited by M. ILyas* (2003) pp. 2.1-2.16.

[7] D. Dharmaraju, A. Roy-Chowdhury, P. Hovareshti, and J. S. Baras, Inora-a unified signaling and routing mechanism for QoS support in mobile ad hoc networks, *Proceedings of International Conference on Parallel Processing Workshops* (Aug. 2002) pp. 86-93.

[8] Y. Dong, T. Yang, D. Makrakis, and I. Lambadaris, Supernode-based reverse labeling algorithm: QoS support on mobile ad hoc networks, *Canadian Conference on Electrical and Computer Engineering, IEEE CCECE* Vol.3 (May 2002) pp. 1368-1373.

[9] A. Veres et al., Supporting service differentiation in wireless packet networks using distributed control, *IEEE JSAC* (Oct. 2001).

[10] I. Gerasimov and R. Simon, A bandwidth-reservation mechanism for on-demand ad hoc path finding, *IEEE/SCS 35th Annual Simulation Symposium* (San Diego, CA, Apr. 2002) pp. 27-33.

[11] I. Gerasimov and R. Simon, Performance analysis for ad hoc QoS routing protocols, *Mobility and Wireless Access Workshop, MobiWac* (2002) pp. 87-94.

[12] A. Gupta and D. Sanghi, QoS support in mobile ad-hoc networks, *IEEE International Conference on Personal Wireless Communications* (Dec. 2000) pp. 340-344.

[13] Y.-K. Ho and R.-S. Liu, On-demand QoS-based routing protocol for ad hoc mobile wireless networks, *Proceedings of the Fifth IEEE Symposium on Computers and Communications (ISCC)* (Jul. 2000) pp. 560-565.

[14] I. Jawhar and J. Wu, A race-free bandwidth reservation protocol for QoS routing in mobile ad hoc networks, *Hawaii international conference on system sciences - HICSS-37* (Big Island, Hawaii, Jan. 2004).

[15] S.-B. Lee, G.-S. Ahn, X. Zhang, and A. T. Campbell, Insignia: An ip-based quality of service framework for mobile ad hoc networks, *Journal of Parallel and Distributed Computing* Vol.60 No.4 (Apr. 2000).

[16] S.-B. Lee and A. T. Campbell, Insignia: In-band signaling support for QoS in mobile ad hoc networks, *Proceedings of 5th international workshop on mobile multimedia communications (MoMuC)* (Berlin, Oct. 1998).

[17] W.-H. Liao, Y.-C. Tseng, and K.-P. Shih, A TDMA-based bandwidth reservation protocol for QoS routing in a wireless mobile ad hoc network, *IEEE International Conference on Communications (ICC)* Vol.5 (2002) pp. 3186-3190.

[18] W.-H. Liao, Y.-C. Tseng, S.-L. Wang, and J.-P. Sheu, A multi-path QoS routing protocol in a wireless mobile ad hoc network, *IEEE International Conference on Networking* Vol.2 (2001) pp. 158-167.

[19] C. R. Lin, Multimedia transport in multihop wireless networks, *IEEE Proceedings on Communications* Vol.145 No.5 (Oct. 1998) pp. 342-346.

[20] C. R. Lin. Admission control in time-slotted multihop mobile networks. *IEEE Journal on Selected Areas in Communications* Vol.19 No.10 (Oct. 2001) pp. 1974-1983.

[21] C. R. Lin and M. Gerla, A distributed control scheme in multi-hop packet radio networks for voice/data traffic support, *IEEE International Conference on Communications (ICC)* Vol.2 (Seattle, Jun. 1995) pp. 1238-1242.

[22] C. R. Lin and M. Gerla, Asynchronous multimedia multihop wireless networks, *Proceedings of IEEE INFOCOM '97* Vol.1 (Apr. 1997) pp. 118-125.

[23] C. R. Lin and C.-C. Liu, An on-demand QoS routing protocol for mobile ad hoc networks, *Proceedings of IEEE International Conference on Networks (ICON)* (Sep. 2000) pp. 160-164.

[24] C. R. Lin and J.-S. Liu, QoS routing in ad hoc wireless networks, *IEEE Journal on Selected Areas in Communications* Vol.17 No.8 (Aug. 1999) pp. 1426-1438.

[25] H.-C. Lin and P.-C. Fung, Finding available bandwidth in multihop mobile wireless networks, *Proceedings of IEEE 51st Vehicular Technology Conference (VTC)* Vol.2 (Tokyo, May 2000) pp. 912-916.

[26] B. S. Manoj and C. S. R. Murthy, Real-time traffic support for ad hoc wireless networks, *The 10th IEEE International Conference on Networks (ICON)* (Aug. 2002) pp. 335-340.

[27] P. Mohapatra, J. Li, and C. Gui, QoS in mobile ad hoc networks, *IEEE Wireless Communications* (Jun. 2003) pp. 44-52.

[28] V. D. Park and M. S. Corson, A highly adaptive distributed routing algorithm for mobile wireless networks, *Proceedings of IEEE INFOCOM 1997* Vol.3 (Apr. 1997) pp. 1405-1413.

[29] C. E. Perkins, *Ad Hoc Networking*, (Upper Saddle River, NJ, USA, Addison-Wesley, 2001).

[30] C. E. Perkins and P. Bhagawat, Highly dynamic destination-sequenced distance-vector (dsdv) for mobile computers, *Proc. of ACM SIGCOMM '94* (Aug. 1994) pp. 234-244.

[31] C. E. Perkins and E. M. Royer, Ad hoc on demand distance vector (aodv) routing, *Internet Draft*, (Aug. 1998).

[32] S.-T Sheu and T.-F. Sheu, Dbase: a distributed bandwidth allocation/sharing/extension protocol for multimedia over IEEE 802.11 ad hoc wireless LAN, *Proceedings of INFOCOM 2001* Vol.3 (Apr. 2001) pp. 1558-1567.

[33] B. Wang and J. C. Hou, Multicast routing and its QoS extension: problems, algorithms, and protocols, *Network, IEEE* Vol.14 No.1 (Jan./Feb. 2000) pp. 22-36.

[34] J. Wang, Y. Tang, S. Deng, and J. Chen, QoS routing with mobility prediction in MANET, *IEEE Pacific Rim Conference on Communications, Computers and Signal Processing (PACRIM)* Vol.2 (Aug. 2001) pp. 357-360.

[35] Z. Wang and J. Crowcroft, Qos routing for supporting resource reservation, *IEEE Journal on Selected Areas in Communications* Vol.14 (1996) pp. 1228-1234.

[36] J. Wu and H. Li, On calculating connected dominating set for efficient routing in ad hoc wireless networks, *Proc. of the Third International Workshop on Discrete Algorithms and Methods for Mobile Computing and Communications* (Aug. 1999) pp. 7-14.

[37] C. Zhu and M. S. Corson, A five-phase reservation protocol (fprp) for mobile ad hoc networks, *Proceedings of IEEE INFOCOM 1998* Vol.1 (Mar./Apr. 1998) pp. 322-331.

[38] C. Zhu and M. S. Corson, QoS routing for mobile ad hoc networks, *Proceedings of IEEE INFOCOM 2002* Vol.2 (Jun. 2002) pp. 958-967.

Topology Management of Hierarchical Mobile Ad Hoc Networks

Mario Gerla and Kaixin Xu
Computer Science Department
University of California at Los Angeles, Los Angeles, CA 90095
E-mail: {gerla, xkx}@cs.ucla.edu

Contents

1 Introduction

The ad hoc wireless networking technology shows great potential and importance in many situations due to its independence of a fixed infrastructure and its instant deployment and easy reconfiguration capabilities. Usually, a mobile ad hoc network (MANET) is assumed to be homogeneous i.e., all mobile nodes in the network share the same random access wireless channel with a single omnidirectional radio. However, a flat ad hoc network has poor scalability [3, 9, 10]. In [10], theoretical analysis implies that even under optimal network layout conditions, the throughput for each node declines rapidly toward zero while the number of nodes is increased. This is proved in an experimental study of scaling laws in ad hoc networks employing IEEE 802.11 radios presented in [9]. The measured per node throughput declines much faster in the real testbed than in theory. These results reflect that a "flat" ad hoc has an inherent scalability problem. Besides the capacity limitation, ad hoc routing protocols also pose a heavy burden to the network. Flooding is usually adopted by routing protocols to search a path or propagate routing information. In large-scale network with mobility, routing overhead will consume a major fraction of the available bandwidth. Thus further limits the scalability of "flat" ad hoc networks.

One way to reduce the control overhead and improve the network scalability is to organize the ad hoc network in some kind of hierarchy, which is usually called as topology management. One well studied way of topology management is clustering including both proactive clustering [1, 7, 13, 14, 18] and passive clustering [6]. The major idea of clustering is to divide the mobile nodes into geographical groups. Each group has one cluster head and some cluster members. Typically a cluster head is in the center of the cluster and the distance between any two nodes in the same cluster is less than a fixed number of hops, which controls the cluster size. The cluster head usually maintains some information of its members. Those nodes which are in multiple clusters are usually referred as gateway nodes. The gateway nodes can connect multiple clusters. The clustering information can be utilized to reduce the size of the routing tables and correspondingly routing control overhead. One typical clustering based routing scheme is the Clusterhead-Gateway Switch Routing (CGSR) [2]. In CGSR, any source and destination node pair can find a path in a format of "Clusterhead - Gateway - Clusterhead - Gateway ..." between them. Data packets are all routed through such Clusterhead-Gateway switch paths. Another typical example is the Hierarchical State Routing (HSR) [4], which is a multi-level clustering based link state routing protocol. It maintains a logical hierarchical topology by using the clustering scheme recursively. More examples can be found in [6, 13].

One similar topology management scheme is the landmark hierarchy as being utilized in the LANMAR routing scheme [5]. Under the assumption of group mobility, mobile nodes are divided into different mobile groups. Nodes in the same group move in the same fashion. Then, in each group, a landmark node is elected to represent the whole group. Apparently this builds some kind of logical hierarchy. Routing across groups can be summarized as only maintaining routes to the landmark node of each group. By maintaining "summarized" routing information, LANMAR reduces both routing table size and control overhead effectively. It greatly improves routing scalability to large, mobile ad hoc networks. Another type of topology management scheme is to build a virtual backbone throughout the whole network such as in [12]. Such a backbone network can be viewed similar to the backbone of the wired network. However, this virtual backbone network is only logical concept. The backbone links are emulated, in most times, by multihop paths.

The topology management schemes described above can be used for simplifying the network protocols and reducing control overhead. However, most of them still assume the "flat" ad hoc network. The capacity limitation of "flat" ad hoc network is still there. To future improve network scalability in large scale ad hoc networks, recently, physically hierarchical ad hoc networks have been proposed to improve the performance of large scale ad hoc networks [8, 12, 17, 19]. In such hierarchical networks, some nodes are more powerful with multiple wireless radios installed. One of the radios usually has longer radio range and higher bandwidth. These radios can be used to connect these powerful nodes to build a high level wireless backbone network, which provide a "short cut" to connect remote nodes. This wireless backbone network is not a fixed infrastructure like the Internet backbone. Instead, nodes consisting of the wireless backbone network are also mobile and perhaps join and leave the backbone network dynamically. Thus, it is also a typical ad hoc network. In this chapter, we refer such a backbone network as a mobile backbone network (MBN). The powerful nodes are named as backbone nodes (BNs).

An inherent challenge of operating such hierarchical ad hoc network architecture is again how to maintain and manage the hierarchical topology. Major difficulty comes from the fact that the topology of the backbone network is very dynamic due to mobility and node failures. The ultimate goal is to achieve a uniform distribution of backbone nodes in the whole field even under mobility and node failures. Static backbone nodes deployment at the beginning cannot achieve this goal. Even they can be distributed very well at the beginning, their topology may be in a very bad shape after mobility. Moreover, when node failures happen, new backbone nodes are required to replace old one. All these problems require good topology management of the wireless backbone network.

In this chapter, we introduce two schemes for managing the wireless backbone network topology. The first scheme is a backbone node election algorithm. The main idea is to deploy redundant backbone capable nodes. Here backbone capable nodes mean those mobile nodes which physically have the capability to play the role as backbone nodes. Only a limited number of backbone nodes then are elected among these candidates. Others are kept as backup. When one backbone node is destroyed or moves out from a certain area, if there is any backbone capable node available in that area, a new backbone node will be elected to replace the old one. In this way, we can dynamically reorganize the backbone network topology. Another scheme is called backbone nodes automatic repositioning scheme. It aims at achieving uniform distribution of backbone nodes by letting the backbone nodes automatically coordinate their positions with each other. Combining both schemes together, we can then maintain a good backbone network topology. It in turn can improve the network performance, which is proved via simulations.

This chapter is organized as following. We first give an brief overview of the mobile backbone network architecture in section 2. Then in section 3 and 4, we introduce two topology management schemes, which are the backbone node election algorithm and the backbone node automatic repositioning scheme. We also discuss (in section 5) that how the routing scheme can be extended to efficiently utilize the hierarchical network structure. Performance evaluations via simulation are given in section 6 and we conclude the chapter in section 7.

2 Ad Hoc Network with Mobile Backbones

The hierarchical ad hoc network structure discussed in this chapter is called mobile backbone network (MBN). It is a hierarchical network in which a set of nodes functionally more capable than the ordinary nodes form the backbone. The basic scenario consists of a large numbers of mobile nodes deployed over a large area. Among these, the backbone nodes (BN) have

Figure 1: Illustration of an Hierarchical Mobile Ad Hoc Network.

the ability of forming multilevel backbone networks using long range radios. Usually, radios at each backbone level use some form of channel separation (e.g., antenna directivity, different codes, different frequencies, or combinations thereof) in order to minimize interference across levels. Radios in the same level share the same frequency and channel resources. Unlike the wired network, the nodes in the mobile backbone network are also moving, thus the backbone topology is dynamically changing. In many scenarios such as the battlefield, the hierarchical structure is an inherent feature of the application. Different units have different communication devices and capacities. For example, the wireless radios installed in military vehicles have a more ample energy supply and thus are more powerful than those carried by the dismounted soldiers. Unmanned Aerial Vehicles (UAVs) and even satellites can be used for providing higher level and broader reach connections. Figure 1 illustrates a three level hierarchy where the first level supports ground communications among soldiers; and second and third level are implemented using tanks and UAVs respectively. In this chapter, most of our discussions and simulations are based on a two level hierarchical architecture. However, the topology management schemes can be easily extended to multi-level hierarchical networks.

Hierarchical ad hoc networks have great potential in real time constrained applications, especially in the digitized battlefield. However, the backbone design and topology management of such a network are quite challenging if the nodes are mobile. One issue is how to deploy the backbone nodes. The main difficulties are mobility and BN failures. Using a clustering scheme to elect the BNs is a natural choice. Clustering has been widely used to form logically hierarchical networks [1, 18] and to partition a large scale network into small groups. However, a drawback of current clustering schemes is cluster instability, as indicated in many papers such as [1]. Conventional clustering schemes work effectively only in networks with very low mobility or no mobility at all, such as the sensor networks. Instability of the clusters and frequent changes of BNs introduce high routing O/H and make the hierarchy difficult to manage. In this chapter, we will introduce a clustering scheme to achieve good stability. In addition to using clustering for managing the backbone network topology, we also introduce another scheme called backbone node automatic repositioning to achieve uniform distribution of the BNs in the whole field. Combining the two schemes together, we can manage the mobile backbone network in a very efficient way. The two schemes are introduced and investigated in the rest of sections.

3 Mobile Backbone Node Election

To deploy the backbone node, the simplest way is to pre-assign backbone nodes and scatters them uniformly across the field at initialization. However, such a static deployment has two main problems. First, the BNs are constantly moving. Thus after some time, some BNs may congregate in small geographical areas, creating congestion; while other areas may be depleted of BNs altogether. This certainly is not a good scenario. The second concern is fault tolerance. BNs may fail or even be destroyed (a likely event considering the emergency applications envisioned for MANETs). New BNs should be deployed to replace the defunct ones. Static deployment cannot fulfill these requirements. One solution is to deploy some redundant backbone capable nodes (i.e., nodes with long range radios) and to dynamically elect a proper subset to BNs. When one BN is destroyed or moves out of a certain area, a new BN will be selected from the backbone capable node pool. If two backbone nodes move near to each other, one of them will give up its backbone role. The backbone node election is completely distributed and dynamic. It must result in a backbone node distribution that reflects the distribution of ordinary nodes. A Distributed Clustering algorithm is the most common approach to this problem.

3.1 Random Competition based Clustering

Many clustering schemes have been proposed in the literature [1, 7, 13, 14]. Among them, the Lowest ID (LID) and Highest Degree (HD) algorithms are widely used due to their simplicity. The details of the two algorithms can be found in [7, 14]. Most of these work mainly focused on how to form clusters with good geographic properties such as minimum overlap of clusters etc. However, stability is probably the most critical property in applications involving mobility. This is because clustering is often used to support hierarchical routing. In particular, for the hierarchical structure, the stability of the backbone nodes is important, and it directly depends on the stability of the clustering algorithm that elects them.

Here we introduce a clustering scheme called Random Competition based Clustering (RCC) [19], which targeting stability, simplicity and light overhead. The main idea is that any node that doesn't belong to any cluster, can initiate a cluster formation by broadcasting a packet to claim itself as a cluster head. The first node, which broadcasts such a packet, will be elected as the cluster head by its neighbors. All the neighbors, after hearing such a broadcast, give up their right to be a cluster head and become members of this cluster. Cluster heads have to periodically broadcast a "cluster head claim packet" (CHCP) to maintain their role. Since there is a delay between CHCP broadcast and reception by neighbors, several neighbor nodes may simultaneously broadcast CHCPs. To reduce such concurrent broadcasts, a random timer is introduced. Each node defers by a random time before its cluster head claim. If it hears another cluster head claim during this random time, it gives up its broadcast. The idea of "first claim node wins" was first proposed in the passive clustering scheme in [6]. Due to the specific limitations imposed by "passive clustering" no timers were used in [6]. However, the RCC scheme is "active clustering". The introduction of an explicit random timer is necessary to reduce conflicts. Of course, the random timer reduces, but cannot completely eliminate concurrent broadcasts. When a concurrent broadcast is detected, node ID resolves the conflict. The node with lower ID becomes the cluster head.

The Random Competition based Clustering (RCC) scheme is more stable than conventional clustering schemes such as LID and HD. In the LID scheme, when the cluster head hears a node with a lower ID, it will give up its cluster head role. Similarly, in the HD scheme, when a node with more neighbors appears, the cluster will also be reformed. Due to node mobility, such things may happen very frequently. In RCC, one node gives up its cluster head position only when another cluster head moves near to it. Since cluster heads are usually at least two hops away, clusters formed by RCC are much more stable. The low control overhead of our scheme is clear. In the lowest ID and highest degree clustering schemes, each node has to have complete neighbor information. In our scheme, only the cluster heads need to broadcast a small control packet periodically. All other nodes just keep silent.

Usually clustering schemes are single hop based, that is the cluster head can reach all members in one hop. This is not suitable for backbone node election. We want to approximately control the number of elected BNs. To achieve this, the clustering schemes must be extended to form K-hop clusters. Here, K-hop means that a cluster head can reach its members in at most K hops. Adjusting the parameter K can control the number of elected cluster heads. To extend RCC scheme from single hop to multihop clustering, each node stamps hop distance (to its cluster head) in the cluster head claim packet and forwards it to its neighbors. A mobile node will select the nearest cluster head within its K-hop scope to be its cluster head. When there is no cluster head within its K-hop scope, a backbone node capable node (after deferring some random time) claims itself as a cluster head. In multihop clustering, the probability of concurrent cluster head claims is high due to the higher latency for propagating cluster head claim packet K-hop away. The random defer time plays an important role here.

3.2 Stability of the Backbone Nodes

We use GloMoSim [20], a packet level simulator specifically designed for ad hoc networks, to evaluate the stability of the proposed Random Competition based Clustering (RCC) algorithm. We compare its stability with that of the Lowest ID (LID) and Highest Degree (HD) algorithms. Since we are targeting large-scale networks, 1000 mobile nodes are deployed. The field is a 3200mX3200m square. Each mobile node has an IEEE 802.11b wireless radio with transmission range of 175m. The DCF mode of IEEE 802.11 MAC is used and channel bandwidth is set to 2 Mbps. Node mobility model is random waypoint mobility [11]. Simulation time of each run is 6 minutes.

The stability of clusters includes two parts, stability of cluster head and stability of cluster members. We define two metrics, average lifetime of a cluster head and average membership time of a cluster member. The average lifetime of a cluster head is defined as the average time period during which one node plays the role as a cluster head uninterrupted. The average membership time is the average time that one mobile node remains in a cluster. These two metrics fully reflect the stability of clusters. In a MBN, average lifetime of a cluster head is exactly the average lifetime of a BN. In our simulations, we only implement the basic clustering scheme without considering the "gateway" node selection feature as in [7, 14] etc. The

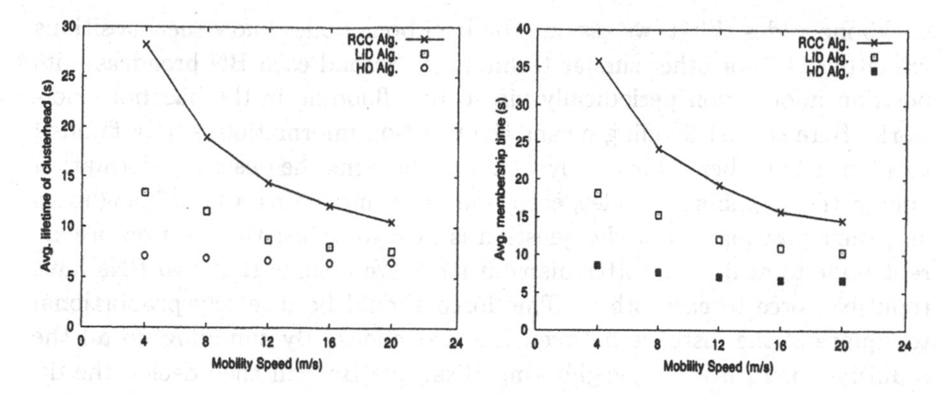

Figure 2: Average lifetime of CH   Figure 3: Average membership time

clustering scope is 1 hop. Simulation results are given in Figure 2 and Figure 3.

From Figure 2 and Figure 3, we can see that the RCC algorithm is more stable than the LID and HD algorithms under both low mobility and high mobility. Stability of the HD algorithm is the worst. This is due to the fact that the degree of one node is changing very frequently under mobility. In our experiments, we have assumed that every node has capability to be a cluster head. In reality, only a small fraction (e.g. 10-25%) of the total mobile nodes are backbone capable. Thus, we expect clusters to be more stable as an established cluster head has fewer challengers.

4 Backbone Node Automatic Repositioning

The backbone node election scheme discussed in previous section provides good robustness and reliability of the mobile backbone network. The clustering algorithm focuses on the stability of backbone nodes, thus stability of the backbone network. However, the distribution of those elected backbone nodes may not be uniformly scattered in the whole field. To further optimize the topology of the backbone network, in this section, we introduce a backbone node automatic repositioning (BNAR) scheme to re-coordinate the positions of backbone nodes to achieve good distribution in the whole field.

The major idea of the BNAR scheme is to let each backbone node adjust its position according to the position information it learns from neighboring backbone nodes. First, we assume the backbone nodes know their positions via either GPS or other similar techniques. Second each BN broadcasts its position information periodically via scoped flooding in the backbone network. Here scoped flooding means the position information is only flooded to a limited number of hops away. After exchanging the position information among the neighboring nodes, each node now maintains a list of neighbors and their positions. Now the question is how to adjust the position of current node to achieve a better distribution? We assume that two BNs have repulsive force to each other. This force should be inversely proportional to square of the distance between the two nodes. By summing up all the repulsive forces from all neighboring BNs, one BN can then decide the direction and speed it should move. Such mobility will then result in a better distribution of BNs.

Now, let's give a formal description of the algorithm. Suppose node A with position (x_a, y_a) has a neighbor list with m nodes as $(N_1, N_2, ..., N_i, ..., N_m)$. The position of any neighbor N_i is (x_i, y_i). The repulsive force from node N_i to node A is given as $|\overrightarrow{f_i}| = \frac{K_1}{D_i^2}$. Here K_1 is a constant and D_i is the distance between node A and node N_i, which is given as $D_i = \sqrt{(x_a - x_i)^2 + (y_a - y_i)^2}$. Sum the repulsive forces from all nodes in the neighbor list, we get the $\overrightarrow{f} = \overrightarrow{f_1} + ..., +\overrightarrow{f_i}, +..., +\overrightarrow{f_m}$. The direction of $\overrightarrow{f}$ is then the direction node A should move to. For simplicity, the moving speed of node A is calculated as $|\overrightarrow{v}| = K_2 * |\overrightarrow{f}|$. Here K_2 is another constant parameter. The values of K_1 and K_2 can be adjusted to achieve better performance in different environment.

The above algorithm does not consider the boundary of the field. However, in nearly all simulation experiments, the field size is fixed with boundaries. If the boundary is not considered, actually many BNs will be forced to move to the edges. This certainly is not the good distribution we are seeking for. To take the field boundary into account, we assume the edges of the fields also have repulsive forces to each BN. This force is computed the same way as the repulsive force between two BNs. Just the distance now is only the vertical or horizontal distance from the BN to the edge. Thus, each BN has to consider additional 4 repulsive forces from the edges of the field (Here, we assume all BNs have the same height. Thus only 4 edges need to be considered.).

The proposed backbone node automatic repositioning scheme is fully distributed. Each node makes its mobility decision based on the information it receives from neighboring nodes. In the implementation, a BN only periodically checks its neighbor list and update its mobility speed and direction. This algorithm for calculating the mobility vector emulates the particle diffuse procedure within a sealed space.

5 Routing in Mobile Backbone Network

After the backbone network is well organized, we need to apply a proper routing protocol to operate it efficiently. The routing scheme in the MBN has some requirements: it must be able to exploit the high level backbone links, enhancing throughput and delay with respect to scheme without a backbone. It must do so without compromising (in fact, possibly enhancing) scalability and fault tolerance. In fact, considering the emergency recovery, unfriendly or even hostile environments where ad hoc networks are deployed, the backbone nodes can very possibly become disabled or may fail to operate. Maintaining connectivity in the face of backbone node failures is a strong requirement. Thus, the addressing and routing scheme cannot be totally "dependent" on the health of the backbone. To meet the challenges of our extremely volatile environment, we extend the Landmark Ad Hoc Routing (LANMAR) [5] to operate in the MBN. We call this solution Hierarchical LANMAR Routing (H-LANMAR). The details of H-LANMAR are presented in the following subsections.

5.1 Landmark Ad Hoc Routing (LANMAR) Overview

LANMAR [5] is an efficient routing protocol designed for ad hoc networks that exhibit group mobility. Namely, one can identify logical subnets in which the members have a commonality of interests and are likely to move as a "group". The logical grouping is reflected in the address used within the ad hoc network, namely the two field address $< GroupID, HostID >$. LANMAR uses the notion of landmarks to keep track of such logical groups. Each logical group has one node serving as a "landmark". The landmark node is dynamically elected. The routes to landmarks are propagated throughout the network using a distance vector mechanism (in this study, we assume DSDV). In addition to landmark distance vector propagation, LANMAR relies also on a local, myopic routing algorithm (in this work, we use Fisheye State Routing (FSR) [15], with limited scope; but, any other proactive routing scheme could work). Within the local scope, LANMAR thus runs link state routing. For nodes outside of the scope, only landmark distance vectors are broadcast. In local FSR routing, each node periodically exchanges in-scope topology information with its immediate neighbors. Updates carry the sequence numbers assigned by the sources. To the Fisheye update, the source also piggybacks a distance vector of all landmarks. Thus, in LANMAR each node has detailed topology information about nodes within its scope and has a distance and routing vector to all landmarks.

When a node needs to relay a packet to a destination that is within its local scope, it uses accurate routing information available from the Fisheye Routing Tables. The packet will be forwarded directly. Otherwise, the packet will be routed towards the landmark corresponding to the destination's logical subnet, carried in the packet header. When the packet arrives within the scope of the destination, it is routed to it directly (possibly without going through the landmark).

5.2 Hierarchical Landmark Ad Hoc Routing (H-LANMAR)

LANMAR can be well integrated into the MBN by virtue of the fact that it is itself logically hierarchical. Routing information to remote nodes is summarized by landmarks. Now, we will extend such a logical hierarchical structure to utilize the physical hierarchy. In the original LANMAR scheme, we route the packet toward the corresponding remote landmark along a long multi-hop path. In the hierarchical MBN, we can route the packet to the nearest BN, which then forwards it through a chain of MBN links to a remote BN near the remote landmark. Finally, the remote BN sends the packet to the remote landmark or directly to the destination if it is within its scope. This will greatly reduce the number of hops. The procedure is illustrated in Figure 4. We can see that by utilizing the backbone links, the 8-hop path is reduced to be 4 hops long, a great improvement!

We extend the LANMAR routing protocol so that it can take the "short cut" described above. First, all mobile nodes, including ordinary nodes and BNs, are running the original LANMAR routing via the short-range radios. This is the foundation for falling back to "flat" multi-hop routing if BNs fail. Second, a BN will broadcast the landmark distance vectors to neighbor BNs via the backbone links. The neighbor BNs will treat this packet as a normal landmark update packet. Since the higher level paths are usually shorter, they will win over (and thus replace) the long multi-hop paths in the level 1 network. From landmark updates the ordinary nodes thus learn the best path to the remote landmarks, including the paths that utilize the backbone links.

One important feature of our routing scheme is reliability and fault tol-

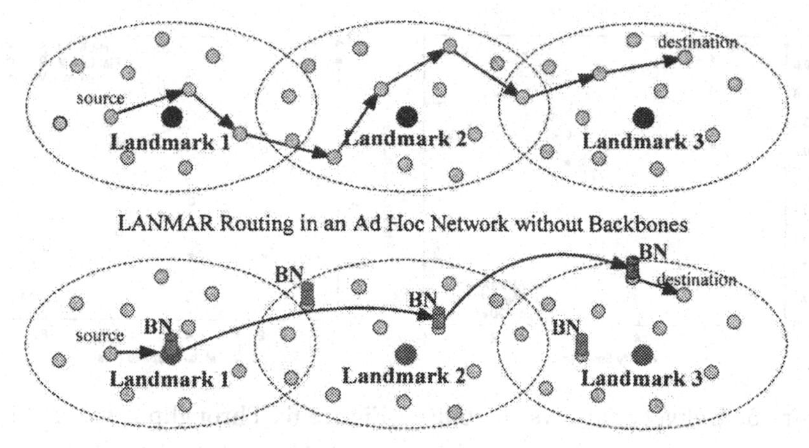

Figure 4: Illustration of H-LANMAR routing in a MBN.

erance. The ordinary nodes are prevented from knowing the backbone links explicitly. The backbone links are indirectly learned via BN routing broadcasts. Now, suppose a BN of one group is destroyed by enemies, the shorter paths via this BN will expire. Then new landmark information broadcasted from other nodes will replace the expired information. Thus, in the worst case, routing in this group goes back to original landmark routing while other groups with BNs can still benefit from backbone links among themselves. When all backbone capable nodes are disabled, the whole network becomes a "flat" ad hoc network running the original level 1 LANMAR routing, which can still provide connectivity, yet at lower performance.

6 Performance Evaluation

6.1 Performance Evaluation of Backbone Node Election

In this section, we present simulation results to compare the H-LANMAR in the MBN with the original LANMAR routing and AODV routing [16] in a "flat" ad hoc network. The purpose of these experiments is to show that how H-LANMAR running on top of MBN can improve the network performance effectively. Same network scenario as in previous experiments is used and channel bandwidths of the "short range" and "long range" radios are set to 2Mbps and 11Mbps respectively. The scope of backbone election

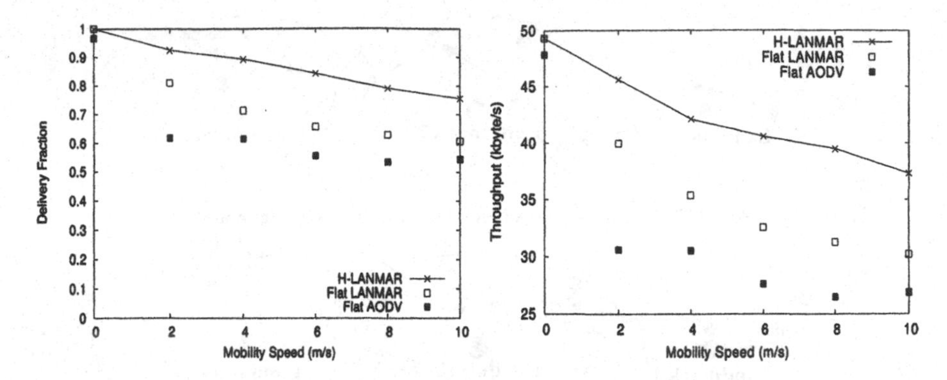

Figure 5: Delivery ratio vs. mobility Figure 6: Throughput vs. mobility

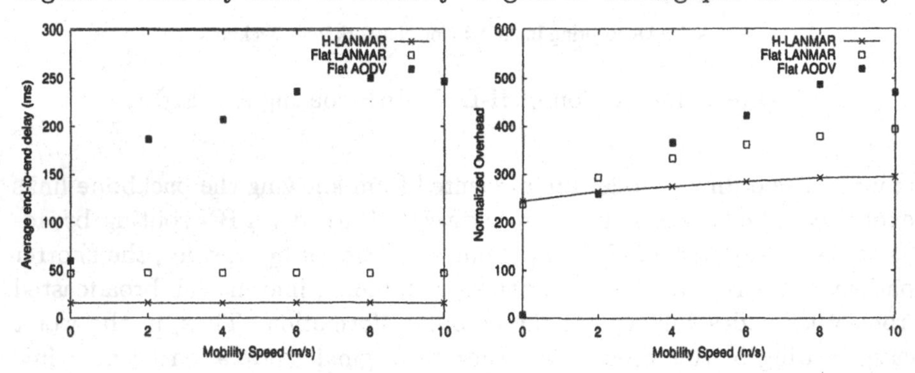

Figure 7: Packet delay vs. mobility Figure 8: Routing OH vs. mobility

is fixed to be 2. 30 randomly selected CBR pairs are used to generate traffic. We increase the node mobility from 0m/sec to 10m/sec to compare the performance. Results are given from Figure 5 to Figure 8.

In Figure 5, the delivery fraction of H-LANMAR clearly outperforms "flat" LANMAR and AODV as mobility increases. Without mobility, all three protocols have delivery fraction nearly equal to 1. This is due to the fact that in a stationary network the routing information in the node routing table is always accurate. Only few packets are dropped on the way to destinations. Note that the CBR traffic load was chosen so as not to saturate the network. However, when the nodes are moving, routing information tends to become obsolete very rapidly. By utilizing the backbone links, H-LANMAR can propagate new routing information very quickly and efficiently and keep the routing table more up-to-date than the other schemes. This is way H-

LANMAR can achieve a high delivery fraction in high mobility while the other two degrade quickly. Similar results, this time in terms of network throughput, are reported in Figure 6.

Figure 7 shows average delay as a function of mobility speed. The average end-to-end delay of AODV increases rapidly with mobility speed. This is due to the on-demand routing maintenance feature of AODV. With increased mobility speed, path interruptions and expirations are more frequent. AODV delays packets in intermediate queues as it searches for new paths. In contrast, LANMAR and H-LANMAR are proactive, thus the average delay (of the packets that actually get delivered to destination) is not significantly affected by speed. H-LANMAR further reduces the delay by using backbone network.

Figure 8 gives the normalized routing overhead (NRO) of the three protocols. The normalized routing overhead is defined as the number of routing packets used in order to route one data packet successfully. In low mobility or no mobility, the routing overhead of AODV is much smaller than LANMAR and H-LANMAR. In fact, AODV re-computes a route only when it expires because of lack of user traffic. Thus, its NRO is very small. However, with increasing mobility, the frequent link breaks and path expirations cause the overhead of AODV to increase sharply. As a result, the NRO increases very quickly. This is an indication that AODV has a scalability problem in large-scale, mobile ad hoc networks. Compared with AODV, the overhead of LANMAR and H-LANMAR is only minimally affected by mobility.

6.2 Performance Evaluation of Backbone Node Automatic Repositioning

Now, we investigate how the backbone node automatic repositioning (BNAR) scheme can further improve the network performance by re-shaping the backbone network topology. The simulation configurations are exactly same to the experiments in previous subsection except the BNAR scheme is applied. Only H-LANMAR routing is adopted. We compare the network performance with and without BNAR for different fraction of backbone capable nodes. The results are shown from Figure 9 to Figure 11.

In Figure 9, we observe that the data packet delivery ratio is increased greatly by applying the BNAR scheme. This is due to the fact that backbone nodes are now more uniformly distributed to help delivering packets. Also we observe that the delivery ratio is also increase along with the increase of the fraction of backbone capable nodes. This is because the more the

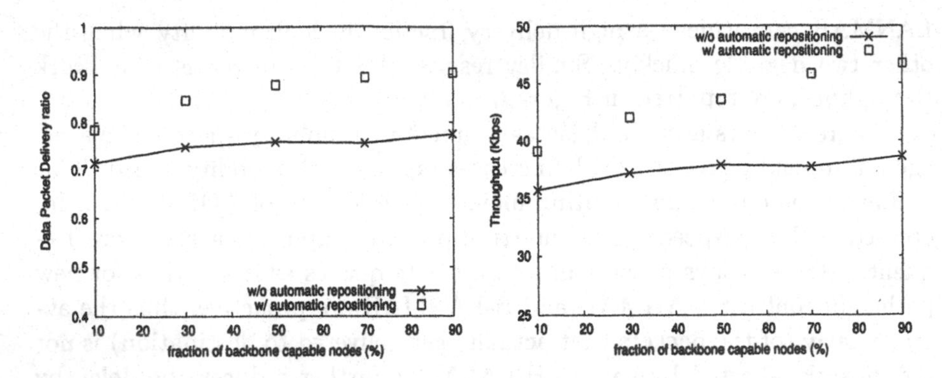

Figure 9: Delivery ratio with BNAR. Figure 10: Throughput with BNAR.

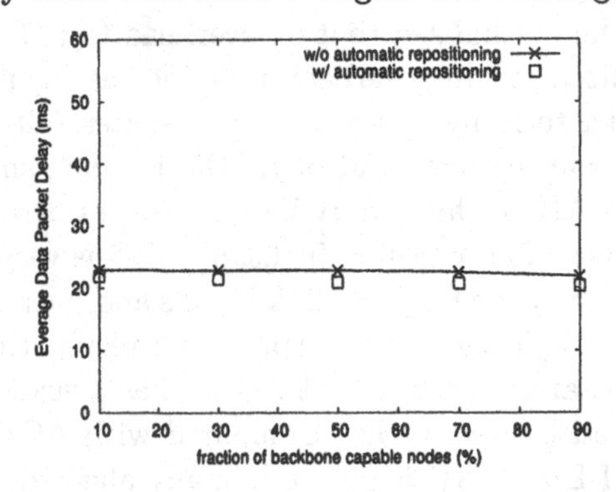

Figure 11: Average delay with BNAR.

number of backbone capable nodes, the better the backbone node election algorithm can build the backbone network. However, since the backbone election algorithm doesn't guarantee the uniform distribution of the elected nodes, the re-coordinating of backbone nodes can still help improving the network performance. Only when the two schemes are combined, we can then get a good distribution of backbone network, while still keeping the good stability of the network hierarchy. The performance improvements are also confirmed in Figure 10, where the total throughput of the CBR connections also gets increased when the BNAR scheme is applied. Figure 11 shows the average delay of data packets with and without the BNAR scheme adopted. We observe that BNAR scheme reduces the average data packet delay in some degree.

7 Conclusion

Building hierarchical ad hoc network is a promising approach to provide scalability to future large scale ad hoc networks. However, due to node mobility and node failures, the topology management of the wireless backbone network is not as simple as in the wired networks. In this chapter, we introduced two schemes to help managing the hierarchical network structure. The goal is to achieve good distribution of backbone nodes as well as robustness, reliability and stability of the backbone network under node mobility and node failures. Simulation results show that topology managed hierarchical ad hoc network can improve the network performance significantly.

References

[1] S. Banerjee and S. Khuller, A clustering scheme for hierarchical control in multi-hop wireless networks, *Proceedings of IEEE INFOCOM'01* (Apr. 2001).

[2] C.-C. Chiang and M. Gerla, Routing and multicast in multihop, mobile wireless networks, *Proceedings of IEEE ICUPC'97* (Oct. 1997).

[3] M. S. Corson, Flat scalability - fact or fiction?, *ARO/DARPA Workshop on Mobile Ad Hoc Networking* (Mar. 1997).

[4] X. H. G. Pei, M. Gerla and C. C. Chiang, A wireless hierarchical routing protocol with group mobility, *Proceedings of IEEE WCNC'99* (Sep. 1999).

[5] M. Gerla, X. Hong, and G. Pei, Landmark routing for large ad hoc wireless networks, *Proceeding of IEEE GLOBECOM 2000* (Nov. 2000).

[6] M. Gerla, T. Kwon, and G. Pei, On demand routing in large ad hoc wireless networks with passive clustering, *Proceedings of IEEE WCNC'00* (Sep. 2000).

[7] M. Gerla and J. T. Tsai, Multicluster, mobile, multimedia radio network, *ACM-Baltzer Journal of Wireless Networks* Vol.1 No.3 (1995)

[8] D. L. Gu, G. Pei, M. Gerla, and X. Hong, UAV aided intelligent routing for ad hoc wireless networks, *Proceedings of IEEE WCNC'00* (Sep. 2000).

[9] P. Gupta, R. Gray, and P. R. Kumar, An experimental scaling law for ad hoc networks, *http:black1.csl.uiuc.edu~prkumar* (May 2001).

[10] P. Gupta and P. R. Kumar, The capacity of wireless networks, *IEEE Transactions on Information Theory* Vol.46 No.2 (Mar. 2000).

[11] D. B. Johnson and D. A. Maltz, Dynamic source routing in ad hoc wireless networks, *Mobile Computing, edited by T. Imielinski and H. Korth, Chapter 5* (1996).

[12] Y. Ko and N. H. Vaidya, A routing protocol for physically hierarchical ad hoc networks, *Technical Report 97-010* (Sep. 1997).

[13] P. Krishna, N. H. Vaidya, M. Chatterjee, and D. K. Pradhan, A cluster-based approcah for routing in dynamic networks, *Proceedings of ACM SIGCOMM Computer Communication Review* (1997).

[14] C. R. Lin and M. Gerla, Adaptive clustering for mobile networks, *IEEE Journal on Selected Areas in Communications (JSAC)* Vol.15 No.7 (Sep. 1997).

[15] G. Pei, M. Gerla, and T. W. Chen, Fisheye state routing in mobile ad hoc networks, *Proceeding of ICDCS 2000 Workshops* (Apr. 2000).

[16] C. E. Perkins and E. M. Royer, Ad-hoc on-demand distance vector routing, *Proceedings of IEEE WMCSA'99* (Feb. 1999).

[17] R. Sanchez, J. Evans, and G. Minden, Networking on the battlefield: Challenges in highly dynamic multi-hop wireless networks, *Proceeding of IEEE MILCOM'99* (Oct. 1999).

[18] P. Sinha, R. Sivakumar, and V. Bharghavan, Enhancing ad hoc routing with dynamic virtual infrastructures, *Proceedings of IEEE INFOCOM'01* (Apr. 2001).

[19] K. Xu, X. Hong, and M. Gerla, Landmark routing in ad hoc networks with mobile backbones, *Journal of Parallel and Distributed Computing* Vol.63 (2003).

[20] X. Zeng, R. Bagrodia, and M. Gerla, GloMoSim: a library for parallel simulation of large-scale wireless networks, *Proceedings of PADS'98* (May 1998).

Efficient Resource Discovery in Wireless AdHoc Networks: Contacts Do Help

Ahmed Helmy
Department of Electrical Engineering
University of Southern California, Los Angeles, CA 90089
E-mail: helmy@usc.edu

Contents

1 Introduction

Resource discovery is an essential element in the design of infrastructure-less networks. Many classes of multi-hop wireless networks, including mobile ad hoc networks (MANets) and sensor networks, are designed as infrastructure-less, unattended, rapidly-deployable networks. In these cases, each network node does not have prior knowledge of the resources available in the network. Hence, resource discovery is imperative to the design of ad hoc and sensor networks. In general, a resource can be any type of service or capability, such as nodes with high energy, processing power or storage, multiple interfaces, printing capability, or sensing capability. Also a resource may be shared information, as in file-sharing, publish-and-subscribe, storage-retrieval, and querying for sensed data.

Resource discovery in wireless ad hoc and sensor networks is a challenging problem mainly due to the following reasons. First, the lack of infrastructure, where there are no well-known servers in a pre-defined network structure as in DNS. Second, the highly dynamic nature of ad hoc and sensor networks that leads to frequent changes in the network topology and resource availability and locations. Network dynamics arise mainly due to the characteristics of wireless channels, in terms of fading, interference and variability in the loss rates, in addition to possible node movement in mobile networks. Third, scarcity of energy and communication bandwidth in such networks necessitates the design of new resource discovery protocols qualitatively different from those designed for the wired Internet. In the wired Internet protocols were not designed for energy and communication constrained environments. Moreover, in many cases wireless networks may scale up to thousands of nodes rendering the discovery problem even more challenging.

Service discovery may occur between the querier and the resource node directly, according to the client-server model. Another alternative is to use the directory lookup model where the target resource node registers itself (or its service) in the directory. A node interested in the stored data may perform a lookup in the directory before issuing the request. In either model, an efficient distributed search protocol is necessary. The design and analysis of search protocols in wireless ad hoc networks is the topic of this chapter.

Efficiency of a resource discovery protocol may be measured using different metrics. Some of the most meaningful metrics in the context of wireless ad hoc networks include communication overhead, energy consumption, delivery success rate and response delay. In general, bandwidth in wireless networks is a scarce resource. In addition, due to the broadcast nature of the wireless medium, collisions, due to medium contention, may further decrease the throughput of the network. Furthermore, nodes in ad hoc networks are, in general, battery powered and hence are energy constrained. Communication is a main consumer of energy, during both transmission and reception of packets. Hence, reducing communication overhead and overall energy consumption is one of the main goals of ad hoc networking protocols.

There are several design approaches for distributed search protocols that may apply to resource discovery. In general, a resource discovery architecture in wireless ad hoc and sensor networks may be categorized as 'location-aware' or 'location-free'. Location-aware architectures require that node (or resource) location information be widely available in the network. Such architectures usually use geographic or trajectory routing to forward the updates or queries and include geographic rendezvous mechanisms (such as GLS [7], Rendezvous Regions [55] [39] [53], and GHT [33]) and trajectory advertisement schemes (such as TBF [54], and [56] [57]). These architectures may be efficient when (and if) location information is available. On the other hand, location-free architectures do not require or use location information. In this chapter we shall present and study location-free resource discovery architectures, and will not be discussing location-aware architectures.

Here we provide a simple classification of the main common location-free search techniques, and provide an overview of each of those techniques, along with a discussion on their applicability and performance in ad hoc networks. Specifically, an overview is given for the following schemes.

1. Flooding-based techniques, including (a) simple flooding (or expanding ring search), (b) scoped flooding and (c) efficient reduced broadcast,

2. Ad hoc on-demand routing (with caching),

3. Hierarchical architectures, including (a) cluster-based hierarchy, (b) landmark hierarchy (c) dominating set (or backbone) approaches, and

4. Hybrid (loose hierarchy) routing, including (a) zone routing and (b) contact-based architectures.

This is not meant to provide an exhaustive list for the large body of literature that exists in this area. Rather, it is a representative subset of the main approaches. Then we shall present an elaborate discussion of the design and evaluation of the contact-based architecture as an illustrative case study and as a promising architecture for efficient resource discovery in large-scale ad hoc networks. A comparative simulation study is given at the end of the chapter.

2 Overview of Resource Discovery Approaches

We address the problem of resource discovery in infrastructure-less wireless mobile ad hoc networks. Hence, architectures that require infrastructure (e.g., DNS) are not suitable for our problem. Centralized approaches are neither robust nor scalable.

Perhaps one of the simplest search schemes is flooding. Flooding each request throughout the network enables the search for the resource sought in every node in the network. Simple flooding causes every node to forward the request packet until all the nodes in the network receive it. In wireless networks, flooding leads to what is called a 'broadcast storm' [23] with many packets sent in a relatively short period, potentially leading to collisions and congestion. Several variants of flooding attempt to alleviate the expensive cost of simple flood. Scoped-flooding (or expanding ring search) attempts to perform the search in expanding rings and stops when the request is satisfied. Reduced broadcast techniques use heuristics to suppress flooding redundancies.

Several on-demand ad hoc routing protocols build upon flooding to obtain up-to-date routes, but use caching to take advantage of the history of route requests to avoid unnecessary flooding. The efficacy of the cache depends on the cache spread, which in turn affects the cache hit rate, and the cache validity ratio. The cache hit and validity ratios depend on dynamics of the cached information. In case of highly dynamic scalable networks the cache efficacy may decrease, forcing more request flooding.

To provide a scalable architecture, some approaches use a hierarchical scheme based on dynamically formed 'clusters'. The clusters are re-formed with topology dynamics due to node failure and mobility. One common method to form clusters is to elect a dominant set of nodes that reach all other nodes in the network within 1 hop. This scheme attempts to guarantee request reception at each node in the network for every request.

A hybrid approach is used in the zone routing protocols, in which a proactive (*push*) mechanism is used intra-zone and a reactive (*pull*) mechanism is used inter-zone. In this approach each node establishes its zone independently forming a loose hierarchy. Flooding requests between the zone periphery nodes (called *borders*) is used in the inter-zone reactive phase in a mechanism called *bordercasting*. Bordercasting attempts to achieve full coverage of the network by querying the border nodes. The contact-based architectures use the concept of zones, but instead of using bordercasting, only contact-nodes that lie out-of-zone are elected and queried on-the-fly. Contacts may be elected in an energy-aware fashion to achieve load balancing and to extend the network lifetime [1].

3 Resource Discovery Approaches

3.1 Flooding-based Approaches

Flooding is a simple, commonly used technique for resource discovery. In simple flooding a message is sent to all nodes in the network with every discovery request. A mechanism to prevent loops is usually employed, where each node stores the request and querier IDs and forwards the query only once. In general, if there are N nodes in the network then each node gets to transmit the packet except the target node; i.e., $N-1$ transmissions. Over wireless channels when a node transmits a packet during flooding, it is broadcast to all neighbors of that node. If there are on average g neighbors per node - where g is called the average node degree - then we get $g.(N-1)$ receptions per flooding. Put differently, for every link AB (between nodes A

[1]Several other resource discovery systems have been developed for ad hoc networks including [58] [59] [60] [61]. For brevity we do not discuss those systems.

and B) the flooded message will traverse twice; once when the node A sends to B and again when B rebroadcasts, so number of receptions is $\sim 2L$, where L is the number of links in the network. Energy is expended in both transmissions and receptions. As one might expect, it is usually not necessary for every node in the network to transmit the packet in order to achieve full network coverage (in which all nodes in the network receive the packet). Hence, simple flooding may lead to unnecessary redundant transmissions. For frequent requests flooding may incur significant communication overhead, and hence may be undesirable for wireless networks.

Expanding_ring search (ERS) techniques (or scoped-flooding) are also commonly used for discovery. ERS uses repeated flooding with a limit on the number of flooding hops (TTL). If the resource is not found in a trial, then this hop limit (TTL) increases for subsequent trials by an increment. A simple version of ERS may use a constant increment of '1'. Variants of ERS may be derived, however, by varying the increment, either to a constant greater than 1 or to a variable such as exponential in the number of trials. ERS usually performs better than simple flooding when the resource is relatively close to the querier. However, ERS may be quite inefficient when the querier-target pairs are randomly distributed the network diameter is high, which tends to be true in many wireless networks due to clustering of nodes.

Reduced_broadcast techniques [23] [62] use heuristics to reduce the redundancies of flooding and conserve communication. The main idea is to exploit node density to reduce redundant transmissions, sometimes at the expense of reducing network coverage. The heuristics are used to estimate when message rebroadcasts are likely to be effective. These schemes may be quite effective when the amount of broadcast redundancy is high due to high node density. In situations where the wireless network is not highly dense, the effect of reduced broadcast is quite limited. The aim of these heuristics is to deliver 'broadcast' messages that should be received by all nodes in the network; a goal sometimes different from that of resource discovery. Some broadcast reduction heuristics include (i) probabilistic forwarding, (ii) count of message receptions, (iii) node distance, or (iv) node locations.

In probabilistic flooding, each node rebroadcasts the message with probability p. When $p = 1$, the scheme degenerates into flooding.

The counter-based scheme takes advantage of the following observation; a node may receive the flooded message x times, and the additional network coverage obtained by rebroadcasting a message decreases with increase in x. Hence, after receiving a number of messages, it may be desirable for a node to suppress its message rebroadcast if the expected added coverage is very low. So, in the counter-based scheme a node maintains a message counter (x) and a counter threshold (C_{th}). Upon first receipt of the message a node waits a random time during which it counts other receptions, when $x > C_{th}$ then the rebroadcast is suppressed.

The distance-based scheme also attempts to suppress rebroadcasts that are expected to achieve very low coverage, but it performs its calculations based on distance. If a message is received from a nearby node, then there is a low added coverage achieved by rebroadcast. In this scheme, a node waits a random time and maintains the least distance (d) to the nodes from which it received messages. Each node also has a distance threshold set at D_{th}. If $d < D_{th}$ then the broadcast is suppressed.

A related scheme is the location-based scheme. Knowing the locations of the sending nodes enables the receiving node to estimate the added area coverage, a. If a is greater than a certain area threshold (A_{th}) then rebroadcast, otherwise suppress.

The location-based scheme is the most effective among these heuristics, since many redundant rebroadcasts maybe achieved without affecting message reachability significantly. It also works well for various node distributions, but requires location information. The counter-based scheme is simple (does not require location information) and achieves noticeable reduction in number of redundant rebroadcasts if the network is dense (where the broadcast redundancy is high).

3.2 Ad Hoc On-demand Routing

Routing protocols in ad hoc networks can be generally classified as proactive (table-driven) or reactive (on-demand). Previous studies [43] have shown that on-demand routing protocols with caching (e.g., DSR [5], AODV [4]) achieve better throughput and overhead performance better than proactive protocols in terms of throughput and overhead especially with node mobility. On-demand protocols employ mechanisms for route caching to avoid unnecessary flooding of route requests and robust route repair mechanisms to deal with frequent route breaks.

Performance studies of ad hoc routing that usually use long-lived assigned connections mainly capture the route repair phase of the routing protocols, while the route discovery phase is invoked only during the initial setup period of the connections. However, resource discovery protocols may exhibit significantly different behavior, where the route discovery/setup phase (and caching) becomes the dominant factor affecting performance while route maintenance/repair is very rarely triggered. Caching is used in on-demand (reactive) ad hoc routing [5, 4] to alleviate the effects of flooding. Caching can be quite efficient, when the cache hit ratio is high and the cache is valid. The effectiveness of this approach thus depends on the cache validity, which in turn is a function of the network and information dynamics and request patterns. In cases of mobile networks and frequent un-correlated short requests, however, one might expect the cache performance to degrade as the on-demand routing degenerates with high probability into flooding.

To gain better understanding of the effect of caching on protocol overhead, we build a caching model for the on-demand DSR approach and evaluate it. Earlier studies on on-demand ad hoc routing protocols investigated relatively small (40-100 node) networks, with long lived connections [5] [40] [41] [42] [43] [44]. We instead study cache performance for larger scale networks and for resource discovery.

3.2.1 Caching Model

Our caching model follows the dynamic source routing (DSR) design [5] [40]. A source looking for a target (or a destination) triggers a route request ($RREQ$) on demand. The process of route request is illustrated in Figure 1. First, the source (Q) looks up its own local cache for a route to the target (T). If local cache is not found, then the source sends a query to its first hop neighbors and they perform cache lookup. If a cached route is not found, or if the found cache does not result in positive response from the target (e.g., due to invalidity of the cache), then the source floods the route request throughout the network. The target replies to requests from distinct neighbors to create multi-paths, and intermediate nodes with cached route to T, respond to the request. The replies (from the target and intermediate nodes) traverse the reverse pathes to the source, and nodes along the route (and their neighbors) cache the route information (i.e., aggressive caching). When a cached route is used and is found to be invalid or out-of-date, it is attached to the flooded route request to invalidate all copies of that route in the network.

Now, let us define the cache hit ratio as the fraction of cache lookups that are found, and the cache validity ratio as the fraction of cache hits that are valid. Let the local cache hit ratio be p_1, and the local cache validity ratio be q_1. Similarly, let neighbor cache hit ratio be p_2, and the neighbor cache validity ratio be q_2. Furthermore, let the validity ratio for the routes obtained and used by flooding be q_r. Note that if the flooded request reaches the target T (i.e., destination) causing a reply to be send to Q, and if that reply is used, then it should be valid. There may be other situations where

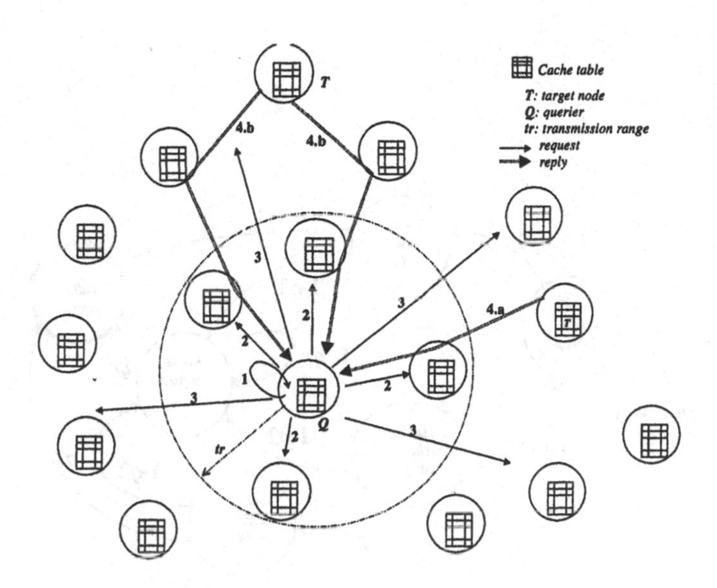

Figure 1: DSR-like routing model: a querier node Q issues a request for a target resource T. The request process progresses as follows: 1. Q performs local lookup, 2. If a cached route to T is not found then request a lookup from 1-hop neighbors (within transmission range), 3. If a cached route is not found (or is invalid) then flood a request throughout the network, 4.a. Intermediate nodes with cached route to T reply to Q, 4.b. The target T replies to requests returning multiple paths to Q.

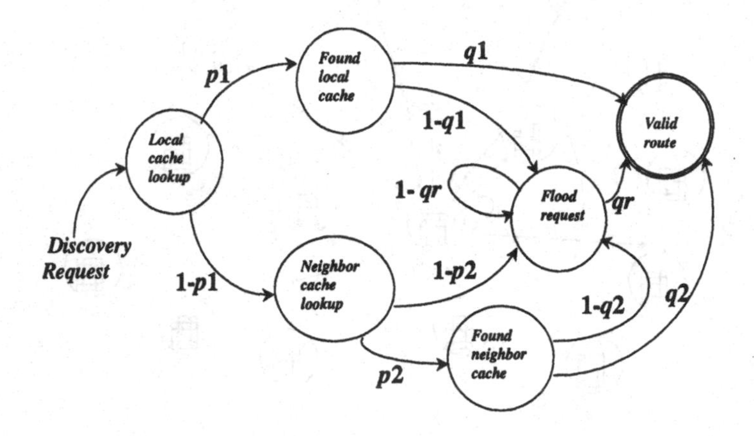

Figure 2: State diagram for the probabilistic decision chain for on-demand routing in DSR-like mechanisms.

nodes en-route to the destination have a cached route for T, in which case the flooded request may not reach T, and in which case there is a probability that the replies may be invalid. In these situations another request may be flooded to obtain a valid route. The state diagram for this process and the state transition probabilities are given in Figure 2.

From the previous figures we can get an estimate of the overhead (or cost) incurred. For now we define cost in terms of requests and replies transmitted and forwarded. In later analysis we shall study the cost in terms of energy consumption. The three main procedures of the route discovery mechanism are I. local cache lookup, II. neighbor cache lookup, and III. flooding the request.

I. **Local cache lookup** does not incur any communication overhead.

II. **Neighbor cache lookup** incurs one request transmission and replies from neighbors with a route cache. If g is the average node degree, and α is the average fraction of neighbors that respond to the request ($0 \leq \alpha \leq 1$), then number of replies is $\alpha.g$. Let the cost of neighbor cache lookup be NC, then $NC = 1 + \alpha.g$.

III. **Request flood** incurs $\sim (N-1)$ request transmissions, and replies from the target T for multiple routes equal to the number of T's neighbors forwarding the request; i.e., $\sim g$. Each reply traverses L hops from T to Q, where L is the average path length in hops between the querier and the target. The flood will also trigger replies from $\beta.N$ intermediate nodes, each traversing $\gamma.L$ hops, where β is the average fraction of intermediate nodes responding to the request and γ is the fraction of the average path length for path lengths for the path from Q to an intermediate node. Let the cost of flooding be CF, then we get

$$CF \sim (N-1) + g.L + \beta.N.\gamma.L$$

As was mentioned previously, there may be a need for repeated floods if no valid route is obtained. Assuming, for simplicity, that the subsequent floods are independent with each having an average of qr success rate, then we get a geometric distribution and the overall (expected) cost of flooding becomes CF/q_r.

From the state machine we get an expression for the overall overhead (Ov) per discovery for on-demand routing, as follows.

$$Ov = (1-p_1).NC + (1-(p_1q_1 + p_2q_2 - p_1p_2q_2)).CF/q_r$$

3.2.2 Simplifications for Resource Discovery

Several minor modifications may apply to ad hoc on-demand routing for resource discovery. Since the resource discovery request will be sent in one-shot (i.e., in a very short transfer) there is no need to store multiple paths for future route maintenance, and only one route to the target T is needed. Hence, we can eliminate the multiple responses from the target and the responses from the intermediate nodes, thus reducing the discovery overhead. This would result in $b = 0$, $q_r = 1$ and $CF = (N - 1) + L$. If we consider medium to large scale networks where $N \sim 100$ to 1000 nodes or more, and consider a reasonable range for $g \sim 6$ to 12 neighbors, then we can ignore CN with a small margin of error. Also, if we combine effects of local and neighbor cache lookups simply as p for the cache hit ratio and q for the cache validity ratio, such that $p.q = p_1q_1 + p_2q_2 - p_1p_2q_2$, we then get:

$$Ov \sim (1 - pq).(N + L)$$

It is clear that the overhead depends heavily on the cache hit and validity ratios. Let us define the cache efficacy ($CE = pq$) as the combined effect of cache hit and validity ratios. But what are the main factors affecting CE? In the following subsection caching performance is evaluated and analyzed over a variety of network size and mobility degrees to attempt to extract those factors.

3.2.3 Simulation and Analysis of Caching

We observe the performance of the on-demand routing caching scheme through simulation. Networks with 40 to 2000 nodes were simulated with various mobility degrees. Each node moves using a "random waypoint" model with no pause time. Each node selects a random value between $[0,V_{max}]$ m/s and a random destination to which it moves with constant speed. This process is repeated whenever a node reaches its destination. V_{max} was varied from 0 to 60m/s. Querier-target pairs were chosen randomly and 1000 queries were performed in each run with 10 queries per sec. A cache warm-up period was allowed before measurements were taken in each run. Each data point represents an average of 10 simulation runs with different random seeds. The average node degree was kept almost constant for all topologies at 8 neighbors per node, in 1kmx1km network. The results are given in Figure 3 and Figure 4. Figure 3 shows the cache efficacy for various mobility desgrees and network sizes. For very small scale networks (40-100 nodes) the efficacy is relatively high ($\sim$ 50-70%) especially for low mobility cases. As the number

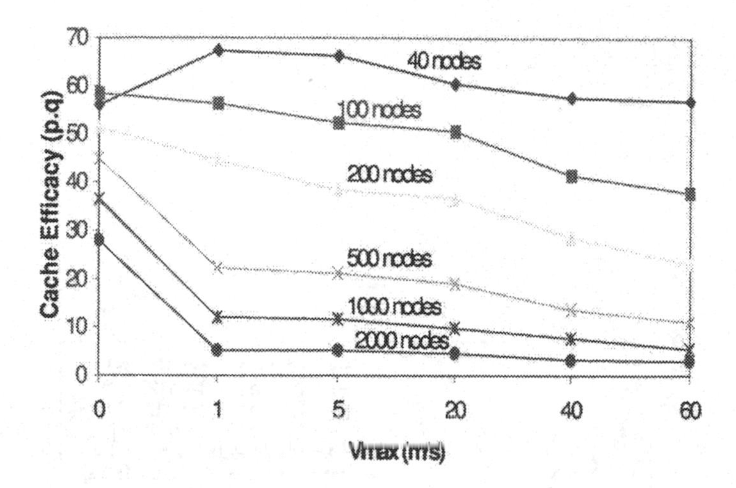

Figure 3: The cache efficacy with various velocities and various network sizes. The cache performance degrades drastically with scale of the network and with (even very low) mobility.

of nodes increases, however, the cache efficacy drops dramatically, even for very low mobility, to $\sim$ 10% for 1000 nodes and to $\sim$ 5% for 2000 nodes. Figure 4 gives a closer look at the cache metrics. It is apparent that the effect of network size is more significant than the effect of mobility. The cache hit ratio (p) drops from $\sim$ 73% (for 40 nodes) to $\sim$ 30% (for 2000 nodes). The more drastic drop occurs in the valid cache ratio (q), from $\sim$ 92% (for 40 nodes) to $\sim$ 14% (for 2000 nodes), which brings the overall cache efficacy ($p.q$) down.

For moderate to large-scale networks (above 1000 nodes) the performance of on-demand routing with caching approaches flooding, where the on-demand routing protocol resorts to flooding more than 90% of the time due to cache misses or invalid cache hits.

3.3 Hierarchical Approaches

For scalability, several hierarchical approaches have been proposed [20, 21, 45]. Many hierarchical architectures are ***cluster*-based**, in which each clus-

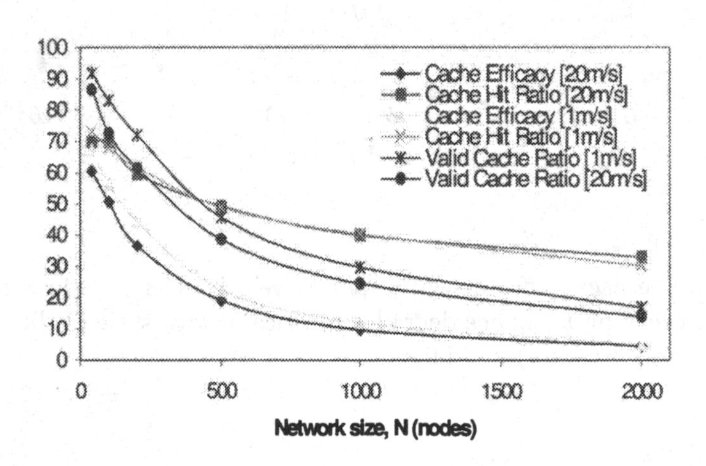

Figure 4: Cache hit ratio (p), valid cache ratio (q) and cache efficacy ($p.q$) with the network size for 1m/s and 20m/s. The cache efficacy with various velocities and various network sizes. The cache performance degrades drastically with scale of the network and with (even very low) mobility.

ter (or group) of neighboring nodes elects a cluster-head to relay traffic to the other clusters. One advantage of this approach is that a request may be forwarded to the cluster heads (via intermediate nodes) without flooding the whole network. A disadvantage, however, is that a cluster-head may become a single-point-of-failure or a point of traffic concentration and a potential bottleneck. The **landmark hierarchy** [6] [14] [15] uses landmarks as directions for routing, but does not use landmarks as communication relays between clusters. This alleviates traffic concentration at the landmarks and reduces the risk of single-point of failure. However, the highest level landmark needs to periodically flood its information throughout the network, which may be unsuitable for wireless networks. One major concern in these hierarchical approaches is their reliance on complex coordination mechanisms, for election, promotion and demotion. Hence, in highly dynamic environments, such as wireless ad hoc networks, complex hierarchical approaches are susceptible to major re-configuration with node failure, sleep schedules and mobility, leading to serious performance degradation.

One common method used for clustering is based on **dominating sets** (DS) [24] [25] [46] [47]. A dominating set of nodes in a network is a subset of the nodes in the network such that each node is either in that set or is a neighbor of a node in the set. The problem of finding the minimum dominating (MDS) set has been proven to be NP-complete. Several heuristics have been proposed to approximate the optimal solution. The proposed solutions provide various trade-offs between the establishment (and maintenance) of the dominating set (sometimes called backbone) and the cost of broadcast. A good survey on these schemes is provided in [46]. Some protocols use a connected dominating set (CDS) as shown in Figure 5, while others propose a weakly-connected DS ($WCDS$) as shown in Figure 6. $WCDS$ schemes lead to a smaller number of clusters and fewer nodes in the backbone than CDS schemes and thus incur less overhead to setup. Dominating set approaches, in general, guarantee (in theory) that the network will be fully covered. By contrast, the heuristics used for reduced broadcast cannot guarantee full network coverage.

In general, DS-based approaches may operate in two-steps or one-step (on-the-fly). The two-step protocols (e.g., [47]) conduct the DS election (or backbone setup) in one phase, then use another phase to perform the discovery. Because the two-step protocols setup and maintain a backbone they are subject to re-configuration with mobility. On the other hand, the single-step protocols, such as dominant pruning as in [24][25], conduct the dominating set election while performing the discovery, and hence are more resilient to mobility effects. In general, cluster-based approaches using dominating

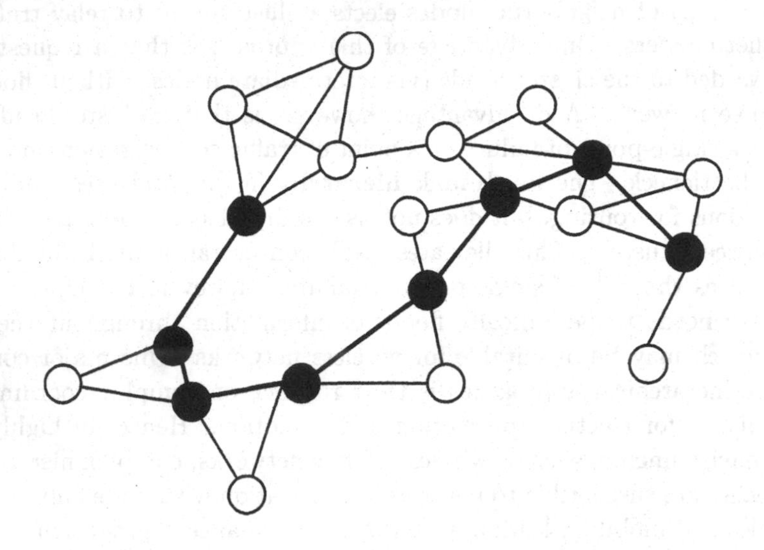

Figure 5: Example connected dominating set, black nodes belong to the backbone constituting a dominating set.

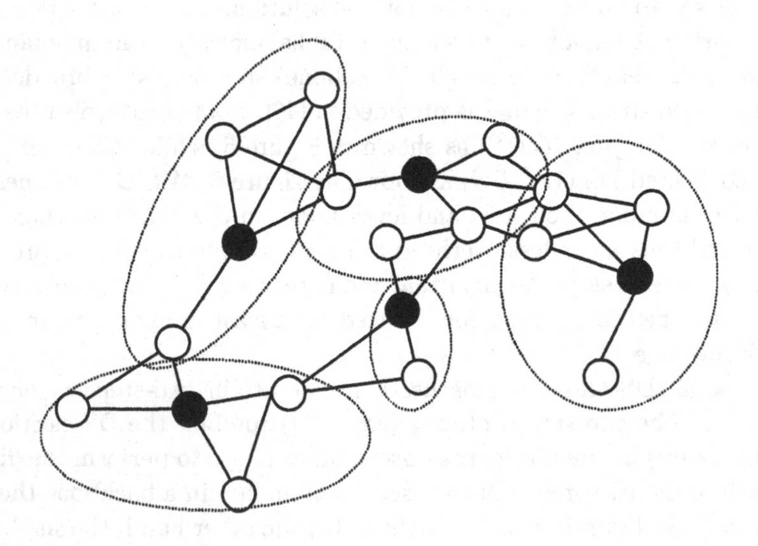

Figure 6: Example weekly connected dominating set.

sets attempt to achieve complete network coverage by ensuring that every node receives every request. These approaches may be quite effective when broadcast redundancy is high (i.e., in highly-dense networks). For networks in which node distribution is not highly-dense, however, the effects of overhead using these approaches is quite limited, and perhaps hybrid approaches should be investigated.

3.4 Hybrid, Loose Hierarchy

Hybrid schemes use a mix of proactive and reactive schemes in an attempt to establish a reasonable overhead balance. Nodes exchange periodic and triggered information with neighbors within a limited number of hops to form zones. The proactive protocol is used for intra-zone updates, while reactive, on-demand, request forwarding is used for inter-zone discovery. The zones form a loose hierarchy because each node maintains its zone independently, and no major re-configuration is incurred with mobility. Unlike dominating set schemes, hybrid approaches, in general, do not deliver every request to all nodes. Rather they attempt to achieve complete network information coverage by forwarding the request to nodes that (collectively) have information about network routes or resources. In this section we discuss two hybrid approaches (a) the zone routing protocol, and (b) the contact-based architectures.

3.4.1 The Zone Routing Protocols (ZRP)

The zone routing protocol (ZRP) [9][10][11][12][13] uses a hybrid approach, where link state is used intra-zone and on-demand routing (border-casting) is used inter-zone. A feature of zone routing is that, unlike cluster-based approaches, a zone is node-specific, and no complex coordination is used for cluster-head or landmark election. In zone routing each node transmits its information up to R hops away, where R is called the zone radius. Based on this message transmission, each node independently collects information from its neighbors and forms its own view of the network, called a zone. Nodes at exactly R hops away are called border nodes. Figure 7 shows the concept of a zone and independent zone construction by network nodes.

In ZRP the querier, Q, sends the request to its borders, and the borders send it to their borders, and so on. Query control mechanisms are used to reduce redundant querying. Requests are broadcast (or multicast) hop by hop and are recorded by nodes along the path. Query detection mechanisms (called QD-1 and QD-2) specify that intermediate nodes along

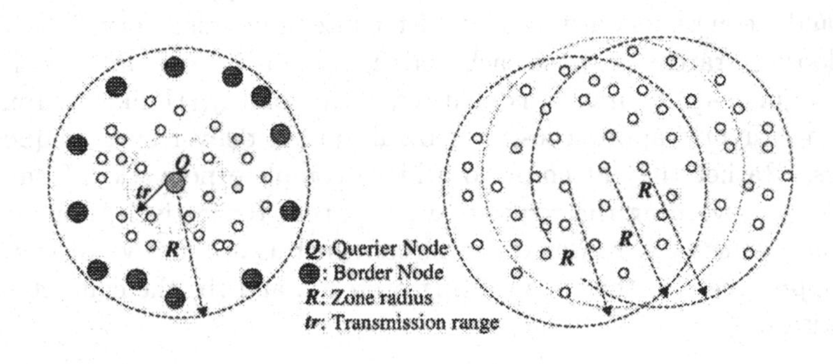

Figure 7: Concept of zone radius, R (in hops), border nodes, and per-node zone in hybrid routing with loose hierarchy.

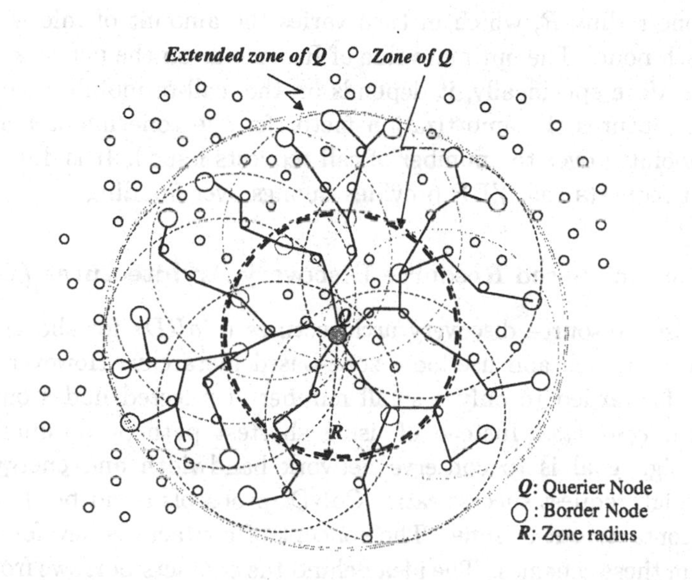

Figure 8: ZRP bordercasting with query early detection and control, various zone for the border node of Q are shown using different colors for illustration. Bold lines indicate the paths traversed by the bordercast messages.

the forwarding path (and their neighbors) record the request information. Upon receiving a request sent to a border that has been previously visited, the intermediate node terminates such request. The intermediate node has knowledge of the previously visited borders in its zone by maintaining intra-zone information of up to $2R$-1 hops (called the extended zone). Hence, the redundant request can be terminated early. This scheme is called early termination (*ET*). Illustration of the basic operation of ZRP and border-casting is given in Figure 8. ZRP design attempts to achieve an overhead balance between intra-zone maintenance and inter-zone discovery by varying the zone radius R, which in turn varies the amount of information to store in each node. The optimal value of R depends on the network size and dynamics. More specifically, it depends on the 'call-to-mobility ratio' that in essence captures the amortization factor for the zone maintenance cost (due to mobility) over the number of call requests issued. It is during those bordercast requests that ZRP provides savings over flooding.

3.4.2 Contact-based Resource Discovery Architectures *(CoRDs)*

Contact-based resource discovery architectures (*CoRDs* for short) employ the hybrid approach and use loose zone-based hierarchy. However, the requests are forwarded to only a small number of selected nodes out of the zone, called *contacts*. Instead of using shortest path or optimal routes, CoRDs design goal is to conserve network bandwidth and energy, while achieving high request success ratio. CoRDs protocols avoid border-casting by using contacts out-of-zone. The concept of contacts is key for efficient discovery in these schemes. The idea behind the contacts borrows from *small worlds* [1] [17]. Unlike relational or random graphs, wireless networks are spatial graphs (in which links are a function of distance, among other factors) that tend to be highly clustered, leading to very high degree of separation. For a node, contacts are a few nodes outside of the zone that act as short cuts to transform the wireless network into a small world and hence reduce the average degrees of separation between the querier and the target. When a request is made, a contact-selection protocol is invoked. Contact selection mechanisms aim to reduce zone-overlap and to elect contacts that increase the coverage of the search. The search proceeds according to several possible policies; single-shot, or multiple-expanding trials. Several protocols have been designed for contact-selection, including *CARD* [48] [49], *MARQ* [50], and *TRANSFER* [51]. The first two architectures, *CARD* and *MARQ*, use a pro-active approach that selects and maintains contacts. *CARD* uses zone-edge information to select useful contacts, while *MARQ* exploits mobility by choosing contacts moving away from the zone. The third architecture, $TRANSFER$, on the other hand, uses a re-active approach, by choosing contacts dynamically, on-the-fly, when the request is issued. The reactive nature of this protocol reduces the maintenance overhead and is more resilient to network dynamics. In the rest of this chapter the contact-based architecture will be presented as an elaborate case study for the design, evaluation and analysis of an efficient resource discovery protocol for large-scale ad hoc wireless networks. The contact selection mechanism presented use the dynamic contact selection mechanism, as in $TRANSFER$ [51].

4 Contact-based Resource Discovery Architectural Overview, Design and Evaluation

In the contact-based architecture, each node in the ad hoc network keeps track of a number of nodes in its vicinity within R hops away. This defines the zone of a node. The zone is maintained using a proactive localized link state protocol. Each node chooses its zone independently, and hence no major re-configuration is needed when a node moves or fails. There is no notion of cluster head, and no elections that require consensus among nodes. A neighbor discovery protocol is used by which each node identifies nodes 1 hop away (through periodic beacons). The link state protocol provides neighbor information to other nodes in the zone. Typically the number of nodes in the zone is small (less than 100 nodes). As part of the zone information each node keeps routes to nodes and pointers to resources in its zone. Nodes exactly R hops away are called *borders*. When a querier node Q (potentially any node in the network) issues a resource discovery request, it first checks to see if the resource (or destination) is in its zone. If not, then it seeks the assistance of a number of contacts (NoC) outside the zone, as follows. First, a request is issued to NoC (say 3) of Q's borders (R hops away). Each border, B, receiving the request would in turn select another node, C, at r hops away to which to forward the request. C is called a contact node. To increase the search efficiency, C should have low zone overlap with Q. Proper setting of the parameter r helps to reduce such overlap. Contact nodes act as short cuts that bridge between disjoint zones. This helps to reduce the degrees of separation between Q and the target nodes. Degrees of separation in this context refer to the number of intermediate nodes from the querier node to the target. The main architecture is shown in Figure 9, where the querier node Q chooses three of its borders, B_1, B_2, B_3 to which to send a request message. Each of the borders in turn chooses one contact

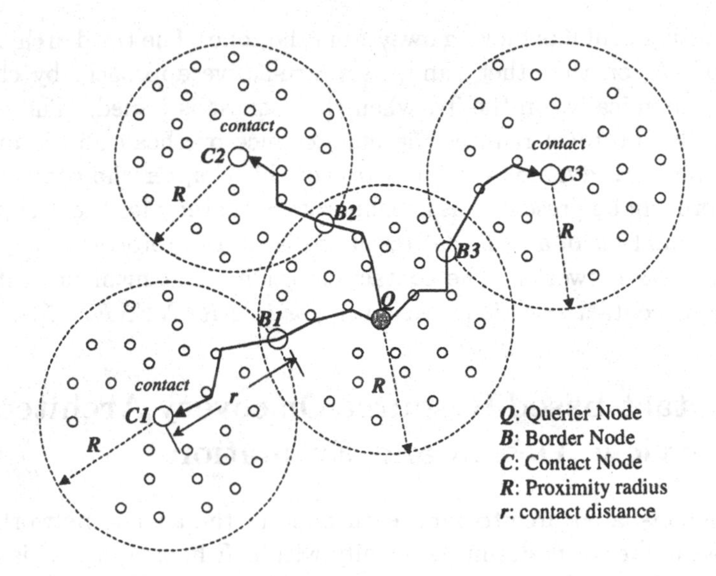

Figure 9: Each node in the network has a zone of radius R hops. A querier node, Q, sends a request through a number of its borders equal to the number of contacts (NoC), in this case $NoC = 3$. Each border node, B_i, chooses one of its borders, C_i, to be the direction for forwarding the request r hops further until it reaches the contact. The contacts are at most $(R+r)$ hops away from Q. In this example $r = R = 3$.

at r hops away to which to forward the request. C_1, C_2, and C_3 represent the contacts. The number of borders (and subsequently contacts) chosen, NoC, and the contact distance (r hops) are design parameters. If $r = R$ then the contact is a border of a border of Q. Questions regarding setting the design parameters, such as number of contacts (NoC), contact distance (r), and zone radius (R), will be presented in the evaluation section. First, we describe the contact selection scheme.

4.1 Contact Selection and Search Policies

This section describes the contact-selection protocol and the notion of levels of contacts. Then presents various search policies.

4.1.1 Contact Selection Protocol

The main purpose of a contact node is to act as a short cut to increase the view of the network by searching for the target in uncovered parts of the network. Hence, it is important for a contact to have a zone that does not overlap significantly with that of the querier node, Q, or the other contacts of Q. However, contacts do not know about each other, and do not know their shortest distance to the querier (remember that contacts are outside of the querier's zone). To address this problem an algorithm is used to reduce zone overlaps, thus increasing coverage and reducing search overhead. The first kind of overlap occurs between the contact's zone and the querier's zone. To reduce this overlap the request is directed out of the querier's zone. One simple approach to try to achieve this is for the border node to randomly choose one of its borders to which to forward the request. This, however, often leads to significant overlap with the querier's zone rendering the contact ineffective and the query success rate becomes low. Another simple approach is for the border node to avoid sending the request through the node from which it was received. However, wireless networks have a high clustering coefficient[2] [1][17]. This means that, on average, there is relatively high probability that the neighbors of a neighbor of B are also neighbors of B. Therefore, it is not sufficient to avoid only the previous hop since there may still be a good chance that the border may forward the request through nodes that belong to Q's zone. This is illustrated in Figure 10 (a), where the border node B receives the request from node L (the previous hop), and forwards it to contact C_1 through node x. Node x is a neighbor of node L and is within Q's zone, and hence would lead to a contact less than $R+r$ hops away. In many cases the contact chosen this way may have a zone heavily overlapping with Q's zone. The problem in forwarding the request outside of Q's zone to a useful contact is the loss of direction for the forwarded message at the border of the zone (since Q knows only about nodes R hops away). To achieve a sense of direction without location information, a mechanism is designed that uses information about the neighbors of B's previous hop, L, as explained next. A querier node, Q, sends a request to NoC of its borders. Consider one of those borders, B. Let node L be the last hop before B on that path. Note that B is R hops away from Q, and L is R-1 hops away from Q. All L's neighbors (including B) are 1 hop away from L, and hence are at most R hops away from Q. That is, all L's neighbors

[2] The clustering coefficient (cc) measures the probability that neighbors of a node are also neighbors of each other. In [17] it was shown that for wireless networks cc=0.58 (high clustering) for settings similar to our study.

are within Q's zone. As was mentioned before, due to high clustering many of L's neighbors (all of which are in Q's zone) may also be B's neighbors. Hence, B should attempt to avoid forwarding the request through any of L's neighbors. As illustrated in Figure 10 (b), B avoids L's neighbors (x, y, z) and is able to forward the request to a contact, C_2, that has significantly less zone overlap with Q than C_1 does. This scheme is called the zone overlap reduction (POR) scheme. Note that the above examples used $r = R$ for illustration. In cases where r is not equal to R, POR is used to select a border for B that provides direction for choosing the contact, this is called the direction border. If $r < R$ then POR is performed by B and then the contact is selected between B and its direction border. If $r > R$ then the direction border needs to perform POR again to find its own direction border, and so on. POR is performed $\lceil r/R \rceil$ times at each chosen border.

The second type of overlap occurs between zones of contacts. To reduce this overlap the querier node, Q, attempts to select borders to which it has disjoint routes. This is done using the zone information (with no extra overhead). If NoC borders are chosen by the end of this procedure then Q sends the request to the chosen borders. Otherwise, borders are chosen with minimum route overlap (i.e., with different 2nd hop nodes, then 3rd hop nodes, etc.). Otherwise, new borders are chosen randomly until NoC borders are chosen. This scheme does not guarantee non-overlap between contacts' zones, but performs quite efficiently during requests, as we shall show. We call this scheme the route overlap reduction (ROR) scheme. The POR and ROR mechanisms are performed as part of the query forwarding process, i.e., 'on-the-fly'. Reduced broadcast heuristics or dominating set schemes may be used to reduce the overhead of the zone maintenance.
Power-Aware Contact Selection: The contact-selection criteria can (and in fact should) take power into consideration. Information about power levels in nodes and rate of energy consumption may be piggybacked upon the proximity exchange messages. During the contact selection mechanisms (i.e., POR and ROR) nodes with low remaining energy are given low selection priority. To achieve this the following power-aware algorithm is used. A node that is selecting a border or contact first applies the POR and ROR rules as described above. If the selected node(s) (or any of the nodes en route to the selected node(s)) has less than a power threshold (Pth_1) of remaining power, then the selection process is repeated for $Pth_2 < Pth_1$, so on, until the selection is made. Hence, this algorithm selects the most energy-capable nodes that satisfy the non-overlap rules of POR and ROR.

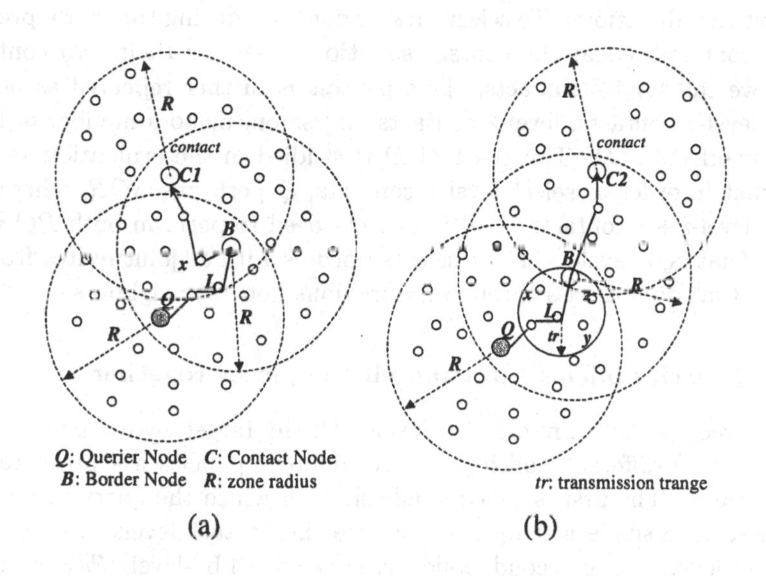

Figure 10: (a) The border node, B, forwards the request towards its border C_1 via node x. C_1's zone has significant overlap with Q's zone. By only using random forwarding or avoiding only node L (the previous hop) B can easily lose sense of direction and choose a poor contact. (b) By using neighbor information of L, B avoids forwarding the request to L or any of its neighbors (x, y, z), all of which are in Q's zone. Hence, B is more likely to choose a useful contact, C_2. The overlap between C_2's zone and Q's zone is a lot less than overlap between C_1's and Q's zones.

4.1.2 Levels of Contacts - putting the first pieces together

The above contact selection schemes (POR and ROR) provide a mechanism to select NoC contacts that have distances up to $R + r$ hops away from Q. These contacts are called level-1 contacts. To select the level-1 contacts Q performs ROR to reach NoC borders, then those borders (and their respective direction borders, and so on, $\lceil r/R \rceil$ times) perform POR to get the contacts' directions. To select farther contacts during the query process, level-1 contacts repeat the contact selection to choose their own contacts, which we call level-2 contacts. This process is further repeated as needed at the level-1 contacts, level-2 contacts and so on, up to a number of levels called maxDepth, D. The effect of D is studied in the evaluation section. Note that in order to select level-1 contacts, Q performs ROR, whereas to choose level-$i+1$ contacts, level-i contacts need to perform both POR and ROR. That is, a level-i contact selects borders with disjoint routes from its borders that do not pass through its previous hop L's neighbors.

4.1.3 Search Policies - putting all the pieces together

Given a request and a number of levels, D, the target search process may proceed using different policies. Three different policies are presented for target search. The first is called single-shot, in which the querier sends out a request, in a single attempt, to traverse the contact levels in succession, up to D levels. The second policy is called level-by-level (*lbl*), in which the request is sent out in several attempts. The first attempt is performed with level depth of 1. Until and unless the target is found, each subsequent attempt, i, is performed with level depth $d_i = 1 + d_{i-1}$. Attempts continue up to $d_i = D$. The third policy is called *stepsearch* (or simply *step*), and is similar to *lbl* except that increasing the depth occurs in steps instead of increments of 1. For our study we choose an exponential step increase; i.e., $d_i = 2d_{i-1}$.

Single-shot Policy In this policy the request is sent out from the querier node once, in a single attempt. The request is forwarded directly from level-1 contacts to level-2 contacts, up to level-D contacts. In a sense, this policy is analogous to flooding between contacts. To further clarify this policy we give a simple, first order, theoretical estimate of its overhead. These estimates are given only for illustration purposes. At each level-i, the theoretical number of contacts visited is $(NoC)^i$, and the theoretical number of hops traversed is $(R + r).(NoC)^i$. Hence, the number of transmissions is given by $[(R+r).\sum_{i=1}^{D}(NoC)^i]$. This provides a theoretical upper bound. The search employs loop and re-visit prevention mechanisms, the effect of which are not considered in this simple theoretical analysis. After considering these mechanisms, the overhead may be reduced drastically. An example of the single-shot search is given in Figure 11 (a).

Level-by-level (*lbl*) Policy In *lbl* the querier node, Q, may need to send the request several times, in multiple attempts, until the target is reached or D is reached. Starting with 1 level, the number of levels visited in each attempt d is incremented by 1. If the querier does not get a positive response, it initiates another attempt after increasing d. Hence, the number of contacts visited in each attempt is given by $\sum_{i=1}^{d}(NoC)^i$, and the theoretical upper limit on number of transmissions is $[(R+r).\sum_{d=1}^{D}\sum_{i=1}^{d}(NoC)^i]$.

Exponential *Step* Search Policy Step search is similar to *lbl*, except that the number of levels visited in attempt i, d_i, is incremented exponentially; i.e., $d_i = 2d_i - 1$ (e.g., 1,2,4,8..) until the target is found or d_{max} is reached, where d_{max} is the first d_i that satisfies the inequality $2d_{max} > D$ for $D > 2$. (For $D \leq 2$, $d_{max} = D$). For example, if $D = 20$ then $d_{max} = 16$. For the step policy the upper limit on number of transmissions is given by $[(R+r).\sum_{d=1,2,4,8,\ldots}^{d_{max}}\sum_{i=1}^{d}(NoC)^i]$.

An example of *lbl* (or *step*) with $D = 2, R = r = 3$, and $NoC = 3$ is given in Figure 11 (b). Schemes *lbl* and *step* are identical for $D = 2$. Note that the level-1 contacts visited on the first attempt are not necessarily similar to level-1 contacts visited on the second attempt. This is due to the randomization of the first border selection. From Figure 11 this effect is clear, and it results in different policies reaching different parts of the network. It seems, however, that *single-shot* may not reach parts of the network near the querier, but those parts are likely to be reached by *lbl* and *step* due to the randomization (*rotation-like*) effect, as illustrated in Figure 12. Another performance implication due to the different policies is in the request latency. Intuitively, *single-shot* incurs less delivery time than the other policies because it completes its search in a single attempt. *Step* search is expected to complete its search in less number of attempts than *lbl*.

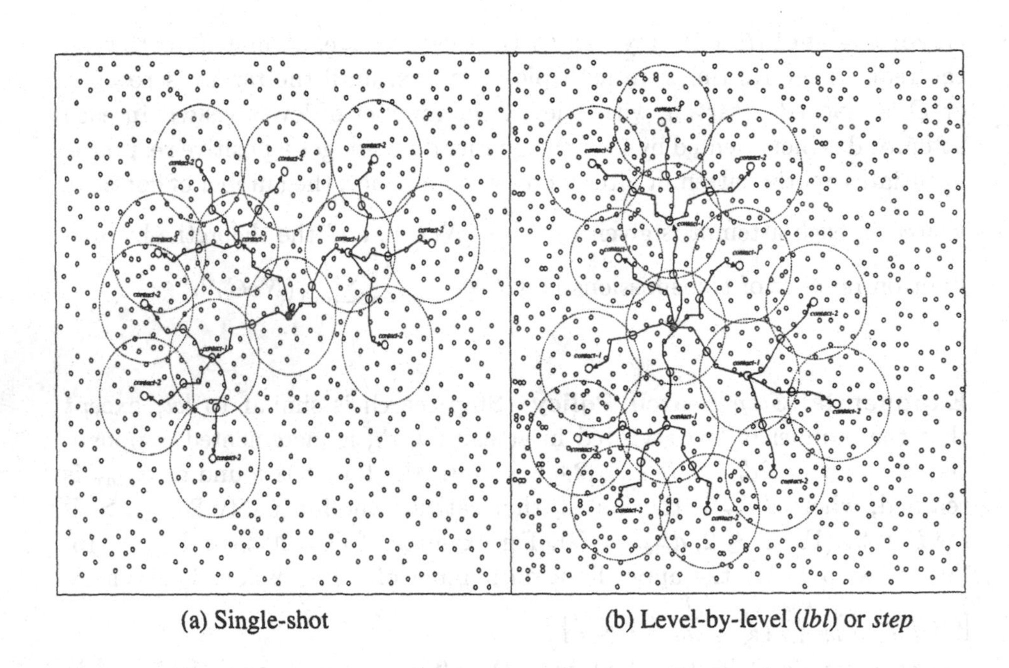

(a) Single-shot (b) Level-by-level (*lbl*) or *step*

Figure 11: Examples of search policies with $D = 2$, $R = r = NoC = 3$: (a) The single-shot policy forwards the request in one attempt reaching level-1 and level-2 contacts (called contact-1 and contact-2), (b) The level-by-level (*lbl*) policy forwards the request in multiple attempts with increasing the visited levels. In the first attempt only '3' level-1 contacts are visited. In the second attempt 3 different level-1 contacts are visited and the request is forwarded to '9' level-2 contacts. It is clear that different policies reach different parts of the network. Single-shot may not be able to achieve good coverage near Q with low NoC.

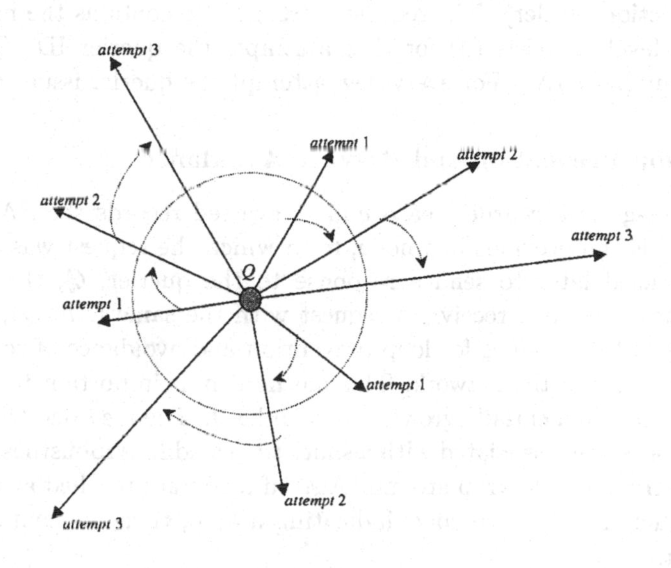

Figure 12: The rotation-like effect between attempts in *step* and *lbl* increases network coverage. In *lbl*, attempt$_i$ reaches the level$_i$ contacts, so on.

4.2 Request Forwarding and Processing

The rules for processing the requests are the same for all of the above policies. This section presents the request processing, forwarding, and loop prevention.

4.2.1 The Request Message

The request message contains the target ID; the node ID or the resource key. The destination-ID in the request message contains the ID of the border node (or the direction border). The request message also contains the maximum number of levels to visit (d) for that attempt, the querier ID (Q) and a sequence number (SN). For every new attempt the querier issues new SN.

4.2.2 Loop Prevention and Re-visit Avoidance

As the message is forwarded, each node traversed records the SN, Q and P, where P is the previous hop node, from which the request was received. P may be used later to send a response to the querier, Q, through the reverse path. If a node receives a request with the same (SN, Q), it drops the request. This provides for loop prevention and avoidance of re-visits to the covered parts of the network. This mechanism is important to keep the overhead from exponentially growing at each level. The recorded (SN, Q, P) is kept as soft state, associated with a short timer, adding robustness against querier failure and SN wrap around. Also, if a contact reached at any level finds the querier in its own zone, indicating a loop, then the contact drops the request.

4.2.3 Search, Processing and Forwarding

A contact (or border) receiving the request, first performs a target search in its local zone information. If the target is found, the request is delivered and a response is forwarded on the reverse path (if needed), with each node forwarding the response to its recorded previous hop, P. Otherwise, further processing is performed as follows. In order for a recipient of a request message to determine which functions to perform, and whether it is a contact, two fields are included in the request message; level-count and the hop-count. Initially, the level-count is set to d and the hop-count set to $(R+r)$. Hop-count is decremented and checked at every hop: If hop-count reaches '0', then the receiving node acts as a contact. A contact decrements the level-count and resets the hop-count field to $(R + r)$. If level-count reaches '0' the contact drops the request. If level-count is not '0', the contact selects NoC borders (using POR and ROR), and sends the request to those borders. If the hop-count is not '0', and the current node ID is same as the destination ID of the request message, the receiving node acts as a border node. It selects a direction border (using POR), and sends the request towards it. Otherwise, the request is simply forwarded to the next hop to the destination. The request message is unicast hop by hop, it is not broadcast hop by hop. This has an impact on the energy consumed at each hop. In broadcast, all the sender's neighbors consume energy to receive the message, whereas in unicast only the intended recipient consumes full reception energy, after a handshake for channel reservation, other neighbors may go into idle/sleep mode.

5 Evaluation and Comparison

In this section we study the various design parameters for $TRANSFER$. In addition, we compare several resource discovery approaches including flooding, expanding ring search[3], ZRP, dynamic dominating set scheme and reduced broadcast. Particularly, for $TRANSFER$, the following questions are addressed: (1) How many contacts (NoC) to choose? (2) What is the best contact distance (r)? (3) What should be the maximum depth (D) for the search? (4) How should we set the zone radius (R)? (5) What is the best search policy, ***single-shot***, ***lbl*** or ***step***? (6) How does replication affect the protocol performance? and (7) Is there a specific combination of settings that performs well for a wide variety of networks? The main performance metrics include energy consumption due to communication overhead and the request success rate. Note the trade-off between success rate and overhead; the more the success rate the more the overhead and vice versa. In order to balance these conflicting goals a penalty is introduced for request failures. Any failure beyond an acceptable level will be recovered using flooding. Hence, the scheme used in this study is contact-based search, if failed then fallback to flooding. Since this penalty is quite expensive it will be natural for the best performing parameters to avoid resorting to flooding by achieving a very high request success rate using contacts only.

[3]Several variants of expanding ring search - with various constant and exponential TTL increments - were studied and were found to perform worse than flooding due to the large network diameter of wireless networks. For clarity of presentation we omit results for the expanding ring search.

5.1 Simulation Setup

Extensive network simulations are used to investigate the design space parameters and evaluate the performance of the contact-based protocols under various settings of r, NoC, and D. The overall communication overhead for the architecture is evaluated. This overhead consists of two components: (a) zone establishment and maintenance, and (b) per request (or per query) overhead. Each node keeps track of other nodes in its zone. To keep storage requirements and zone overhead at a reasonable limit, the number of nodes per zone is limited to 100 nodes. This limit is achieved for all simulated networks by setting $R = 3$. The transmission range (tr) is taken as 110m. A wide range of networks is studied, as shown in Table 1. The area of the network is varied to maintain network connectivity, and to keep the zone nodes under 100 (for zone radius of $R = 3$). N nodes are randomly placed in a square of 'l m xl m'.

Nodes	Area (mxm)	Node Degree	Border Nodes	Zone Nodes
200	1000x1000	7.6	15.1	35
500	1400x1400	8.9	20.5	44.8
1000	2000x2000	9.1	21.7	46.8
2000	2800x2800	9.7	24.7	52.9
4000	3700x3700	11	30.3	62.2
8000	4800x4800	13	38.8	77.8
16000	6500x6500	14.3	44.6	88.2
32000	9200x9200	14.3	45	88.9

Table 1. Networks used in the simulation. Nodes are initially randomly distributed. Number of border and zone nodes are given for $R = 3$.

For mobility the random way point model is used, where a destination is chosen randomly and a velocity is chosen randomly from $[0, V_{max}]$. Once the destination is reached, another random destination is chosen, so on. V_{max} was varied from 0 to 60 m/s [4]. In the simulations the energy is measured taking into account the 802.11 MAC-layer handling of both hop-by-hop transmitted and received packets.

[4]The random way point model is commonly used in evaluating wireless ad hoc routing protocols. There are, however, recent studies [63, 64, 65, 66, 67, 68, 69, 70, 71, 72] that show the inadequacy of this mobility model to capture some important mobility characteristics. In the future we plan to evaluate our protocols using a richer set of mobility models. However, we expect performance of our query protocols to be relatively robust to mobility patterns due to its reactive (on-the-fly) hierarchy establishment.

Hop-by-Hop Communication Energy Model The energy consumed when a request is forwarded at each hop is due to packet transmission at the sender and packet reception at the recipient(s). Depending on the mode of the message, whether unicast, multicast or broadcast, the number of actual recipients vary. Recipients include the intended recipients and other nodes (within the transmission range) that are in receive state. In general, a wireless node may be in one of three power states: (i) transmit state, (ii) receive state, or (iii) idle/sleep state. The power expended in each of these states varies drastically. Also, the overall power consumed is a function of the duration of stay in any of these states (mainly a function of the packet size). The amount of energy consumed during the transmission of a request packet is denoted by E_{tx}. Similarly, E_{rx} refers to the energy of request reception. If a message is broadcast, it is received by all neighbors. The average number of neighbors per node is the average node degree (g). For a unicast message, there is a small handshake phase to inform the neighbors of the impending transmission. In IEEE 802.11 the CSMA/CA is used with handshake and medium reservation. The handshake involves broadcast of small RTS and CTS messages causing the neighbors to sleep until the end of transmission. The power consumption due to handshake is denoted by E_h. Hence, the hop-by-hop energy consumption is given as follows:

- Energy consumed by a unicast message (E_u):

$$E_u = E_{tx} + E_{rx} + E_h = E_{tx}(1 + f + h),$$

where $f = E_{rx}/E_{tx}$ and $h = E_h/E_{tx}$.

- Energy consumed by a broadcast message (E_b):

$$Eb = E_{tx} + g.E_{rx} = E_{tx}(1 + f.g),$$

where g is the average node degree. This study used f=0.64, and h=0.1[5]. The simulator differentiates between (hop-by-hop) unicast and broadcast messages and applies the energy model accordingly. The energy is presented in E_{tx} units. The results of the simulations are discussed next. The first part of the results discusses the effect of r, NoC, and D on the performance of the

[5]The power consumption numbers were based on reasonable averages of data from Lucent, Cisco and 3Com 802.11b wireless cards. For the unicast case, a short RTS/CTS handshake reserves the channel for data transmission, other nodes within radio range backoff for the duration of the transmission and go to sleep/idle mode. The handshake consumes a small fraction (h) of the actual transmission energy (E_{tx}). This fraction depends on the transmitted packet size. A reasonable (on the high side) estimate of h is~ 10%.

different search policies. For this set of simulations the 1000 node topology in Table 1 is used. The query overhead for $TRANSFER$ is insensitive to mobility because of its on-the-fly contact selection and query mechanisms. The results are shown for $V_{max} = 20m/s$. The second part of the results presents scalability analysis and comparisons. Each data point represents an average of 10 simulation runs with different random seeds. Querier-target pairs were chosen randomly. 1000 such queries were performed in each run; i.e., a total of 10,000 queries (or requests) for each data point. First present the overhead per request (hereafter referred to as overhead per query) is presented, then the zone overhead, and finally the overall overhead.

5.2 Overhead per Query

The overhead per query is affected by the various design parameters. Here we investigate the effect of the contact distance (r), the number of contacts (NoC), the maximum depth (D), and the degree of replication. This study aims to understand the performance trends of the different $TRANSFER$ policies with the various parameters.

5.2.1 Effect of Contact Distance (r)

Several experiments were conducted with various NoC and D. Only partial results are shown to illustrate the trend, using $NoC = 3$ and $D = 33$ in a 1000 node network. Figure 13 shows the effect of varying r and clearly indicates favorable settings for the different search policies. In general, as r grows, the contacts' location extends farther away from the querier's zone. For single-shot policy, as r increases a consistent increase in the energy consumed per query is observed. This is due to a drop in the request success rate. Remember that drop in success rate translates into fallback to flooding, which consistently produces more energy consumption. The drop in success rate is due to reduced coverage of areas near Q's zone or the contacts' zones. Hence lower values of r ($0 \leq r \leq 2$) are preferred for single-shot. On the other hand, for lbl and step policies, the trend is different. Due to the contact rotation-like effect between attempts, *lbl* and *step* can still maintain good coverage with increasing r up to a certain distance. Hence, high request success rate is achieved with less energy consumption due to fallback to flooding. Further increase in r generally leads to more energy consumption due to drop in success rate. At very low values of r (e.g., $r \leq 2$), although *lbl* and *step* achieve high success rate, they also incur added overhead due to zone overlap between Q and level-1 contacts (and in general between level-i

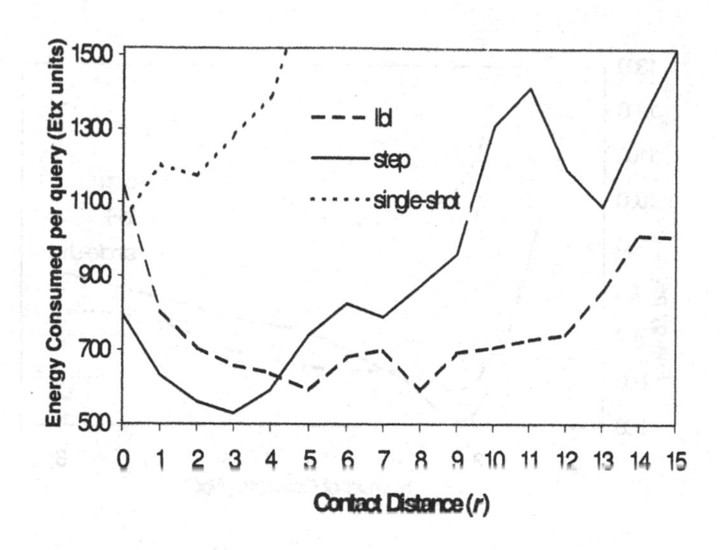

Figure 13: Effect of Contact Distance (r).

contacts and their respective level-$i+1$ contacts). This overlap reduces with increase in r, with the best values around 3-8 hops (3 being best for *step*, 5 and 8 being best for *lbl*).

5.2.2 Effect of Number of Contacts (NoC)

To understand the effects of NoC on the different policies different favorable settings of r based on our previous analysis were evaluated. Results in Figure 14 are shown for $r = 2$ (for *single-shot*), $r = 8$ (for *lbl*) and $r = 3$ (for *step*). For all policies, a very low number of contacts ($NoC < 3$) incurs high energy consumption due to fallback to flooding because of low success rate. Increasing NoC increases success rate until almost all requests succeed then an increase in overhead is observed due to additional (unnecessary) search branches with increase in NoC. For *lbl* and *step* the best setting is at $NoC = 3$, while for *single-shot* the best setting being $NoC = 4$ mainly due to the inability of 3 contacts to establish complete coverage near Q's zone and the contacts' zones.

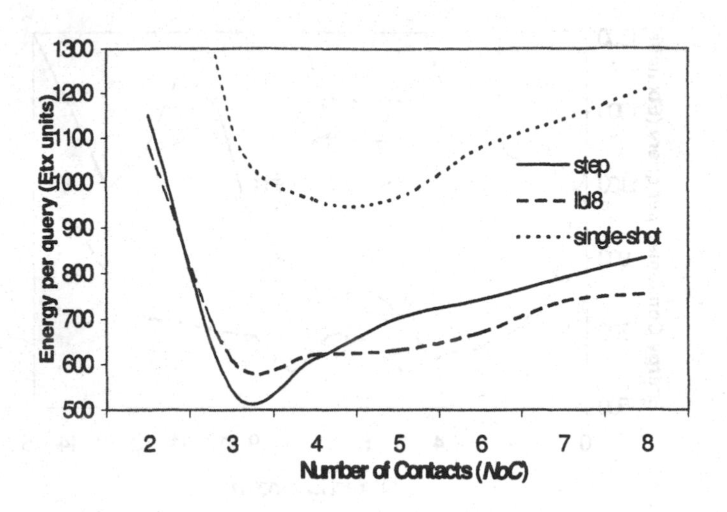

Figure 14: Effect of Number of Contacts (NoC).

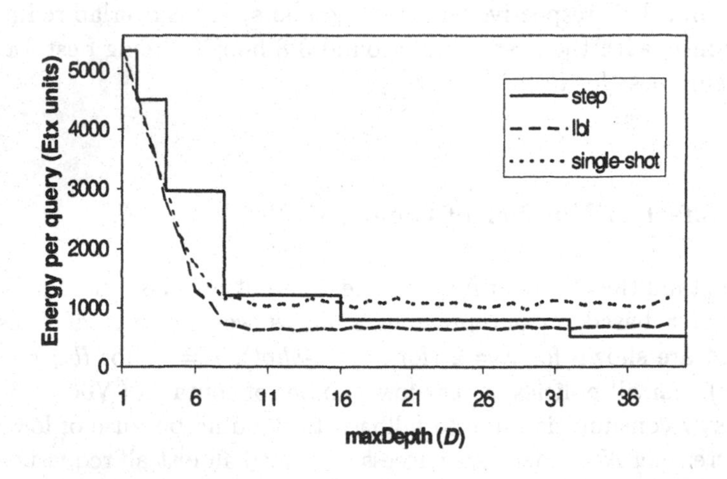

Figure 15: Effect of maximum depth (D).

5.2.3 Effect of Maximum Depth (D)

Using favorable settings for r and NoC the effect of increasing the maximum contact depth, D, is investigated. Results in Figure 15 show that increasing D generally decreases the power consumption by increasing the success rate and subsequently reducing fallback to flooding. It is not the case that increasing D exponentially increases the number of contacts visited. Although the number of potential contacts grows, loop prevention drastically reduces the number of visited contacts. After certain values of D (10 for *lbl*, 13 for *single-shot* and 33 for *step*) most requests (above 97.5%) become successful and energy consumption saturates. Note that $D = 33$ for step translates into a maximum of 6 attempts. Increase in D does not necessarily translate into increase in number of attempts. The average number of attempts (for $D > 10$) is 3.1 attempts for *step*, 4.0 for *lbl*, and of course 1 for *single-shot*. For larger networks we expect this number to rise and D required for high success rates is expected to rise as well.

5.3 Scalability Analysis of Total Overhead

In this section we investigate how the energy consumption grows with the increase in network size (as in Table 1). There are two main overhead components for $TRANSFER$: (a) query overhead, and (b) zone maintenance. We analyze scalability of query overhead, zone overhead, and total overhead.

5.3.1 Scalability of $TRANSFER$ Query Overhead

Parameter setting for this experiment was based on earlier analysis. For *single-shot* results are presented for two settings: (a) $r = 2$, $NoC = 4$, (b) $r = 3$, $NoC = 3$. The maximum depth, D, was increased to 65 to achieve better success rate for *single-shot*. For *step* and *lbl* we used $D = 33$ and $NoC = 3$. For *step* we used $r = 3$, and for *lbl* $r = 8$. Results are presented in Figure 16. For all network sizes it is clear that the *step* policy achieves the best performance (with success rate of 97.5% or better without flooding for all network sizes). *lbl* incurs more overhead than *step*. *single-shot* exhibits an interesting behavior, for $NoC = 4, r = 2$ it achieves between 90-96% success rate for sizes below 4000, then the success rates go above 97.5% from 4000 nodes and on, consistently rising with increase in N. For *single-shot* with ($NoC = 3$, $r = 3$) lower success rates (82-89%) are reached for sizes below 4000, the success rate increases to (94-97%) for 4000-8000 nodes. After 8000 nodes this setting achieves 97.5% and above success rate. It is interesting that *single-shot* with this (3,3) setting achieves less success rate

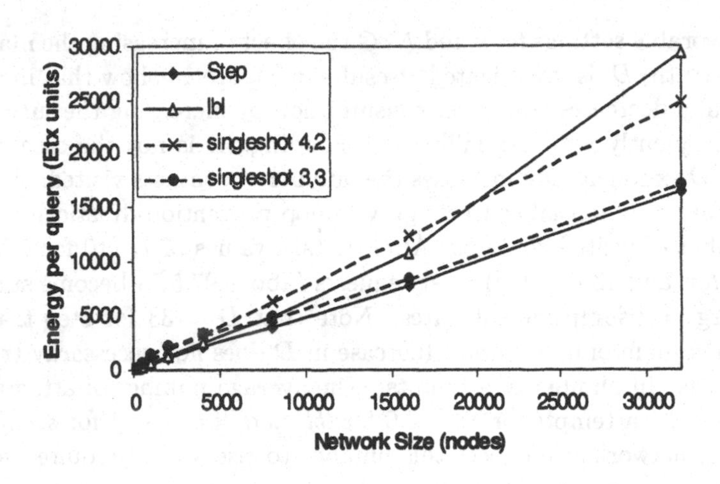

Figure 16: Scalability of query overhead for $TRANSFER$ policies.

than the previous (4,2) setting for small networks, yet there is a cross-over point at 4000 nodes after which $single - shot(3,3)$ performs better than $single - shot(4,2)$ and approaches performance of *step*. This is shown more clearly in Figure 17, which shows the query overhead ratio of *step* to the others. There are two reasons for this trend: a) for lower NoC *single-shot* incurs less overhead, b) with the increase in number of nodes there are more branches to search, giving more chance to cover, at higher contact-levels, what was not covered at lower contact-levels (near Q), thus increasing the success rate for contact-based search and decreasing fall-back to flooding.

Latency Analysis: Figure 18 shows the trend for average number of attempts with increase in nodes. The *single-shot* average is always$\sim$ 1, and the largest increase occurs for *lbl* (reaching 13.7 attempts for 32,000 nodes). *Step* scales reasonably well, with 5.2 average attempts for 32,000 nodes. Based on this analysis, we observe that *lbl* provides no advantage over *single-shot* or *step*. *Step* provides the best performance in terms of energy consumption, and possesses desirable scaling characteristics in terms of overhead and latency. *single-shot* exhibits the best latency among these policies and may be set to achieve good performance at higher scale. One desirable feature

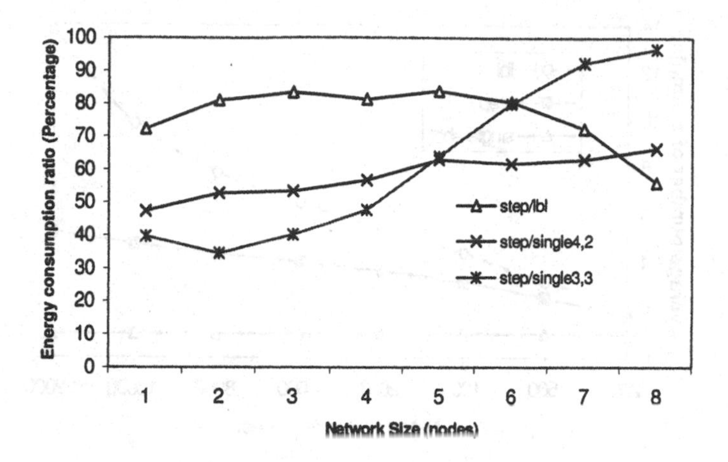

Figure 17: Query overhead of various $TRANSFER$ policies relative to step.

of step, however, is its persistent good performance over a wide spectrum of network sizes, with the setting ($R = 3, NoC = 3, r = 3, D = 33$ [max attempts=6]). We use these settings in the rest of the simulations.

As mentioned before, the forwarding and processing rules are the same for all policies. Depending on the query requirement different policies may be used for different kinds of requests. This is simply achieved by setting the right parameters in the request message. For example, to implement *single-shot*, the querier sets the maximum level of contacts to visit (d) to the maximum depth (D) and performs a single attempt.

5.3.2 Comparison (Query Overhead)

We compare the approaches of flooding, on-demand routing with cache, i.e., DSR-like (we refer to this as ODC), reduced broadcast (location-based scheme as described in[23]), dynamic minimum dominating set (MDS) (dominant pruning as in [24][25]), ZRP and $TRANSFER$. For illustration we briefly explain flooding overhead. For flooding, in a network of N nodes, the request is transmitted by $N-1$ nodes. We get $E_{flood} = (N-1).E_{tx}(1+f.g) = E_{tx}(N+2Lf)$, for large N, where L is the number

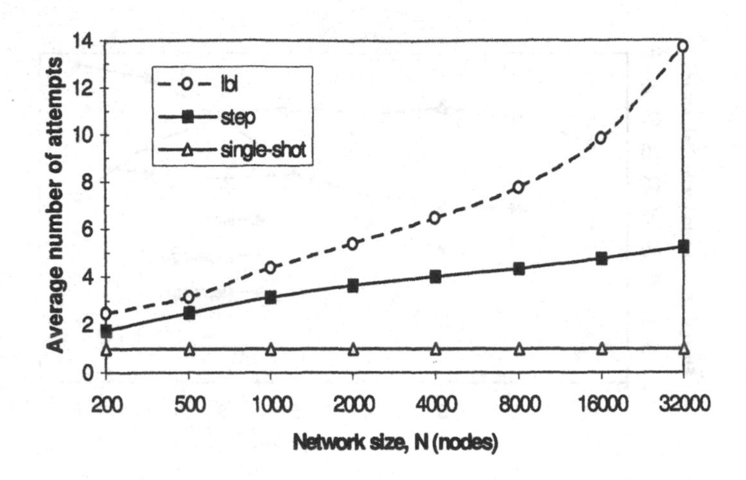

Figure 18: Average attempts per query for $TRANSFER$ policies.

of links in the network, g is the node degree, and $g = 2L/N$. Figure 19 shows the results for query overhead for $V_{max} = 20m/s$ (similar trends were observed for other velocities). Noticeable improvement is observed in performance by using contacts, especially in large-scale networks. This is due to: (i) decrease in number of transmitted packets per query, and (ii) avoiding broadcast and using hop-by-hop unicast messages.

Figure 20 shows overhead of the various schemes relative to flooding. Performance of ODC degrades with increase in network size, and it approaches flooding in large-scale networks. Reduced broadcast ranges from 55-70% while MDS ranges from 48-55% of flooding overhead. ZRP's performance varies from 33.5% of flooding in small networks to 46% at larger scale. The *step* policy has the best performance with 7.4% in small networks to 5.8% of flooding in large networks.

Per-node Energy Analysis The above analysis presented the total energy (per-query) consumed by the whole network. Such analysis does not show the energy distribution in the network, which would be more representative of the network lifetime and potential for partitioning. In this analysis several experiments are conducted to compare the energy distribution in the

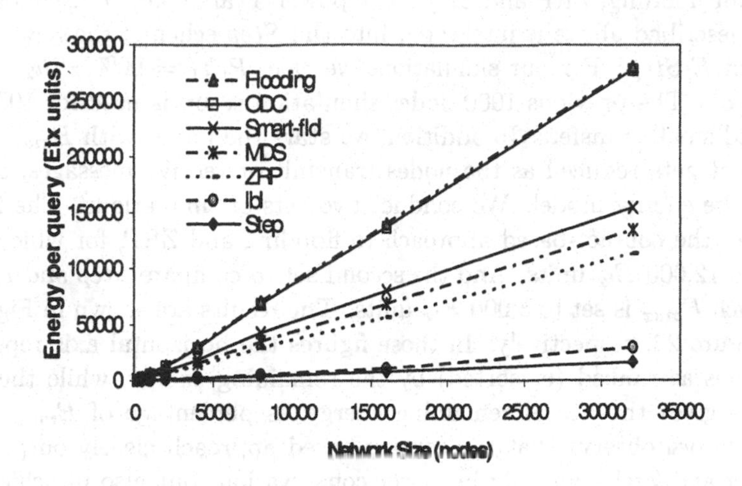

Figure 19: Comparing resource discovery schemes using energy consumption.

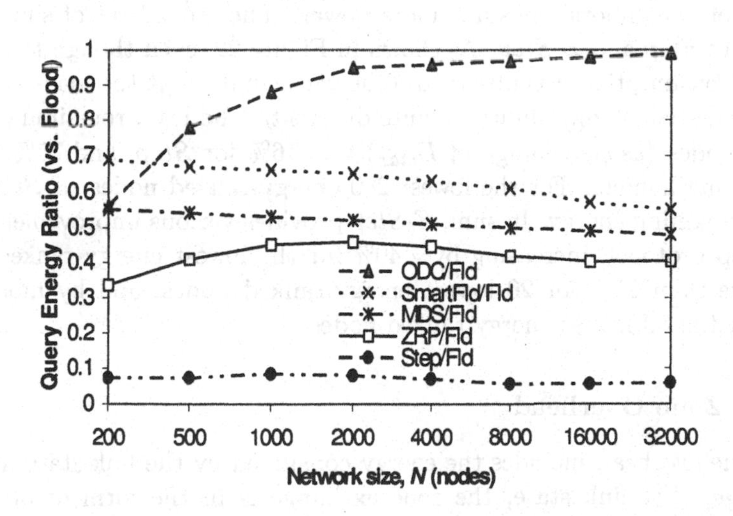

Figure 20: Query overhead of the various schemes normalized by flooding.

nodes for flooding, ZRP and *Step*. The power-aware contact selection algorithm described above is integrated into the *Step* scheme; the new scheme is called *E-Step*. For our simulations we take $Pth_1 = 90\%, Pth_2 = 80\%$, and so on. The previous 1000 nodes simulation setup is used for 1000 randomized small transfers. In addition, we start the nodes with E_{max} energy level that gets reduced as the nodes transmit or receive messages, according to the energy model. We conduct two sets of simulations. The first to compare the contact-based approach to flooding and ZRP, for which E_{max} is set to 12,000 E_{tx} units. And the second set to compare *Step* and *E-Step*, for which E_{max} is set to 3,000 E_{tx} units. The results are shown in Figure 21 and Figure 22, respectively. In those figures the horizontal axis represents the nodes as ranked (or sorted) by the remaining energy, while the vertical axis gives the actual remaining energy as percentage of E_{max}. From Figure 21 we observe that the contact-based approach clearly outperforms flooding and ZRP, not only in power conservation, but also in achieving a balanced power consumption between network nodes. For this set of simulations, the remaining energy for the lowest energy-ranked nodes was as follows (as a percentage of E_{max}): for flooding 1%, for ZRP 42%, for *Step* 80%, and for *E-Step* 90%. For flooding we notice a staircase distribution for the remaining energy, this is due to the fact that all the nodes participate in the flooding for every query and nodes having higher node degree (i.e., more neighbors) consume more power. The second set of simulations compares *Step* and *E-Step*. As shown in Figure 22, even though the overall energy consumption was observed to be very similar but the load-balancing (or energy-balancing) ability is quite different. The least remaining energy in the nodes (as percentage of E_{max}) was 16% for *Step*, and 55% for the *E-Step* mechanism. For the lowest 280 energy-ranked nodes, *E-Step* gives more remaining energy. In sum, *E-Step* provides various improvements over the Step protocol, increasing by$\sim$ 40% for the lowest energy-ranked node, by more than 25% for 20 lowest energy-ranked nodes, and by more than 15% for the 50 lowest energy-ranked nodes.

5.3.3 Zone Overhead

The zone overhead includes the energy consumed by the link state message exchange. For link state, the zone exchange is in the form of broadcast messages by each node, up to R hops away. This exchange increases linearly with mobility (with more link changes). So, this overhead is normalized with respect to mobility using $Z(R)$. The zone overhead is also a function of the number of nodes in the zone. Figure 23 shows $Z(R)$ for $TRANSFER$ and

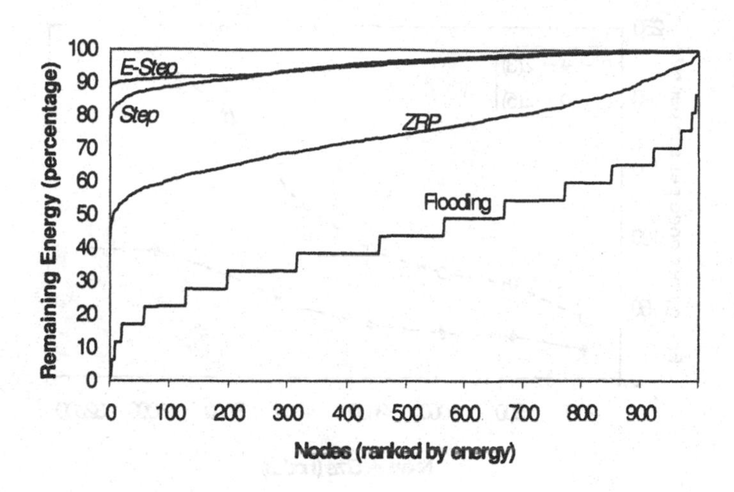

Figure 21: Remaining energy for Flooding, ZRP, *Step* and *E-Step* (E_{max} = 12,000).

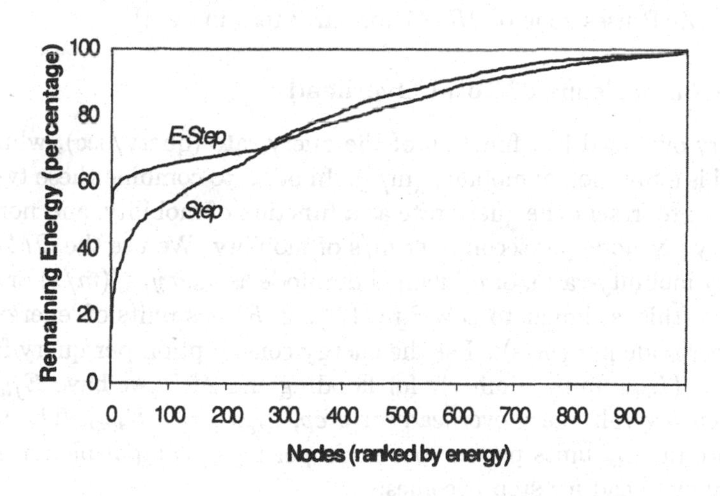

Figure 22: Remaining energy for *Step* and *E-Step* ($E_{max} = 3,000$).

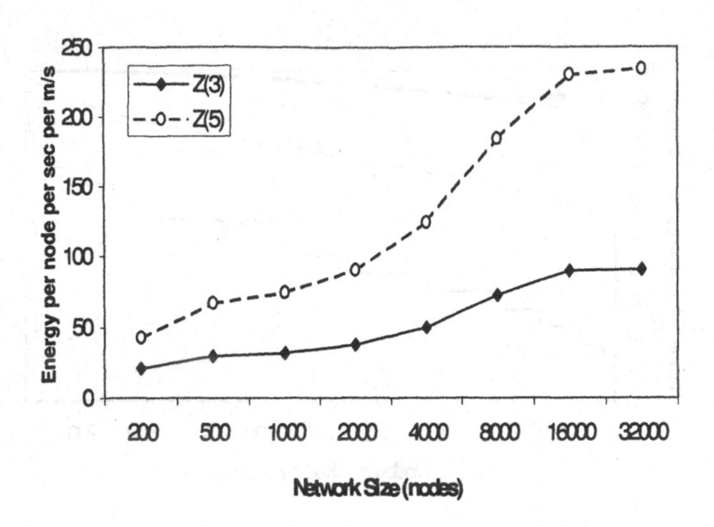

Figure 23: Normalized Intra-zone overhead for the basic proximity $R = 3$, $Z(3)$ and the extended zone of $2R - 1$, $Z(5)$.

for ZRP. (ZRP uses zone of $2R - 1$ for early termination).

5.3.4 Comparisons of Total Overhead

The query overhead is a function of the query rate (query/sec), while zone overhead is a function of mobility (m/s). In order to combine these two components we represent the query rate as a function of mobility, and normalize the energy per node per second per m/s of mobility. We use the QMR metric (query-mobility-ratio, or q) defined per node as $query/s/(m/s)$ or simply query/km (this is similar to CMR in [12]). $Z(R)$ has units of 'energy (E_{tx}) per sec per node per (m/s)'. Let the energy consumption per query for *step* be E_{step} in (E_{tx}) units. Similarly for flooding and ZRP, we have E_{flood} and E_{ZRP}. The overall query overhead for *step*, $E_{Qstep} = q.E_{step}$. The units of E_{Qstep} are in 'E_{tx} units per sec per node per m/s', compatible with Z(R). The total overhead for step becomes:

$$E_{Tstep} = Z(R) + E_{Qstep} = Z(R) + q.E_{step}.$$

For flooding, no zone overhead is incurred, so $E_{Tflood} = E_{Qflood} = q.E_{flood}$. For ZRP the zone (or intra-zone) overhead is incurred for $2R - 1$

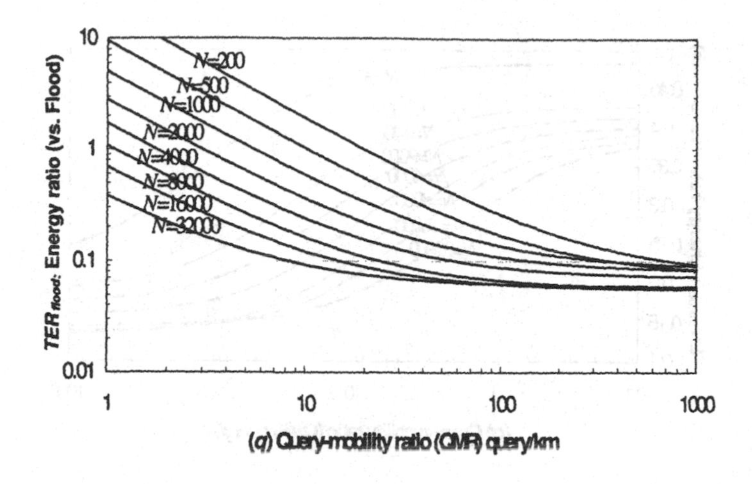

Figure 24: Total energy ratio vs. flood (TER_{flood}).

hops, hence $E_{TZRP} = Z(2R-1) + E_{QZRP} = Z(2R-1) + q.E_{ZRP}$. For brevity, we evaluate the total energy-consumption ratio, TER, of *step* to flooding and ZRP. We get:

$$TER_{flood} = \frac{E_{TStep}}{E_{Tflood}} = \frac{Z(R)+q.E_{step}}{q.E_{flood}},$$

and

$$TER_{ZRP} = \frac{E_{TStep}}{E_{TZRP}} = \frac{Z(R)+q.E_{step}}{Z(2R-1)+q.E_{ZRP}}.$$

Figure 24 shows TER_{flood} as function of the QMR (query-mobility ratio) q (query/km). We note that a logarithmic scale was used for q to resolve the rapid drop in the total energy-consumption ratio.

For very low values of q (1-10 query/km) and small to medium network sizes (200-4000 nodes) flooding performs better. This is due to the very low number of queries triggered as compared to the intra-zone maintenance overhead[6]. Note that, zone-based protocols perform well when the intra-

[6] We suspect that a scenario of very low q, indicating relatively inactive nodes, is unlikely in large-scale ad hoc networks. A more likely scenario is that when the nodes are inactive for extended periods of time, they may go to sleep or 'off' mode and not participate in intra-zone exchange. Maintaining zone information without being active is not desirable.

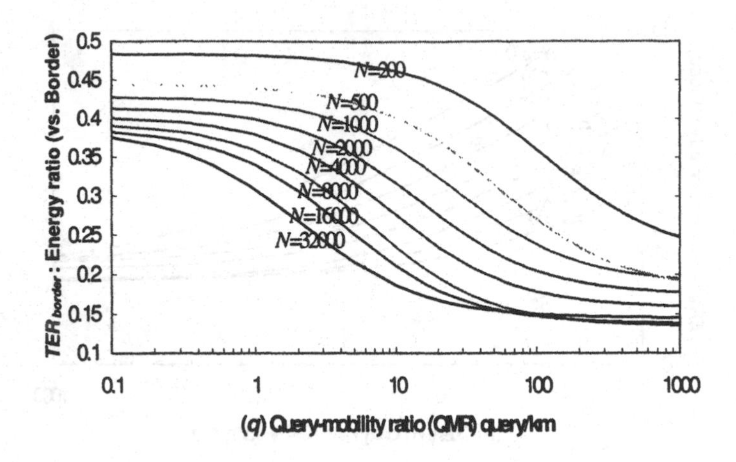

Figure 25: Total energy ratio vs. ZRP (TER_{ZRP}).

zone overhead is amortized over a reasonable number of queries in order to achieve overall gain. It is clear that for medium to large-scale networks and for medium to high rate of queries, $TRANSFER$ has a significant advantage over flooding, where TER_{flood} approaches 5% for large networks. We now turn to Figure 25 to analyze the trends in TER_{ZRP}. We notice a trend different from that for TER_{flood}, mainly because ZRP is also a zone-based approach and incurs more intra-zone overhead by using the extended zone of radius $(2R-1)$. Effect of the extended zone is clearest for small QMR where the intra-zone overhead has the dominant effect, whereas for high QMR the effect is mainly due to the query overhead. For a small network (200 nodes) and for low q, we get $TER_{ZRP}\sim$ 48%, while for high q, TER_{ZRP} is just below 25%. For medium to large-scale networks (500-32000 nodes) and for low q, TER_{ZRP} ranges from 37% to 44%, and for high q, TER_{ZRP} ranges from 13% to 20%. Hence, the best gains for $TRANSFER$ can be observed for higher values of QMR, where TER_{ZRP} approaches 14% for large networks.

6 Conclusions

In this chapter, we provided an overview of the main approaches for resource discovery in wireless networks. The problem of resource discovery is becoming more challenging and pressing as more emerging infrastructure-less networks are designed and implemented. The various approaches described in this chapter included flooding-based approaches, including expanding ring search and reduced broadcast heuristics, on-demand ad hoc routing, hierarchical cluster-based and dominating set schemes, and hybrid loose hierarchy schemes including zone routing and contact based architectures. Reduced broadcast approaches and dominating set approaches attempt to reduce the redundancies of flooding but attempt to all discovery requests to all nodes in the network. On-demand ad hoc routing utilizes information caching to reduce effects of flooding, but is greatly affected by network dynamics that cause cache invalidation. Hybrid approaches, such as zone and contact-based routing use the proactive approach within the zones, and the reactive approach for out-of-zone discovery. Zone routing uses bordercasting, while contact-based schemes select out-of-zone contact nodes to avoid bordercasting. After conducting an extensive comparative study between the above schemes, it was observed that hybrid approaches are useful when the cost of zone maintenance is amortized over a reasonable number of requests, otherwise flooding-based approaches work better. In general, if power control is used such that only active nodes participate in zone construction and maintenance, then hybrid approaches seem to scale best. In particular the contact based approach is promising as it performs well for a fixed set of parameters over a relatively wide array of scenarios and network sizes.

7 Acknowledgements

The author acknowledges the useful feedback provided during discussions with the wireless networking laboratory research group at USC. Specifically the author thanks Shao-Cheng Wang for feedback on overhead analysis and cache modeling for DSR, and Karim Seada for comments on contact selection protocols. Ahmed Helmy was supported by the NSF CAREER Award 0134650, and research grants from Intel and Pratt&Whitney Institute for Collaborative Engineering (PWICE).

References

[1] D. Watts and Strogatz, Collective dynamics of small-world networks, *Nature* (1998).

[2] J. Kleinberg, Navigating in a small world, *Nature* Vol. 406 (Aug. 2000).

[3] C. E. Perkins and P. Bhagwat, Highly dynamic destination-sequenced distance vector routing (DSDV) for mobile computers, *ACM CCR* (Oct. 1994).

[4] C. E. Perkins and E. M. Royer, Ad-hoc on-demand distance vector routing, *Proceedings of 2nd IEEE Workshop on Mobile Computing Systems and Applications* (Feb. 1999).

[5] D. B. Johnson and D. A. Maltz, Dynamic source routing in ad-hoc wireless networks, *Mobile Computing* (1996) pp.153-181.

[6] P. Guangyu, M. Gerla and X. Hong, LANMAR: landmark routing for large scale wireless ad hoc networks with group mobility, *ACM Mobihoc* (2000).

[7] J. Li, J. Jannotti, D. Couto, D. Karger and R. Morris, A scalable location service for geographic ad hoc routing, *ACM Mobicom* (2000).

[8] L. Blazevic, S. Giordano and J.-Y. Le Boudec, Anchored path discovery in terminode routing, *IFIP-TC6 Networking* (May 2002).

[9] M. Pearlman and Z. Haas, Determining the optimal configuration for the zone routing protocol *IEEE JSAC* (Aug. 1999) pp. 1395-1414.

[10] Z. Haas and M. Pearlman, The Zone Routing Protocol (ZRP) for ad hoc networks, *IETF Internet draft for the Manet group* (Jun. 1999).

[11] Z. Haas and M. Pearlman, The performance of query control schemes for the zone routing protocol, *ACM SIGCOMM* (1998).

[12] Z. Haas and M. Pearlman, ZRP: A hybrid framework for routing in ad hoc networks, in *Ad Hoc Networks* (Addison Wesley, 2001).

[13] Z.J. Haas and M.R. Pearlman, The perforamnce of query control schemes for the zone routing protocol, *ACM/IEEE Transactions on Networking*, (Aug. 2001).

[14] P. F. Tsuchiya, The Landmark hierarchy: a new hierarchy for routing in very large networks, *ACM CCR* Vol. 18 No. 4 (Aug. 1988) pp. 35-42.

[15] S. Kumar, C. Alaettinoglu and D. Estrin, SCOUT: scalable object tracking through unattended techniques, *IEEE ICNP* (2000).

[16] J.J. Aceves and M. Spohn, Bandwidth-efficient link-state routing in wireless networks, *Ad Hoc Networks* (Addison Wesley, 2001).

[17] A. Helmy, Small worlds in wireless networks, *IEEE Communications Letters* Vol. 7 No. 10 (Oct. 2003) pp. 490-492.

[18] T.-W. Chen and M. Gerla, Global state routing: a new routing scheme for ad-hoc wireless networks, *Proc. IEEE ICC* (1998).

[19] S. Murthy and J.J. Garcia-Luna-Aceves, An efficient routing protocol for wireless networks, *ACM Mobile Networks and App. Journal, Issue on Routing in Mobile Communication Networks* (Oct. 1996).

[20] C.-C. Chiang, Routing in clustered multihop, mobile wireless networks with fading channel, *Proc. IEEE SICON* (Apr.1997).

[21] Q. Zhang, B. Li, J. Liu, K. Sohraby and W. Zhu, Resource discovery in mobile ad hoc networks, in *Handbook on Ad Hoc Wireless Networks* (CRC Press, 2002).

[22] T. Clausen, P. Jacquet, A. Laouiti, P. Muhlethaler, A. Qayyum and L. Viennot, Optimized link state routing protocol, *IEEE INMIC* (2001).

[23] S. Ni, Y. Tseng, Y. Chen and J. Sheu, The broadcast storm problem in a mobile ad hoc network, *ACM Mobicom* (1999).

[24] H. Lim and C. Kim, Flooding in wireless ad hoc networks, *Computer Communications Journal* Vol. 24 No. 3-4 (2001) pp. 353-363.

[25] A. Laouiti, A. Qayyum and L. Viennot, Multipoint relaying: an efficient technique for flooding in mobile wireless networks, *HICSS* (2002).

[26] L. Breslau, D. Estrin, K. Fall, S. Floyd, J. Heidemann, A. Helmy, P. Huang, S. McCanne, K. Varadhan, Y. Xu and H. Yu, Advances in network simulation (ns), *IEEE Computer* Vol. 33 No. 5 (May 2000) pp. 59-67.

[27] H. Gupta, S. Das and Q. Gu, Connected sensor cover: self-organization of sensor networks for efficient query execution, *ACM MobiHoc* (2003).

[28] C. Intanagonwiwat, R. Govindan and D. Estrin, Directed diffusion: a scalable and robust communication paradigm for sensor networks, *ACM MobiCom* (Aug. 2000).

[29] M. Chu, H. Haussecker and F. Zhao, Scalable information-driven sensor querying and routing for ad hoc heterogeneous sensor networks, *Int'l Journal High Performance Computing Applications* (2002).

[30] R. Govindan, J. Hellerstein, W. Hong, S. Madden, M. Franklin and S. Shenker, The sensor network as a database, Technical Report 02-771, Computer Science Department, UCLA, (Sep. 2002).

[31] P. Bonnet, J. E. Gehrke and P. Seshadri, Querying the physical world, *IEEE Personal Communications* Vol. 7 (Oct. 2000).

[32] P. Bonnet, J. Gehrke and P. Seshadri, Towards sensor database systems, *Mobile Data Management* (2001).

[33] S. Ratnasamy, B. Karp, L. Yin, F. Yu, D. Estrin, R. Govindan and S. Shenker, GHT - a geographic hash-table for data-centric storage, *First ACM WSNA Workshop* (2002).

[34] W.R. Heinzelman, J. Kulik and H. Balakrishnan, Adaptive protocols for information dissemination in wireless sensor networks, *ACM MobiCom* (Aug.1999) pp. 174-185.

[35] W.R. Heinzelman, A. Chandrakasan and H. Balakrishnan, Energy-efficient communication protocol for wireless microsensor networks, *33rd International Conference on System Sciences (HICSS)* (2000).

[36] D. Braginsky and D. Estrin, Rumor Routing Algorithm For Sensor Networks, *First WSNA Workshop* (Sep. 2002).

[37] N. Sadagopan, B. Krishnamachari and A. Helmy, The ACQUIRE mechanism for efficient querying in sensor networks, *First IEEE ICC SNPA Workshop* (May 2003).

[38] B. Karp and H. Tung, Greedy perimeter stateless routing for wireless networks, *ACM MobiCom* (2000).

[39] K. Seada and A. Helmy, Rendezvous regions: a scalable architecture for service provisioning in large-scale mobile ad hoc networks, *ACM SIGCOMM, Refereed poster* (2003).

[40] D. B. Johnson, D. Maltz and J. Broch, DSR: the dynamic source routing protocol for multihop wireless ad hoc networks, in C. Perkins (ed.) *Ad Hoc Networks*, (Addison Wesley, 2001) pp. 139-172.

[41] Y. Hu and D. Johnson, Caching strategies in on-demand routing protocols for ad hoc wireless networks, *ACM MobiCom* (2000).

[42] Y.-C. Hu and D. B. Johnson, Ensuring cache freshness in on-demand ad hoc network routing protocols, *Proc. POMC Workshop on Principles of Mobile Computing* (Oct. 2002) pp. 25-30.

[43] J. Broch, D. A. Maltz, D. B. Johnson, Y.-C. Hu and J. Jetcheva, A performance comparison of multi-hop wireless ad hoc network routing protocols, *ACM MobiCom* (1998).

[44] D. A. Maltz, On-demand routing in multi-hop wireless ad hoc networks, Ph.D. Thesis, School of Computer Science, Carnegie Mellon University, Pittsburgh, PA, (May 2001).

[45] M. Steenstrup, Cluster-based networks, in C. Perkins (ed.) *Ad Hoc Networks* (Addison Wesley, 2001) pp. 75-138.

[46] Y. Chen and A. Liestman, Approximating minimum size weakly-connected dominating sets for clustering mobile ad hoc networks, *ACM MobiHoc* (Jun. 2002).

[47] U. Kozat and L. Tassiulas, Service discovery in mobile ad hoc networks: an overall perspective on architectural choices and network layer support issues, *AdHoc Networks Journal* (Elsevier, In press, 2003).

[48] A. Helmy, S. Garg, P. Pamu and N. Nahata, Contact based architecture for resource discovery (CARD) in large scale MANets, *IEEE/ACM IPDPS Int'l Workshop on Wireless, Mobile and Ad Hoc Networks (WMAN)* (Apr. 2003) pp. 219-227.

[49] A. Helmy, S. Garg , P. Pamu and N. Nahata, CARD: a contact-based architecture for resource discovery in ad hoc networks, *ACM Baltzer Mobile Networks and Applications (MONET), Special issue on Algorithmic Solutions for Wireless, Mobile, Ad Hoc and Sensor Networks*, (2004).

[50] A. Helmy, Mobility-assisted resolution of queries in large-scale mobile sensor networks (MARQ), *Computer Networks Journal Elsevier Science, Special Issue on Wireless Sensor Networks* Vol. 43 No. 4 (Nov. 2003) pp. 437-458.

[51] A. Helmy, TRANSFER: transactions routing for ad-hoc networkS with efficient energy, *IEEE GLOBECOM* (Dec. 2003).

[52] J. Liu, Q. Zhang, J. Zhang, W. Zhu and B. Li, A novel framework for QoS-aware resource discovery in mobile ad hoc networks, *IEEE ICC* (2002).

[53] K. Seada and A. Helmy, Rendezvous regions: a scalable architecture for service location and data-centric storage in large-scale wireless networks, *IEEE/ACM IPDPS 4th Int'l Workshop on Algorithms for Wireless, Mobile, Ad Hoc and Sensor Networks (WMAN)* (Apr. 2004).

[54] D. Niculescu and B. Nath, Trajectory-based forwarding and its applications, *ACM MobiCom* (2003).

[55] A. Helmy, Architectural framework for large-scale multicast in mobile ad hoc networks, *IEEE International Conference on Communications (ICC)* Vol. 4 (Apr. 2002) pp. 2036-2042.

[56] J. Tchakarov and N. Vaidya, Efficient content location in wireless ad hoc networks, *IEEE International Conference on Mobile Data Management (MDM)* (Jan.2004).

[57] I. Aydin and C.-C. Shen, Facilitating match-making service in ad hoc and sensor networks using pseudo quorum, *IEEE ICCCN* (2002).

[58] D. Doval and D. O'Mahony, Nom: Resource location and discovery for ad hoc networks, *Med-hoc-Net* (2002).

[59] V. Verma, S. Helal, N. Desai and C. Lee, Knoark: a service discovery and delivery protocol for ad hoc networks, *IEEE WCNC* (2003).

[60] O. Ratsimor and D. Chakrabotry, Allia: alliance-based service discovery protocol for MANETs, *ACM Mobile Commerce Workshop* (2002).

[61] D. Chakraborty, A. Joshi, T. Finin and Y. Yesha, GSD: a novel group-based service discovery protocol for MANETs, *MWCN* (2002).

[62] B. Williams and T. Camp, Comparison of broadcasting techniques for mobile ad hoc networks, *ACM MobiHoc* (2002).

[63] F. Bai, N. Sadagopan and A. Helmy, The IMPORTANT framework for analyzing the impact of mobility on performance of routing for ad hoc networks, *AdHoc Networks Journal - Elsevier Science* Vol. 1 No. 4 (Nov. 2003) pp. 383 - 403.

[64] N. Sadagopan, F. Bai, B. Krishnamachari and A. Helmy, PATHS: analysis of PATH duration statistics and their impact on reactive MANET routing protocols, *The Fourth ACM International Symposium on Mobile Ad Hoc Networking and Computing (Mobihoc)* (Jun. 2003) pp. 245-256.

[65] F. Bai and A. Helmy, A survey of mobility modeling and analysis in wireles adhoc networks, Book chapter in submission to Kluwer Academic Publishers and USC Technical Report (Feb. 2004).

[66] F. Bai and A. Helmy, The IMPORTANT framework for analyzing and modeling the impact of mobility in wireless adhoc networks, Book chapter in submission to Kluwer Academic Publishers and USC Technical Report (Feb. 2004).

[67] D. Son, A. Helmy and B. Krishnamachari, The effect of mobility-induced location errors on geographic routing in ad hoc networks: analysis and improvement using mobility prediction, *IEEE Wireless Communications and Networking Conference (WCNC)* (Mar. 2004).

[68] Y. Lu, H. Lin, Y. Gu and A. Helmy, Towards mobility-rich performance analysis of routing protocols in ad hoc networks: using contraction, expansion and hybrid models, *IEEE International Conference on Communications (ICC)* (Jun. 2004).

[69] C. Shete, S. Sawhney, S. Herwadka, V. Mehandru and A. Helmy, Analysis of the effects of mobility and node density on the grid location service in ad hoc networks, *IEEE International Conference on Communications (ICC)* (Jun. 2004).

[70] J. Yoon, M. Liu and B. Noble, Sound mobility models, *ACM MobiCom* (Sep. 2003).

[71] J. Yoon, M. Liu and B. Noble, Random waypoint considered harmful, *IEEE Infocom* Vol. 2 (Apr. 2003) pp. 1312-1321.

[72] A. Jardosh, E. Belding-Royer, K. Almeroth and S. Suri, Towards realistic mobility models for mobile ad hoc networks, *ACM Mobicom* (Sep. 2003).

Hybrid Routing Protocols for Mobile Ad-hoc Networks

Lan Wang and Stephan Olariu
Department of Computer Science
Old Dominion University, Norfolk, VA 23529
E-mail: {wang_l, olariu}@cs.odu.edu

Contents

1 Introduction

A mobile ad-hoc network (MANET) is a collection of mobile nodes that communicate using a wireless medium forming an autonomous network. In addition to attending to its own business, each node also acts as a router, forwarding packets on behalf of other nodes. Examples of applications of MANET include: tactical operations, search-and-rescue missions, law enforcement, and virtual classrooms, among many others. Compared to wireline networks and to more traditional wireless networks (such as cellular networks), MANET has the following distinguishing characteristics: (1) lack of pre-existing infrastructure, (2) potential for rapid node mobility, and (3) all communications are carried over a bandwidth-constrained wireless medium.

The effectiveness of a distributed routing algorithm depends on the accuracy of the topology information available to it. Unfortunately, rapid changes in the topology of MANET can easily cause the distribution of up-to-date routing information to saturate the scarce network bandwidth. Substantial research effort has been devoted to balance this tradeoff in order to achieve better performance in terms of high throughput, low control overhead, short delay, low energy consumption, scalability, etc.

The emerging consensus [3, 4, 14] is that there is no single basic MANET routing protocol that operates efficiently under all the network conditions. Considering diverse MANET applications where mobility, traffic, size and density may vary significantly, different tradeoffs have to be made in different situations. An ideal protocol should be able to combine the strengths of basic protocols and adapt their behavior at the appropriate time and for the appropriate scope of the network. This motivates the study of hybrid MANET routing protocols. In this chapter, we present a survey of routing protocols for MANET with a focus on such solutions that integrate various routing approaches in a hybrid framework and adapt to the changing network dynamics to achieve better performance in a wide variety of environments.

The remainder of this chapter is organized as follows. After a quick review on the basic proactive/reactive MANET routing protocols in Section 2, Section 3 surveys recent hybrid MANET routing protocols. Subsection 3.1 reviews *cluster-centric* hybrid routing protocols where the clustering of the nodes in the MANET is leveraged for the design of efficient routing protocols. Subsection 3.2 reviews a number of alternatives to the cluster-centric view, that we refer to as *node-centric*. One of the important node-centric routing protocols is the classic Zone Routing Protocol that is discussed in some detail. Next, Section 4 presents the details of our new hybrid node-centric routing protocol – TZRP. Subsection 4.1 begins by motivating the need to decouple concerns about traffic characteristics and mobility in ZRP. Subsection 4.2 presents a high-level overview of TZRP. The details of TZRP are discussed in Subsection 4.3. Subsection 4.4 presents our simulation results that show that TZRP outperforms ZRP. Finally, Section 5 offers concluding remarks.

2 MANET Routing Protocols: a Quick Review

Numerous routing algorithms targeted at small-to-medium MANET have been proposed in the literature. Traditionally, they can be classified as either *proactive* (table-driven) or *reactive* (on-demand) protocols. The key distinction between them is *when* routing information is acquired and maintained about specific destination nodes. In pure proactive routing protocols, consistent and up-to-date routing information to all nodes is maintained at each node, whereas in reactive routing protocols, routes are created and maintained only when desired by the source node.

A *reactive* routing protocol typically has two components: route discovery and route maintenance. The goal of route discovery is to find a new route to the destination node. Once a route is found, route maintenance mechanisms are used to validate an active route, do local repair when the route breakage is detected, and to notify the source node of the route failure when necessary. DSR [15] and AODV [22] are typical examples of reactive routing protocols. They share similar basic route discovery and route maintenance mechanisms, but differ in where and how the routing information is stored: DSR is a *source routing* protocol while AODV is a distance vector-based hop-by-hop routing protocol. When a source node s needs to reach a destination d, it floods a Route Request (RREQ) packet into the network. Each forwarding node appends its own address to the RREQ (in the case of DSR) or adds the route to s in its routing table (in the case of AODV). When a RREQ reaches the destination (or a node with a cached route towards the destination), a Route Reply (RREP) packet is sent back to s following the reverse path traced by RREQ. Each node forwarding the RREP adds the route to d in its routing table (in the case of AODV). When s receives the RREP, a source route to d is available from RREP (in DSR), or routing table entries are established at all the nodes along a route from s to d (in AODV). Once a route is found, it is used to forward data packets from the source to the destination as long as it is still valid, even if it is no longer optimal due to node mobility. When a route is broken, the source node is notified by a Route Error (RERR) packet. It can then attempt to use some other route to the destination already in its cache or can invoke route discovery again to find a new route. By virtue of source routing, routing loops can be easily detected and eliminated in DSR, while AODV attains loop-free routing by using a destination-based sequence number.

To minimize the flooding overhead, in AODV the RREQ messages are initiated with increasing time-to-live (TTL) value. If no RREP is received, a source node issues a new RREQ for the destination based on an exponential back-off algorithm. This is called *expanding ring search.* LAR [17] demonstrates how location information provided by GPS can be exploited to reduce the number of nodes involved in RREQ flooding.

TORA [19] is designed to minimize the control overhead in the route maintenance stage by localizing reaction to topological changes. The idea is to establish a destination-rooted directed-acyclic graph (DAG) in the form of a relative *height* value at each involved node during the flood-based route discovery stage. This DAG is used during the route maintenance stage based on *link-reversal* algorithm to establish new routes quickly in response to link breakage. Different from both AODV and DSR, TORA requires node coordination and reliable control message exchanges among neighbors.

In *proactive* routing protocols, a lot of work was done to adapt existing wireline network routing protocols, such as Distance Vector (DV) and Link State (LS), to MANET. DSDV [21] is one of the earliest routing algorithms for MANET. Its key advantage over traditional DV protocols is that it uses sequence number to guarantee loop-freedom. Link-state routing algorithm is also adapted for MANET because of its faster convergence and better support for QoS compared to the DV-based approaches. In this direction, the major focus is how to reduce the control overhead by disseminating Link-State Update (LSU) throughout the network more efficiently and how to reduce the generation frequency and propagation scope of LSU.

For example, OLSR [12] reduces the control overhead compared to traditional LS based on *Multi-Point Relays* (MPR). A node's MPRs are a minimum subset of its 1-hop neighbors that can cover all of its 2-hop neighbors. In OLSR, only MPR nodes further forward LSU, and the size of a LSU is reduced since it only includes a node's links to its MPR neighbors, instead of the whole neighborhood list as in a traditional LSU. In the *source-tree*-based approaches (STAR [5] and TBRPF [18]), each node maintains a shortest path tree to every other node in the network. A node computes its source tree based on its own links and the source trees reported by neighbors. A node generates a LSU only when its source tree changes. In TBRPF, a node reports only the changes to a part of its source tree in LSU based on estimating which links in the tree are actually used by its neighbors. In STAR, instead of generating a LSU whenever the source tree changes, a node generates a LSU only when a long-term routing loop may happen or network membership changes (called *Least Overhead Routing*).

There is an on-going debate on whether a proactive or reactive approach is best for MANET. In terms of *routing table size*, a proactive protocol has to maintain entries for all the nodes in the network, hence does not scale well to large networks. By contrast, a reactive routing protocol only maintains routing information to the active communicating nodes. In terms of *delay*, proactive protocols have a route to the destination readily available whenever it is needed, while reactive protocols suffer from longer route acquisition latency due to the on-demand route discovery. In terms of *bandwidth consumption*, reactive routing protocols are generally considered to have lower control overhead. But when new routes have to be found frequently, the flooding of RREQ may cause significant overhead. In addition, route optimality cannot be achieved in such protocols. Instead, a path is used as long as it is valid, although it may no longer be optimal. This means that the amount of bandwidth wasted due to the sub-optimality of paths may become excessive when the session-to-mobility ratio is high. Proactive routing protocols have long been considered as having more control overhead than reactive approaches. However, this is based on the somewhat unfair assumption that a proactive routing protocol necessarily has to provide optimal paths and all node must have a consistent view of the network. In fact, as demonstrated by STAR, by relaxing the route optimality and the consistent view constraint, proactive protocols can potentially be designed with the same level of control overhead as reactive protocols. In a sense, this flexibility of balancing the tradeoff between routing overhead and path optimality is an advantage of proactive approaches over reactive approaches since reactive protocols intrinsically lack the ability to achieve path optimality. Essentially, the tradeoff between routing control overhead and path optimality is a key problem when comparing the performance of different routing protocols for bandwidth-constrained MANET [32].

3 Hybrid MANET Routing Protocols

Ideally, a hybrid routing protocol should have the following properties: (1) *efficient*: the protocol should choose suitable basic components and integrate them organically to achieve better performance than any single component; (2) *adaptive*: the protocol should be able to dynamically adjust the extent of each component to achieve different desired performance goals under different network conditions; such adaptation mechanisms generally require a clear mapping between performance metrics and hybridization parameters; (3) *simple*: the hybridization should be lightweight, avoiding excessive control overhead.

There is a large design space for hybridization between various basic proactive and reactive protocols. In this section, we review several hybrid MANET routing schemes proposed in the literature. We classify these schemes into *cluster-centric* and *node-centric*. In the cluster-centric schemes, explicit clusters are formed and maintained as efficient control structures for abstracting and propagating routing states, and the boundary of clusters is the switching point between different routing strategies. In the node-centric schemes, no special effort is spent on forming/maintaining explicit control structures, and each node makes use of routing information to its neighbors (as well as to some other special nodes) to switch between different routing strategies.

3.1 Cluster-Centric Protocols

A natural way of implementing hybrid routing is to organize the network into a collection of node groups (clusters), and to adopt different routing strategies for intra- and inter-cluster traffic, respectively. Indeed, hierarchically organizing a network is a well-studied problem in large-scale wireline networks and has been shown to be effective in minimizing the size of routing tables, thus optimizing the use of network resources. In the case of MANET, partitioning nodes into clusters can have other benefits as well: the clustering control structure not only makes a large network seem smaller but, more importantly, can make a highly dynamic network appear less dynamic, essentially hiding mobility.

Many cluster-centric routing schemes have been proposed in the literature [11, 13, 23, 24, 33]. Among them, the $(\alpha, t) - cluster$ framework [23] is directly targeted at maintaining an effective cluster topology that adapts to node *mobility* to achieve hybrid routing that balances the tradeoff between proactive and reactive strategies according to the temporal and spatial dynamics of the network. Specifically, the $(\alpha, t) - cluster$ framework dynamically organizes mobile nodes into clusters in which the probability of path availability α can be bounded for a period of time t. Since α establishes a lower bound on the probability that a given cluster path will remain available for time t, it controls the cluster's inherent stability. For a given α (stability level), the role of t is to manage the cluster size, essentially controlling the balance between route optimality and cluster maintenance overhead.

Routing is achieved utilizing a dynamic two-level hierarchical strategy consisting of pure proactive routing (DSDV or OLSR) and least-overhead proactive routing (STAR) operating at each level. Each node maintains two routing tables. The level-1 routing table consists of one entry for each destination node within the same cluster and of one entry for each neighboring cluster, indicating the next-hop nodeID along the optimal path to the corresponding destination. The level-2 routing table consists of one entry for each cluster in the network, indicating the next clusterID along the current active path toward the corresponding destination cluster, which can be resolved to a next-hop nodeID using the level-1 routing table. The level-2 protocol requires that one node (the node with the lowest nodeID in each cluster) generate an update on behalf of its cluster. When a level-2 update is generated, it is flooded to all the nodes in each neighboring cluster, but not transmitted beyond the neighboring clusters. Using the strategy of STAR, every node maintains level-2 topology information.

To forward a packet to a desired destination, a source node must first use a *location management* protocol to discover the current clusterID associated with the destination node. This binding procedure is similar to a reactive route discovery process, and the associated overhead is a common problem to all the cluster-centric schemes. However, in the $(\alpha, t) - cluster$ framework, the level-2 information maintained is used to infer a broadcast tree to forward a request to every cluster only once. Furthermore, each request need only be processed by one node in each intermediate cluster, and if the target is discovered along a given subtree, early termination of the query thread on that subtree is easily achieved. Finally, the request provides binding information directly to the target of the request. Consequently, the response can be sent directly to the source of the request via unicast routing.

The main contribution of the $(\alpha, t) - cluster$ framework is that it emphasizes the importance of mobility in cluster formation and develops a formal mobility model to characterize link and path availability. If both α and t are sufficiently large, then $(\alpha, t) - clustering$ tends to identify groups of nodes moving together. However, both the definition of $(\alpha, t) - cluster$ and the associated clustering algorithms described in [23] need to be refined for working effectively in a general MANET. In [24], two major modifications to the $(\alpha, t) - clustering$ framework are proposed: (a) In a cluster, the *pairwise* (α, t)-reachability is considered to be too restrictive, hence the cluster is redefined so that (α, t)-reachability is only required between a potential joining node and the *parent node* of the cluster; (b) A node does not leave a cluster until the cluster becomes disconnected. These modifications do solve the convergence problem of the cluster maintenance algorithm, but also compromise the clustering goal and introduce new problems that need to be further investigated. For example, choosing a *good* parent node becomes a critical problem.

The $(\alpha, t) - cluster$ framework clearly demonstrates both the benefits and challenges of a cluster-centric hybrid routing protocol. Generally, the hierarchical clustered MANET forces a tight coupling between routing and clustering. It is a very challenging task to determine which combination of routing and clustering algorithms is the most appropriate for a particular network.

3.2 Node-Centric Protocols

In cluster-centric routing protocols the formation and maintenance of the clustering infrastructure needs explicit exchanges of special-purpose control messages. By contrast, in the *node-centric* routing schemes discussed in this sub-section, each node makes use of an implicit control structure that is naturally associated with itself: the area that consists of all the nodes reachable in k hops from it (i.e. its k-hop neighborhood). Such a structure is constructed and maintained as a by-product of exchanging regular routing information among nodes. No special-purpose algorithm is needed. Such a structure can be considered to be an *implicit cluster.*

3.2.1 Fisheye Routing and FSLS

The *fisheye* [16] routing concept is based on the observation that nodes that are far away do not need to have complete topological information in order to make a good next hop decision. Given an approximate view of the distant parts of the network, a node can forward a packet in the *proper* direction toward the destination. As the packet makes progress toward the destination, the view of the destination's region becomes more accurate, providing for more precise packet forwarding. This suggests that propagating every LSU over the network may not be necessary. The fisheye technique is used in FSR [7, 11] and DREAM [1] (using location information provided by GPS). This class of approaches is further generalized and analyzed in FSLS [31]. In FSLS, the reduction of control overhead is achieved both in space (by limiting the propagation scope of LSUs) and in time (by limiting the time between successive LSU generations). Specifically, a node wakes up every $2^{i-1} * t_e$ (i = 1,2,3,...) seconds and transmits a LSU with TTL set to s_i if there has been a link state change in the last $2^{i-1} * t_e$ seconds. The choice of a good set of values $\{s_i\}$ is determined by the traffic pattern. Assuming a uniform traffic distribution among all nodes in the network, the values of s_i that achieve the best balance between proactive overhead and route sub-optimality is derived in HSLS [31]. In these *fuzzy* proactive protocols there is a higher chance for short-term loops caused by routing inconsistencies due to different views of the network at different nodes. Currently, there is no mechanism in FSLS to avoid such loops. These loops are detected and removed by means of TTL expiration.

3.2.2 ZRP

The Zone Routing Protocol (ZRP) [8] provides a hybrid routing framework that is locally proactive and globally reactive. The goal is to minimize the sum of the proactive and reactive control overhead. In ZRP, a node proactively propagates LSU to all the nodes in its k-hop neighborhood, where k is called *Zone Radius*. Thus, each node has an up-to-date view of its *routing zone*, that is, all the nodes and links in the node's k-hop neighborhood. The routing zone nodes that are at a distance of exactly k hops are called *peripheral nodes*. Each node has its own associated routing zone (hence, its own set of peripheral nodes), and routing zones of neighboring nodes overlap heavily.

ZRP is hybrid not only in that it adopts pure proactive routing for intra-zone traffic and reactive routing for inter-zone traffic but, more importantly, the zone structure maintained for proactive routing is exploited in the reactive routing process through a mechanism called *bordercasting*. Rather than blindly broadcasting a node's route request to all its neighbors, bordercasting directs the request to peripheral nodes only. Using the zone topology maintained, each peripheral node decides whether to reply to the request or to further forward it to its own peripheral nodes.

The heavily overlapping neighboring routing zones can lead to query duplication and backward propagation. To alleviate the problem, special query control mechanisms (*Query Detection* and *Early Termination*) [8] are used to identify those peripheral nodes that have been covered by the route query (i.e. that belong to the routing zone of a node that already has bordercast the query) and to prune them from the bordercast tree. This encourages the query to propagate outward, away from its source and away from covered regions of the network.

The latest version of bordercasting [9] works as follows. When a node receives the first copy of a RREQ, (a) if the node is not an intended recipient of the RREQ, it is implied that the node's own routing zone has been covered by other bordercasting nodes. Thus, the node marks its entire routing zone as *covered* and discards the query; (b) if the node is an intended recipient of the query, it proceeds to process the query: if the node knows a route to the destination, it forwards the query to the destination; otherwise, the node forwards the query to 1-hop neighbors that span its *uncovered* peripheral nodes in its bordercast tree. After forwarding the RREQ, the node marks its entire routing zone as *covered*.

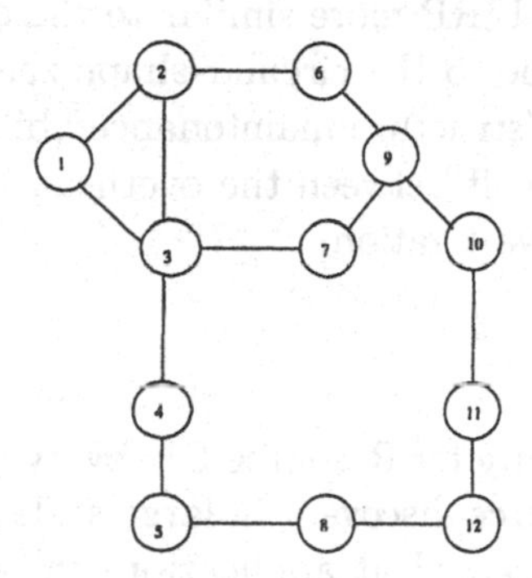

Figure 1: *Bordercasting in ZRP.*

The *efficiency* (the number of forwarding nodes compared to flooding) and *effectiveness* (the query success ratio compared to flooding) of bordercasting depends on traffic pattern and the instantaneous network topology. Consider the network topology shown in Figure 1, and assume *ZR*=2. For a query from node 1 to node 9, the query is terminated at nodes 2 and 3 because both 2 and 3 have the destination inside their routing zone. In this case, most of the nodes of the network (to wit, nodes 4–12) are not involved in the propagation of the query. However, when node 2 wants to find a route to node 8, the query propagation involves most of the nodes in the network (except nodes 5, 8, 12) before a route is found.

The optimal zone radius value is dynamically adjusted using *Min Searching* and *Traffic Adaptive Estimation* [8]. For example, if a node notices that the ratio of proactive overhead to reactive overhead during a certain interval exceeds a certain threshold, zone radius is reduced; if the ratio is lower than a certain threshold, zone radius is increased. By adjusting the globally-uniform zone radius, a good balance between proactive and reactive control overhead can be achieved and the total routing control overhead is minimized.

In recent work under the ZRP framework [29], it is argued that using a uniform zone radius throughout the whole network is sub-optimal. Instead, having independent zone radii allows each node to distributedly and automatically configure its optimal zone radius, hence fine performance tuning can be achieved. Such an Independent Zone Radius (IZRP) approach makes ZRP much more complicated since each node now has to know which nodes have a demand for its LSU. This information is generally acquired by requiring each node to broadcast periodically an explicit *Zone Building Packet* to all the members in its zone. In fact, exchanging such explicit control messages makes IZRP more similar to the cluster-centric approach, and ZRP's simplicity due to the circular-shape zone structure and the implicit zone membership/structure maintenance ability by LSU exchanges is compromised. The tradeoff between the overhead and benefit of this IZRP scheme needs further investigation.

3.2.3 CARD

Contact-based architecture for Resource Discovery (CARD) [10] is proposed as a framework for resource discovery in large-scale MANET. In the context of routing, CARD is targeted at applications in which most of the traffic consists of short flows and small transactions [10]. In such applications, the cost of discovering routes is usually the dominant factor instead of the data transfer as in long flows. As a result, CARD strives to minimize the control overhead during route discovery instead of finding and maintaining shortest paths.

In CARD, each node proactively maintains the routing information in its R-hop neighborhood (called *vicinity*) and keeps track of all the nodes that are exactly R hops away from itself (called *edge nodes*). The vicinity and edge nodes are similar to ZRP's zone and peripheral nodes, respectively. The key difference is that in addition to the above information, each node maintains paths to a few distant nodes called *contacts*. The underlying motivation is that, based on the *small world* concept, these contacts can help find a route to distant destinations more efficiently.

The selection and maintenance of contacts is the key mechanism of the CARD framework. Theoretically, each node should maintain as few contacts as possible to cover as many nodes outside of the source node's vicinity as possible. This is, to some extent, equivalent to the source-dependent minimum k-dominating set problem (a k-dominating set is a subset D such that each node is within k-hops of a node in D). In practice, several heuristics are proposed to provide maximum increase in reachability with the addition of each new contact by minimizing the overlapping between contacts.

A source node s selects its contacts one by one. To select a contact, s sends a *Contact Selection Query* (CSQ) control packet to one of its edge node. The edge node further forwards the packet to a randomly chosen neighbor. The receiving node decides whether to become a contact for s by checking for overlap with s's vicinity, the vicinities of all the already selected contacts and the vicinities of s's edge nodes. If there is no overlap, then the node is selected as a contact. If the node fails to become a contact, it further forwards the control packet. If the packet reaches a node whose distance to s exceeds a predetermined *Maximum Contact Distance* (r), the packet is returned to the last sender (*backtracking*). A contact is searched in such a depth-first way until one is found. Note that since searching a contact may involve backtracking, and multiple contacts are searched sequentially, this contact selection phase tends to be time-consuming.

Once a contact c is selected and the route from s to c is established, this route has to be validated periodically. If the route is broken, local recovery is used to try to salvage it. If salvage fails or the length of this route exceeds a certain threshold, the contact is considered lost. If the number of contacts falls below a certain threshold, new contacts are selected.

When source node s needs a path to destination d, it first checks whether d exists in the vicinity. If not, s sends a *Destination Search Query* (*DSQ*) control packet to its contacts. The *Depth of Search* (D) field in *DSQ* controls the levels of contacts queried. By doing such a sequential expanding ring search, CARD avoids the complicated query control mechanisms as in ZRP. The tradeoff is a longer route acquisition latency (when the destination is far away) adding to the already long delay in the reactive route search approach.

CARD provides a wide range of adjustable parameters to achieve fine tuning for various desired performance goals. However, determining and adjusting the optimum values and combinations of vicinity size, number of contacts, maximum contact distance, and maximum depth of search is a challenging problem.

Another hybrid routing protocol based on the *small world* concept is described in [2] in the context of *position-based routing* [35].

3.2.4 LANMAR, Netmark, and SHARP

ZRP and CARD make no special assumptions about individual nodes. However, in many practical applications some nodes enjoy special properties that happen to be relevant to routing.

LANMAR [26, 36] is designed for MANET that exhibit *group mobility*. A landmark node is selected for a group of nodes that are likely to move together. A *scope* is defined such that each node would typically be within the scope of its landmark node. Each node propagates link state information corresponding only to nodes within its scope, and distance vector information for all landmark nodes. When a node needs to send a packet to a destination within its scope, the local link state routing table is used directly. Otherwise, the packet will be routed towards the landmark nodes of the destination. When the packet arrives within the scope of the destination, it is routed using local link state tables, without necessarily going through the landmark node.

In [6, 28], a node-centric hybrid routing protocol is proposed based on the assumption that some special nodes in a MANET are more popular than others. In this protocol, a *hot-spot* node is called a *netmark*. Paths between netmarks and ordinary nodes are maintained proactively, while routes between ordinary nodes are set up on demand. It is worth noting that the key difference between netmark-based and landmark-based protocol is that a node is not assumed to be associated with any node in the netmark protocol. Hence, in a large MANET where a netmark is only known within a given number of hops from itself, when a node needs to communicate with a destination beyond its scope, the packet can no longer be directed to a specific node as in LANMAR.

SHARP [27] is also predicated on the existence of *hot-spot* nodes. A *proactive zone* is defined around each hot-spot node. A node-specific zone radius determines the size of a given proactive zone. Each node at a distance not exceeding the zone radius is a member of the proactive zone of that hot-spot node Ni. Nodes within Ni's proactive zone maintain routes proactively only to the central node Ni. The nodes that are not in the proactive zone of a given destination use reactive routing protocols to establish routes to that node. It is interesting to note that SHARP's proactive zone is far more lightweight than ZRP's routing zone.

The reactive component of SHARP is AODV with the optimization mechanisms of route caching and expanding ring search. The proactive component of SHARP is adapted from TORA [19]. The idea is that in the proactive zone centered at node Ni, a directed acyclic graph (DAG) that is rooted at Ni and is consisted of all the nodes in the proactive zone is built and maintained constantly. The proactive component has two procedures: *DAG construction* and *DAG update*. During DAG *construction*, the center node sends a construction control packet, which is further forwarded by the other nodes in the proactive zone. During this forwarding process, each node is assigned a *height* value. The height is the distance of the node from the center. A data packet arriving at a node is transmitted along the downstream link to the neighbor with the lowest height. During DAG *update*, with link failures, as long as there is a downstream link, a node does not take any special action since a route to the center is still available (although not necessarily shortest). Only when all the downstream links at a node have failed, the node reverses the orientation of its upstream links by choosing a new height greater than the height of all its neighbors and broadcasting a new update control packet. Each node receiving this update packet records the new height of this neighbor and, if necessary, adjusts the orientation of its own upstream links and initiates a new update control packet.

Compared to DAG construction, the DAG update procedure introduces less control overhead. However, the routes maintained may deviate significantly from the shortest path; and, with the movement of nodes, a node's path to the center node may have to involve nodes that just moved into the proactive zone, especially in the less dense networks. To deal with these situations, the DAG construction procedure has to be invoked periodically. The more frequently the DAG construction procedure is invoked, the more proactively shortest paths are maintained at the expense of more control overhead.

Each node continually monitors network characteristics including average lifetime of immediate links and average node degree. This information is sent to the destination node periodically. The destination node also locally maintains statistics about the data traffic that it has received. Using this information, each destination independently computes the optimal proactive zone radius to bound routing control overhead, loss rate, or delay jitter.

3.3 Comparison of Different Hybrid Routing Protocols

In this section we have reviewed various hybrid routing schemes proposed in the recent literature, focusing on their motivations, various explicit/implicit structures maintained, choices of basic components, and hybrid routing methodologies.

To recap, in cluster-centric protocols, explicit clusters are formed and maintained as routing units. The clustering constraint includes node locality and group mobility. The task of creating and maintaining such clusters generally involves significant overhead in the face of mobility. By contrast, node-centric approaches can provide some extent of scalability without involving too much overhead. However the lack of explicit control structures may lead to inefficiency for abstracting and propagating routing states.

We have to point out that there is no fundamental differences between these protocols. For example, CARD can be considered as a generalization of LANMAR or Netmark if mobility-group leaders or hot-spot nodes are chosen to be contacts. LANMAR can be considered as a special case of either Netmark (in a small network) or $(\alpha, t)-clustering$ (in a large network with group mobility). This similarity suggests the possibility of further hybridizations between these protocols.

4 TZRP - A Two-Zone Hybrid Routing Protocol

The main goal of this section is to propose a novel hybrid routing protocol - the Two Zone Routing Protocol (TZRP) - developed by the authors as an extension of the Zone Routing Protocol.

4.1 Motivation

Although Zone Routing Protocol (ZRP) provides an elegant and powerful hybrid routing framework, the choice of the sepcific proactive or reactive protocols used therein is of key importance. In fact, the *bordercasting* mechanism - the key component of ZRP- has some very important implications on ZRP's IARP (IntrA-zone Routing Protocol) component: the IARP must be able to provide up-to-date topology information of the routing zone.

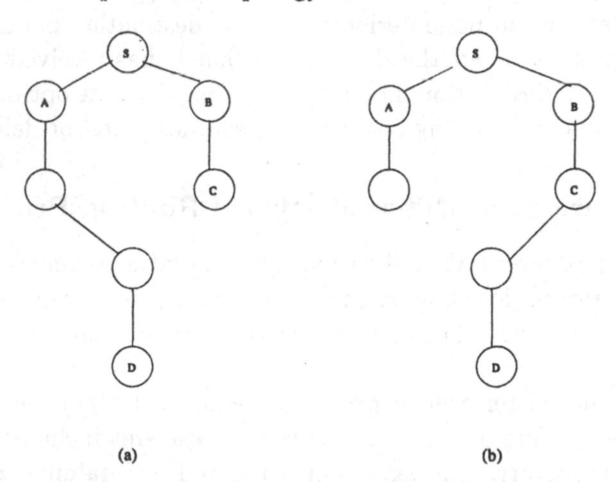

Figure 2: *Inaccurate zone topology information can lead to query failure.*

Consider the scenario illustrated in Figure 2, and assume a zone radius of 3. The actual topology is shown in Figure 2(a). However, due to an IARP that fails to provide accurate zone topology information, the topology perceived by nodes S, A, and B is that of Figure 2(b). When S wants to find a route to D, it constructs its bordercast tree. S prunes A from its RREQ receiving set and sends the RREQ only to B. B further forwards it to C, which has no choice but to terminate the query thread. Hence the query procedure fails. Since the source S has to wait for an amount of time proportional to the expected network diameter before realizing the query failure and trying again, such a query failure can cause significantly longer route acquisition latency.

This example illustrates the importance of the freshness and consistency of the IARP information maintained at each node. Indeed, bordercasting in ZRP requires an IARP that converges very fast, so that the distance vector variants and the long-timer-based link state variants are generally not suitable to work as IARP. By contrast, the event-driven link-state approach is the ideal choice. However, in the bandwidth-limited MANET, frequent topology changes make a pure event-driven implementation infeasible. For this reason, most link-state approaches are implemented in a timer-based fashion [11, 31], whereby a LSU is sent out only at some specific intervals. The smaller the interval, the shorter the convergence time, and the better approximation of an event-driven Link State routing can be achieved.

As discussed in [20], both mobility and traffic pattern influence routing control overhead, hence the optimum configuration of the zone radius. For a high-mobility scenario, the event-driven IARP incurs a very large proactive control overhead. This drives towards a smaller zone radius. However, reducing the zone radius also reduces the initial hit ratio since more nodes will be outside of the node's immediate knowledge. For example, assume that node X's zone (with a radius of 5) is divided into 4 areas: A, B, C, and D. Using an IARP that can approximate event-driven LS reasonably well, some nodes in area A and C are moving so fast that a lot of LSUs are generated. These LSUs are received by X, and when such proactive traffic is too large, X will decrease its zone radius to, say, 4. The result is that X no longer maintains proactively routing information to its 5-hop neighbors, even though these 5-hop neighbors are quite stable with respect to X. When X needs to find a route to one of these nodes, global bordercasting is required. Note that, although ZRP has several mechanisms to terminate a query thread as early as possible [8], asymptotically, once the query goes out of the initial zone, at least half of the network will still be flooded [30]. Consequently, bordercasting is still an expensive procedure compared to an immediate available route, hence should be avoided as much as possible.

Basically, the single zone structure of the original ZRP framework is intended to serve a dual purpose simultaneously as far as reducing routing control overhead is concerned: (a) it maintains routes to nearby nodes proactively so that local traffic can be routed immediately; in scenarios featuring *traffic locality*, this can result in a significant reduction in reactive control overhead since it avoids global search as much as possible; (*b*) it provides a structure that can be exploited to achieve efficient flooding (bordercasting) when a global search is necessary. The key problem with this framework is that although the *accurate* topology information of the circular shape (instead of any other shape) zone is necessary for purpose (*b*), such information is not necessary for purpose (*a*).

In fact, bordercasting is used in finding a route to a destination whose location is *unknown* to the source. This implies that bordercasting features a *global* behavior and the protocol must ensure that a query passes through even the *weakest* part of the network and reaches the destination's zone. Hence, inaccurate topology information used by any bordercasting node to prune its bordercast tree may terminate a query prematurely, causing a bad global effect. On the other hand, taking advantage of traffic locality to reduce routing overhead is a *local* behavior. FSLS[31], FSR, and GSR[11] have demonstrated that reduced frequency and accuracy in LSU generation and propagation work well in determining the next hop to a distant node. This is based on the observation that past routing information to a distant node provides a good approximation for the current route to that node.

Understanding the requirements for information accuracy of different components of a hybrid MANET routing protocol like ZRP is important since accurate topology is intrinsically expensive to maintain in MANET and hence should be limited to small scope. The high sensitivity to mobility renders the zone structure in ZRP less useful as a means of adapting to changing traffic patterns when mobility becomes high. This motivated us to find a companion structure that works when high mobility forces the zone radius to be small, to achieve fine tuning of the total routing control overhead.

4.2 Basic Idea of TZRP

In outline, the basic idea of TZRP is as follows: each node X maintains two zones, both with X as the center node. One is called the *Crisp Zone*, with radius ZR_c, the other is called the *Fuzzy Zone*, with radius ZR_f. Generally, we have $ZR_c \leq ZR_f$. Node X maintains proactively up-to-date topology of its Crisp Zone; however, X does not have to know the exact topology of its Fuzzy Zone. Instead, a *Fuzzy-sighted*-like proactive routing protocol [31] is employed as the IARP in node X's Fuzzy Zone.

In a low-mobility scenario, where topology changes are infrequent, a large Crisp Zone can be maintained with little proactive overhead. In such a case, we have $ZR_c = ZR_f$, which is the same as the original ZRP. In a high-mobility scenario where it is too costly to maintain a large Crisp Zone, ZR_c is reduced to a smaller size. However, since the control overhead to maintain the Fuzzy Zone is long-timer based and largely independent of the node's mobility pattern, a large ZR_f can be kept, hence the traffic locality benefit is still preserved to a great extent due to fuzzy proactive routing.

Essentially, TZRP aims to decouple the protocol's ability to adapt to traffic pattern from the ability to adapt to mobility. The Crisp Zone is used to balance the influence of mobility on routing control overhead, and the Fuzzy Zone is used to balance the influence of traffic pattern on routing control overhead. By adjusting these two radii independently, a lower total routing control overhead can be achieved. Thus, TZRP can be considered as a general MANET routing framework that can balance the tradeoffs between pure proactive, fuzzy proactive, and reactive routing approaches more effectively.

4.3 Details of TZRP

In this subsection, we describe TZRP in detail. The key difference from the original ZRP is that our IARP component explicitly distinguishes between event-driven IARP and timer-based IARP. Specifically, we use a short-timer-based implementation to approximate Crisp IARP, and we use a long-timer-based implementation to achieve Fuzzy IARP.

(A) *Generation and propagation of LSU.*

A node X *wakes up* every T_{short} seconds, checks its link state table, and determines whether there is a need to send a LSU:

- If during the last T_{long} seconds, X has not sent any LSU with $TTL = ZR_f - 1$, then X sends a LSU with $TTL = ZR_f - 1$; otherwise,
- If there was any change in X's 1-hop neighborhood (i.e. a new neighbor is found or an existing neighbor is lost) during the last T_{short} seconds, then X sends a LSU with $TTL = ZR_c - 1$; otherwise,
- If X finds that there is a new node in its ZR_c-hop neighborhood during the last T_{short} seconds, then X sends a LSU with $TTL = ZR_c - 1$.

The structure of the Crisp Zone is exploited during the propagation of a LSU. When $ZR_c \geq 2$, each node X maintains its shortest paths to every 2-hop neighbor by exchanging LSUs with TTL=1. X uses the minimum number of 1-hop neighbors to cover all the 2-hop neighbors using one of the known greedy heuristics [34]. The selected 1-hop neighbors form a *forwarding set* for X. X includes this forwarding set information in each LSU it generates or forwards. When a node Y receives a LSU, it first checks whether the LSU is received for the first time. If not, Y discards this LSU; otherwise, Y integrates this LSU into its link state table. Then, if the LSU's TTL=1, Y discards it; else Y checks whether itself appears in the forwarding set of the LSU; if *yes*, Y decrements the TTL field of the LSU, calculates its forwarding set, appends this forwarding set to the LSU, and forwards it; if *no*, Y just discards the LSU.

(B) *Computing Crisp/Fuzzy IARP route.*

When a node receives a LSU with a larger sequence number generated by node S, it deletes all the existing entries with S as the source in the current link state table, and then inserts the link state entries contained in the LSU just received. Since LSU is generated in an event-driven way in the Crisp Zone, the entries derived from Crisp Zone LSUs can always be updated on time by the latest LSUs received. However, the entries derived from Fuzzy Zone LSUs may not be updated on time by the originator since the Fuzzy Zone LSU generation is based on a long timer and the originator node may move out of the fuzzy zone. Consequently, these entries age and are deleted from the link state table after a predetermined period, say, T_{long}. When there is a route to be resolved, the intra-zone routing table is re-computed based on the latest link state table. Specifically, node X uses Dijkstra's algorithm to compute a shortest path from X to any other node of which it is aware. All the shortest paths with length of exactly ZR_c hops constitute X's *bordercast tree*, which is used in bordercasting as described later. Note that Fuzzy IARP route entries only intend to work as a heuristic to increase the initial query hit ratio, hence the probability of routing failure using such an entry is higher than a Crisp IARP route. In such a fuzzy route failure case, the source node X will receive a RERR. Upon receiving a RERR, if node X finds that the corresponding entry is a fuzzy entry, it deletes all the LSUs starting from the intended destination so that the fuzzy information will not be used again until the next LSU originated from the node is received.

(C) *Bordercasting.*

If the destination node is unreachable from X through either a Crisp IARP route or a Fuzzy IARP route, then a reactive route query *bordercasting* procedure is invoked. We follow the latest version of BRP described in [9]. Specifically, X sends a RREQ to all its 1-hop neighbors which are direct children of X in the *bordercast tree* constructed above. Only those children nodes can further forward the RREQ. A sending node appends its forwarding set in the RREQ. Upon receiving the first copy of a RREQ, a node can determine whether it is a forwarding node by checking the forwarding set information piggy-backed in the RREQ. If a node finds that it is not in the forwarding set, it simply discards the RREQ.

A node Y in the forwarding set processes the RREQ as follows. If there is an IARP route from Y to the query destination with length not longer than ZR_c (hence a Crisp IARP route), Y unicasts the RREQ to the destination, which then sends a RREP back to the query source, indicating that a route to the destination has been found. Otherwise, node Y constructs its bordercast tree in the following way: First, it computes the shortest path tree with X as the root. All the nodes that are ZR_c or fewer hops from X are marked as *covered*. Second, Y computes the shortest path tree with Y as the root, and all the uncovered nodes as leaves. The paths with length exactly ZR_c hops constitute Y's bordercast tree. Then Y appends its forwarding set in the RREQ and further forwards it. Finally, Y marks all nodes that are ZR_c or fewer hops from Y as *covered*.

During the bordercasting procedure, routing information is created and maintained at the involved nodes. We adapt AODV as the IERP (IntEr-zone Routing Protocol). That is, during the propagation of RREQ, a backward routing entry towards the query source is established at each forwarding node; during the propagation of RREP, a forward routing entry towards the query destination is established at each forwarding node. The destination sequence number is used in the similar way as in AODV to prevent routing loop. Unlike AODV, zone information is used for route maintenance. When a link breakage on an active route is detected, the upstream node checks whether the destination node can be reached through any alternative Crisp/Fuzzy IARP route. If so, the route is locally-repaired successfully; otherwise, a RERR is sent back to the source as in AODV.

4.4 Simulation Results and Discussion

We have simulated the TZRP protocol using the *ns-2* simulator [25]. In our simulations, we have N=200 nodes, each of which has a radio transmission range of T_r=250m and transmission rate of 2Mbps. Initially the nodes are distributed uniformly at random in an area A which is either a square or a rectangle. The nodes move according to the *random way-point model*; in all our simulations, we set the pause time to zero and each node always moves at the fastest speed V. The values of A and V vary in different scenarios, as illustrated in Table 1. The node density, D, in Table 1 is calculated as $D = \frac{N*\pi*T_r^2}{A}$ and corresponds to the expected degree of a node in the underlying graph. Each simulation begins at time 0 and ends at time 190s. We begin collecting statistical data of various control packets starting at t=10s until the end of the simulation.

Scenario	A (m×m)	V (m/s)	D
1	2000×2000	10/20	9.81
2	2500×2500	10/20	6.28
3	4000×1000	10/20	9.81
4	5000×1250	10/20	6.28

Table 1: *Simulation scenarios*

Our protocol relies on periodic HELLO beacons to detect link formations and breakages. The HELLO beacons are sent every 0.1s, and the number of tolerable missed HELLOs is 2. In addition, MAC-layer link breakage detection is enabled. In our simulation, we use a simplified MAC layer in the sense that we assume the MAC-layer broadcast is reliable and is not impaired by collisions. We understand that this is not the case for the family of IEEE 802.11 MAC protocols. Indeed, there are complicated interactions between ZRP and the underlying MAC layer. For example, if a LSU or RREQ transmission is translated into multiple reliable MAC-layer unicasts, then the number of control overhead introduced by bordercasting can be overwhelming; on the other hand, if a HELLO, LSU, or RREQ transmission is translated into a single unreliable MAC-layer broadcast, then ZRP's behavior becomes more unpredictable since either of these messages can be lost. We note that, in fact, for the purpose of our simulation an ideal MAC is highly desirable as it helps us focus on understanding and analyzing the behavior of bordercasting without being distracted by heavy cross-layer interactions.

We extended the scenario generation tool in *ns-2* to generate traffic based on a given flow distance distribution. By controlling the flow distance, we can clearly identify whether a flow is intra-zone or inter-zone for a specific scenario. This enables us to determine whether the reduction in total routing control overhead is more attributed to traffic locality or to efficient bordercasting as the zone radius increases. In addition, the flows generated in this way can have a large chance to be between connected nodes.

4.4.1 Sensitivity of Bordercasting to IARP Timer

The goal of this set of simulations is to demonstrate the influence of the IARP timer on the effectiveness of bordercasting. In our implementation, if a RREP is not received within 0.4s after the first RREQ for a query is issued, the source node resends the RREQ and doubles its waiting time. After *three* failed attempts, the query is dropped. We calculate the query success ratio at each attempt. Combining this number with the *route acquisition latency* gives us a sufficiently good indication of the effectiveness of bordercasting. Note that when calculating the route acquisition latency, only the successful queries are considered.

A (m×m)	V (m/s)	Average Route Length (hops)
2000×2000	10	6.3
2000×2000	20	6.3
2500×2500	10	7.2
2500×2500	20	7.5
4000×1000	10	8.2
4000×1000	20	8.3
5000×1250	10	10.2
5000×1250	20	9.9

Table 2: *Traffic pattern: average route length of queries*

For this set of simulations, we generate 2000 queries during the 3-minute simulation time for each scenario, and we examine those flows whose distance between the source and destination is at least 5 hops at the instant when the flow is generated at the source. The average route length for each scenario is shown in Table 2. In this set of simulations, we make the duration time of each flow short, and each flow has only one packet to send. The intention is to isolate the effects of various possible route maintenance optimizations and to focus on the route discovery procedure only. Also, we use pure timer-based IARP (i.e. without the optimization of propagating LSU using forwarding set) since we want to focus on IARP's influence on bordercasting only. We study the behavior of bordercasting for various values of zone radius, and the results featured in Figures 3, 4, 5, and 6 correspond to a value $ZR = 3$.

Figures 3 – 6 indicate clearly that *node density*, *node mobility*, and *Crisp Zone IARP timer* are key factors that have a significant influence on the effectiveness of bordercasting.

Notice that when node density is high (as illustrated in Figure 3), node mobility has relatively little influence on route acquisition latency and query success ratio of bordercasting. This is because a large number of threads are generated for a single query, and although some of them lose their directions

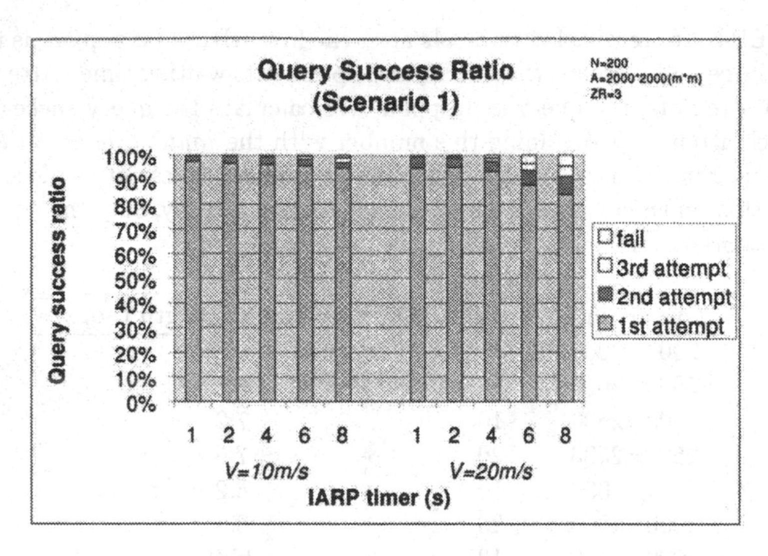

(a) Scenario 1, Query success ratio

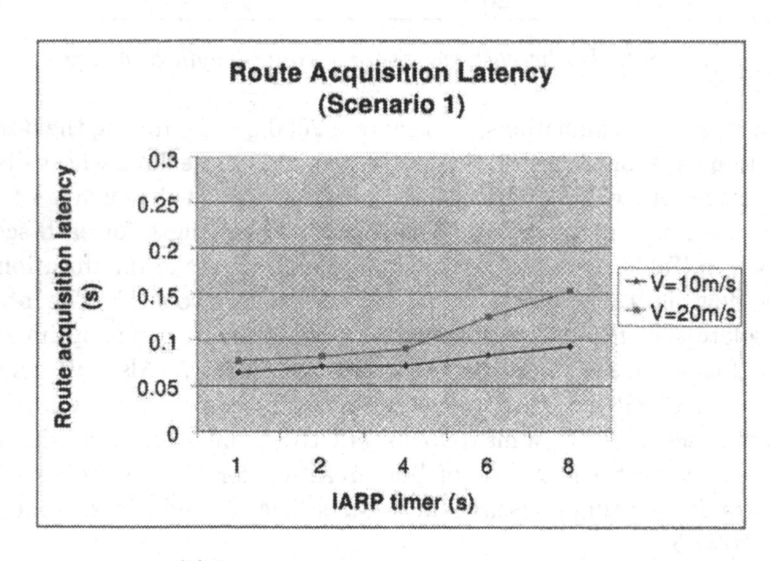

(b) Scenario 1, Route acquisition latency

Figure 3: *The influence of IARP timer on bordercasting: scenario 1*

and are terminated prematurely due to the inaccuracy of topology information when mobility is high and/or IARP timer is long, the probability that

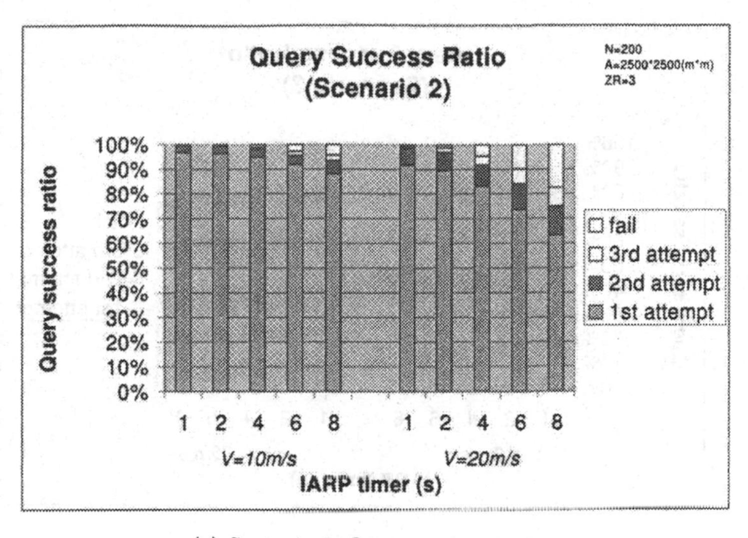

(a) Scenario 2, Query success ratio

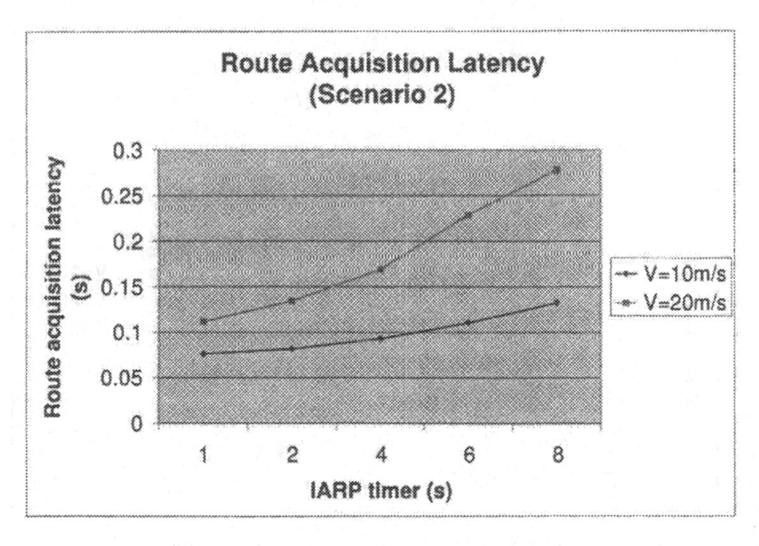

(b) Scenario 2, Route acquisition latency

Figure 4: *The influence of IARP timer on bordercasting: scenario 2*

at least one thread survives and reaches the destination is still high. As a result, the query success ratio is relatively stable, and the route acquisition

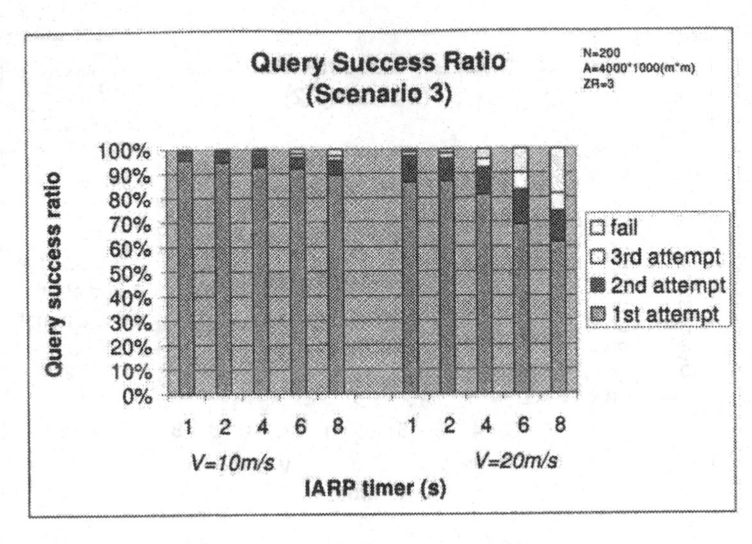

(a) Scenario 3, Query success ratio

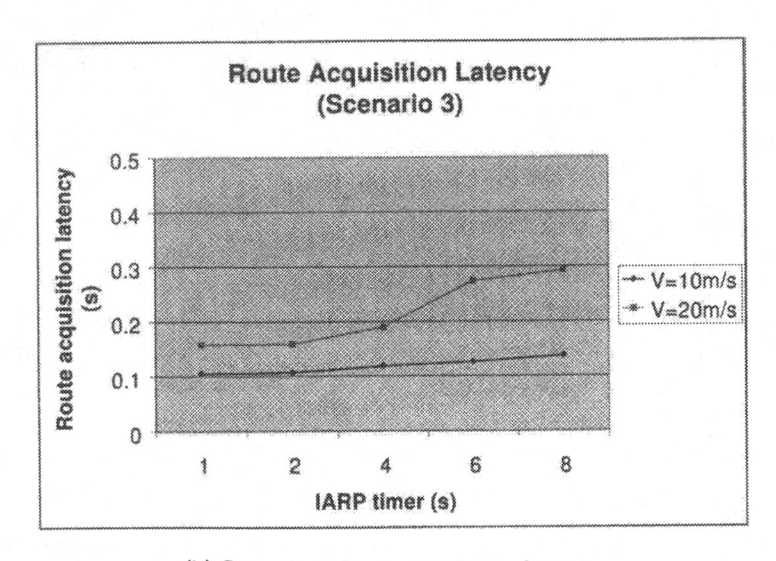

(b) Scenario 3, Route acquisition latency

Figure 5: *The influence of IARP timer on bordercasting: scenario 3*

latency only increases slightly.

However, as the node density decreases (see Figure 4), fewer threads

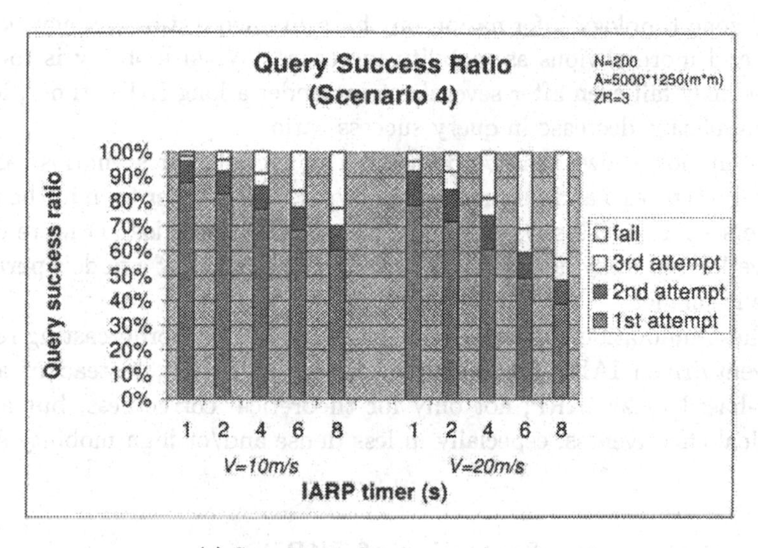

(a) Scenario 4, Query success ratio

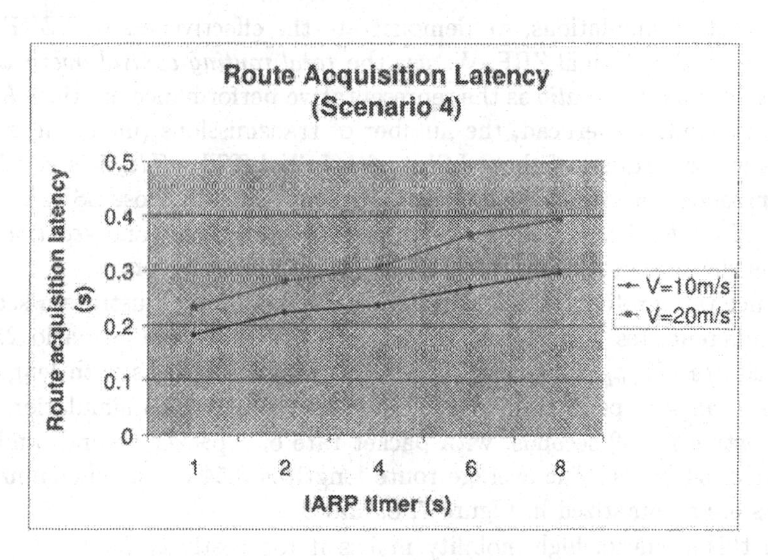

(b) Scenario 4, Route acquisition latency

Figure 6: *The influence of IARP timer on bordercasting: scenario 4*

are generated for each query, and the number of alternative routes to a destination decreases as well. In this case, the influence of the accuracy of the zone topology information on the route acquisition latency becomes more and more obvious as mobility increases. When mobility is too high, queries may fail even after several retries under a long IARP timer, leading to a significant decrease in query success ratio.

Simulation shows that bordercasting in rectangular scenarios (as illustrated in Figures 5 and 6) is more sensitive to IARP timer than in the case of square scenarios. This is because in rectangular scenarios (1) more queries involve further-away destinations, and (2) the number of a node's peripheral nodes is smaller.

These simulation results clearly demonstrate that bordercasting requires an event-driven IARP (or a short-timer-based IARP), instead of a long-timer-based fuzzy IARP, not only for theoretical correctness, but also for practical effectiveness, especially in less dense and/or high mobility scenarios.

4.4.2 Performance Evaluation of TZRP

In this set of simulations, we demonstrate the effectiveness of TZRP compared with the original ZRP. We use the *total routing control overhead* and *data packet delivery ratio* as the representative performance metrics. Among the total control overhead, the number of transmissions (including generation and forwarding) of those LSUs with initial TTL=ZR_c-1 is considered *pure proactive overhead*, the number of transmissions of those LSUs with initial TTL=ZR_f-1 is considered *fuzzy proactive overhead*, and *reactive overhead* is the sum of RREQ, RREP, and RERR transmissions.

Since this protocol is an extension of ZRP, we only illustrate a scenario that differentiates TZRP from the original ZRP. We use Scenario 2, with $V = 20m/s$, $T_{short} = 1s$, and $T_{long} = 8s$. Here, 300 flows with length not longer than 4 hops are generated between 10s and 150s simulation time, each lasting for 30 seconds, with packet rate of 1 packet/second, and each packet of 64 bytes. The average route length is 3.54 hops. The simulation results are summarized in Figure 7, 8, and 9.

In this scenario, high mobility makes it too costly to maintain a large Crisp Zone, which is reflected in Figure 7 as a significant increase in the pure proactive overhead when ZR increases by 1. By comparison, we can notice from Figure 8(a)(b) that the increase of the fuzzy proactive overhead is much less drastic as the Fuzzy Zone radius increases. Hence, by reducing ZR_c and keeping a large ZR_f, TZRP achieves a better balance between proactive and reactive control overhead than the original ZRP, as shown in Figure 9.

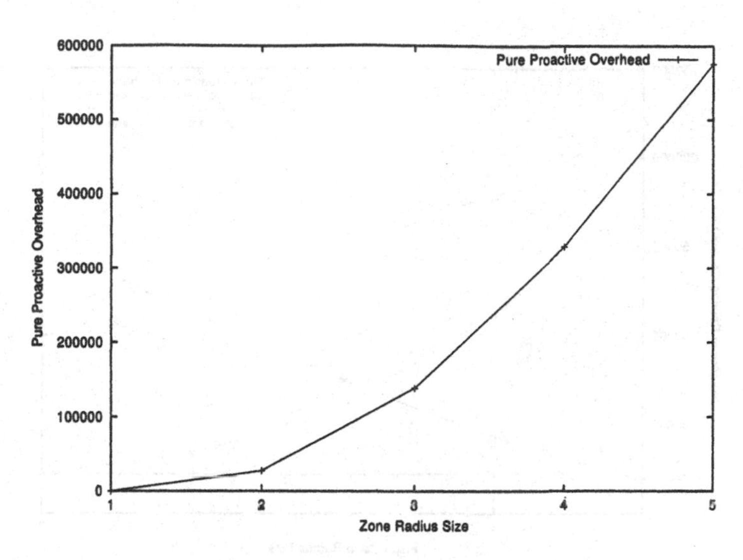

Figure 7: *Pure proactive routing control overhead*

$ZR/(ZR_c, ZR_f)$	Packet Delivery Ratio
1	87.7%
2	89.9%
3	91.3%
4	93.4%
5	**95.4%**
(2,3)	92.6%
(2,4)	93.9%
(2,5)	94.5%
(2,6)	95.1%
(2,7)	**95.3%**
(3,4)	94.4%
(3,5)	**95.4%**
(3,6)	95.9%
(3,7)	96.2%

Table 3: *Packet delivery ratio under different zone radius settings*

Table 3 indicates that the data packet delivery ratio of TZRP with (ZR_c,ZR_f) is larger than that achieved by ZRP with ZR=ZR_c. Yet another important lesson learned from this table concerns the effect of the interplay between the Crisp and Fuzzy Zone sizes. Consider the boldface entries in

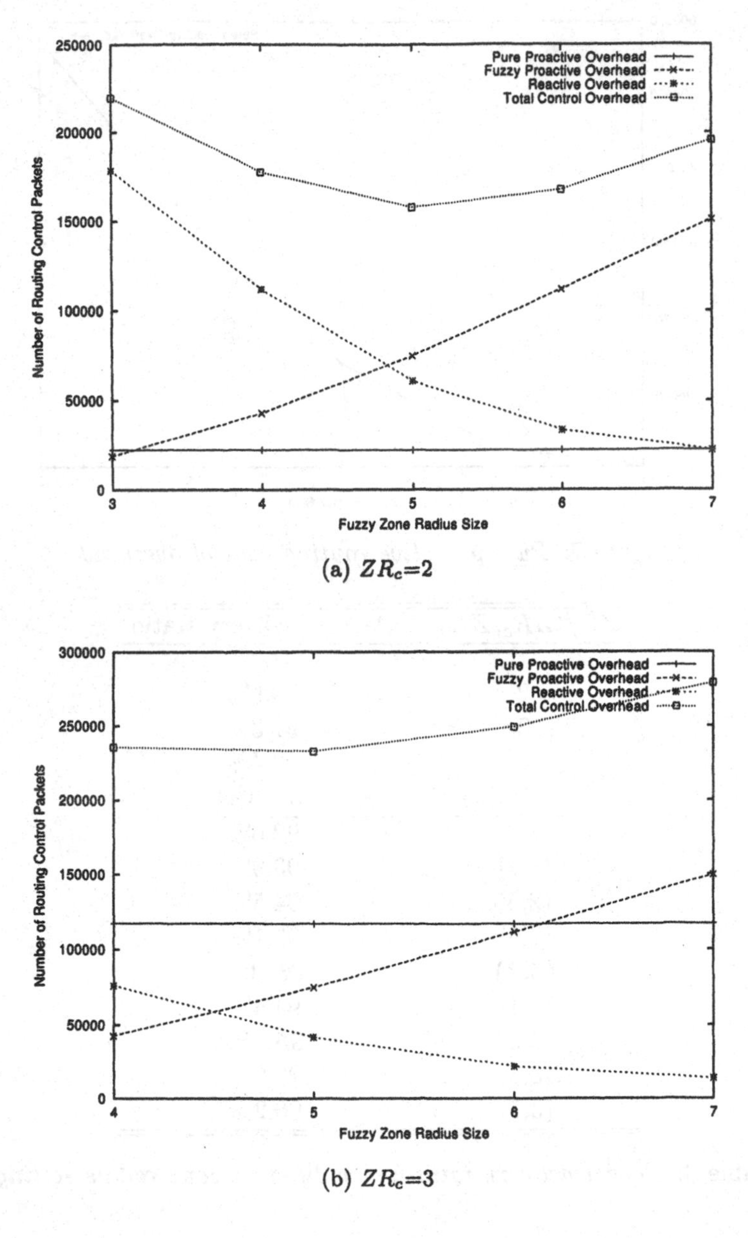

(a) ZR_c=2

(b) ZR_c=3

Figure 8: *Fuzzy proactive routing control overhead*

this table. In ZRP a packet delivery ratio of 95.4% commands a ZR of 5. In

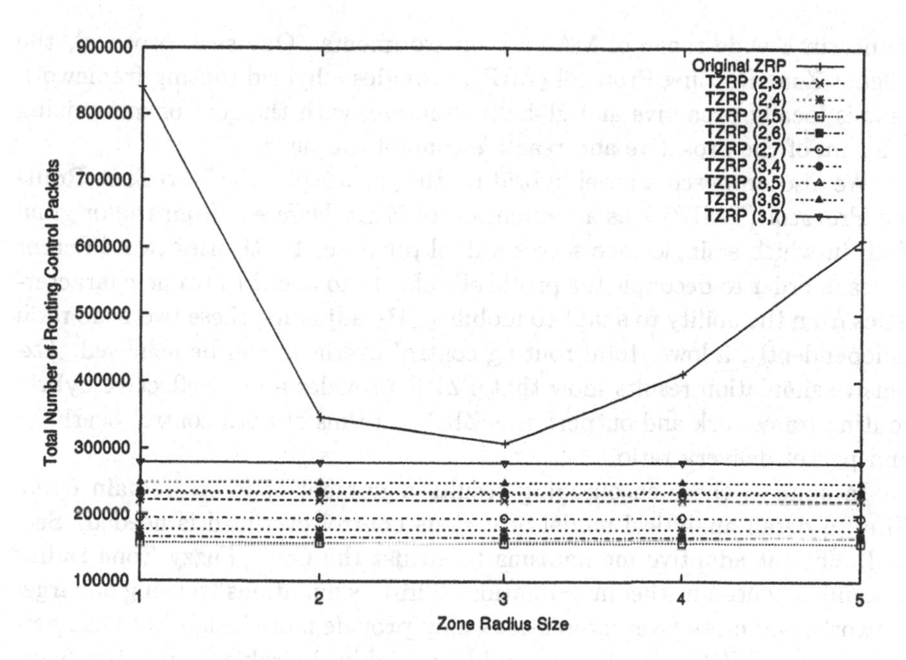

Figure 9: *Total routing control overhead*

TZRP the same level of performance is obtained with either a Crisp Zone size of 2 and a Fuzzy Zone size of 7 or with a Crisp Zone size of 3 and a Fuzzy Zone size of 5. At the same time, as shown in Figure 9, for ZR=5 the control overhead of ZRP is almost three times larger as that of TZRP(2,7) or TZRP(3,5). Note that in our simulation, zone is used for route maintenance, and if an alternative Crisp/Fuzzy route to the destination cannot be found, then the route cannot be repaired locally, and the data packet is simply dropped. Since TZRP has more topology information for use at hand, the probability of such data packet dropping is smaller. This, together with the simulation result in Subsection 4.4.1, shows that although a long-timer-based Fuzzy IARP cannot work effectively to guide bordercasting, it does play an effective role in avoiding reactive overhead and guiding a packet to make a good next-hop decision to the destinations inside the Fuzzy Zone.

5 Concluding Remarks

In this chapter, we have reviewed a class of hybrid MANET routing protocols. By integrating suitable proactive and reactive components to adapt to changing network conditions, a hybrid protocol can provide better performance in a wide range of MANET environments. One such protocol, the elegant Zone Routing Protocol (ZRP), provides a hybrid routing framework that is locally proactive and globally reactive, with the goal of minimizing the sum of the proactive and reactive control overhead.

We also proposed a novel hybrid routing protocol – the Two-Zone Routing Protocol (TZRP) – as an extension of ZRP. Different from the original ZRP in which a single zone serves a dual purpose, TZRP uses two different zones in order to decouple the protocol's ability to adapt to traffic characteristics from the ability to adapt to mobility. By adjusting these two zone radii independently, a lower total routing control overhead can be achieved. Extensive simulation results show that TZRP provides a more effective hybrid routing framework and outperforms ZRP in terms of both control overhead and packet delivery ratio.

In spite of these encouraging results a number of issues remain open. First, a detail analytical model of various control overhead is needed. Second, efficient adaptive mechanisms to adjust the Crisp/Fuzzy Zone radius dynamically need further investigation. Third, simulations focusing on larger networks and cross-layer interactions may provide more insight into the performance of TZRP. Finally, it would be highly desirable to use the fuzzy information inherent in the Fuzzy Zone in a form of fuzzy bordercasting. This promises to be an exciting area for further work.

References

[1] S. Basagni, I. Chlamtac, V.R. Syrotiuk and B.A. Woodward, A distance routing effect algorithm for mobility (DREAM), *Proc. ACM/IEEE MOBICOM* (Dallas, Texas, Oct. 1998) pp. 76-84.

[2] L. Blazevic, L. Buttyan, S.G.S. Capkun, J.P. Hubaux and J.Y.L. Boudec, Self-organization in mobile ad-hoc networks: the approach of terminodes, *IEEE Computer Communications Magazine* Vol.39 No.6 (Jun. 2001).

[3] J. Broch, D.A. Maltz, D.B. Johnson, Y.-C. Hu and J. Jetcheva, A performance comparison of multi-hop wireless ad hoc network routing protocols, *Proc. ACM/IEEE MOBICOM* (Dallas, Texas, Oct. 1998) pp. 85-97.

[4] S. R. Das, C. E. Perkins, E. M. Royer and M. K. Marina, Performance comparison of two on-demand routing protocols for ad hoc networks, *IEEE Personal Communications Magazine – Special Issue on Ad hoc Networking* Vol.8 No.1 (2001) pp. 16-28.

[5] J.J. Garcia-Luna-Aceves and M. Spohn, Source-tree routing in wireless networks, *Proc. 7-th IEEE International Conference on Network Protocols (ICNP)* (Toronto, Canada, Oct./Nov. 1999).

[6] J.J. Garcia-Luna-Aceves, Flow-oriented protocols for scalable wireless networks, *Proc. 5-th ACM international workshop on Modeling analysis and simulation of wireless and mobile systems* (Atlanta, Georgia, USA, 2002).

[7] M. Gerla, Fisheye state routing protocol (FSR) for ad hoc networks, *IETF MANET Internet Draft*, draft-ietf-manet-fsr-03.txt, work in progress, (Jun. 2002).

[8] Z.J. Haas and M.R. Pearlman, The performance of query control schemes for the zone routing protocol, *ACM/IEEE Transactions on Networking* Vol.9 No.4 (2001) pp. 427-438.

[9] Z. J. Haas, M. R. Pearlman, P. Samar, The bordercast resolution protocol (BRP), *IETF MANET Internet Draft*, draft-ietf-manet-zone-brp-02.txt, work in progress, (Jul. 2002).

[10] A. Helmy, S. Garg, P. Pamu and N. Nahata, Contact based architecture for resource discovery (CARD) in large scale MANETs, *Proc. International IEEE/ACM Parallel and Distributed Processing Symposium* (Nice, France, Apr. 2003) pp. 219-227.

[11] A. Iwata, C. Chiang, G. Pei, M. Gerla and T. Chen, Scalable routing strategies for ad hoc wireless networks, *IEEE Journal on Selected Areas in Communications, Special Issue on Ad-Hoc Networks* Vol.17 No.8 (1999) pp. 1369-1379.

[12] P. Jacquet, P. Muhlethaler and A. Qayyam. Optimized link-state routing protocol, *IETF MANET Internet Draft*, draft-ietf-manet-olsr-11.txt, work in progress, (Jul. 2003).

[13] M. Joa-Ng and I-Tai Lu, Peer-to-peer zone-based two-level link state routing for mobile ad hoc networks, *IEEE Journal on Selected Areas in Communications*, Vol.17 No.8 (1999) pp. 1415-1425.

[14] P. Johansson, T. Larsson, N. Hedman, B. Mielczarek and M. Degermark, Scenario-based performance analysis of routing protocols for mobile ad-hoc networks, *Proc. ACM/IEEE MOBICOM* (Seattle, Aug. 1999) pp. 195-206.

[15] D. B. Johnson, D. Maltz, Y. Hu and J. Jetcheva. The dynamic source routing protocol for Mobile Ad Hoc Networks. *IETF MANET Internet Draft*, (Feb. 2002).

[16] L. Kleinrock and K. Stevens, Fisheye: a lenslike computer display transformation, *Technical report* (UCLA, Computer Science Department, 1971).

[17] Young-Bae Ko and Nitin H. Vaidya, Location-aided routing (LAR) in mobile ad hoc networks, *Proc. ACM/IEEE MOBICOM* (Dallas, Texas, Oct. 1998) pp. 66-75.

[18] M. Lewis, F. Templin, B. Bellur and R. Ogier, Topology broadcast based on reverse-path forwarding, *IETF MANET Internet Draft*, (work in progress, Nov. 2002).

[19] V.D. Park and M.S. Corson, A highly adaptive distributed routing algorithm for mobile wireless networks, *Proc. IEEE INFOCOM* (Kobe, Japan, Apr. 1997) pp. 1405-1415.

[20] M.R. Pearlman and Z.J. Haas, Determining the optimal configuration for the zone routing protocol, *IEEE Journal on Selected Areas in Communications, special issue on Ad-Hoc Networks*, Vol.17 No.8 (1999) pp. 1395-1414.

[21] C.E. Perkins and P. Bhagwat, Highly dynamic destination-sequenced distance-vector routing (DSDV) for mobile computers, *SIGCOMM Symposium on Communications Architectures and Protocols* (London, UK, Sep. 1994) pp. 212-225.

[22] C. E. Perkins and E. M. Royer, Ad-hoc on-demand distance vector routing, *Proc. 2-nd IEEE Workshop on Mobile Computer Systems and Applications* (Feb. 1999) pp. 90-99.

[23] A. B. McDonald and T. Znati, A mobility based framework for adaptive clustering in wireless ad-hoc networks, *IEEE Journal on Selected Areas in Communications, Special Issue on Ad-Hoc Networks* Vol.17 No.8 (1999) pp. 1466-1487.

[24] A. B. McDonald and T. Znati, A dual-hybrid adaptive routing strategy for wireless ad-hoc networks, *Proc. IEEE Wireless Communications and Networking Conference (WCNC'2000)* (Chicago, IL, Sep. 2000) pp. 1125-1130.

[25] The Network Simulator - NS-2. http://www.isi.edu/nsnam/ns.

[26] G. Pei , M. Gerla and X. Hong, LANMAR: landmark routing for large scale wireless ad hoc networks with group mobility, *Proc. 1-st ACM International Symposium on Mobile and Ad Hoc Networking and Computing* (Boston, Massachusetts, Nov. 2000).

[27] V. Ramasubramanian, Z.J. Haas, and E.G. Sirer. SHARP: A hybrid adaptive routing protocol for mobile ad hoc networks, *Proc. 4-th ACM International Symposium on Mobile Ad Hoc Networking and Computing* (Jun. 2003).

[28] S. Roy and J.J. Garcia-Luna-Aceves, Node-centric hybrid routing for ad hoc networks, *Proc. 10th IEEE/ACM International Symposium on Modeling, Analysis and Simulation of Computer and Telecommunications Systems-Workshops (MASCOTS 2002)* (Fort Warth, Texas, Oct. 2002).

[29] P. Samar, M.R. Pearlman, and Z.J. Haas, Hybrid routing: the pursuit of an adaptable and scalable routing framework for ad hoc networks, *The Handbook of Ad Hoc Wireless Networks* (CRC Press, 2003).

[30] C. Santivanez, Asymptotic behavior of mobile ad hoc routing protocols with respect to traffic, mobility and size, *Technical Report* TR-CDSP-00-52, Communications and Digital Signal Processing Center, ECE Dept., Northeastern University (Boston, MA, 2000).

[31] C. Santivanez, R. Ramanathan, and I. Stavrakakis, Making link-state routing scale for ad hoc networks, *Proc. 2-nd ACM International Symposium on Mobile Ad Hoc Networking and Computing* (Long Beach, CA, Oct. 2001).

[32] C. Santivanez, A.B. McDonald, I. Stavrakakis and R. Ramanathan, On the scalability of ad hoc routing protocols, *Proc. IEEE INFOCOM* (New York, Jun. 2002).

[33] M. Steenstrup, Cluster-based networks, Chapter 4 in C.E. Perkins (ed.) *Ad Hoc Networking* (Addison-Wesley, Reading, MA, 2000).

[34] I. Stojmenovic, M. Seddigh, and J. Zunic, Dominating sets and neighbor elimination based broadcasting algorithms in wireless networks, *IEEE Transactions on Parallel and Distributed Systems* Vol.13 No.1 (2002) pp. 14-25.

[35] I. Stojmenovic, Position based routing in ad hoc networks, *IEEE Communications Magazine* Vol.40 No.7 (2002) pp. 128-134.

[36] P.F. Tsuchiya, The landmark hierarchy: a new hierarchy for routing in very large networks, *ACM SIGCOMM Computer Communication Review* Vol.18 No.4 (1998) pp. 35-42.

Localization in Wireless Ad Hoc Networks

Dmitri D. Perkins, Ramesh Tumati, Hongyi Wu, and Ikhlas Ajbar
The Center for Advanced Computer Studies
University of Louisiana at Lafayette
E-mail: {perkins, rxt2909, wu}@cacs.louisiana.edu

Contents

1 Introduction

Node localization, position estimation, and geolocation are all terms that are widely used to describe the process of estimating the position or location of a mobile node (MN) with respect to some spatial coordinate system[8]. In this chapter, we discuss the challenges and highlight current research developments in the area of node localization in wireless ad hoc networks[1] [2, 6].

1.1 What Are Ad Hoc Networks?

As communication devices become more intelligent and detached from wired networks, researchers are envisioning a truly ubiquitous computing environment that will allow users to communicate from anywhere and at anytime. Wireless ad hoc networks [6]—an emerging network architecture with several unique characteristics, are part of this vision. Ad hoc networks are infrastructureless self-organizing, peer-to-peer, and rapidly deployable [9, 5, 22]. They are comprised of wireless nodes, which can be deployed anywhere, and must cooperate in order to dynamically establish communications using limited network management and administration [7]. Nodes in an ad hoc network may be highly mobile, or stationary, and may vary widely in terms of their capabilities and uses [15]. The primary objectives of this *new* network architecture are to achieve increased *flexibility*, *mobility* and *ease of management* relative to infrastructured wireless networks. This is achieved by eliminating the need for fixed base stations (BSs) (as in cellular networks and wireless LANs); thereby, enabling instant infrastructure wherever ad hoc nodes are activated, and eliminating many of the constraints to node mobility that are imposed by a fixed network. Due to their inherent flexibility, ad hoc networks have the potential to serve as a ubiquitous wireless infrastructure, capable of interconnecting thousands of devices [23] and supporting a wide range of networking applications. It is hoped that ad hoc networks will emerge as an effective complement to infrastructured LANs

[1] The term *wireless ad hoc network* includes both mobile ad hoc (MANETs) and sensor ad hoc networks.

(wired and wireless), and even wide-area mobile networking services, such as Personal Communication Systems (PCS).

In an ad hoc network environment, the transmission range of each mobile node is limited and variable due to numerous system and environmental factors, including transmission power, receiver sensitivity, noise and other channel effects, namely, path-loss, shadow fading, Raleigh fading, Doppler shift, and interference. Node mobility may exacerbate several of these capacity limiting effects. Furthermore, signal range may be limited by design in order to increase system throughput by minimizing channel access contention [27], and to increase battery lifetime by minimizing transmission power. In general, a node's transmission range is neither fixed, nor symmetric—it demonstrates temporal and spatial variability. Consequently, the wireless links of an ad hoc network are not fixed entities—their status changes over time and is dependent on the relative spatial location of the nodes, transmitter and receiver characteristics, and the signal propagation properties of the environment. These wireless links not only represent wireless end points, as in infrastructured wireless networks, they represent the network topology itself. Thus, as nodes move freely, the topology of an ad hoc network changes dynamically.

A set of five (5) properties have been identified, which are the basis for the many challenges faced by the design and implementation of ad hoc networks [9]:

1. There is no centralized authority for network control, routing or administration.

2. Network devices, including user terminals, routers, and other potential service platforms are free to move rapidly and arbitrarily in time and space.

3. All communication, user data and control information, are carried over the wireless medium. There are no wired communications links.

4. Resources, including energy, bandwidth, processing capacity and memory, that are relatively abundant in wired environments, are strictly limited and must be preserved.

5. Mobile nodes that are end points for user communications and process user applications must act cooperatively to handle network functions, mostly notably routing, without specialized routers.

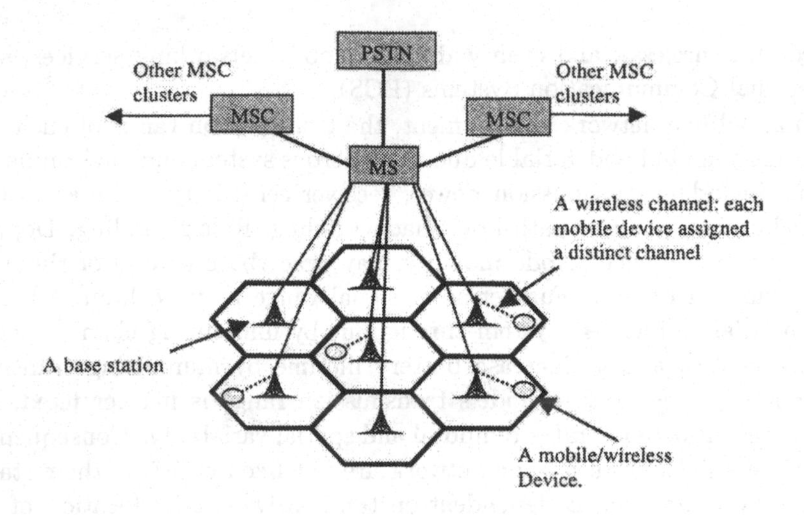

Figure 1: A traditional cell-based network

The challenges stemming from the properties enumerated above affect every aspect of system design and performance—from issues related to physical and medium access control (MAC) layer design, to network-layer issues including routing, addressing and mobility management, to application layer issues, and, of particular interest here, mobile node localization.

Unlike infrastructured networks (see Figure 1), ad hoc networks cannot rely on dedicated and centralized base stations, access points or routers/switches to forward traffic across fixed network segments between mobile users. Furthermore, direct communications between all nodes is infeasible due to limited transmission range and node mobility. As such, store-and-forward packet routing is required over multi-hop wireless paths. The mobile nodes themselves must cooperate in order to dynamically maintain routes and forward traffic on behalf of other nodes (see Figure 2). In order to maintain communications subject to *router* mobility and the subsequent dynamic status of the wireless network links, the routers must implement adaptive algorithms that are responsive to the changes in the network topology, without over-utilization of network resources.

1.2 Why Localization in Ad Hoc Networks?

Ad hoc networks find their applications in places where there is a tight constraint on the network setup-time. They can be used for search and rescue

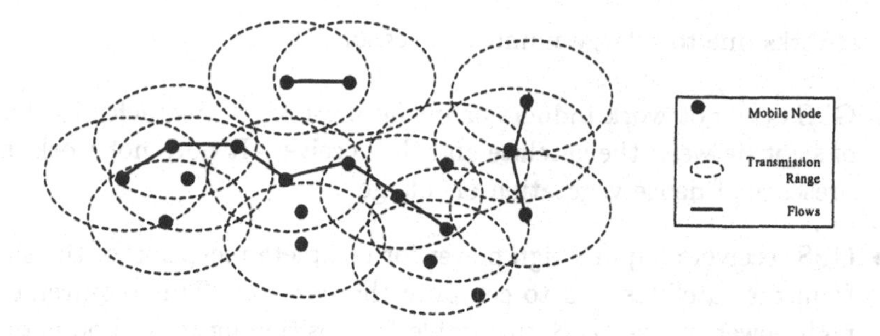

Figure 2: A mobile ad hoc networking environment

operations, conference meetings, tracking and monitoring in hostile environments, environmental control and monitoring, monitoring vehicular traffic in congested areas of a city and for providing security in shopping malls, parking lots and other public places. Most of these applications require the nodes to know their position information to accomplish their respective tasks and the networking tasks efficiently. For example, suppose a large-scale sensor ad hoc network has been deployed in some remote location and has the job of reporting hazardous conditions and events of special interest (e.g., blast, movement, etc.). The sensor node should not only report the existence of a hazard or the occurance of an event, but should also report the location. Thus, the location or position information of mobile nodes in wireless ad hoc networks is necessary and can support location-aware applications (e.g., monitoring, tracking, and location-based services), location-based routing [13, 4], coverage management [16] and collaborative signal processing. The questions is how will all the nodes obtain accurate estimates of their current location in some specified coordinate system.

1.3 Why Not GPS?

The Global Positioning System (GPS) is a worldwide radio-navigation system formed from a constellation of 24 satellites and their ground stations [12]. GPS uses the satellites as reference points to effectively calculate the positions of ground nodes. Some of the real world applications of GPS include location estimation, tracking, navigation , mapping and providing timing services. To use GPS, a node must be equipped with a GPS receiver which is responsible for estimating the absolute position of the node in the global coordinate system. Though GPS makes it possible to provide a wide range of positioning services, it is not a completely viable solution for ad hoc networks due to a large number of reasons [1].

- GPS does not work indoors or in the presence of obstacles in the line of sight between the satellite and the receiver. It does not work in the presence of dense vegetation or foliage.
- GPS receivers require high power for complete reception of the signals from the satellites and to compute the position. This requirement of high power makes GPS unsuitable for positioning in ad hoc networks which have strict energy constraints.
- The cost of attaching a GPS receiver to every node in the ad hoc network, which could consist of thousands of nodes in sensor ad hoc networks, is prohibitive.
- Position errors with GPS are still around 3 to 30m, which is a considerable error for dense sensor ad hoc networks.

1.4 Designing Localization Algorithms for Ad Hoc Networks

In this section, we discuss some general assumptions, design issues, and goals for ad hoc localization algorithms. The unique characteristics (i.e., infrasturctureless, wireless multi-hop routing, frequently changing topologies, and mobile routers) of ad hoc networks precludes the direct application of existing infrastructure-based localization approaches and presents unique challenges in designing localization schemes for ad hoc networks. As such, the non trivial nature of the position estimation problem in ad hoc networks has attracted significant research interests and several algorithms have been proposed in recent years. For example, a key presumption of the infrastucture-based localization schemes (to be discussed in Section3) is that the mobile node is always a single hop from several reference nodes (e.g., a stationary base station aware of its location) stations). This assumption is not necessarily valid, as ad hoc network architecture is primarily an autonomous self-organizing multihop network. That is, to access a reference node, an unknown node may need to transmit a signal via several intermediate nodes.

1.4.1 Assumptions

The localization algorithms proposed for ad hoc networks are typically based on several key assumptions as described below.

1. **Multi-hop Routing:** Base stations and access points are not present in ad hoc networks and, as such, the mobile nodes must cooperate to deliver packets between distant nodes. Generally, ad hoc localization algorithms assume the existance of a routing protocol which can be used to deliver packets to special nodes that can assist in the localization process.

2. **Coordinate System:** A localization algorithm can estimate the positions of the nodes in either a global coordinate system or a relative coordinate system. In a global coordinate system, the positions of all the nodes will be based on a global reference. In relative coordinate system, the positions of the nodes are based on some locally relevant coordinate system. Typically, algorithms based on a global coordinate system assume a small fraction of nodes in the network are aware of their absolute location, for example via GPS.

1.4.2 Design Issues and Goals

Location systems must satisfy certain properties to be used with infrastructureless ad hoc networks. The required properties include:

1. **Distributed operation:** In ad hoc networks, the location estimation procedure should be distributed rather than centralized i.e, nodes are responsible for computing their positions. In centralized approach a single node is reponsible for computing the positions of all the nodes in the networks. This approach fails if the central node fails resulting in a collapse of the location system.

2. **Low response time:** The position estimation procedure must have a fairly low response time as the ad networks are characterized by their dynamic topologies. The location system should enable the nodes in the network to estimate their positions on fly as the network topology changes.

3. **Infrastructureless:** Ad hoc networks are generally characterized by their lack of or reduced dependance on infrastruture support. So the location systems should not require extensive infrastruture support or pre planning to be used with ad hoc networks.

4. **Energy efficient:** Communication and computation are the main sources of power consumption. The nodes in the ad hoc networks are energy constrained as they are not connected to a continuous power

source and, therefore, must utilize their energy in an efficient manner to prolong the lifetime of the node. So the communication and computation overhead involved in the computation of position of the node must be minimum to prevent the battery of the node from draining soon. Also the signaling complexity involved in finding the location should be minimum when a part of the network changes.

5. **Scalable:** Ad Hoc networks range from small local area networks to large-scale sensor networks. So depending on the application, the location system must be able to scale from smaller networks to large scale networks like sensor networks.

6. **Accurate:** The positions estimated by the localization systems must be accurate and consistent irrespective of the changing topologies and other environmental factors. The level of accuracy and consistency of the position estimates generally depend on the application for which the localization system is being used. In general, the accuracy of the position estimates must be comparable with the nodes communication range.

2 How Node Localization Works

In this section, we describe the general process of node localization in wireless networks. Let us consider a network comprising two types of nodes: *reference nodes* and *unknown nodes. Reference nodes* are those nodes aware of their own (x, y)-coordinates in a specified coordinate system while *unknown nodes* are unaware of their location. Reference nodes may have obtained their coordinates via manual configuration, via GPS, or by some other means. The goal of a localization algorithm is to use the reference nodes[2] to somehow estimate the position coordinates of unknown nodes.

Localization schemes for wireless ad hoc networks can be classified as either *range-based* or *connectivity-based* depending on what information (e.g., distance and angle measurements or proximity contraints) is used to provide position estimation. Range-based techniques attempt to perform position estimation by estimating the distance or direction of an unknown node with respect to several reference nodes and can be further classified as distance-based or direction based. Using the known position of reference nodes and

[2]Position estimation requires a minimum of three reference nodes in a two-dimensional coordinate system or four in a three-dimensional coordinate system. A two-dimensional space is assume in this chapter, but the concepts easly extend to higher dimensions.

the estimated distances from the unknown node to several reference nodes, the (x, y)-coordinates of the unknown can be determined. For example, the coordinates (x_m, y_m) of an unknown node m can be determined using characteristics of the signals transmitted between several reference nodes and the unknown m. In order to determine (x_m, y_m), the distance and direction (or both) of the unknown node m must be estimated by several of the reference nodes. Alternatively, the distance and direction of several references nodes can be estimated by the unknown node m such that it can calculate its own position. Distances can be estimated using properties of the received signals such as signal strength and time of arrival. Directions can be determined from the angle of arrival of the received signal.

Connectivity-based methods do not attempt to estimate the distances or directions of an unknown node to several reference nodes. Instead the location of an unknown node is inferred by its proximity to several reference nodes. The remainder of this section provides an overview of the general mechanisms involved in the range-based and connectivity-based approaches.

2.1 Range-based Methods

Typically, range-based methods employ a two-phase process to provide position estimation for unknown nodes. Phase I is the ranging phase where distance or direction estimation is performed. In Phase II, the ranging information obtained in Phase I is used to compute the (x, y)-coordinates of an unknown node. Typically, some form of trilateration or triangulation is used in Phase II.

2.1.1 Distance-based Techniques

Signal Strength Method

This method can be used for estimating the distances of an unknown node to reference nodes based on signal strength. This method can only be used with radio signals. Here the distance is estimated from the path loss using radio propagation models. By knowing the transmitted power and the received power, the effective path loss can be calculated. The distance between the transmitter and the receiver can then be obtained by converting the obtained path loss into distance using the appropriate path loss model that well represents the environment through which the radio signal propagates. This method is highly unpredictable as the signal suffers from multipath, fading and shadowing effects. The RSSI measurements rely heavily on the environmental interference and they are also non linear. This method can have measurement errors up to 50% of the range of the nodes making it less suitable for most applications, which require very accurate position estimates. So this method is normally used along with other techniques to get the distances.

Time of Flight Method

In this method, time of flight of the signal between the unknown node and the reference node is used to estimate the distance between them. Using the time of flight and the known speed of the signal , the distance can be computed using speed distance relationship. This method can be used with RF, acoustic, infrared and ultrasound signals. The only drawback with this approach is that complex hardware is needed for perfect synchronization when only RF signal is used. On the other hand acoustic signals have low frequency and low speed due to which the cost and complexity of the hardware needed for these signals is low. But acoustic signals cannot be used outdoors and have lower penetration than RF signals. So for this method both RF and ultrasound signals are employed. The radio signals are used for synchronization as they have much higher speed when compared to the ultrasound and ultrasound is used for estimating the ranges by taking the time of flight into account. Timing methods produce precise results and are more robust when compared to the signal strength methods.

Position Calculation

Distances between an unknown node and reference nodes obtained in the previous phase can now be used to compute the position of the unknown node. Depending on the number of distance measurements obtained, localization can be performed using trilateration or multilateration.

Trilateration: This technique is used when the distances of the node to three reference points are known. The position of the node is the intersection of the circles as shown in Fig.3. The center of each circle is a reference node and the radius is its distance to the node whose position is to be estimated. Figure 3 also gives the set of equations which are used for computing the position of the node (x, y) using trilateration.

Multilateration: This technique is used when the distances of the node to more than three reference nodes are available. Here, an estimate of the position is obtained by reducing the difference between the actual distances and the estimated distances between the unknown node and the reference

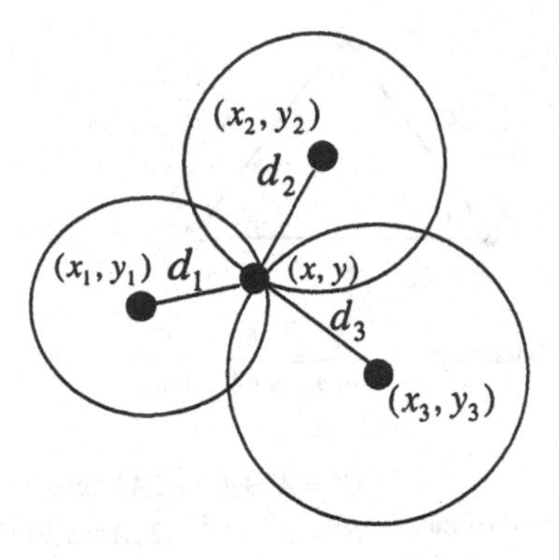

$$\begin{bmatrix} (x-x_1)^2 + (y-y_1)^2 \\ (x-x_2)^2 + (y-y_2)^2 \\ (x-x_3)^2 + (y-y_3)^2 \end{bmatrix} = \begin{bmatrix} d_1 \\ d_2 \\ d_3 \end{bmatrix}$$

Figure 3: Trilateration

nodes. The difference between the actual and estimated distances between the unknown node j and the reference node i can be expressed as shown in equation 1, where d_{ij} is the estimated distance between the unknown node j and reference node i.

$$f_i(x, y) = \sqrt{(x_j - x_i)^2 + (y_j - y_i)^2} - d_i \tag{1}$$

Finally, minimum least squares estimate is used here to get the optimal position estimate. The position of the unknown node is obtained by finding the minimum of the function defined in (2).

$$F(x, y) = \sum_{i=1}^{n} f_i(x, y)^2 \tag{2}$$

2.1.2 Direction-based Techniques

Similar to the distance-based methods, the position of an unknown node can be determined in two phases: *angle estimation* and *position calculation*. In the phase one, angular bearings of the unknown node relative to at least three reference nodes are estimated. From the angles obtained in the previous phase the position is estimated using triangulation. Phased antenna arrays can also be used for determining the angles of the nodes relative to

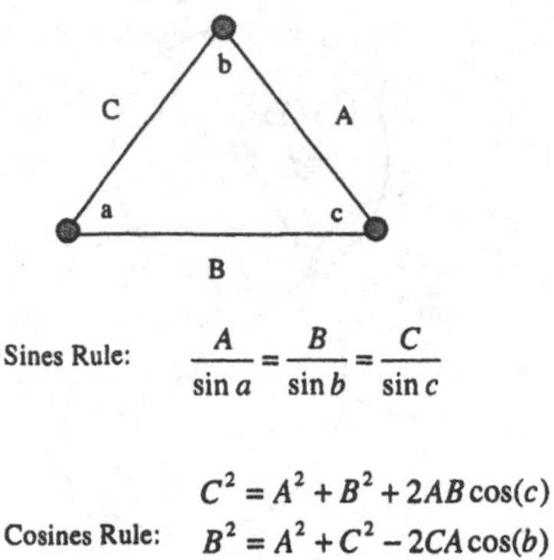

Figure 4: Triangulation

the reference nodes. Here each reference node must have an array of antennas separated by a known measure. Using time taken by the signal to reach the various antennas and the known separation measure between the antennas, it is possible to get the angle from which the emission originated. So the unknown node location can be found by using several reference nodes and the angles of the unknown node relative to the reference nodes. This is a centralized approach and the position estimation is done at the cell site. This method cannot be used indoors because of multipath effects and it also requires costly antenna arrays at the cell sites.

Triangulation: Using the angle estimates from Phase I, triangulation can now be used determine a position estimate for the unknowm node using trigonometry laws of sines and cosines as shown in Figure 4.

2.2 Connectivity-based Approach

In connectivity-based (also called range-free) localization systems, unknown nodes find their locations based on their proximity to the reference nodes. Here the unknown nodes estimate connectivity relationships to sufficient number of reference nodes instead of estimating distances or angles. The unknown node obtains the connectivity constraints to the reference nodes from the communication with them. Finally the unknown node solves the proximity constraints to get its position estimate. For example, let us suppose a node is in the communication range of four reference nodes as shown in Figure 5. In connectivity-based location systems the position of the node

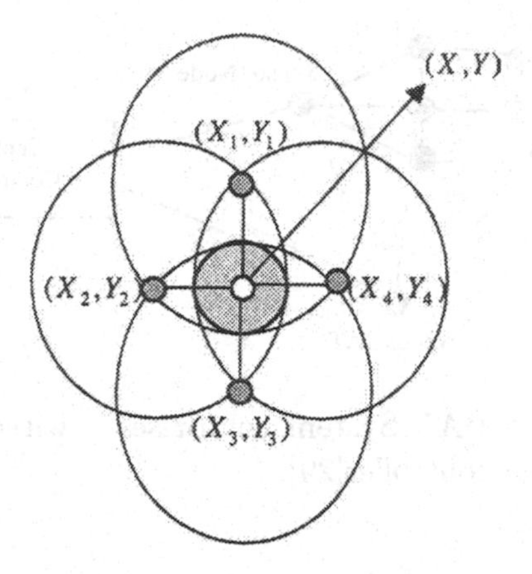

Figure 5: Example of connectivity-based node localization

can be estimated as the geometric center of the quadrilateral formed by the reference nodes or it can be estimated as the geometric center of the common(shaded) region for all the four reference nodes. Here the communication between the node and the reference nodes are transformed into relationships and these relationships are solved to estimate the position of the node. This class contrasts with the range-based localization methods, where the unknown node estimates distances or angles to the reference nodes from the communication with them and computes the position using one of the aforementioned position calculation techniques.

3 Localization Techniques in Infrastructured Systems

In the next section, we will examine the localization algorithms designed for wireless ad hoc networks. However, before that, it is perhaps instructive to briefly consider existing position estimation systems that have been designed for infrastructure-based networks such as wireless LANs and cellular networks, further motivating the need for new algorithms for the ad hoc networking paradigm. Many localization, monitoring and tracking schemes have been developed for infrastructured environments. These systems include Active Bat[29], Active Badge [28], RADAR [3], SpotON [11], and Cricket[21]. Typically, these location estimation systems use a central server which monitors mobile nodes via a wired network of sensors or base stations. The systems require that the person or node being tracked be equipped

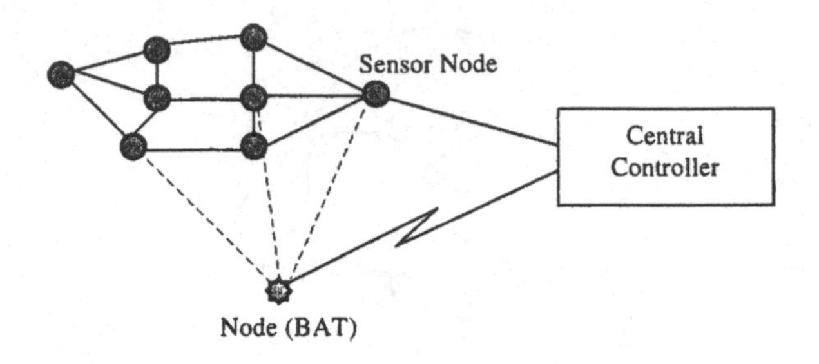

Figure 6: The Active BAT System comprises a wired network of sensors connected to a central controller[29].

with a small communication device. Communication between the sensors and tracking device is reported to a server which can then provide position estimation or tracking information.

As an example let us consider the infrastructure-based position estimation system Active BAT. Active Bat is an ultrasonic location estimation system, which can be used for large-scale deployments in indoor environments. The main advantages of Active Bat are that it requires less power, wireless, less expensive and provides precise position estimates and orientation of the nodes. An active bat system consists of a network of sensors mounted on a ceiling and connected to a central controller. To determine the location of an unknown node, the central controller sends radio signals to the sensors and the node (see Figure 6). Upon receiving the radio signal from the central controller, the BAT attached to the unknown node transmits ultrasound signals which is received by the nearby sensors. The sensors, then, estimate their respective distances to the unknown node using time of arrival of the signal and the speed of ultrasound and forward the distance information to the central controller where trilateration or multilateration is performed to compute the position of the node. The main drawback of this approach is that it requires significant amount of wiring between the sensors and the central controller making it unsuitable for ad hoc networks.

Localization is also relevant in cellular networks, which are infrastructured-based wide area networks which enables a mobile node (MN) to communicate over long distances. In these networks the coverage area is divided into cells. Each cell has a set of mobile nodes and a base station through which nodes in one cell communicate with each other and with nodes in other cell. Similar to wireless LANS, a MN is always a single hop from a central controller (e.g., base station). MNs in a cellular environment can use fixed base stations to estimate their current location. Further, a positioning service can determine the location of a mobile node(subscriber) and provide location-based information such as nearby restaurants or other services. In fact, by 2005, FCC (E-911) is requiring mobile telephone operators to be capable of accurately determining (with in a few meters) the positions of 911 callers.

4 Localization Algorithms for Ad Hoc Networks

This section contains short descriptions of several proposed localization algorithms for ad hoc networks. The algorithms differ along multiple dimensions, including range measurement technique, scalability, position error, and communication and computational requirements, coordinate system, and the infrastructure support required. Table 1 provides a characterization of the algorithms described in this section.

4.1 Range-based Algorithms

4.1.1 Ad-Hoc Localization System (AHLoS)

The Ad Hoc Localization System (AHLoS), proposed by Savvides and Srivasatava in [1], is a distributed localization algorithm for sensor ad hoc networks. AHLoS assumes is reference nodes and uses the a fraction of reference nodes are present in the networl and uses an interative process to allow unknown nodes to estimate their positions in a global coordinate systems. The ToA technique is the primary distance estimation method used in AHLoS.

AHLoS employs an iterative process and works as follows. Upon deployment, each reference node broadcasts its location which is received by all neighboring unknown nodes. Using the ToA technique, an unknown node measures the distance from the neighboring reference nodes. Using these distance measurments and the location information recieved from the reference nodes, the neighboring unknowns can estimate their own location. Once an unknown node estimates its position, it becomes a reference node and broadcast its own position information to neighboring unknown nodes. This process is then repeated until the positions of all unknown nodes with three or more reference nodes are estimated.
Atomic Multilateration: In the basic case, an unknown node 0 uses atomic multilateration to compute its (x_0, y_0)-coordinates if the unknown node has at least three neigboring reference nodes. Figure 7a illustrates a topology for which atomic multilateration can be used. A key component of AHLoS is

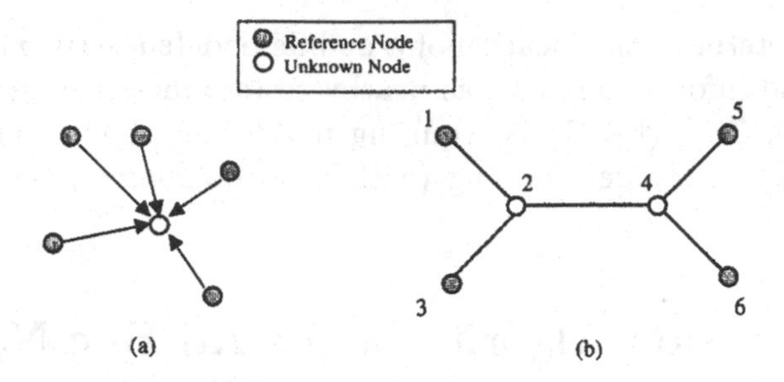

Figure 7: Examples of Multilateration [1].

a Maximum Likelihood estimation procedure which is used to estimate the position of nodes by minimizing the differences between the measured distances and estimated Euclidean distances. The measured distance between an unknown node 0 and a neighboring reference node i is also obtained by using an ultrasound signal. Now, an error estimate for reference node i can be obtained using equation 3, where x_0 and y_0 are the estimated coordinates for unknown node 0, and t_{i0} and s are the propagation time and speed, respectively, of the ultrasound signal as it travels from reference node i to unknown node 0.

$$f_i(x_0, y_0, s) = s \times t_{i0} - \sqrt{(x_i - x_0) - (y_i - y_0)} \qquad (3)$$

If unknown node 0 has at least three reference nodes, a Maximum Likelihood estimate of node 0's position is obtained by taking the minimum mean square estimate of a system of equations as follows, where α is the weight applied to each equation. Please see [1] for more details.

$$F(x_0, y_0, s) = \sum_{i=1}^{N} \alpha_i^2 f_i(x_0, y_0, s)^2 \qquad (4)$$

Collaborative Multilateration: In an ad hoc network, it is certainly possible that the case will arise where an unknown node will not have at least three neighboring reference nodes. In this case, a procedure called collaborative multilateration [1] is used. In collaborative multilateration an unknown node estimates its position by using non-neighboring reference nodes which are multiple hops from the unknown node. Figure 7b illustrates a topology for which collaborative multilateration can be applied. Here a unique position estimate can be computed for nodes 2 and 4 since they each have two neighboring reference nodes and each has the other as a neighboring node.

The primary advantages of this localization scheme are it incurs less cost, can operate indoors and does not require pre-planning or expensive infrastructure support. The main drawback of iterative multilateration is the error accumulation that results from the use of unknown node that estimates their positions as beacons. Further, in some cases, collaborative multilateration will not yield a unique position. This localization scheme produces satisfactory results for small networks but it is not scalable to large networks. The network connectivity, node density and beacon node placement are crucial for this scheme to enable most of the nodes to estimate their locations. :subsubsectionAd-hoc Positioning Algorithm (APS) The APS algorithm [19] is a distributed hop-by-hop positioning algorithm. It does not require special infrastructure or setup, provides global coordinates, and requires recomputation only for moving nodes. It provides approximate location for all nodes in a network where only a fraction of nodes have self-location capability. At the beginning, reference nodes flood their location to all nodes in the network. Using one of propagation methods proposed in this algorithm, nodes estimate their distances to reference nodes. From the estimated distances and known positions of the reference nodes, the position estimate is determined using trilateration, which is a form of triangulation.

Three propagation methods are proposed in this algorithm, which can be used by the nodes to determine their distances to the reference nodes depending on the characteristic of the network.

DV-Hop Using the DV-Hop propagation method, each node maintains a list of the reference nodes and its distances to them in number of hops. The reference points broadcast their position information to their neighbors in a packet containing a hop count field. This hop count field is initially set to 0. Nodes that receive the packets from the reference nodes store the position information of reference nodes within 1 hop. Then they increment the hop count field in the packet by 1 and rebroadcast it to their neighbors. After some time, each node has the position information of all reference nodes and the distances to them in hops. After a reference node receives the locations of the other reference nodes and distances to them in hops, it computes correction (the average distance for one hop) that is defined as the sum of distances from this reference node to other reference nodes divided by the sum of number of hops to other reference nodes (Figure 8). This correction is propagated through the network by controlled flooding. Usually, nodes in the network use the correction of the closest reference point and compute the distances to the reference nodes using the hop counts they have to them.

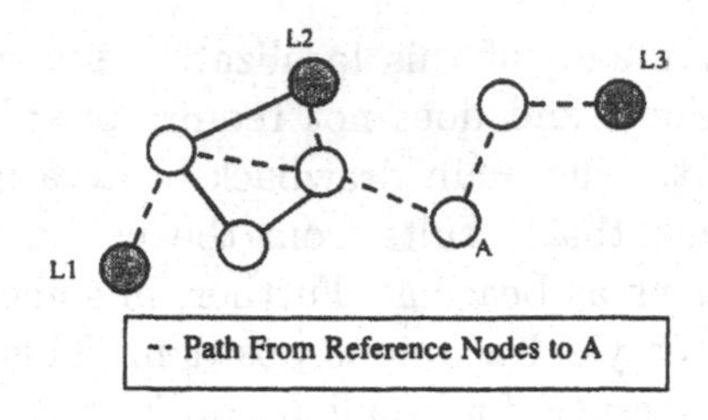

Figure 8: DV-hop correction example $L_i, i = 1, 2, 3$ are reference nodes A is unknown node[19].

A node in the network can then estimate its position using the distances by performing trilateration. The advantages of DV-Hop are its simplicity, robustness, and immunity to measurement errors. It works well in dense and regular topologies, but for sparse or irregular networks, the accuracy decreases. Another drawback of DV-Hop is that it works only for isotropic networks.

DV-Distance The DV-Distance propagation method is similar to DV-Hop except that the distance between the nodes is measured using signal strength and the cumulative distance to the reference node is propagated rather than the number of hops. This is more efficient than the DV-Hop, because not all the hops have the same size. The drawback of this propagation method is that it is sensitive to measurement errors.

Euclidean In Euclidean propagation method, each node in the network needs to receive messages from at least two neighbors that know their distance to the reference point, and to each other (Figure 9). Each node calculates the distance to the reference node by applying Pythagoras generalized theorem. Euclidean method provides good accuracy with the increase in the number of GPS enabled nodes. It works well for non-isotropic networks, but at the expense of greater communication overhead. It is also sensitive to range errors.

APS is distributed, does not require infrastructure support, enables the nodes to estimate their positions in the global coordinate system and requires computation only for the moving nodes. DV based algorithms work well for most purposes and they have a low signaling complexity. The position error in the position obtained by APS is always less than one radio hop from the true location, but at the expense of the number of operations that trilateration requires.

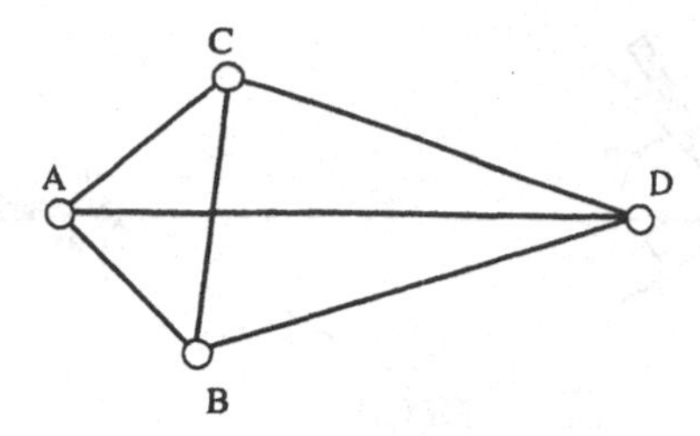

Figure 9: Euclidean propagation method[19].

4.1.2 Robust Positioning Algorithm

Robust Positioning algorithm [25] is a distributed algorithm for determining the positions of the nodes in an ad hoc wireless sensor network. This algorithm is separated into two phases: start-up phase and refinement phase. The start-up phase addresses the problem of small number of reference nodes that are often scattered over a large area. In the start-up phase, the Hop-TERRAIN algorithm (similar to DV-hop) is used. The Hop-TERRAIN is run once at the beginning of the algorithm. It finds the number of hops from an unknown node to each of the reference nodes in the network, and then multiplies this number of hops by the average hop distance to estimate the distance between the node and each of reference nodes. Then, using triangulation, the node can get its estimated position. The triangulation consists of solving a system of linearized equations ($Ax = b$) by means of a least squares algorithm. Of course, the resulting position estimate is not accurate, but it provides an initial point for refinement.

In phase II, an iterative refinement algorithm that focuses on increasing the accuracy of the position estimate as much as possible is used. A node broadcasts its position estimate, receives the positions and corresponding range estimates from its neighbors, and performs lateration to determine its new position. In many cases the constraints imposed by the distances to the neighboring locations will enforce the new position towards the true position of the node. When the position update becomes small, refinement stops and reports the final location. The refinement algorithm is simple, but it is not clear what the condition of convergence is and how accurate the final solution will be. Refinement is improved by solving the system ($wAx = b$) instead of solving ($Ax = b$), where w is the vector of confidence weights.

This algorithm is distributed and is scalable to large networks. It achieves position errors of less than 33% in a scenario with 5% range measurement

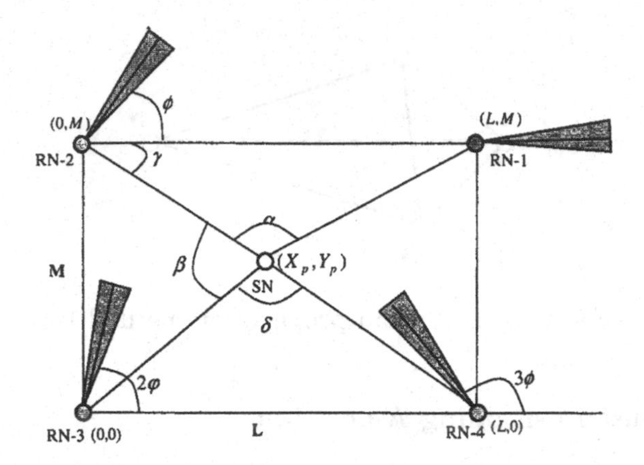

Figure 10: The model of rotating directional reference signals from 4 reference nodes, located on the corners of the sensor network area. Also shown is a sensor node SN with its angular bearings with respect to the 4 reference nodes[18].

error, and 5% reference points. This algorithm requires the nodes to have a high degree of connectivity (around 8-10) and reasonable fraction of reference nodes placed on the edges of the network. Hop-TERRAIN fails for highly irregular network topologies where the variance in actual hop distances is very large.

4.1.3 A Directionality-based Localization Scheme

In [18], Nasipuri and Li present a distributed localization scheme for static sensor ad hoc networks based on an angle of arrival estimation technique. The proposed localization scheme requires a set of reference nodes that are capable of covering the entire network area by powerful directional wireless transmissions. The network model for this algorithm consists of a large number of sensor nodes, which are located in random but fixed locations and a central processing and control unit. The network consists of at least three fixed wireless stations or reference nodes. Figure 10 shows a network model with four reference nodes RN1, RN2, RN3, RN4 and an unknown sensor node SN.

The reference signal is a radio signal on a narrow directional beam, which rotates with a constant angular speed. The reference signals from different reference nodes should be different from each other, which is achieved by using different carrier frequencies. There is constant angular speed between the directional beams from the reference nodes. All reference nodes are wired and controlled by a central controller, so it is possible to achieve phase synchronization and identical angular speeds in the reference signals. A rotational directional beam is implemented here by mechanically rotating the directional antenna. In this algorithm, the unknown node notes the times of the arrival of the reference signals from the reference nodes. Based on the times of arrival of the reference signals, its angular bearings with the reference nodes are computed. Using the angular bearings the position of the unknown node is computed using triangulation. The errors in the position estimates usually occur due to the non-zero width of the directional beam and due to the multiple signals generated by multipath reflections. The error due to the non-zero width of the reference signal can be corrected by considering the time only when the strength of the signal is maximum. The errors due to reflections can be corrected by finding common solution from a number of solutions obtained by using at least four reference nodes and considering three at a time. The amount of computations and the difficulty of resolving the multiplicity of solutions increase exponentially with the number of multipath reflections received. This algorithm is may not scalable. Although, the computation for the positions is performed at the node itself but it requires significant cost for setting up antennas (at reference nodes) used for transmitting rotating directional beams.

4.1.4 A Localized Algorithm

In [17], the authors propose a distributed localization algorithm based on the collaborative location discovery similar to AHLoS. This proposed algorithm differs from AHLoS as it specifies the order in which the nodes should estimate their positions such that the location errors are reduced. Here a node does not accept the result of its first trilateration as its position and it continues to accept location information from other nodes and adjust its position estimate. If a node has more than three reference nodes, then it can perform more than one trilateration and the location of the node would be the center of mass of all the generated locations. The priority for selecting the unknown nodes that should be first to estimate their positions is as follows:

- Nodes whose positions generated by the trilateration process are more consistent measured by the variance of the locations relative to the center of mass of the locations.
- Unknown nodes whose majority of neighbors have already accepted

their estimated locations and become reference points for other nodes.

An objective function is used to evaluate the above two properties. Once the node evaluates its objective function it compares it with the values reported by the nodes within a predefined number of hops. If the value of the objective function of the node is lower than the values of all of its n-hop neighbors then the node accepts its position and becomes a reference point. Other nodes start the multilaterations again. This localized algorithm for location discovery can produce results with accuracy similar to that of centralized counterpart while maintaining all the benefits of distributed algorithms

4.1.5 A Self Positioning Algorithm

All of the localization algorithms discussed thus far have assume a network model where a small fraction (≥ 3) of node are reference nodes. The Self Positioning Algorithm (SPA) [24] is a distributed, infrastructure less positioning algorithm for wireless ad hoc networks and does not assume the existence of reference nodes. This algorithm does not require the use of GPS and it enables the nodes to obtain their relative positions in the local space. SPA can be used when GPS cannot be used either because the signal is weak or it is not economically feasible to use it. One drawback of SPA is it can compute the positions of the nodes only in the 2-dimensional space.

The nodes in the network are responsible for finding their own positions using the information from their neighboring nodes. The unknown node computes the distances to the neighboring nodes and uses the position information of the neighbors along with the distances computed to obtain their positions. The distances between the nodes are estimated using Time of Arrival (ToA) technique. Using SPA, mobile nodes estimate their positions in three steps.

Step 1: Initially each node i is responsible for building its own coordinate with itself being the center. The neighboring nodes if possible compute their positions in the local coordinate system. Each node in the network performs the following sequence of steps:

- Detect all the one-hop neighbors
- Compute the distances to all one-hop neighbors
- send the neighbors list and the corresponding distances to all the neighbors.

- Choose two non-collinear neighboring nodes p and q such that node p is one the x-axis and node q has a positive y component with respect to the local coordinate system of the node. The positions of p and q are obtained using some trigonometry laws of sines and cosines.
- The node i can compute the position of its neighboring nodes using the position information of nodes p and q provided the node is also a neighbor to p and q and has distances to them.
- The positions of nodes, which are not neighbors of p and q, are computed by using the positions of node i and at least two other nodes for which the positions are already computed.

The set of nodes whose locations are computed in the local coordinate system of node i is referred to as local viewset connectivity (LVS). The size of LVS depends on the choice of p and q. So p and q must be chosen such that the number of nodes that computed positions must be more.

Step 2: After the nodes build their own local coordinate systems, there will be a number of coordinate systems each having a different direction. So during this step, the directions of all the coordinate systems are adjusted to obtain a single direction and it becomes the direction of the network coordinate system. This can be accomplished by choosing one of the nodes in the network as center and using its coordinate system as the network coordinate system. So all the coordinate systems of the remaining nodes are transformed in the direction of the network coordinate system of the node by rotation or mirroring or using both.

Step 3: All the nodes in the local view set of node i already have their positions computed in the coordinate system of node i. During this step the positions of all the remaining nodes in the network must compute their positions in the transformed network coordinate system. This is illustrated in Figure 11.

Let us suppose that node i is considered as the network center so all the coordinate systems of the nodes must be aligned to the direction of node i's coordinate system. After all the nodes align their coordinate systems in the direction of the network coordinate system, every node must compute its position in the network coordinate system. In figure 11, let us suppose that node i is considered as the network center. So node k aligns its coordinate system in the direction of the network coordinate system. Now node k knows its position in the network coordinate system, as it is a neighbor to node i

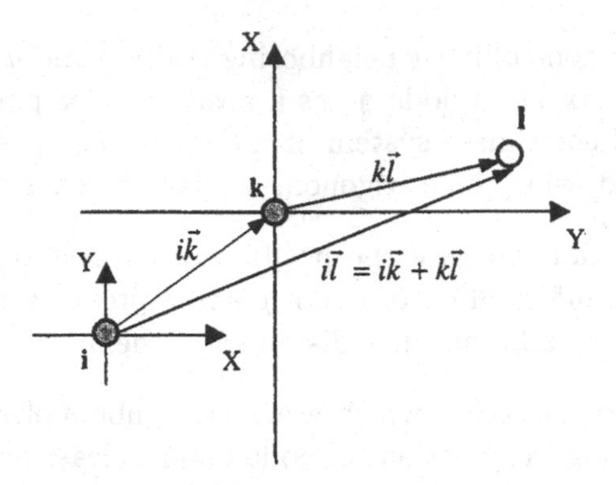

Figure 11: Position computing when the local coordinate systems have the same direction[24].

and it also knows the position of the node l in its own coordinate system. Now the position of node l is obtained from the following relation:

$$\vec{il} = \vec{ik} + \vec{kl} \tag{5}$$

Similarly the positions of 3-hop neighbors of node i can be obtained using the position of node l and their respective positions in the coordinate system of node l. In this way all the nodes in the network can obtain their position estimates in the network coordinate system. Nodes which are not able to build their local coordinate systems can compute their positions using the using the already computed positions of at least three neighbors and performing triangulation.

There are many disadvantages in choosing a local coordinate system of single node i as a network coordinate system. Motion of node i will require all nodes in the network to recomputed their positions in the network coordinate system. So, this approach can be used only for small networks where nodes have low mobility and where disconnection of nodes is not expected.

A more stable approach is to compute the center of the coordinate system as a function of small group of nodes in the network, which are stable and less likely to disappear from the network. This group of nodes is referred to as Location Reference Group (LRG). LRG is chosen such that the density of the nodes in the LRG is highest in the network. In this approach the network center is the geometric center of all the nodes in the LRG. The network direction is computed as the mean value of the directions of all the local coordinate systems of the nodes in the LRG. The larger the LRG the more stable it is and more difficult it becomes to maintain and more costly to compute the center and direction of the network coordinate system.

Initially the node with highest density factor (say in n-hops) becomes the location reference group master. All the nodes in LRG will have the complete information about all the nodes in the LRG. If the node does not have LRG master in its n-hop neighborhood, it starts an initialization timer. If within certain time the node does not receive the new position information issued by the LRG master, it sends a reference signal to all the nodes in n-hop neighborhood and gets their node id, mutual distance and directions of their coordinate systems. It compares the n-hop neighbors list with the list of LRG members. A node that is a LRG member and has the highest number of nodes in its n-hop neighborhood will become the new LRG master and its n-hop neighbors will become the new LRG.

This localization scheme assumes that all the nodes have the same power range. The increase in node connectivity, i.e., increase in radio range, increases the LVS connectivity. Also the larger the LRG is more stable are the network center and direction of the network coordinate system. This localization scheme requires the following to provide better position estimates.

- The power range must be large enough to ensure LVS connectivity.
- The size of LRG must be large enough so as to ensure network center and direction stability.
- The accuracy of range measurements affects the position estimate accuracy. So range error mitigation techniques must be used to provide stable position estimates.

The major drawback of this approach is that the positions of the nodes cannot be mapped onto a global space.

4.2 Connectivity-based Algorithms

4.2.1 Localization from Mere Connectivity

MDS-MAP [26] is a centralized localization scheme for communication networks, which uses connectivity information to estimate the locations of the nodes. This method can also use additional information such as the estimated distances between neighbors or known positions of the beacon nodes if available, for improving the accuracy of the location estimates. The MDS-MAP algorithm is based on multidimensional scaling which is a data analysis technique. The algorithm is robust to measurement error when nodes are placed uniformly throughout the network. This algorithm can estimate the locations of the nodes in global space when the network has at least three refernce nodes or in local space when there are no reference nodes in the network.

MDS solves the problem of estimating the positions of the nodes such that the connectivity information between the nodes is maintained. They are many types of MDS techniques, which are based on whether the similarity data derived from the points in a multidimensional space is qualitative or quantitative. Classical MDS is used in this localization that maintains a single dimensional similarity matrix derived from the points in the multidimensional space. In classical MDS the similarity data is quantitative and the proximities of the objects are treated as distances in a Euclidean space. The goal of metric MDS is to find a configuration of points in a multidimensional space such that the inter-point distances are related to the provided proximities by some transformation. If the proximity information is measured without error in the Euclidean space, the MDS will exactly create the configuration of points. In classical MDS the proximity information is transformed into Euclidean distance using the relation:

$$d_{ij} = a + bp_{ij} \tag{6}$$

The distances are computed so that they are as close to proximities as possible. The distances for which the differences between them and the actual distances is minimum are considered to be close to the proximity information. MDS-MAP can be used under two different scenarios: when only connectivity information is available and when distances between the nodes are available along with the proximity information. The network is represented as a connected in directed graph with vertices V and edges E. Vertices correspond to the nodes and edges correspond to the connectivity information between the nodes. The position of the nodes is estimated by inserting the nodes in the 2 dimensional or 3 dimensional space with the connectivity information between the nodes maintained. This relative map can be transformed into a absolute map if sufficient number of reference nodes are available. MDS-Map consists of three steps.

Step 1: The shortest path between all pairs of nodes computed and these distances are used to construct the distance matrix for MDS. In the connectivity-only cases each node knows the identities of the nodes in its neighborhood but not distances to them. In known-distance cases each node knows its distances to its neighbors. The distance information is modeled as the true distance modeled by the gaussian noise. During this step the distances are assigned to the edges in the connected graph. If the distances between the nodes are known then the measured distances are assigned to the edges. When only connectivity information is available, then all edges are assigned a value of 1. Then a shortest-path algorithm such as Dijkstra or Floyd's algorithm can be applied to obtain the shortest path between all pairs of nodes.

Step 2: Classical MDS is applied to the distance matrix to obtain a relative map of all the nodes.

Step 3: The relative map obtained in the previous step can be transformed into an absolute map if there are at least 3 reference nodes for 2-D space or 4 reference nodes for 3-D space. The relative map is transformed through linear transformations, which include scaling, rotation and reflection. Here the goal is to minimize the sum of the squares of the errors between the true positions of the reference nodes and their transformed positions in the MDS map.

4.2.2 APIT: An Area-based Localization Scheme

An area-based range-free localization scheme, called APIT, is proposed in [10]. APIT is applied to the heterogeneous wireless sensor networks, where a small number of nodes, called reference nodes, are equipped with high-powered transmitters and GPS receivers. Each reference obtains its own coordinates through its GPS receiver and periodically sends out reference messages that include its own location information.

The entire network area is represented by a grid array as shown in Figure 12. The initial value of every grid entry is zero. Upon receiving the reference messages, the regular sensor node (without GPS receiver) performances Point-In-Triangulation (PIT) tests. More specifically, the sensor node chooses three reference nodes from which it receives the reference messages and test whether it is inside the triangle formed by connecting the three reference nodes. If the sensor is inside the triangle, the values of corresponding grid region are incremented by one. Otherwise, the grid area is decremented by one. The PIT test is repeated until all reference nodes combinations are exhausted or the required accuracy is achieved. As a result, the sensor node calculates the center of gravity of the maximum overlapping area (i.e., the area with the highest value) as the estimation of its location.

APIT is a simple approach without the need of accurate range informa-

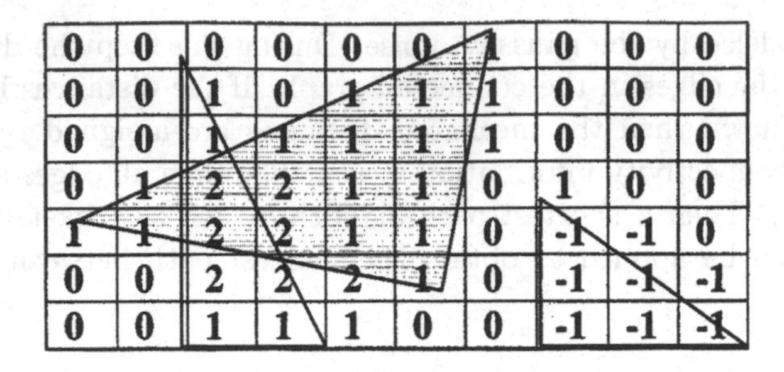

Figure 12: The SCAN approach[10].

tion. The localization in APIT is not highly accurate, but its accuracy is may prove sufficient for many sensor network applications.

4.2.3 GPS-less localization Algorithm

This algorithm [20] is a connectivity based localization method in unconstrained outdoor environments for very small low-cost devices. The algorithm assumes an idealized radio model with the following assumptions: Perfect spherical radio propagation and the same transmission range for all the radios. Also, it asuumes a fixed number of nodes in the network situated at known positions (x_1, y_1) to (x_n, y_n) and are uniformly distributed with overlapping regions. These nodes serve as reference points. They transmit periodic reference signals containing their respective positions. Each mobile node listens for a fixed period of time t and collects all the reference signals it receives from various reference points. Using this information, each mobile computes a connectivity metric CM_i (equation 7 for each reference node i. $Nrecv(i,t)$ is number of reference signals from reference node i that have been received in time t and $Nsent(i,t)$ is the number of references sent by reference node i in time t).

$$CM_i = \frac{Nrecv(i,t)}{Nsent(i,t)} \times 100 \tag{7}$$

The estimated location of a node is the centroid of the set of reference points for which the connectivity metric exceed a specified threshold. To improve accuracy, the number of reference points used to cover the networking grid can be increased, thereby reducing the separation distance and increasing the range overlap between reference points. The separation distance between adjacent reference points for different size grids is shown in Figure 13.

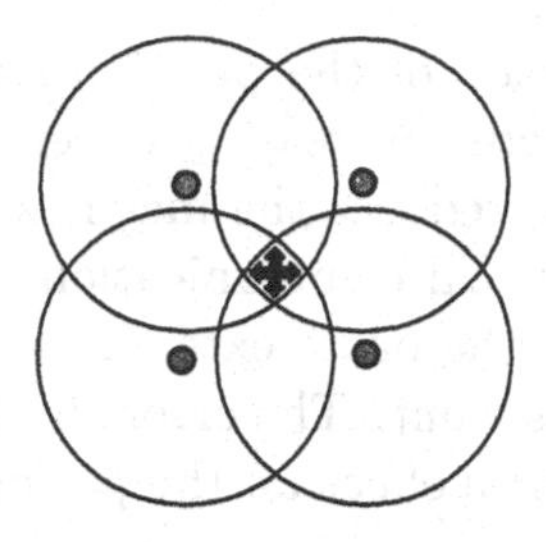
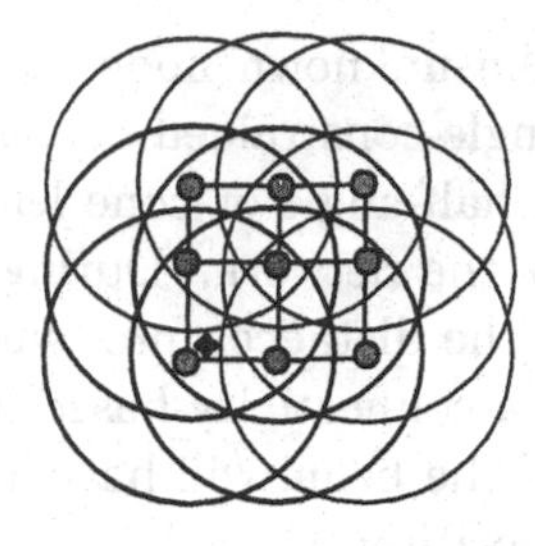

Figure 13: Granularity of Localization Regions vs. Range Overlap[20].

This proposed algorithm cannot be used indoors because of reflection and occlusion. This algorithm is simple, RF based and the accuracy depends on the number of reference points available. It requires no coordination among reference points or sensor nodes, so it is scalable to large, distributed networks. Some of the drawbacks of this approach are the reference signals should be synchronized so as to avoid collisions, high power consumption at the reference nodes and they should be placed uniformly throughout the network. This system is not robust in the event of the failure of the reference nodes, so the network must contain extra reference.

4.2.4 Convex Position Estimation in wireless sensor networks

In [14] the estimation of positions of the unknown nodes in an ad hoc network based on connectivity-induced constraints is described. The position information of the unknown node here is obtained by solving the connection imposed proximity constraints obtained from its communication links to the known nodes. The area of the bounding rectangle decreases as constraints become tighter. The feasible solutions to the position estimation problem are obtained using convex optimization. The problem of translating a sensor network position estimation problem to a linear or semi definite program is proposed to solve it efficiently. This is based on connectivity and pair wise angles between the nodes. This algorithm can be used to obtain a solution of the problem for large networks. The radio-based communications impose radial constraints that can be fixed radius constraint or variable radius constraint. The use of variable radius constraint improves the performance of the positioning algorithm over a fixed radius constraint. Placing the reference nodes on the perimeter of the network can reduce the errors in the position estimates of the unknown nodes. Though the computational overhead is more for placing a rectangular bound around the possible position estimates of the unknown node, accuracy of the position estimate is improved. For angle-constrained connections, decreasing uncertainty through a reduction in half-angle or cone length reduces the mean position estimation error over the network. During optical communication, the angle can be known but the distance upto, which the beam extends, would be necessary to obtain a connectivity-based constraint. The errors in the estimation the distance of the beam will have adverse effect on the performance of the positioning algorithm.

The performance of the algorithm increases with the increase in the connectivity but it also increases the amount of communication overhead, which is incurred from the transmissions of the connectivity information to the central computer. Further, in this localization scheme the position estimation problem is formulated as a linear or semidefinite program, which is based on connectivity information and pair wise angles between the nodes. For radially constrained connections, using a variable radius constraint improves performance over a global fixed radius. The nodes with the known positions must be placed at the corners of the network for the positions to be estimated with less error. This localization scheme incurs a great amount of computational overhead to estimate the positions of the unknown nodes. For angle-constrained connections, the decrease in uncertainty of the cone half-angle or cone length reduces the mean position estimation error over the network. For the angular constraints the distance information is required neglecting which would adversely impact the performance. Finally the performance of the localization scheme with increase in node connectivity.

5 Comparing Ad Hoc Localization Algorithms

In this section, we provide a summary of the algorithms discussed in this chapter. Table 1 provides a comparison of the localization algorithms discussed in this chapter based on several design goals and performance characteristics such as limitations, infrastructure support required, localization methodology (i.e., range-based vs. connectivity-based), distributed vs. centralized operation, coordinate system, accuracy, and cost.

Range-based versus Connectivity-based

A localization algorithm for ad hoc networks can employ either a range based (distance or direction estimation) approach or connectivity-based approach to estimate the position of unknown nodes. Connectivity-based approaches are sometimes called range-free approaches. In range-based localization methods, the range or distance estimates of the unknown node from at least three or more reference nodes are used to compute the node's position. After estimating distances to the reference nodes, the position of the unknown node is computed using trilateration or multilateration or triangulation. Range-free or connectivity-based localization algorithms compute the positions of the unmkown node using only connectivity or proximity information obtained from nearby reference nodes.

Distributed or Centralized

Localization algorithms can work either in a centralized or distributed manner. In centralized approach a node or group of nodes in the network generally known as a central or server node is responsible for estimating the positions of the nodes. A server node is responsible for computing the positions of the unknown nodes from the information it receives from unknown nodes or reference nodes. A server node may also be responsible for periodically computing the positions of the unknown nodes or other important nodes (e.g., nodes which provide a useful service) and broadcasting the information throughout the network. In this approach each node has to send the information obtained from the reference node to a server node, which results in more communication overhead. Also the unknown nodes near the server node may be responsible for continuously propagating information to and from the server node. Hence, a centralized approach may results in uneven load distribution and power consumption where the nodes near the server node suffer more power loss when compared to the other nodes.

In distributed approach, each node in the network is responsible for computing its own position. Here each node uses the information of the reference nodes to estimate their positions. As the node need not transmit its information to the central node the amount of communication overhead with this approach is less than that of centralized approach. As the node is responsible for finding its own position, this approach results in even power distribution between the nodes. Also with this approach the changes in the network are reflected soon when compared to the centralized approach. A distributed approach is also robust to central nodes failures and it enables nodes to estimate their positions as soon as the network topology changes with minimal cost.

Coordinate System

A localization algorithm can estimate the positions of the nodes in either global coordinate system or relative coordinate system. In global coordinate system, the positions of all the nodes will be based on some global reference. Some localization algorithms require the use of reference nodes with known positions (either through GPS or some other means) to enable them to compute the global positions of the unknown nodes. Here, The position information of the reference nodes is used to estimate the position of the unknown node in the global coordinate system. For localization algorithms, which compute the positions of the nodes in global coordinate system, the network is assumed to have at least three reference nodes for 2-dimensional position estimate or at least four reference nodes for 3-dimensional position estimate. Global position estimation algorithms are used for mapping objects on a geographical map, target tracking and monitoring and other location aware applications.

In relative coordinate system, the positions of the nodes are based on some local reference. Some localization algorithms build a local coordinate system by considering a node in the network as a center and defining the direction of the coordinate system. Such algorithms do not require GPS and are considered to be GPS-less. In these algorithms the positions of the nodes are computed relative to the local coordinate system. Relative position estimation algorithms are used to accomplish networking tasks like routing, collaborative signal processing and coverage management.

Infrastructure Support

Localization systems for ad hoc networks can either be infrastructure based or Infrastructure less. Infrastructure based localization systems require some form of network set up for the node positions to be estimated. Infrastructure less localization systems does not require any preplanning or network support for the estimation of nodes positions.

Position accuracy

The positions estimated by the localization systems must be accurate and consistent irrespective of the changing topologies and other environmental factors. The required level of accuracy and consistency of the position estimates generally depend on the application for which the localization system is being used.

Cost

The costs involved in localization systems include the costs for implementing the system on different nodes and infrastructure setup costs.

Table 1: Characteristics of Localization Algorithms

Location System	Positioning Approach	Assumptions	R / C†	L / G‡	Limitations
Self Positioning Algorithm	(1) ToA (2)Triangulation	(1) high degree of node connectivity; (2)size of reference group must be chosen such the network center and direction are stable	R	G	(1) only relative positions can be obatined; (2) positioning accuracy depends heavily on ranging accuracy; (3) can be used only for getting 2 dimensional position estimates.
Ad Hoc Localization Algorithm	(1) TDoA (2) Atomic, Iterative & Collaborative multilateration	(1)Requires high degree of node connectivity; (2) a large number of reference nodes are required; (3)reference node placement is critical.	R	G	(1) can only be used indoors; (2) positioning accuracy depends on ranging accuracy; (3) iterative and collaborative multilateration may cause high error.
Localized location algorithm	(1) TDoA (2) Atomic, Iterative & Collaborative multilateration	(1)high degree of node connectivity; reference node placement can be chosen	R	G	(1) can only be used indoors; (2) positioning accuracy depends on ranging accuracy; (3) iterative and collaborative multilateration may cause high error.
Ad Hoc Position System					
1. DV-Hop	(1) HopCount & average hopsize are used. (2)Lateration	(1)The network must have atleast three beacon nodes.	R	G	accuracy of the position estimates is independent of measurement errors but performs well only for isotropic networks.
2. DV-Distance	(1) Signal strength (2) Lateration		R	G	(1)Accuracy of the position estimate depends on the accuracy of the range measurements between the nodes. (2)Performs well only for isotropic networks.
3. Euclidean	(1) Signal strength (2) Triangulation		R	G	(1)better accuracy, but incurs higher computation and communication overhead. (2)positioning accuracy depends on the accuracy of the range measurements.
Directionality based Estimation	(1) ToA & signal strength (2) Triangulation	(1)At least three reference nodes wired to a central controller placed at the corners of the network; (2) reference nodes are equipped with directional antennas which cover the entire network.	R	G	(1)reference nodes require rotational beams and must cover the entire network; (2) requires infrastructure support (e.g., reference nodes and central controller).
GPS-less localization	(1) Connectivity based	(1)Assumes idealized radio model where all the nodes have the same transmission range. (2)Requires network covering reference nodes	C	G or L	(1)synchronization is required to prevent the collisions between the beacons. (2)accuracy of the position estimates depends on the seperation between the reference nodes and their ranges.
Convex position estimation	(1) Connectivity based	(1)Position estimate is obtained by solving semidefinite or linear programs obtained by using radial or angulkar constraint models.	C	G	(1)still under evaluation, but can potentially yield highly accurate results
Robust Positioning Algorithm	(1) ToA & signal strength (2) Lateration	(1)atleast three reference nodes with known positions should be placed on the perimeter of the network.	R	G	positioning accuracy depends on ranging accuracy, although improved by refinement phase.
Localization using mere connectivity	(1) Connectivity Based	(1)Assumes idealized radio model where all the nodes have the same transmission range. (2) Requires network covering reference nodes	C or R	G or L	(1)Positioning accuracy depends on ranging accuracy; (2)Errors will be propapagated and used throughout network.

† Range-based or Connectivity-based
‡ Local or Global coordinate system

References

[1] C.C. Han A. Savvides and M.B. Srivastava, Dynamic fine-grained localization in ad-hoc networks of sensors, *Proceedings of the 7th annual international conference on Mobile computing and networking* (Jul. 2001) pp. 166-179.

[2] I.F. Akyildiz, W. Su, Y. Sankarasubramaniam, and E. Cayirci, A survey on sensor networks, *IEEE Communications Magazine* Vol. 40 No. 8 (Aug. 2002) pp. 102-116.

[3] P. Bahl and V. N. Padmanabhan, Radar: An in-building rf-based user location and tracking system, *IEEE Infocom* Vol. 2 (Tel-Aviv, Israel, Mar. 2000) pp. 75-784.

[4] S. Basagni, I. Chlamtac, V. R. Syrotiuk, and B. A. Woodward, A distance routing effect algorithm for mobility dream, *Proceedings of the Fourth Annual ACM/IEEE International Conference on Mobile Computing and Networking* (Dallas, TX, Oct. 1998) pp. 76-84.

[5] S. Corson and J. Macker, Mobile Ad hoc Networking (MANET): Routing Protocol Performance Issues and Evaluation Considerations, *Internet-Draft*, (Oct. 1998).

[6] Scott M. Corson, J. P. Macker, and G. H. Cirincione, Internet-based mobile ad hoc networking, *IEEE Internet Computing* Vol. 3 No. 4 (Jul./Aug. 1999) pp. 63-70.

[7] A. Alwan et al., Adaptive Mobile Multimedia Networks, *IEEE Personal Communications Magazine* Vol. 3 No. 2 (Apr. 1996) pp. 34-49.

[8] Dieter Fox, Wolfram Burgard, Frank Dellaert, and Sebastian Thrun, Monte carlo localization: Efficient position estimation for mobile robots, *Proceedings of the Sixteenth National Conference on Artificial Intelligence (AAAI'99)*(Jul. 1999) pp. 343-349.

[9] Z. J. Haas, Panel report on ad-hoc networks, *Mobile Computing and Communications Review* Vol. 2 (1988) pp. 1-15.

[10] T. He, C. Huang, B. Blum, J. Stankovic, and T. Abdelzaher, Range-free localization schemes in large scale sensor networks, *Proceedings of the 9th Annual International Conference on Mobile Computing and Networking* (San Diego, CA, 2003) pp. 81-95.

[11] Jeffrey Hightower, Roy Want, and Gaetano Borriello, Spoton: An indoor 3d location sensing technology based on RF signal strength, *UW-CSE 00-02-02*, (Feb. 2000).

[12] B. Hofmann-Wellenhof, H. Lichtenegger, and J. Collins, *Global Positioning System: Theory and Practice*, (Springer-Verlag, 4th edition, 1997).

[13] Y.B. Ko and N.H. Vaidya, Location-Aided Routing (LAR) in Mobile Ad-Hoc Networks, *Proceedings of the 4th Annual International Conference on Mobile Computing and Networking* (Oct. 1998) pp. 66-75.

[14] Kristofer S. J. Pister Lance Doherty and Laurent El Ghaoui, Convex position estimation in wireless sensor networks, *IEEE Infocom* (Anchorage, Alaska, Apr. 2001) pp. 1655-1663.

[15] P. Lettieri and M. Srivastava, Advances in Wireless Terminals, *IEEE Personal Communications* (Feb. 1999) pp. 6-19.

[16] S. Meguerdichian, F. Koushanfar, M. Potkonjak, and M. B. Srivastava, Coverage problems in wireless ad-hoc sensor networks, *IEEE Infocom* (Anchorage, Alaska, Apr. 2001) pp.1380-1387.

[17] S. Meguerdichian, S. Slijepcevic, V. Karayan, and M. Potkonjak, Localized algorithms in wireless ad-hoc networks: Location discovery and sensor exposure, *The Proceedings of the ACM Symposium on Mobile Ad Hoc Networking & Computing (MOBIHOC)* Vol. 10 (Long Beach, CA, Oct. 2001) pp. 106-116.

[18] Asis Nasipuri and Kai Li, A directionality based location discovery scheme for wireless sensor networks, *ACM International Workshop on Wireless Sensor Networks and Applications (WSNA'02)* (Sep. 2002) pp. 105-111.

[19] Dragos Niculescu and B. R. Badrinath, Ad hoc positioning system (aps), *IEEE Globecom* (San Antonio, Nov. 2001).

[20] John Heidemann Nirupama Bulusu and Deborah Estrin, GPS-less low cost outdoor localization for very small devices, *IEEE Personal Communications Magazine* Vol. 7 No. 5 (Oct. 2000) pp. 24-28.

[21] Anit Chakraborty Nissanka B. Priyantha and Hari Balakrishnan, The cricket location-support system, *Proceedings of the 6th Annual Interna-*

tional Conference on Mobile Computing and Networking (Boston, MA, Aug. 2000) pp. 32-43.

[22] C. E. Perkins, Mobile Ad Hoc Networking Terminology, *Internet Draft*, (Nov. 1998).

[23] R. Ramanathan and M. Steenstrup, Hierarchically-Organized, Multihop Mobile Wireless Networks for Quality-of-Service Support, *Mobile Networks and Applications* Vol. 3 (1998) pp. 101-119.

[24] M. Hamdi S. Capkun and J.P. Hubaux, GPS-free positioning in mobile ad-hoc networks, *Hawaii International Conference On System Sciences* Vol. 9 (Jan. 2001) pp. 9008.

[25] Chris Savarese and Jan Rabaey, Robust positioning algorithms for distributed ad-hoc wireless sensor networks, *USENIX Annual Technical Conference* (Monterey, CA, Jun. 2002).

[26] Yi Shang, Wheeler Ruml, and Ying Zhang, Location from mere connectivity, *The Proceedings of the 4th ACM international symposium on Mobile ad hoc networking and computing* (Annapolis, Maryland, USA, Jun. 2003) pp. 201-212.

[27] H. Takagi and L. Kleinrock, Optimal Transmission Ranges for Randomly Distributed Packet Radio Terminals, *IEEE Transactions on Communications* Vol. 32 No. 3 (Mar. 1984) pp. 246-257.

[28] R. Want, A. Hopper, V. Falcao, and J. Gibbons, The active badge location system, *ACM Transactions on Information Systems* Vol. 10 No. 1 (Jan. 1992) pp. 91-102.

[29] A. Ward, A. Jones, and A. Hopper, A new location technique for the active office, *IEEE Personal Communications* Vol. 4 No. 5 (Oct. 1997) pp. 42-47.

Energy-Efficient Broadcasting in Wireless Mobile Ad Hoc Networks

François Ingelrest and David Simplot-Ryl
IRCICA/LIFL, INRIA futurs
Bât. M3, University of Lille I, France
E-mail: `ingelres@lifl.fr`, `simplot@lifl.fr`

Ivan Stojmenović
Computer Science, SITE
University of Ottawa, Ontario K1N 6N5, Canada
E-mail: `ivan@site.uottawa.ca`

Contents

1 Introduction

Mobile networks have emerged as an important component of networking technology and a very challenging research field [40]. After the initial failure of the *UMTS* concept that would have magnified the old *GSM* network, the future of mobile network is now largely opened to innovative concepts. It seems now clear that the mobile network of tomorrow will be based on a multiple interface notion, with the Internet as the natural unifier. Among all these new concepts, mobile ad hoc networks have a prominent place [21].

1.1 Description of Wireless Ad Hoc Networks

Wireless ad hoc networks are formed by a set of hosts that operate in a self-organized and decentralized manner, forming a dynamic autonomous network through a fully mobile infrastructure. Communications take place over a wireless channel, where each host has the ability to communicate directly with any other one in its physical neighborhood, which is determined by its range. More accurately, when an host emits a message, every other one in its physical neighborhood will receive it. Another important characteristic of wireless transmissions is that when two nodes emit a message simultaneously, their common neighbors will experience a collision, and will not receive the message correctly.

Hosts in ad hoc networks can be fixed or mobile, and every collection of mobile hosts with appropriate interfaces may form a temporary network. No fixed infrastructure is needed, thus every host has to discover its environment when the network is formed. Ad hoc networks have multiple applications in areas where wired infrastructure may be unavailable, such as battlefields or rescue areas. It might also be infeasible to construct sufficient fixed access points due to cost considerations. As an example, constructing a fixed network infrastructure for the duration of an outdoor assembly is not realistic.

Here is a summary of the main characteristics of ad hoc networks:

Dynamic topology Hosts are mobile and can be connected dynamically in any arbitrary manner. Links of the network vary and are based on the proximity of one host to another one,

Autonomous No centralized administration entity is required to manage the operation of the different mobile hosts,

Bandwidth constrained Wireless links have a significantly lower capacity than the wired ones; they are affected by several error sources that result in degradation of the received signal,

Energy constrained Mobile hosts rely on battery power, which is a scarce resource; the most important system design criterion for optimization may be energy conservation,

Limited security Mobility implies higher security risks than static operations because portable devices may be stolen or their traffic may cross insecure wireless links.

Figure 1 shows an example of such an ad hoc network. Hosts can be any mobile devices like laptops, cell phones, *PDA*... They communicate via an air interface and detect themselves in real time. New hosts can suddenly appear and old ones can disappear at any time. The topology of this network is very fragile, it can change at any moment and disconnections are frequent due to the mobility or activity status changes.

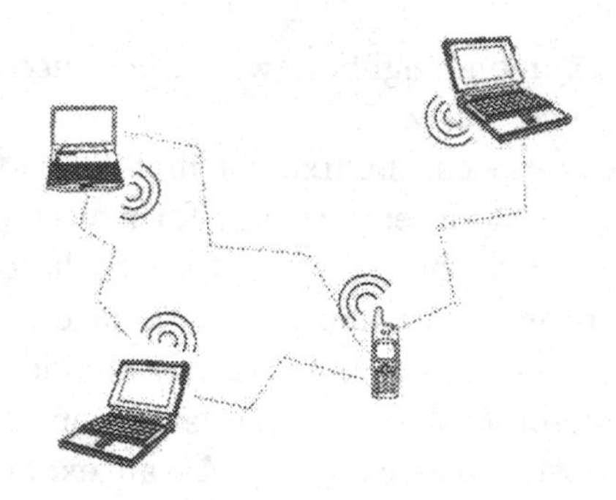

Figure 1: A self-organized network.

1.2 The Broadcasting Task

As already discussed, two hosts in an ad hoc network can communicate directly only if they are in the physical neighborhood of each other, which is determined by a communication range. Due to propagation path loss, the transmission radii are limited, thus communications must take place via a multihop routing.

To establish a connection between two hosts not directly connected, messages must be routed by intermediate hosts, as shown on Figure 2. Hosts A and B are not able to communicate directly, every communication between them must be relayed by an intermediate host C. In this example, when A wants to send a message to B, it is simple for C to relay the message, because it is a neighbor of both of them. In larger networks, with hundreds of hosts, it is much more difficult for an host to find a route to another one, because of the lack of fixed infrastructure.

The traditional method used to discover routes is the dissemination of a

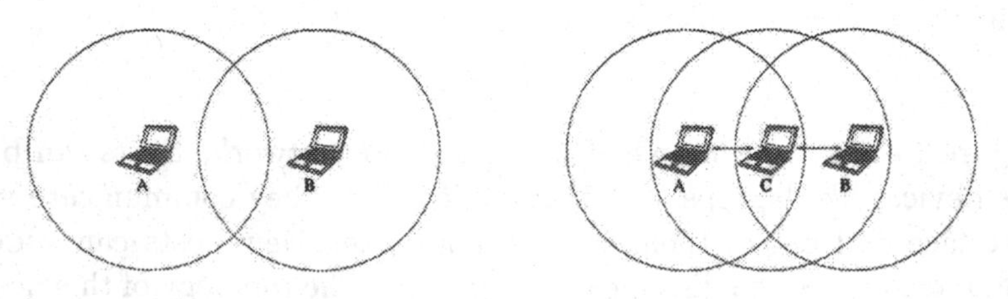

Figure 2: Multihop routing.

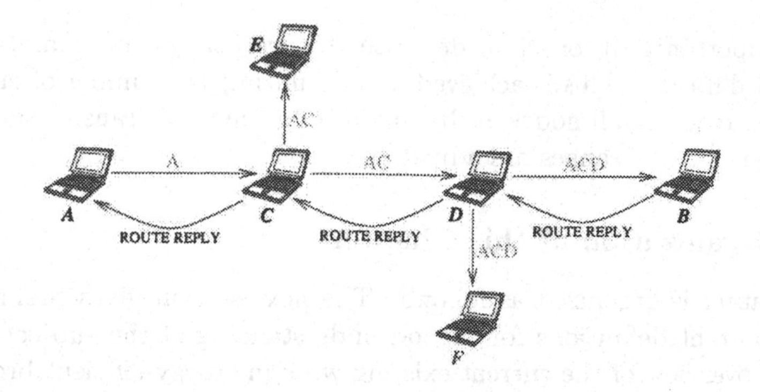

Figure 3: A route request from host A to host B.

message called a *route request*. When an host A wants to find a route to an host B, it sends a route discovery message to its neighborhood. Each of its neighbors adds its name to the message and re-emits it. Every other node in the network does so and finally, the searched host B receives the message with the chain of hosts followed from the source host A. The final step of this process is the emission of a message called a *route reply* by host B to inform A that a route has been found. This procedure is basis for an *IETF* standardized protocol named *DSR* (*Dynamic Source Routing*) [26].

DSR is illustrated by Figure 3. A wants to find a route to B, so it sends a route request message to its neighborhood with its *id* as the initial chain of hosts. C receives this message, writes its name in the chain and re-emits it. A gets the packet, but ignores it (it already knows it and furthermore, it is the one that first emitted it). D does the same as C (as well as E and F) and finally B receives the request. It then just have to answer to A by emitting a route reply. This last message is addressed directly to A, because B knows the route to it. This one is simply written in the route request message, formed by the chain of hosts that relayed the message. In our example, the chain received by B was ACD so it sends the route reply with DCA as the route to follow.

Route discovery is performed by a broadcasting task, while the route reply is an unicast routing operation. Traditional broadcasting used in *DSR* is called a *blind flooding* because every node in the network retransmits once the message, upon receiving the first copy of it, and will consequently ignore further copies of the same message.

The broadcasting task is therefore a fundamental mechanism in route discovery, so the design of an efficient broadcast in ad hoc networks is of prime importance, in order to decrease the overhead, while maintaining a maximal diffusion. This is achieved by minimizing the number of emissions while still reaching all nodes, or by minimizing the total transmission power if the transmission ranges are adjustable.

1.3 Organization of this Chapter

This chapter is organized as follows. The next section gives preliminaries and important definitions for a good understanding of the subject. Sec. 3 gives an overview of the current existing work in energy-efficient broadcasting protocols without range adjustment, where only the number of needed emissions is minimized. Sec. 4 presents energy efficient broadcasting protocols with transmission range adjustments, while Sec. 5 considers broadcasting protocols that use smart antennas, which are able to make directional emissions. Finally, conclusion and direction for future work are given.

2 Preliminaries

2.1 Communication Model

A wireless network can be represented by a graph $G = (V, E)$ where V is the set of nodes (hosts) and $E \subseteq V^2$ the edge set which gives the available communications: (u, v) belongs to E means that u can send messages to v. In fact, elements of E depend on the positions and the communication ranges of the nodes. Let us assume that the maximum communicating range, denoted by R, is the same for all vertices and that $d(u, v)$ is the Euclidean distance between nodes u and v. The set E is then defined as follows:

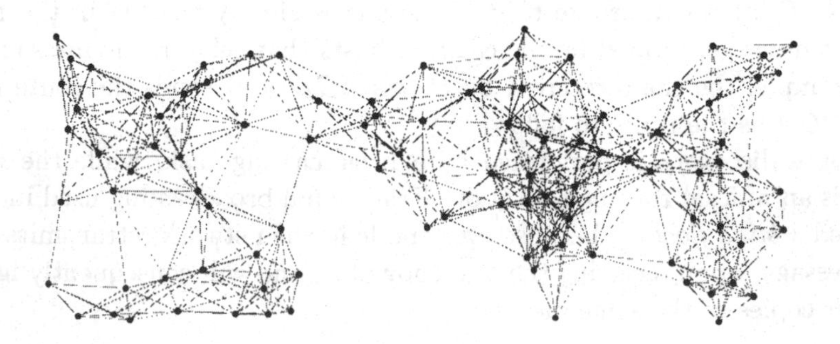

Figure 4: An unit graph with a density of 15 and 100 nodes.

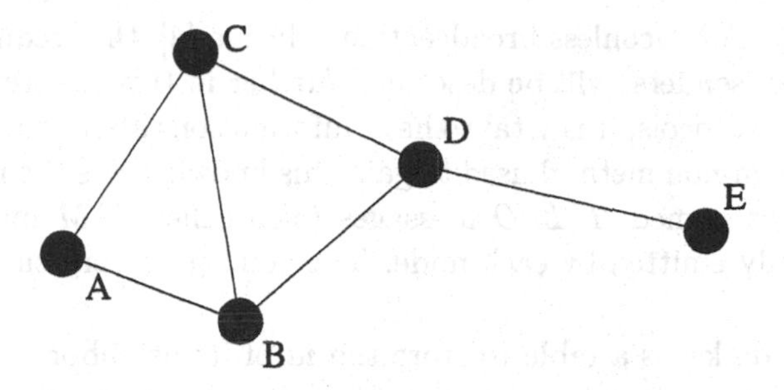

Figure 5: The distance in hops.

$$E = \{(u, v) \in V^2 \mid d(u, v) \leq R\}.$$

So defined graph is known as the *unit graph*, with R as its transmission radius. An example of such a graph is given in Figure 4. Every node $u \in V$ must be assigned an unique identifier (id). We also define the neighborhood set $N(u)$ of the vertex u as

$$N(u) = \{v \mid (u, v) \in E\}.$$

The neighborhood function is naturally extended to set of nodes: for a given subset A of V, we have $N(A) = \cup_{u \in A} N(u)$. The degree of a given node u is the number of nodes in $N(u)$. The density of the graph is the average degree for each node. We also denote by $n = |V|$ the number of nodes in the network.

We measure the distance between two nodes in term of *number of hops*. It is simply the minimum number of links a message has to cross from a source node to reach its destination. In Figure 5, the distance between A and B is one hop, while the distance between B and E is two hops. The one-hop neighborhood of E is $\{D\}$, while its two-hops neighborhood is $\{B, C, D\}$.

2.2 Assumptions

There are a variety of different assumptions that can be made about the operation of ad hoc networks and the amount of knowledge available at each node. The simplest assumption is that there are no control messages, and each node reacts to incoming broadcast message without being aware of its neighborhood. The blind flooding solution works with such a minimal overhead, but is suboptimal due to excessive collision impact from redundant transmissions. A beaconless broadcasting solution [4], that requires position information of senders, will be described further in this chapter.

For most protocols, it is vital to have informations about the neighboring nodes. The common method used to gain this knowledge is the use of special short messages named *HELLO* messages (also called *IAM* messages) that are periodically emitted by each node. The concept is very simple:

- each node keeps a table to store the *id* of its neighbors,
- each node emits periodically a *HELLO* message with its *id*; the frequency used for these messages is generally of 1 second,
- when a node u receives a *HELLO* message from a node v, it adds v to its neighborhood table, or updates the entry if it already exists. v and u are necessarily neighbors, since only physical neighbors of the emitter can receive the message,
- old entries are periodically deleted from the table. When an entry is too old (something like 5 seconds old), it is obvious that the corresponding host is no more in the neighborhood of the node, so the entry is simply removed from the table. The host could have simply moved or be in lack of power.

Many of the algorithms described further need the distance between a node and its neighbors to be applied. The easiest way to compute distances is to know the positions of the nodes, for example by using a location system like the *GPS* (*Global Positioning System*) [27]. Some other mechanisms for positioning can be used, such as *TODA* (*Time Difference Of Arrival*) [36], *AOA* (*Angle Of Arrival*) [34], reception power measurement [2] or phase difference [3]. A survey on location systems is available in [18]. If this kind of system is available, each node includes its position in its *HELLO* messages and thus needed distances can be computed.

Although positioning information may require additional hardware, the latest technological advances are remarkable: a very cheap 7mm x 7mm x 2mm *GPS* receiver now exists. Cartigny and Simplot proposed a software "distance" function denoted by μ [9]. It is defined by:

$$\mu(u,v) = \frac{|N(v) \setminus N(u)|}{|N(u)|}. \quad (1)$$

The numerator is the subtraction between the two sets $N(v)$ and $N(u)$. It is based on the fact that the number of common neighbors of two nodes depends on the distance between these nodes. The higher the distance is, the fewer common neighbors are and the higher the value of μ will be. They also showed that this (non-symmetric) estimation of Euclidean distance is sufficient for broadcasting.

To be able to compute the value of $\mu(u, v)$, a node u needs to know its neighbors, but it also needs to know the neighbors of the node v. To spread this information, *HELLO* messages are used again: when a node emits a message of this type, it includes the list of its own neighbors. Nodes have then to store the neighbors of their neighbors in their table. Typically, this is a two-hops information.

2.3 Energy Model

When emitting a message, a node spends a part of its energy and there are a few energy models used to compute this consumption. In the most commonly used one, the measurement of the energy consumption when transmitting a unit message depends on the range of the emitter u:

$$e(u) = r(u)^{\alpha},$$

where α is a real constant greater or equal than 2 and $r(u)$ is the range of the transmitting node. This model has been used in a few papers, *e.g.* [47].

In reality, however, the model has a constant to be added in order to take into account the overhead due to signal processing, minimum energy needed for successful reception, *MAC* control messages and also possible overhead due to retransmission probability as suggested by Feeney [17]. The general energy consumption formula is:

$$e(u) = \begin{cases} r(u)^{\alpha} + c & \text{if } r(u) \neq 0, \\ 0 & \text{otherwise.} \end{cases} \tag{2}$$

For instance, Rodoplu and Meng considered the model with $e(u) = r(u)^4 + 10^8$ [39]. This last one, also used by Cartigny *et al.* [11], is more realistic as illustrated by Figure 6. With parameters $\alpha = 2$ and $c = 0$, it is clear that the case (b) costs the same energy as (a) by using the Pythagoras theorem. By induction, all illustrated configurations are supposed to have the same energy consumption and can be arbitrary extended.

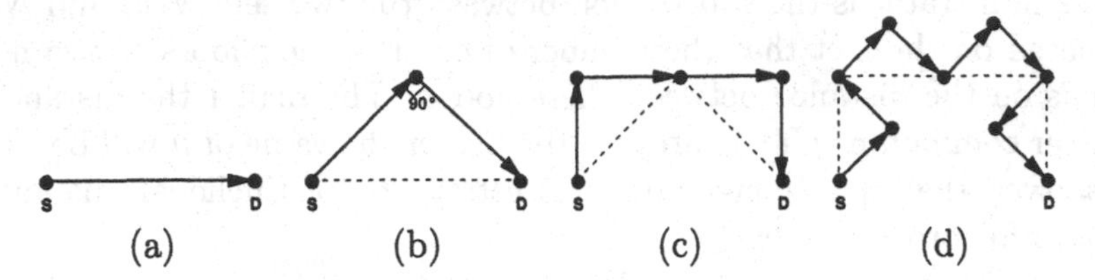

Figure 6: Configurations with same energy consumption for $\alpha = 2$ and $c = 0$.

Sometimes, with the required hardware, it is possible to consider directional emissions, that is a node can choose the angle of its emission. In this case, it is necessary to use a particular version of the formula:

$$e(\theta, r) = \begin{cases} \frac{\theta}{2\pi}(r^\alpha + C_1) + C_2 & \text{if } r \neq 0, \\ 0 & \text{otherwise.} \end{cases} \tag{3}$$

with θ being the chosen angle for the beam, C_1 a constant representing the overhead for correctly positioning the antenna, and C_2 a constant representing the cost of preparing the message for one or more directional transmissions.

3 Energy Efficient Broadcasting Without Range Adjustment

The first obvious idea to reduce the energy consumption is to reduce the number of needed communications to achieve the broadcast. Indeed, not all nodes have to relay the broadcast message to obtain a full coverage of the network, as demonstrated by Figure 7. If the node A sends a broadcast message, its neighbor B is the only needed relay to cover the network. Many protocols have been proposed to minimize the needed number of emissions, that can be grouped into the following families: clustering based, neighbor-elimination based, distance-based, probabilistic, coverage based, and forwarding neighbor based.

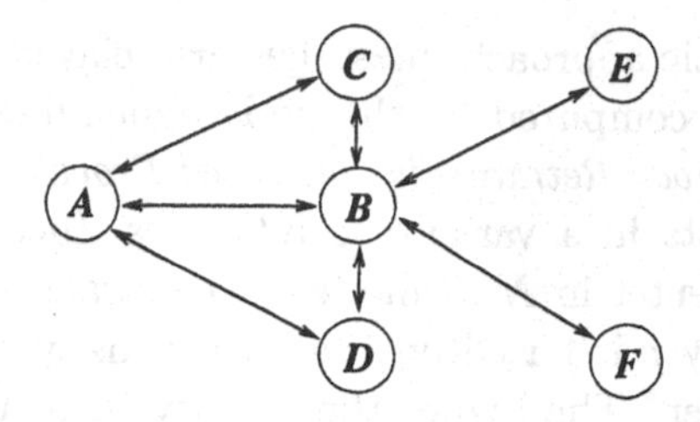

Figure 7: Example where not all nodes have to relay the message.

3.1 Clustering Based Broadcasting Protocols

In a clustering based solution, message is retransmitted once by all clusterheads and border nodes, and it is applicable in conjunction with other methods presented further. Since clustering process has chain effect (local changes may trigger global structure updates), we do not elaborate it further. Jaikaeo and Shen proposed one of variants of such clustering based method to compute a dominating set [25]. Dominating nodes are selected based on their *id*. Each node decides to belong to the set if it has a higher *id* than any of its neighbors, otherwise it associates itself to the nearest dominating set member with highest *id*. Dominating set nodes are then connected to create a backbone. The backbone can also be used for power conservation, with clusterhead nodes deciding which nodes shall go to sleep mode.

3.2 Neighbor Elimination Based Broadcasting Protocols

NES (*Neighbor elimination scheme*) has been independently proposed in [35] and [43]. In this source-dependent scheme, a node does not need to rebroadcast a message if all of its neighbors have been covered by previous transmissions. After each received copy of the same message, each node eliminates, from its rebroadcast list, neighbors that are assumed to have received correctly the same message. If the list becomes empty before the node decides to relay the message, the re-broadcasting is canceled.

3.3 Distance-based and Probabilistic Protocols

In distance-based protocol [33], each node that receives the message for the first time relays it only if the distance between it and the emitter is greater than a fixed threshold.

In the probabilistic approach, messages are relayed with a probability p that can be fixed or computed by the node depending on several parameters. *BRP* (*Border node Retransmission based Probabilistic*) belongs to this family [9]. It consists in a variant of *NES* described above where nodes decide randomly to enter in *NES* mode or to retransmit immediately. The computed probability p is based on the local density of each node, and on distance to the sender. The higher the density is, or the closer the sender is, the lower the probability of relaying will be.

3.4 Coverage and Connected Dominating Set Based Broadcasting

These protocols are based on the computing of a connected dominating subset $S \subset E$ as small as possible, which has to satisfy two characteristics:

- all nodes in the graph are either in S or a neighbor of a node in S, the subset is then called a dominating set,
- it has to be connected.

Several algorithms that compute these kind of sets have been proposed [1, 16]. The latter, called "Generalized rule", is an algorithm that can be applied locally by each node, without any message exchanged with neighboring nodes, solely using the knowledge of the neighborhood. The protocol can be described as follows (this simplified version of the protocol is given in [5]). First, each node checks if it has an intermediate state, that is every node that has at least two neighbors not directly connected is intermediate. Then each intermediate node A constructs a subgraph G of its neighbors with higher *ids*. If G is empty or disconnected then A is in the dominating set. If G is connected but there exists a neighbor of A which is not neighbor of any node from G then A is in the dominating set. Otherwise A is covered and is not in the dominating set. Non-intermediate nodes are never dominant. Dijkstra's shortest path scheme or a depth-first search (*DFS*) can be used to test the connectivity. The procedure apparently has localized maintenance. This procedure is generalized since it allows coverage by any number of neighbors. There exists special cases proposed earlier by Wu and Li [51], where the coverage was restricted to one or two neighbors only.

Figure 8, where a dominating set has been computed, illustrates this algorithm. Black nodes are dominants while white ones are non-dominants. In this example, node 0 is not dominant because all of its neighbors with

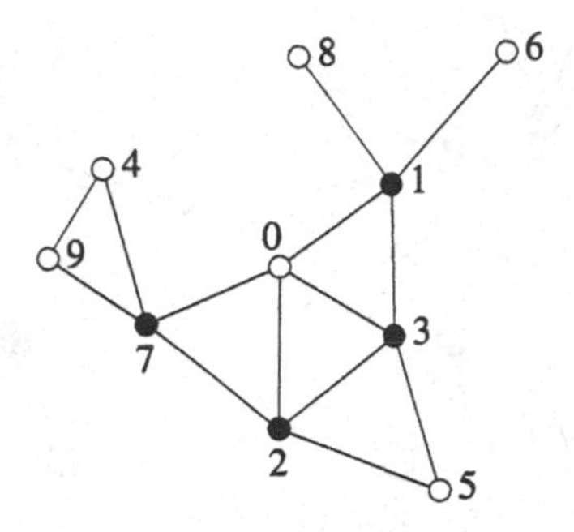

Figure 8: A dominating set computed with Generalized rule.

higher *id* form only one (connected) component, while node 2 is dominant because there exists two components in its neighborhood (*e.g.* $\{3,5\}$ and $\{7\}$) that are not connected. Node 5 is not dominant because it is not intermediate (all its neighbors are directly connected).

The broadcasting protocol with a dominating set is very easy:

- a dominant node that receives the message for the first time relays it, any further receptions are ignored,
- a non-dominant node that receives the message simply ignores it.

Wu and Dai proposed to construct a dominating set by clustering followed by generalized rule on clusterheads [50]. Two versions are proposed. Clusters are created using transmission ranges $\frac{r}{3}$ and $\frac{r}{4}$, while generalized rule on clusterheads is applied with transmission ranges r and $\frac{3r}{4}$, respectively. Significant reductions in the size of connected dominating sets, compared to generalized rule, are reported. However, these reduction come from using clustering operation that is not fully local (decisions made at one part of network have impact on other parts on network, both during construction and maintenance).

3.5 Forwarding Neighbors Based Broadcasting Protocols

Some broadcasting protocols, such as covering based ones, are source-independent, because relays are always the same regardless of the source. With source-dependent protocols, nodes that act as relays are not always the same.

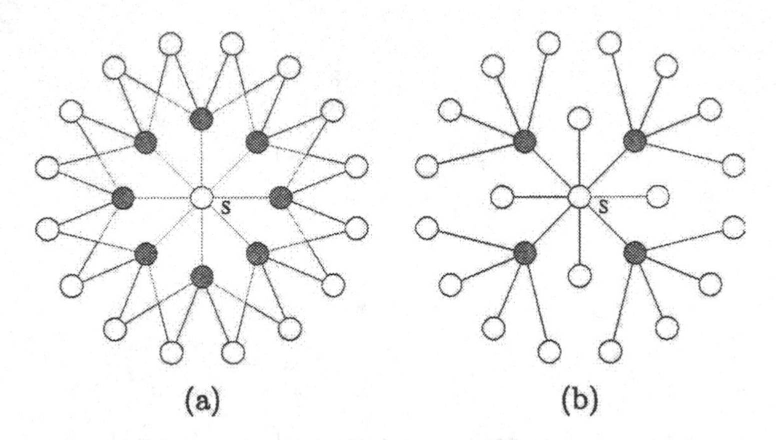

Figure 9: Application of MPR algorithm.

Depending on the protocol used, nodes can select themselves as relays when they receive a message, or they can select their relays within their neighborhood before the emission. Source-dependent protocols balance better the load of the network between the nodes. With dominating sets, relays are always the same, thus hosts that act as relays will quickly lose their energy when several broadcasts are launched. On the other hand, source-independent protocols allow non-dominating set nodes to go to sleep mode without affecting the network operation, thus prolonging the network life considerably.

The *MPR* (*Multipoint Relay*) protocol belongs to the family of forwarding neighbors based broadcasting protocols [37]. It uses a greedy algorithm to compute an optimal selection of neighbors to act as relays, in order to reach every two-hops neighbors. When a node selects some of its neighbors, it forwards its selection with the broadcast packet, thus increasing the traffic. Figure 9 shows an example of *MPR* relays, where S wants to broadcast a message, with black nodes being its relays. In the general case (a), each of its neighbors are relays, while in case (b), S has applied *MPR* to choose them. This protocol is very efficient in terms of energy savings, but unfortunately it is not very resistant to node failures due to the low redundancy of chosen relays. Cartigny *et al.* proposed *RRS* (*RNG Relay Subset*) to address this issue [7]. Moreover, this protocol offers the advantage to allow nodes to select themselves as relays, reducing the size of packets and therefore the number of collisions. *RRS* is based on *RNG* (*Relative Neighborhood*

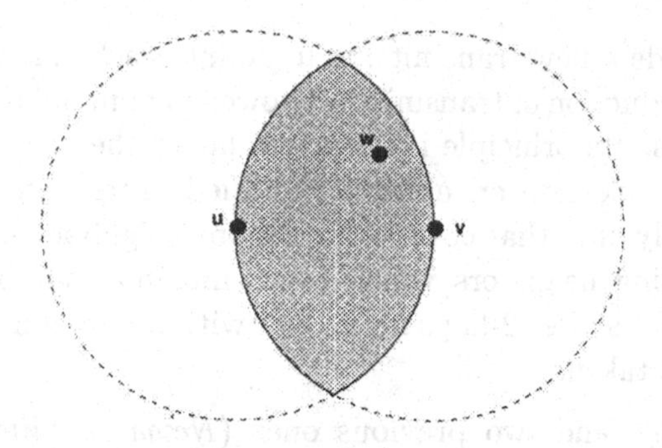

Figure 10: The edge (u, v) is not in RNG because of w.

Graph), which is a geometric concept proposed by Toussaint [45]. The relative neighborhood graph of G, denoted by $RNG(G) = (V, E_{rng})$ is defined by:

$$E_{rng} = \{(u,v) \in G \mid \nexists w \in N(u) \cap N(v), \\ d(u,w) < d(u,v) \wedge d(w,v) < d(u,v)\}$$

This condition is illustrated by Figure 10. The shaded area is the intersection of the two circles centered at u and v with radius $d(u,v)$. An edge (u,v) belongs to the *RNG* if there does not exist a node w in the gray area, which is not the case here, so (u,v) does not belong to the *RNG*. In other words, an edge belongs to the *RNG* if it is not the longest edge in any triangle. In Figure 10, if there exists a neighbor w in the shaded area, uv would have been the longest edge in triangle uvw and therefore would not belong to the *RNG*.

The protocol *RRS* is a variant of *NES* (see Sec.3.2) where nodes survey only a restricted part of their neighborhood: their non-covered *RNG*-neighborhood.

Recently, Kim and Suh proposed a forwarding set selection method and an efficient scheme for broadcasting with transmission power control [28]. The goal of this protocol is not to minimize the total transmission power but to reduce delays at *MAC* layer of forwarding set by avoiding collisions when scheduling transmissions of nodes in forwarding set. Indeed, the node not only decides its forwarding neighbors but makes as well the schedule for their transmissions (with similar iterative scheme, where each next node is either included in an existing transmission spot or a new spot is created for it if it collides with all already scheduled spots) to avoid collisions. An attempt to include a new transmitting node into each existing spot is made by considering reduction of transmission power to minimal necessary to cover designated nodes. Its principle is quite similar to the one of *MPR*, but the forwarding node selection criterion is modified. First, all 1-hop neighbors which are the only ones that cover some 2-hops neighbors are chosen. Then, instead of choosing neighbors which cover maximal number of remaining nodes, a node that shares 2-hops neighbors with minimal number of already selected nodes is taken.

This technique and two previous ones (*Neighbor Elimination Scheme* and *Dominating Sets*) have been unified in a generic protocol proposed by Wu and Dai [49].

3.6 Impact of Realistic Physical Layer

Qin and Kunz [38] considered the impact of a realistic physical layer model on on-demand routing in ad hoc networks. Assuming that each node uses the same transmission radius, in an ideal environment the received signal power only depends on the distance between nodes. With a shadowing model, the received signal power has a Gaussian distribution fluctuation. The signal strength is therefore a function of distance between two nodes (which decides the distribution function) and a random number which is used to choose a number from the distribution. Given two nodes at distance d, there exists therefore probability $f(d)$ that the signal is received correctly. This observation can be used to derive new broadcasting schemes with better expected performance than the existing schemes that assume ideal physical layer. In [41], the following broadcast scheme is proposed. Each node is assumed to know whether or not it belongs to a dominating set, and also knows for each of its neighbors whether or not they belong to a dominating set. Recall that this knowledge can be gained by adding just one bit to any communication between any two neighboring nodes, or by knowing the geographic coordinates of two-hop neighbors of each node. Upon receiving the first copy of a broadcast message, node A will set a timeout, which is short for a node from dominating set, and long for a node which is not in dominating set. Neighbor elimination scheme is applied to eliminate neighbors that are believed to have received the message (that is, have high corresponding probability $f(d)$, where d is their distance from the transmitting node) and are not in dominating set. Neighbors that are in dominating set and are believed to have received the message are not eliminated. They are eliminated only if a message directly from them is received. At the end of the timeout, if the set of neighboring nodes to be covered is not empty, message is retransmitted, otherwise retransmission is canceled. Thus the difference from the existing approach is that each node only recognizes transmissions which are correctly received, when eliminating neighbors. That is, coverage is only accepted from neighbors whose transmission was correctly received. Note that the timeout function does not need to be a constant function, but could depend on the number of received messages, or percentage of node transmission area that has been covered. The distinction is made, however, between timeout for nodes in dominating set and nodes outside it. The dominating set itself may follow any of existing approaches, such as forwarding neighbors (*e.g. MPR*), covering (definitions by Jie Wu *et al.*), or clustering.

3.7 Beaconless Broadcasting

In [4], a beaconless broadcasting method is proposed. All nodes have the same transmission radius, and nodes are not aware of their neighborhood. That is, no beacons or *HELLO* messages are sent in order to discover neighbors prior to the broadcasting process. The source transmits the message to all neighbors. Upon receiving the packet (together with geographic coordinates of the sender), each node calculates the portion of its perimeter, along the circle of transmission radius, that is not covered by this and previous transmissions of the same packet. Node then sets or updates its timeout interval, which inversely depends on the size of the uncovered perimeter portion. If the perimeter becomes fully covered, the node cancels retransmissions. Otherwise, it retransmits at the end of the timeout interval.

3.8 Double-Dominating Sets

Koubaa and Fleurry [29] proposed to enhance reliability of multicasting by requesting that each node is adjacent to at least two clusterheads. This idea is further developed in [42], by defining double dominating sets and double reception based broadcasting, to increase reliability and make a step toward secure broadcasting. Each node X decides not to be in double dominating set if higher priority neighbors make a connected component, and each neighbor of X is neighbor of at least two nodes from the connected component. During broadcasting, the definition can be converted into source-dependent broadcasting, as follows: Node X decides not to re-transmit the message after timeout if all neighbors that transmitted message already, and all neighbors with higher priority together satisfy the property that each neighbor of X is

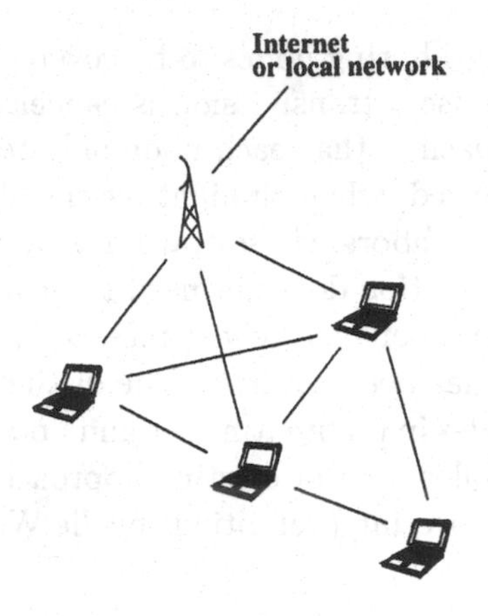

Figure 11: Example of an hybrid network.

a neighbor of at least two of such nodes.

3.9 Broadcasting in Hybrid Networks

Hybrid networks are ad hoc networks with some fixed access points. Correctly used, these ones bring many advantages, mainly in the energy consumption. Indeed, as they are fixed, we suppose they have an unlimited amount of energy, when mobiles use a battery. They can also offer some services that are inaccessible to simple mobiles, like an access to the Internet or a local network. If an access point has such an access, we can assume that every mobile in the network will be able to use it also, thanks to the multi-hop routing. Figure 11 illustrates such a network.

In [22], broadcasting and dominating sets are generalized to hybrid networks. Access nodes, which are assumed to be mutually connected by fast high bandwidth backbone network, are all assumed to be in dominating set, with highest priority. Other nodes then follow given definitions of dominating sets. Let S be the source node of a broadcasting task, with packet arriving at node A. A will retransmit if A is in dominating set and there exists neighbor B of A such that $hc(S,A) < hc(S) + hc(B)$, where $hc(S,A)$ is the number of hops between S and A, and $hc(S)$ and $hc(B)$ are hop counts of S and B to their nearest access nodes. If access nodes are assumed to have significantly larger transmission radius than ad hoc nodes, the condition can be modified to $hc(S,A) < hc(S) + 1$.

3.10 Effects of *MAC* Layer and Mobility

In [48], Williams and Camp classified the broadcast protocols into: blind flooding, probabilistic based, area based, and neighbor knowledge based. They studied twelve broadcast protocols through detailed simulations, particularly focusing on the effects of *MAC* layer and mobility. The comparison is made with same simulator and same parameters, and same scenarios involving bandwidth congestion and topology variations. Results showed that simple protocols like probabilistic do not work well in a congested network. Further, methods using knowledge of neighborhood offer better results than area based methods.

4 Energy Efficient Broadcasting with Range Adjustment

Topology control protocols aim to minimize the needed radius for a transmission at each node, while preserving the connectivity of the network. The connectivity here requires non-directional links (that is any two neighbors in the connected graph must be able to reach each other). In a broadcast oriented variant of the problem, the graph may contain directional edges, and connectivity from source node to all other nodes only is required. When a node has to transmit a message, it does not always have to do it at full power. It is possible to limit the needed range for various reasons (for instance, some neighbors may have already received the message) and thus some energy can be saved. The problem is now to assign transmission ranges to all nodes so that the network remains connected and the sum of selected transmission powers is minimized. This minimum-energy range assignment problem has been shown to be a NP-hard one by Clementi *et al.* [15].

Figure 12 shows the effects of radius adjustment. Case (a) is the unit graph, circles being the areas of communication of each node, which is determined by their original communication range. Case (b) shows the same unit graph, where nodes have adapted their radii to the minimum needed range to reach their furthest neighbor. Finally, case (c) shows the *RNG* of the unit graph with radii adjusted to reach the furthest *RNG*-neighbor of each node. This last case clearly illustrates the fact that the graph becomes directed with radius adjustment, with *D* receiving the communications from *B* without being able to communicate with him.

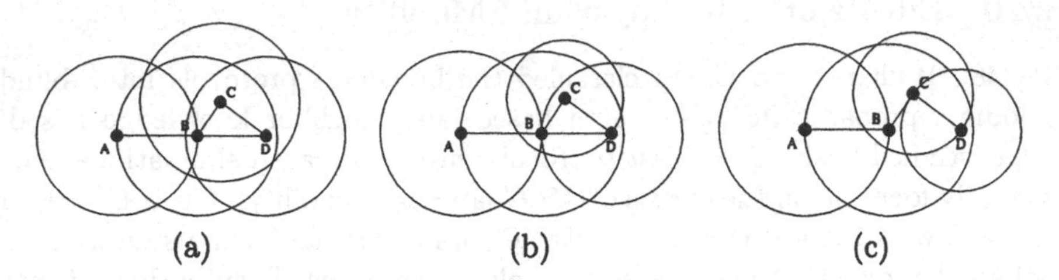

Figure 12: Consequences of radius adjustment.

4.1 Centralized Protocols

Wieselthier *et al.* proposed two centralized greedy heuristics to compute an energy-efficient broadcast tree by assigning a range to each node, called *BLU* and *BIP* [47]. The first one, *BLU* (*Broadcast Least-Unicast-cost*), applies the Dijkstra's algorithm, while the second one, *BIP* (*Broadcast Incremental Power*), is a variant of the Prim's algorithm that uses the broadcast nature of wireless transmissions. Although the authors considered an energy model using a constant c equal to zero, *BIP* fits well in the general model with any other arbitrary value and its performances are the best known ones. Some small improvements have since been proposed but always in a centralized manner and with an energy model using constant $c = 0$ [20, 44, 46].

Wieselthier *et al.* also defined a topology control algorithm based on the *MST* (*Minimum Spanning Tree*) [47], which is used to determine the transmission range of nodes: a node selects the transmission power that permits it to cover all its neighbors in this subgraph. As, by definition, the *MST* is always connected, the graph derived from the new range assignment is also always connected.

4.2 Localized Protocols that Minimize Needed Radii

As centralized solutions cannot really be applied in ad hoc networks without an huge overhead of communications, some localized protocols have been proposed. Cartigny *et al.* [11] proposed a protocol named *RBOP* (*RNG Broadcast Oriented Protocol*) that uses the *RNG* as a connected subgraph instead of the *MST* in the algorithm from Wieselthier *et al.* , the obvious advantage being that the *RNG* can be computed in a totally decentralized manner. Li *et al.* proposed an algorithm to compute a graph named *LMST*

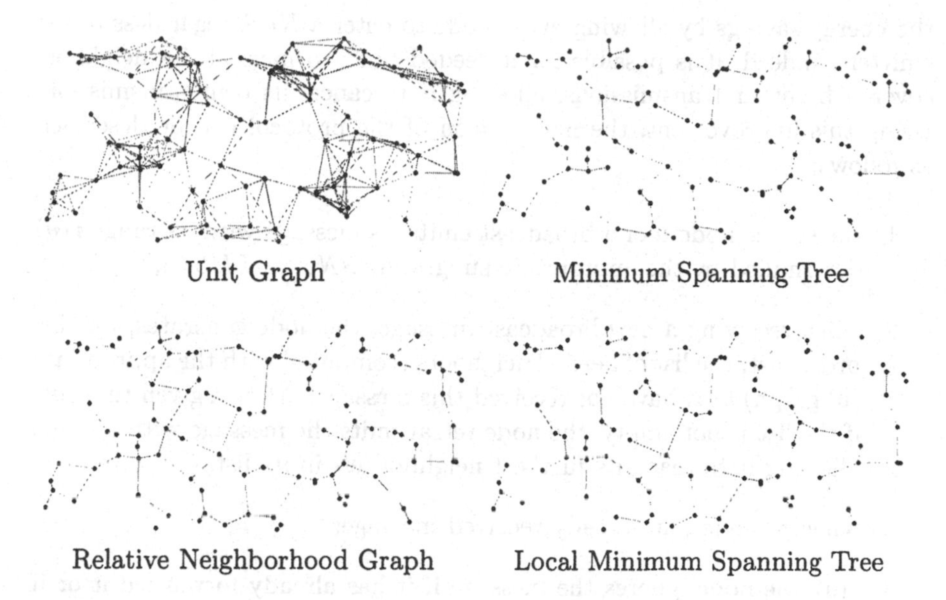

Figure 13: An unit graph with its subgraphs (100 nodes with an average degree of 14).

(*Local Minimum Spanning Tree*) that keeps connectivity [30] and that have since been demonstrated to be a subgraph of the *RNG* [8]. This inclusion proves that *LMST* always performs better than *RNG* when used in this algorithm. This version of *RBOP* is called *LBOP* (*LMST Broadcast Oriented Protocol*) and have been proposed in the same paper [8]. An example of an unit graph and its associated *MST*, *RNG* and *LMST* subgraphs is given in Figure 13. It can also be noticed that *MAC* layer for *LMST*/*RNG* based broadcasting may use acknowledgments from *LMST*/*RNG* neighbors only.

Some improvements to *RBOP* have since been proposed. Indeed, in *RBOP* and *LBOP* protocols, a subgraph is used (respectively *RNG* and *LMST*) to compute a minimal selection of needed neighbors to keep the connectivity. In the description, it is said that only nodes that receive the broadcast message from a neighbor not present in the subgraph enter a *NES* (see Sec. 3.2), otherwise the message is immediately retransmitted. So, the *NES* is used only to ensure a total connectivity, in case of a non-retransmission from a needed node (in case of a collision for example).

This behavior is not always optimal, and it is possible to further enhance the energy savings by allowing every node to enter a *NES*, regardless of the emitter. Indeed, it is possible for a needed node to have all its neighbors covered by other transmissions, allowing it to cancel its own transmission. Using this improvement, the new version of the protocol can be described as follows:

1. the source node u of a broadcast emits its message with its range $r(u)$ determined by the appropriate subgraph (*RNG* or *LMST*),
2. when receiving a new broadcast message, the node generates, for this broadcast, the list of needed neighbors (computed with the appropriate subgraph) that have not received this message. After a given timeout, if the list is not empty, the node retransmits the message with a range allowing it to reach its furthest neighbor left in its list,
3. when receiving an already received message:
   (a) the node ignores the message if it has already forwarded it or if the associated neighborhood list is empty,
   (b) the node removes nodes that are reached by the message from the associated neighborhood list.

This version of the protocol based on the *RNG* is designated by *RBOP-T* (*RNG* Broadcast Oriented Protocol with full Timeout) and the one based on *LMST* in called *LBOP-T*. They have been proposed in [8]

The use of the full *NES* (*e.g.* applied to all nodes in the network) in *RBOP-T* brings a decrease up to 15% in the energy consumption to standard *RBOP*. The substitution of *RNG* by *LMST* in *LMST-T* leads to an improvement of 20% that corresponds to the average degree reduction (from 2.6 for *RNG* to 2.04 for *LMST*). While obviously *BIP* is still the best algorithm with its global knowledge, the overhead of *LBOP-T* is less than 45% with the model $\alpha = 2, c = 0$ and less than 65% with $\alpha = 4, c = 10^8$. This example illustrates the fact that optimized localized algorithms can be very competitive with centralized ones.

Li and Hou independently also proposed to apply *LMST* structure for broadcasting [31], and analytically concluded that multi-hop broadcast is more power efficient than single message with full transmission power when $\alpha \geq 2.2$. This article challenges the generality of such conclusion independently on constant c, which tends to suggest that it is always better to minimize the transmission radius to the minimal one, decided by the structure of *LMST*.

Huang and Shen proposed another algorithm to compute the needed range [19]. Each node broadcasts at each power level and verifies whether the subgraph of all its 1-hop neighbors (when the maximum transmission radius is used) is still connected. Then, each node chooses the minimal range which preserves the connectivity of the subgraph as its transmission radius. The communication overhead that occurs when determining the transmission radius seems significant, although it is limited to 1-hop neighbors, that is, procedure is localized. The advantage of this scheme is that no position information is needed. This method can also be seen as the discrete version of *LMST* structure presented above.

Chen *et al.* proposed a localized minimum energy scheme where each node decides transmission power based on 2-hops information [13]. Upon receiving the first copy of a message, node A sets a timeout and eliminates all neighbors that already received the same message in this or subsequent transmissions by other nodes before timeout expires. All neighbors of current node A are sorted, from furthest to closest, and examined in that order. If, for an examined neighbor B, there exist a common neighbor C such that the power to reach C from A plus the power to reach B from C is lower than power needed for direct transmission from A to B, node A decides not to cover node B, and reduces transmission power and examines, in turn, the next closer node in the list. This continues until a node is found for which this optimization step cannot be performed. The distance to that node is selected for transmission radius and message is retransmitted (unless the set becomes empty, in which case A decides not to transmit at all). Although the scheme is very promising, unfortunately it is compared only with scheme that uses maximum transmission power at each node, not with other existing localized scheme such as [11].

These algorithms are also studied for applications in sensor networks. Cheng *et al.* proved that the problem of assigning transmission power to each sensor, such that the induced topology composed only by bidirectional links is strongly connected, is NP-complete [14]. They proposed two heuristics, one based on selecting the furthest *MST*-neighbor at each node, and the other based on incremental power (adding one node at a time to already constructed connected tree so that the added power is minimized).

Wu and Wu proposed to dynamically reduce the transmission power of each node in the broadcast process based on some observations on the maximal distances from each node to all its neighbors, and based on the

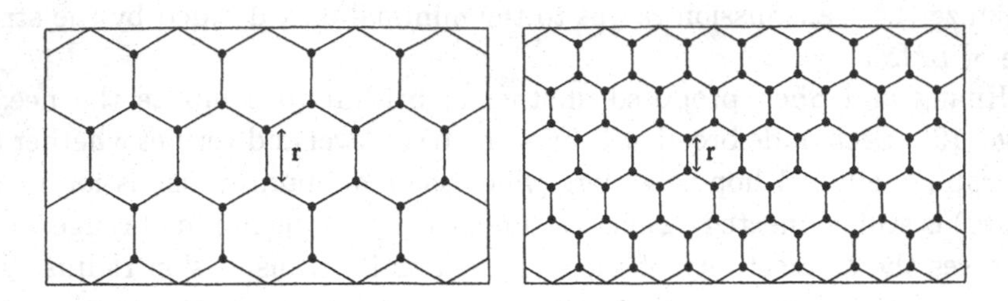

Figure 14: Hexagonal tiling of a surface S for different ranges r.

dominating set status of each node [52]. Nodes which are not in dominating set may still retransmit, and the comparison is made with blind flooding, and methods presented in [53] which were aimed at scheduling node activities, not at minimizing total energy.

4.3 Localized Protocol that Uses a Target Radius

All the presented algorithms that use range adjustment are based on the assumption that the shortest radius is always the optimal one. Indeed, the constant α (see Eq. 2) makes long emissions expensive. Unfortunately, this assertion is not always valid when considering cases with c not equal to zero, because short radii will lead to an increase of needed relays to cover the network and thus to an increase of the energy consumption (the constant c is supported by each of these transmitting nodes). So, it can be assumed that there exists a target radius for a given broadcast that balances the cost of α and c, leading to an optimal energy consumption for a given broadcast. A protocol that uses this idea has been proposed in [23] and is named *TR-LBOP*. Its description goes as follows.

Optimal Radius Value

The value of the optimal radius is computed by considering a geometrical area S on which some nodes that can adjust their radii are placed. To do this, an honeycomb mesh is used, that is the area is divided into several hexagons of side r, where nodes are placed on vertices, with a radius of r. Obviously, the quantity of vertices (*i.e.* nodes) depends of r. Figure 14 shows the tiling for two different values of r.

To find the optimal radius, the value of r that minimizes the total power consumption must be computed. The needed energy is simply computed by considering that each node will have to emit the message once with a radius of r. A high value can be chosen, which would allow nodes to cover a large part of the area with only one emission, or a low value can be chosen, in which case only a small part of the area would be covered. In the first case, only a few emitting nodes would be needed, while in the second one, a lot more of relays would be needed to cover the entire area. A radius that balances between these two behaviors must be found.

The number of hexagons h is computed by using the formula

$$\mathrm{h} \simeq \frac{\text{Surface of the area}}{\text{Surface of an hexagon}} = \frac{2S}{3r^2\sqrt{3}}.$$

As two nodes are placed on each hexagon, the number of nodes $n = 2h$ is then:

$$n = \frac{k}{r^2}, \quad k = \frac{4S}{3\sqrt{3}}.$$

When a broadcast occurs, all these nodes will emit once the message, the power consumption is then:

$$PC(r) = (r^\alpha + c)\frac{k}{r^2}.$$

Minima of this function give optimal radii, depending of the value of α and c. Figure 15 clearly shows that with the hexagonal tiling, when $\alpha = 4$ and $c = 10^8$, the optimal radius is 100. Below this value, there are too many nodes (vertices), making the constant c a problem while a radius greater than 100 makes the constant α a disadvantage.

Given that $\alpha \geq 2$, $c \geq 0$ and $r > 0$, there are only a few possible cases that are enumerated by table 1. It can be noticed that $\alpha = 2$ brings special cases. In the first one ($c \neq 0$), it is better to emit with the maximal possible range, while in the second one ($c = 0$) the chosen radius does not influence the power consumption.

TR-LBOP Protocol

Obviously, the position of nodes cannot be controlled in ad hoc networks, but fortunately radii can be modified, making possible a control of the topology. Roughly, chosen radii will have to be as close as possible from the optimal one, while still guaranteeing the original connectivity of the network.

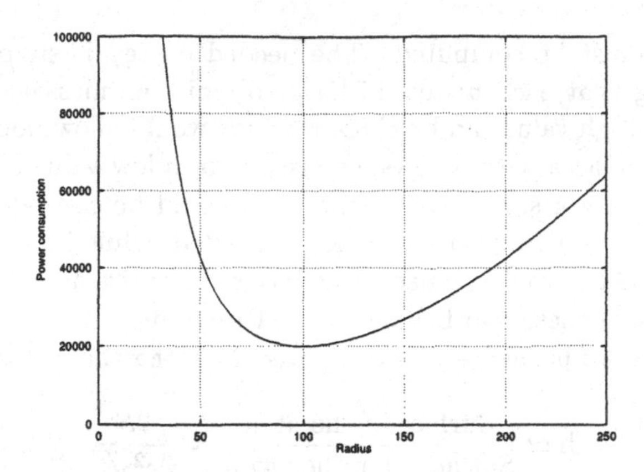

Figure 15: Power consumption vs. chosen radius with $\alpha = 4$ and $c = 10^8$

LBOP is used as a based broadcast protocol because of its advantages: it uses a localized algorithm, the subgraph *LMST* keeps the connectivity and is composed only by small edges. Its only disadvantage is to try to minimize transmission radii, which is not always an optimal behavior. To include the concept of optimal radius, some parts of the protocol have been modified, so that each node increases its radius up to the optimal one when a retransmission is needed. This variant of *LBOP* is called *TR-LBOP* for *Target Radius LMST Broadcast Oriented Protocol* and can be described as follows.

Each node manages two lists L and L' during the *NES*. The first one, L, stores the neighbors needed to keep the connectivity of the network. As *LBOP* uses the *LMST*-subgraph, L is defined by:

R_{opt}	$c = 0$	$c \neq 0$
$\alpha = 2$	-†	R
$\alpha > 2$	0	$\sqrt[\alpha]{\frac{2c}{\alpha-2}}$.

† when $\alpha = 2$ and $c = 0$, the radius has no influence, we use 0 as an arbitrary value.

Table 1: Theoretical optimal emitting radius R_{opt} for each α and c.

$$\forall u \in V \quad L(u) = N_{LMST}(u).$$

The list L' stores every other neighbors of the node, and is defined by:

$$\forall u \in V \quad L'(u) = N(u) \setminus L(u).$$

During the *NES*, a node u will monitor every transmission that occurs in its neighborhood, and each time one of its neighbor receives the broadcast message, its removes from the corresponding list (L if it is a $LMST$-neighbor, L' otherwise) this node. Of course, the node u can immediately remove the node from which it gets the message and their common neighbors.

When the timeout is up, two cases can happen:

- The list L is empty, in which case the node does not need to re-emit the message to keep the network connectivity, so the retransmission is canceled, as with *LBOP*.
- There is at least one node in L. In this case, the node u has to rebroadcast the message to reach the nodes left in L.

In the second case, when the retransmission is needed, the node will have to support the cost of the constant c in the energy model (regardless of its radius). So, as explained previously, it can be clever to increase the needed radius up to the optimal one, when it is possible. Two values, D_L and $D_{L'}$, are used. The first one, D_L is defined by:

$$D_L = \max\{d(u,v) \mid v \in L(u)\},$$

with $d(u,v)$ being the Euclidean distance between u and v. The second one, $D_{L'}$ is defined by:

$$\begin{aligned} D_{L'} &= \{d(u,v) \mid v \in L(u) \cup L'(u) \wedge \\ \delta_{uv} &= \min\{\delta_{uw} \mid w \in L(u) \cup L'(u)\}\}, \end{aligned}$$

with $\delta(u,v)$ being defined by

$$\delta_{uv} = |d(u,v) - R_{opt}|.$$

In other words, the chosen distance is the length of the edge between the node u and its non-reached neighbor which is the nearest one from R_{opt}. The final chosen radius is simply:

$$r(u) = \max\{D_L, D_{L'}\}:$$

This modification leads to a situation where nodes mostly emit with a radius as close as possible to R_{opt}. The increased number of reached neighbors is balanced with the full neighbor elimination scheme of *LBOP*, so the number of relays does not increase dramatically.

The results were obtained by simulating 500 broadcasts in randomly generated static networks composed by 200 nodes with a maximum radius set to 250 meters. They showed that there really exists an optimal radius for which the global energy consumption is minimal. *TR-LBOP* improves *LBOP* by about 10% with the use of this optimal radius, for the density of 50 and using a constant $c = 10^8$. This improvement becomes greater as c increases, while the overhead with *BIP* always stays under 50%. That is, *TR-LBOP* follows *BIP* on all ranges, while the overhead of *LBOP* becomes greater as c increases. The experimental optimal radius observed for $\alpha = 4, c = 10^8$ seems to be around 80 meters, which is near to the theoretical value of 100 meters (see Figure 15), the small difference can be attributed to various border effects. The consideration of the non-zero value of the constant c leads a good improvement of *LBOP*, approaching the performances of *BIP*.

In an upcoming paper, we further explore this idea by using connected dominating sets in the topology control step [24]. This way, nodes that are not dominant can go into sleep mode and thus reduce the energy consumption, instead of being active (Nodes in *TR-LBOP* that enter a *NES* are active and consume some energy).

5 Energy Efficient Broadcasting with Directional Antennas

There exists another method to further improve energy savings when using smart antennas, that are able to make directional emissions. Recently, these antennas have been designed to allow the transmission of a packet from a given node in a particular direction, with signal being sent only inside an angle (or cone in 3-D) oriented in any desired direction. Omnidirectional emissions are then just special cases with an angle of 360°, consuming the full energy possible, as the smaller the angle is, the smaller the energy consumption will be. For this directional case, special protocols must be used to take this possibility into account and thus take advantage of it.

5.1 Centralized Protocols

Wieselthier *et al.* proposed two extensions of *BIP* for directional antennas [47]. The first one is called *RB-BIP* (*Reduced Beam BIP*) and uses one-to-one communication model (with minimal angle) to join neighbors in the *BIP* tree. Because of its construction, the *BIP* spanning tree is exactly the same as the tree constructed by the *MST* algorithm, which is confirmed in experiments. The second improvement is named *D-BIP* (*Directional BIP*). Each node can send only one message, so they have to change the angle of the beam or increase their range to join more neighbors. The tree construction is different from *RB-BIP*, because the protocol has to decide, at each step, if it is better to extend the beam and/or the range of a node, or to add a new communication beam for a neighboring node. This decision is made with respect to the energy consumption of the two algorithms. Hence, the natural tendencies of *D-BIP* are to favor transmissions with large radii and beam angles and to avoid retransmissions by every node (this is especially valid if the constants C_1 and C_2 have a high value, see Equ. 3).

5.2 Localized Protocols

As all these variants of *BIP* are still centralized, Cartigny *et al.* proposed an adaptive localized variant of *DRBOP* (*Directional RBOP*) [10] which takes advantage of both one-to-one and one-to-many schemes [12]. In fact, in a sparse network, it is easy to see that one-to-one communication model with narrow beams fits better than large beams. At contrary, in a dense network, one-to-many communication model fits better since it allows to cover several neighbors with a single beam. In order to fulfill the need of adaptive algorithm, the authors give a one-to-one broadcast protocol called *DLBOP* and a one-to-many broadcast protocol called *OM-DLBOP*.

The first one, *DLBOP* (*Directional LBOP*), is a variant of *DRBOP* where the *RNG* is replaced by the *LMST*. Hence, each node u sends to its *LMST*-neighbors v an unicast message, with a narrow beam of fixed angle β and a range $d(u,v)$. We can notice that *NES* is not applicable here since packets sent to a neighbor by a narrow beam are normally not received by other *LMST*-neighbors. However, a sending node can include in its message the list of neighbors it is supposed to send. Hence by increasing the size of packets, a neighbor elimination scheme can be achieved.

The energy consumption of the *DLBOP* protocol with directional antennas can be derived by summing the power consumption of one-to-one messages from each node. Because the degree of each node is approximately 2 for the *LMST* subgraph (including the local forwarder of the broadcast message) we can expect that each node will broadcast in average one unicast message. Let us denote by d_{lmst} the average distance between *LMST*-neighbors. The energy consumption of the *DLBOP* protocol for n nodes is approximately:

$$E_{DLBOP} = n \times e(\beta, d_{lmst}).$$

The one-to-many variant of *DLBOP*, denoted by *OM-DLBOP* (*One-to-Many Directed LMST Broadcast Oriented Protocol*), consists of sending a single variable angle beam instead of several narrow beams. A node which decides to retransmit the message (because of *LMST*-neighbor elimination scheme reason) uses a single beam with an appropriate angle which allows it to reach its non-covered *LMST*-neighbors. To increase energy savings, it can be useful to extend the range in order to avoid excessive retransmissions that can be expensive if constants C_1 or C_2 are not equal to zero. It is shown that for a given angle γ, there exists an optimal angle. These angles are given by Table 2.

We are now in position to describe *OM-DLBOP* algorithm. We choose to send beams with angle $4\pi/3$ which minimizes the overlap communication zone and provides a good coverage of the neighborhood as illustrated by Figure 16. The angle is positioned symmetrically with respect to the line uv as shown in the same figure where v is the current sender and u the node which sends the broadcast message to v. We denote by $R_{opt}(4\pi/3)$ the function which computes the optimal radius, by taking into account the

	$C_1 = C_2 = 0$	$C_1 \neq 0 \vee C_2 \neq 0$
$\alpha = 2$	constant No $R_{opt}(\gamma)$	monotone decreasing $R_{opt}(\gamma) = R_{max}$
$\alpha > 2$	monotone increasing $R_{opt}(\gamma) = 0$	minimal at $r(\gamma) = \sqrt[\alpha]{\frac{2C_1 + \frac{4\pi C_2}{\gamma}}{\alpha - 2}}$ $R_{opt}(\gamma) = min(r(\gamma), R_{max})$

Table 2: Behavior of total energy consumption function and optimal radius.

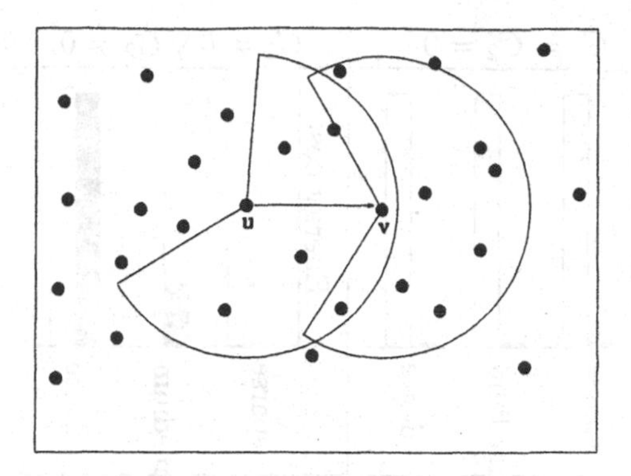

Figure 16: A *OM-DLBOP* broadcast.

chosen angle $4\pi/3$. After applying *NES* restricted to *LMST*-neighbors, a node v which decides to retransmit computes its transmitting angle and range as follows:

- let A be the set of uncovered neighbors and $B \subseteq A$ the set of uncovered *LMST*-neighbors.

- The node u computes the set of nodes closer than $R_{opt}(4\pi/3)$:

$$A' = \{v \in A \mid d(u,v) \leq R_{opt}(4\pi/3)\}.$$

  The "goal" of u is to reach nodes of $C = A' \cup B$, *i.e.* nodes closer than $R_{opt}(4\pi/3)$ and *LMST*-neighbors. If $C_1 = C_2 = 0$, the optimal radius cannot be evaluated. In this case, we consider $R_{opt}(4\pi/3) = 0$. This implies that $A' = \emptyset$ and that the goal of the node is limited to its *LMST*-neighbors.

- The node calculates the angle θ needed to cover C and the distance d to the furthest node of C. If $\theta < \beta$ then set $\theta = \beta$. If C is empty, the retransmission is canceled.

- If $d > R_{opt}(\theta)$ then send θ-beam with d range. Otherwise, the node sets the range of θ-beam (without modifying the orientation) in order to reach all nodes of A closer than $R_{opt}(\theta)$ (thus the selected radius is generally somewhat lower than $R_{opt}(\theta)$).

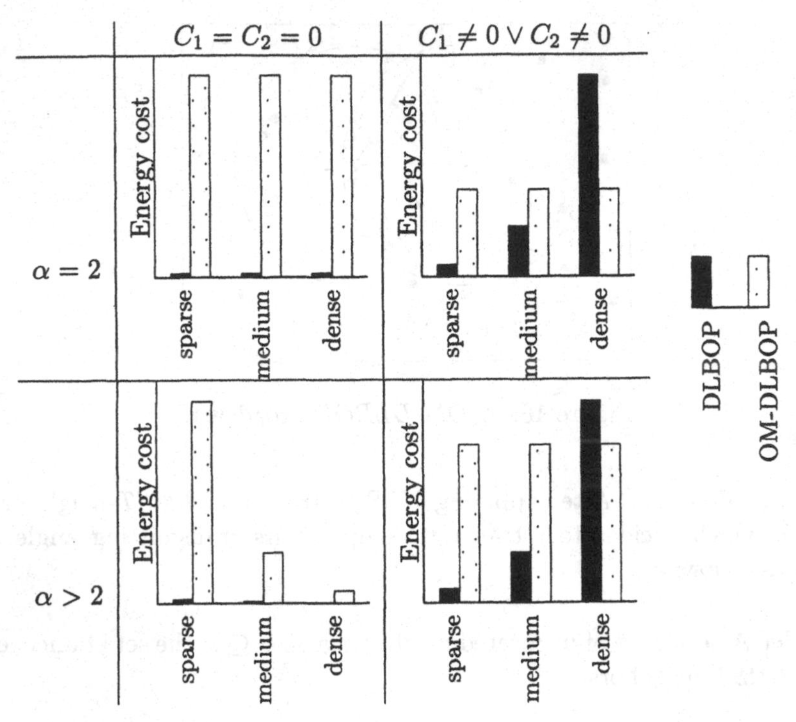

Figure 17: Behavior of one-to-one and one-to-many variants of DLBOP.

5.3 Adaptive Directed *LMST* Broadcast Oriented Protocol (*A-DLBOP*)

It seems that both approaches are valid. To illustrate this fact, Figure 17 shows the behaviors of both protocols for sparse, medium and dense networks and for the different energy models. In this comparison, *OM-DLBOP* energy consumption is evaluated by considering an ideal $4\pi/3$ tessellation.

From this observation, we describe the *A-DLBOP* (*Adaptive Directed LMST Broadcast Oriented Protocol*) algorithm which combines one-to-one and one-to-many communication models [12]. Each node which receives the broadcast message starts a *NES* limited to its *LMST*-neighbors (for narrow beam transmissions, *NES* has limited effectiveness and we suppose that a transmitting node includes the *id* of nodes it targets in all its transmissions). At the end of the timeout period, the node has to choose between one-to-one or one-to-many communication models. For a given node u, the decision algorithm is the following one:

- Let A be the non-covered neighbors set and $B \subseteq A$ the non-covered $LMST$-neighbors set. We denote by A' the set of nodes belonging to A closer than $R_{opt}(4\pi/3)$. As previously, if $C_1 = C_2 = 0$ we consider that $R_{opt}(4\pi/3) = 0$. The "goal" of the node u is to cover the set $C = A' \cup B$. If the set C is empty, the retransmission is canceled.

- The communication model choice is made from a comparison of the energy consumption needed to cover C:

  One-to-one communication: in a flooding over the subset C with one-to-one communication model, each node retransmits the message to its non-covered $LMST$-neighbors. On average, each node has only one non-covered $LMST$-neighbor. Hence the energy consumption with one-to-one communication model using β-beams can be evaluated by:

$$E_{1-to-1} = |C| \times e(\beta, d_{lmst}).$$

  The node u can estimate the average $LMST$ edge length d_{lmst} from the average distance between itself and its $LMST$-neighbors (it may also use more precise average distance in its local minimum spanning tree):

$$d_{lmst} \simeq \frac{1}{|B|} \sum_{v \in B} d(u, v).$$

  One-to-many communication: let θ be the angle needed to cover C (if $\theta < \beta$ we consider that $\theta = \beta$) and d the distance between node u and the furthest node of C. The energy consumption of a single beam which covers C is:

$$E_{1-to-many} = e(\theta, d).$$

- If $E_{1-to-1} < E_{1-to-many}$, then the node u decides to use the one-to-one communication model and sends a β-beam to each of its $LMST$-neighbors (nodes of B).

- Otherwise, the node u decides to use one-to-many communication model and sends a β-beam to cover nodes of C. If $d < R_{opt}(\theta)$, the range of the beam is increased in order to reach nodes of A closest than $R_{opt}(\theta)$.

To resume this process, each node makes its decision based on the comparison $|C|.e(\beta, d_{lmst}) < e(4\pi/3, d)$, where C is the number of nodes covered by a single $4\pi/3$ beam. Alternatively, but leading to a similar conclusion, each node may compare $k.e(\beta, d_{lmst}) < a.e(4\pi/3, R_{opt})$, where k is the number of nodes in the circle of radius R, and a is the ratio between the area of the circle of radius R and the $4\pi/3$ beam of radius R_{opt}. Since, $|C|$ is approximately equal to k/a, it is basically the same comparison.

Experimentations were done by simulating broadcasts in randomly generated networks of 500 nodes, a minimal angle beam of $\pi/9$ and a target radius varying from 10 meters to 250 meters. They showed that the conversion between the two models is beneficial, offering significant energy savings, for full range of densities of networks. The hybrid protocol *A-DLBOP* is able to take the best decision depending on its local density (its number of neighbors). Since each node locally independently chooses between two behaviors, it is possible that, in the same network for the same broadcasting task, different nodes make different choice between the two. The localized protocol *A-DLBOP* gives very good results compared with centralized *D-BIP* protocol. For instance, when $C_1 = C_2 = 0$ the energy overhead is about 30%. The additional spending for other energy model ($C_1 \neq 0$ or $C_2 \neq 0$) grows significantly with C_1 and C_2. For instance, with Rodoplu model [39] ($\alpha = 4$, $C_1 + C_2 = 10^8$), the energy overhead against *D-BIP* is about 500% but the energy reduction against blind flooding is about 95%. The interesting fact is that *A-DLBOP* always gives better results than *DLBOP* and *OM-DLBOP*. Hence, the protocol *A-DLBOP* is an adaptive broadcast protocol where decision is made locally by each node.

6 Conclusion

Broadcasting is an important task in ad hoc networks. In this chapter, we presented several broadcasting methods designed to guarantee delivery to all nodes connected to the source (providing the *MAC* layer is ideal), while maximizing energy savings. We have considered different assumptions and scenarios, depending on the available hardware (such as *GPS* receiver, radius adjustment option, and directional emissions). The techniques described here can also be used for solving similar problems. One of such problems is area coverage in sensor networks. Typical applications of sensor networks are regular reports such as temperature monitoring, where monitoring stations have to gather information from sensors. Monitoring center typically broadcasts its position and request to all the sensors in the network. To prolong network life, most sensors shall be placed into sleep mode. Active sensors should cover the monitoring area. It turns out that covering area can be done with similar methods as covering neighbors, as described in detail in [6]. The final step is to report events using reverse broadcast trees, as described in [5]. A solution based on energy efficient *MPR* method is given in [32]. We anticipate more research on the currently very active research problem of broadcasting in the near future.

References

[1] C. Adjih, P. Jacquet, and L. Viennot, Computing connected dominated sets with multipoint relays, (Technical Report 4597, INRIA, Oct. 2002).

[2] P. Bahl and V.N. Padmanabhan, Radar: an in-building rf-based user location and tracking system, *Proceedings of the IEEE Infocom 2000* (Tel Aviv, Israel, Mar. 2000).

[3] A. Benlarbi-Delaï, D. Simplot, J. Cartigny, and J.-C. Cousin. Using 3d indoor microwave phase sensitive stereoscopic location system to reduce energy consumption in wireless ad hoc networks, In *Proceedings of the Smart Objects Conference (sOc'2003)* (Grenoble, France, May 2003).

[4] J. Carle, D. Simplot, and I. Stojmenović, Area coverage based beaconless broadcasting in ad hoc networks. (In preparation, 2003).

[5] J. Carle and D. Simplot-Ryl. Energy efficient area monitoring by sensor networks. *IEEE Computer Magazine* (to appear, 2004).

[6] J. Carle, D. Simplot-Ryl, and I. Stojmenović, Sensor area coverage, (In preparation, 2003).

[7] J. Cartigny, F. Ingelrest, and D. Simplot, Rng relay subset flooding protocols in mobile ad hoc networks, *International Journal of Foundations of Computer Science (IJFCS)* Vol.14 No.2 (Apr. 2003) pp. 253-265.

[8] J. Cartigny, F. Ingelrest, D. Simplot-Ryl, and I. Stojmenović, Localized lmst and rng based minimum-energy broadcast protocols in ad hoc networks, *Ad hoc Networks*, (To appear, 2004).

[9] J. Cartigny and D. Simplot, Border node retransmission based probabilistic broadcast protocols in ad hoc networks, *Proceedings of the 36th Annual Hawaii International Conference on System Sciences (HICSS-36)* (Hawaii, USA, Jan. 2003).

[10] J. Cartigny, D. Simplot, and I. Stojmenović, Localized energy efficient broadcast for wireless networks with directional antennas, *Proceedings of the Mediterranean Ad Hoc Networking Workshop (MedHocNet'02)* (Sardegna, Italy, Sep. 2002).

[11] J. Cartigny, D. Simplot, and I. Stojmenović, Localized minimum-energy broadcasting in ad hoc networks, *Proceedings of the IEEE Infocom 2003* (San Francisco, USA, Apr. 2003).

[12] J. Cartigny, D. Simplot-Ryl, and I. Stojmenović, An adaptive localized scheme for energy efficient broadcasting in ad hoc networks with directional antennas, (To appear, 2003).

[13] X. Chen, M. Faloutsos, and S.V. Krishnamurthy, Power adaptive broadcasting with local information in ad hoc networks, *Proceedings of the IEEE International Conference on Networks and Protocols (ICNP'03)* (Atlanta, USA, Nov. 2003).

[14] X. Cheng, B. Narahari, R. Simha, M.X. Cheng, and D. Liu, Strong minimum energy topology in wireless : Np-completeness and heuristics, *IEEE Transactions on Mobile Computing* Vol.2 No.3 (2003) pp. 248-256.

[15] A.E.F. Clementi, A. Ferreira, P. Penna, S. Perennes, and R. Silvestri, The minimum range assignment problem on linear radio networks, In *European Symposium on Algorithms (ESA 2000)* (Sarbrüken, Germany, Sep. 2000) pp. 143-154.

[16] F. Dai and J. Wu, Distributed dominant pruning in ad hoc networks, In *Proceedings of the IEEE International Conference on Communications (ICC'03)* (Anchorage, AK, USA, May 2003).

[17] L. Feeney, An energy-consumption model for performance analysis of routing protocols for mobile ad hoc networks, *ACM Journal of Mobile Netwoks and Applications* Vol.3 No.6 (2001) pp. 239-249.

[18] J. Hightower and G. Boriello, Location systems for ubiquitous computing, *IEEE Computer* Vol.38, (Aug. 2001) pp. 57-66.

[19] Z. Huang and C.-C. Shen, Distributed topology control mechanism for mobile ad hoc networks with swarm intelligence, In *Proceedings of the ACM International Symposium on Mobile Ad Hoc Networking and Computing (MOBIHOC'03)* (Annapolis, MD, USA, Jun. 2003).

[20] R.J. Marks II, A.K. Das, M. El-Sharkawi, P. Arabshahi, and A. Gray, Minimum power broadcast trees for wireless networks: Optimizing using the viability lemma, In *Proceedings of the International Symposium on Circuits and Systems (ISCAS 2002)* (Scottsdale, USA, 2002) pp. 245-248.

[21] M. Ilyas (ed.), *The Handbook of Ad Hoc Wireless Networks*, (CRC Press, 2003).

[22] F. Ingelrest, D. Simplot, and I. Stojmenović, Routing and broadcasting in hybrid ad hoc multi-hop cellular and wireless internet networks, (In preparation, 2003).

[23] F. Ingelrest, D. Simplot, and I. Stojmenović, Target transmission radius over lmst for energy efficient broadcast protocol in ad hoc networks, (To appear, 2003).

[24] F. Ingelrest, D. Simplot-Ryl, and I. Stojmenović, Design of a target radius based broadcast protocol using dominating sets in ad hoc networks, (In preparation, 2004).

[25] C. Jaikaeo and C.-C. Shen, Adaptive backbone-based multicast for ad hoc networks, *Proceedings of the IEEE International Conference on Communications (ICC'02)* (New York City, NY, USA, Apr. 2002).

[26] D.B. Johnson, D.A. Maltz, and Y.-C. Hu, The dynamic source routing protocol for mobile ad hoc networks (DSR), (Internet Draft, Work-in-progress, draft-ietf-manet-dsr-09.txt, Apr. 2003).

[27] E.D. Kaplan, *Understanding GPS: Principles and Applications*, (Artech House, 1996).

[28] S.J. Kim and Y.J. Suh, Efficient broadcast schemes with transmission power control in mobile ad hoc networks, (Submitted, 2003).

[29] H. Koubaa and E. Fleury, On the performance of double domination in ad hoc networks, *Proceedings of the Mediterranean Ad Hoc Networking Workshop (MedHocNet'03)* (Mahdia, Tunisia, Jun. 2003).

[30] N. Li, J. C. Hou, and L. Sha, Design and analysis of an mst-based topology control algorithm, *Proceedings of the IEEE Infocom 2003* (San Francisco, USA, Apr. 2003).

[31] N. Li and J.C. Hou, BLMST: A scalable, power-efficient broadcast algorithm for wireless sensor networks, (Submitted, 2003).

[32] J. Lipman, P. Boustead, J. Chicharo, and J. Judge, Resource aware information collection (RAIC) in ad hoc networks, *Proceedings of the Mediterranean Ad Hoc Networking Workshop (MedHocNet'03)* (Mahdia, Tunisia, 2003).

[33] Sze-Yao Ni, Yu-Chee Tseng, Yuh-Shyan Chen, and Jang-Ping Sheu, The broadcast storm problem in a mobile ad hoc network, In *Proceedings of the International Conference on Mobile Computing and Networking (MobiCom'99)* (Seattle, USA, Aug. 1999) pp. 151-162.

[34] D. Niculescu and B. Nath, Ad hoc positioning system (APS) using aoa, *Proceedings of the IEEE Infocom 2003* (San Francisco, USA, Apr. 2003).

[35] W. Peng and X.C. Lu, On the reduction of broadcast redundancy in mobile ad hoc networks, In *ACM MobiHoc 2000* (Boston, Massachusetts, USA, Aug. 2000) pp. 129-130.

[36] N. Priyantha, A. Chakraborty, and H. Balakrishnan, The cricket location-support system, *Proceedings of the International Conference on Mobile Computing and Networking (MobiCom 2000)* (Boston, USA, Aug. 2000).

[37] A. Qayyum, L. Viennot, and A. Laouiti, Multipoint relaying for flooding broadcast messages in mobile wireless networks, In *Proceedings of the Hawaii International Conference on System Sciences (HICSS'02)* (Jan. 2002).

[38] L. Qin and T. Kunz, On-demand routing in manets: The impact of a realistic physical layer model, *Proceedings of the International Conference on Ad-Hoc, Mobile, and Wireless Networks (ADHOC-NOW'03)* (Montreal, Canada, Oct. 2003).

[39] V. Rodoplu and T. H. Meng, Minimum energy mobile wireless networks, *Proceedings of the IEEE International Conference on Communications (ICC'98)* Vol.3 (Atlanta, USA, Jun. 1998) pp. 1633-1639.

[40] I. Stojmenović, *Handbook of Wireless Networks and Mobile Computing*, (John Wiley & Sons, 2002).

[41] I. Stojmenović, The impact of a realistic physical layer on broadcasting in ad hoc networks, (To be published, 2003).

[42] I. Stojmenović, Reliable broadcasting by double domination, (In preparation, 2003).

[43] I. Stojmenović and M. Seddigh, Broadcasting algorithms in wireless networks, In *Proceedings of the International Conference on Advances in Infrastructure for Electronic Business, Science, and Education on the Internet SSGRR* (L'Aquila, Italy, Jul./Aug. 2000).

[44] T.Chu and I.Nikolaidis, Energy efficient broadcast in mobile ad hoc networks, *Proceedings of the International Conference on Ad-Hoc, Mobile, and Wireless Networks (ADHOC-NOW'02)* (Toronto, Canada, 2002) pp. 177-190.

[45] G. Toussaint, The relative neighborhood graph of finite planar set, *Pattern Recognition* Vol.12 No.4 (1980) pp. 261-268.

[46] P.-J. Wan, G. Calinescu, X.-Y. Li, and O. Frieder, Minimum energy broadcast routing in static ad hoc wireless networks, *ACM/Kluwer Wireless Networks* (2002).

[47] J. Wieselthier, G.D. Nguyen, and A. Ephremides, On the construction of energy-efficient broadcast and multicast trees in wireless networks, *Proceedings of the IEEE Infocom 2000* (Tel Aviv, Israel, 2000) pp. 585-594.

[48] B. Williams and T. Camp, Comparison of broadcasting techniques for mobile ad hoc networks, *Proceedings of the ACM International Symposium on Mobile Ad Hoc Networking and Computing (MOBIHOC'02)* (Lausanne, Switzerland, Jun. 2002) pp. 194-205.

[49] J. Wu and F. Dai, A generic distributed broadcast scheme in ad hoc wireless networks, *Proceedings of the International Conference on Distributed Computing Systems (ICDCS'03)* (Providence, Rhode Island, USA, May 2003) pp. 460-468.

[50] J. Wu and F. Dai A distributed formation of a virtual backbone in ad hoc networks using adjustable transmission ranges, *Proceedings of the International Conference on Distributed Computing Systems (ICDCS'04)* (Tokyo, Japan, Mar. 2004).

[51] J. Wu and H. Li, A dominating-set-based routing scheme in ad hoc wireless networks, *Proceedings of the International Workshop on Discrete Algorithms and Methods for Mobile Computing and Communications (DIALM)* (Seattle, USA, Aug. 1999) pp. 7-14.

[52] J. Wu and B. Wu, A transmission range reduction scheme for power-aware broadcasting in ad hoc networks using connected dominating sets, *Proceedings of the 2003 IEEE Semiannual Vehicular Technology Conference (VTC2003-fall)* (Oct. 2003).

[53] J. Wu, B. Wu, and I. Stojmenović, Power-aware broadcasting and activity scheduling in ad hoc wireless networks using connected dominating sets, *Wireless Communications and Mobile Computing* Vol.4 No.1 (Jun. 2003) pp. 425-438.

Energy-Efficient Multicast Protocols

Sandeep K. S. Gupta and Bin Wang
Department of Computer Science and Engineering
Ira A. Fulton School of Engineering at Arizona State University, Tempe, AZ 85287
E-mail: {Sandeep.Gupta, Bin.Wang}@asu.edu

Contents

1 Introduction

The goal of a *multicast* communication (also known as one-to-many/many-to-many communication) service is to efficiently deliver messages originating from a single or multiple sources to multiple recipients. Multicast communication primitives are useful in many distributed systems and applications where multiple parties are involved in tasks such as collaboration, resource and/or information sharing, and co-ordination. In this chapter we are mainly interested in applications which involve distribution of same information to multiple recipients from a single source or multiple sources. A multicast can be viewed as a *selective broadcast* in the sense that only a subset of the system may be interested in receiving the broadcast messages. Although a simple way to implement a multicast operation is to use system wide broadcast and do filtering at the end systems, this can be a significant waste of resources, such as network bandwidth and processing. The degree of inefficiency depends upon the sparsity of the interested systems. At the other extreme a multicast can be implemented by sending a unicast message to each interested system. This may be efficient way to implement multicast for a sparse distributed group but it is a severe waste of resources for moderately sized groups since the message may need to unnecessarily travel a common portion of path to multiple receivers several times wasting bandwidth. This also increases the latency of the multicast message delivery by sequentializing the message delivery to multiple recipients. In systems where bandwidth is at premium and/or low latency is desired multicast services employ multicast delivery structures such as a multicast tree (or mesh for better fault tolerance). The problem of building an optimal multicast tree is akin to the widely known NP-complete Steiner tree problem [1, 2]. Since efficient implementation of multicast operation has the potential to significantly improve application or system performance, understandably many heuristic protocols exist to construct efficient multicast distribution structures.

In this chapter we are mainly concerned with multicast in Wireless Ad hoc NETworks (WANETs). In such networks battery-powered nodes, in addition to performing processing functions, route messages hop-by-hop using wireless transceivers. The optimization goal for constructing a multicast distribution structure in a WANET becomes even more challenging. Not only the bandwidth consumption needs to be optimized to conserve the scarce and shared wireless bandwidth, but also, and perhaps more importantly, consumption of battery power (or energy) (which in some cases, e.g. in some sensor networks, is non-renewable) needs to be minimized in order to maximize the *lifetime* of the network - the duration for which sufficient number of nodes in the system have ample energy to provide the desired service. In case the nodes are also mobile the applicability of maintaining a multicast distribution structure is contingent upon the cost of maintaining the distribution structure in face of disruption due to frequent mobility of nodes.

In this chapter, we present some solutions for constructing energy-efficient multicast distribution trees for WANETs. We begin with describing general techniques for conserving energy needed for communication and describe metrics for modeling cost of energy consumption which can be used to guide optimization of energy consumption for multicasting in wireless networks.

2 Energy-Efficient Communication Techniques

Compared with wired device, one of the greatest limitations of wireless devices is finite power supplies. Thus, any form of communication involving wireless devices need to be as energy-efficient as possible. In this section, we discuss some basic techniques for energy-efficient communication in WANETs. Further, we introduce the basic metrics and cost models used in constructing energy-efficient multicast distribution trees.

2.1 Energy Consumption Model for Wireless Communication

There are three major causes of energy depletion in a node: 1) energy expended for RF propagation; 2) energy expended in the transmitting hardware for operation such as encoding and modulation; and 3) energy expended in the receiving hardware for operations such as demodulation and decoding. For simplicity, we assume that the energy expended for transmission and reception is the same for all the nodes in the system and denote them by E^T and E^R, respectively. We neglect any energy consumption that occurs when the node is simply "on" (idling), although it would be easy to incorporate it into this model. The minimum energy cost (per bit) needed by node i to transmit a packet to an adjacent node j, $E_{i,j}$, is modeled as:

$$E_{i,j} = E^T + K(\max\{p_{min}, r_{i,j}^{\alpha}\}), \qquad (1)$$

where $r_{i,j}$ is the Euclidean distance between i and j, K is a constant dependent upon the properties of the antenna, and α is a constant which is dependent on the propagation losses in the medium, and p_{min} is the minimum transmission power required to send message to an arbitrarily near node (this accounts for the fact that the

$r^{-\alpha}$ dependence is only in the far-field region of the transmitting antenna) [3][4]. Depending upon the relative contribution from the distance-dependent and distance independent energy consumption factors, the radios can be categorized as: *long range radios* and *short range radios*. For long range radios, E^T is much smaller than E^{RF}_{max} (E^{RF}_{max} is the energy cost (per bit) for a node using maximum transmission power). But for short range radios (such as sensor networks), the distance independent part is the dominant factor e.g. $E^T \approx 4E^{RF}_{max}$ for some short range radios [5, 6]. A node can control its communication range by controlling the transmission power. Hence, the connectivity of network depends on the transmission power. We use wireless power control model in [6]: Each node can choose its power level p, where $0 \leq p \leq p_{max}$.

2.2 Basic Techniques for Conserving Energy

In general, there are four basic techniques for energy-efficient communication [7].

1. The first technique is to turn-off non-used transceivers to conserve energy e.g. PAMAS protocol [8].
2. The second technique is scheduling the competing nodes to avoid wastage of energy due to contention. This can reduce the number of retransmission and increase nodes' lifetime by turning off the non-used transceivers for a period of time. For example, a base station in a infrastructure based wireless network can broadcast a schedule that contains data transmission starting times for each node as in [9].
3. The third technique is to reduce communication overhead, such as defer transmission when the channel conditions are poor [10].
4. The fourth technique is to use power control to conserve energy. The transmit power P_t needed to reach a node at a distance of d is proportional to d^{α}, where $\alpha(\geq 2)$ is a transmission medium dependent constant. Hence, a node can adjust its transmission power to a level which is sufficient to reach the receiving node. This has the added advantage of reducing interference with other on-going transmissions.

In this chapter, we will mainly discuss the approaches using power control technique for energy-efficient multicast.

2.3 Wireless Multicast Advantage

A network-wide multicast distribution tree is a spanning tree which "covers" all the multicast group nodes. A *source-based multicast tree* is rooted at a source node.

On the other hand a *group-shared tree* is rooted at a *core node* (also known as rendezvous node) [11, 12]. Although all the leaf nodes in a multicast tree are necessarily multicast group nodes, the intermediate nodes (called *forwarding nodes*) may or may not be multicast group nodes. The message is distributed to all the multicast group members by flooding it on the distribution tree, i.e. the source node forwards its message to all its tree neighbors (children in a source based tree) and a forwarding node forwards the message to all its tree neighbors except the one from which it received the message. Hence, the flooding of a message on the multicast tree gets broken down into several *local forwarding* operations.

The energy consumed for a local forwarding operation (as well as the connectivity in the wireless network) depends largely on the transmission power at the nodes. Assume that each node can dynamically select its transmission power level p^{RF}, where $0 \leq p^{RF} \leq p_{max}$. Let $p_{i,j}^{RF}$ be the minimum power needed to transmit a packet over the link between nodes i and j. The power level information can be obtained from the link layer using power-control techniques [13]. Therefore, the total power expenditure of node i, when forwarding a packet to another node j, $p_{i,j}$ is:

$$p_{i,j} = \begin{cases} p_{i,j}^{RF} + p^T & \text{if } i \text{ is the source node,} \\ p_{i,j}^{RF} + p^T + p^R & \text{otherwise.} \end{cases} \tag{1}$$

Since a leaf node in a multicast tree only receives data without forwarding it further to any other node, its power expenditure is simply equal to p^R in our model.

The power consumption model can be transformed to energy consumption model by introducing time. We use τ to denote the time period for a node to transmit a bit. We assume that this quantity is the same at all the nodes. Then, the total energy (per bit) expenditure of node i, when forwarding a packet to node j, is

$$E_{i,j} = \begin{cases} E_{i,j}^{RF} + E^T & \text{if } i \text{ is source node;} \\ E_{i,j}^{RF} + E^T + E^R & \text{otherwise,} \end{cases} \tag{2}$$

where $E_{i,j}^{RF}$ is the energy cost (per bit) of the link between nodes i and j for a packet transmission, $E_{i,j}^{RF} = \tau p_{i,j}^{RF}$, E^T is energy cost (per bit) of electronics and digital processing, and E^R is energy cost (per bit) at the receiver side. In this chapter, we use $E_{i,j}^{RF}$ as the energy cost of wireless link from node i to node j. Also we assume that link costs are symmetric, i.e. $E_{i,j}^{RF} = E_{j,i}^{RF}$.

The local forwarding operation can exploit the broadcast property of wireless communication. This not only may conserve bandwidth (the message needs to be forwarded only once) and may reduce latency (the message can be simultaneously forwarded to multiple nodes), but also presents an opportunity to conserve energy. For example, consider that a (forwarding or source) node i needs to forward data to its neighbors, say, nodes j and k (see Figure 1) which may be its tree neighbors in a multicast tree. Then, by transmitting data at the power level $\max\{p_{i,j}, p_{i,k}\}$, node i can simultaneously send packets to both nodes j and k. Compared to wired networks, this reduces power consumption for forwarding in a wireless network from sum of power consumed for each forwarding link ($p_{i,j} + p_{i,k}$) to maximum of power required over all the forwarding links ($\max\{p_{i,j}, p_{i,k}\}$). This resource (bandwidth, energy, and time) saving advantage of wireless networks over wired networks is termed as **wireless multicast advantage** [3].

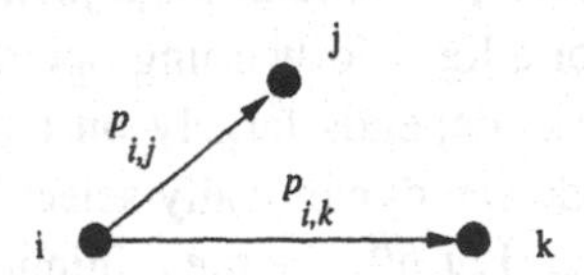

Figure 1: Wireless Multicast Advantage. $p_{i(j,k)} = \max\{p_{i,j}, p_{i,k}\}$, instead of $p_{i,j} + p_{i,k}$.

In summary, wireless medium is a broadcast medium, and one time local transmission can possibly reach all the neighbors. Power control allows a node to conserve energy by transmitting a packet at power level which is sufficient to reach only its tree neighbors (as opposed to all the neighbors) except the one it received the message from. Further, nodes which use more power can reach more nodes reducing the number of forwarding nodes needed to reach all the multicast group members (and hence the network wide multicast tree may have smaller depth). However there is a flip side to this. The node itself will consume energy at a greater rate, hence dieing (running out of energy) faster. Further, more the transmission power used by a node more interferences it causes with other simultaneous transmission[1]. This will reduce the number of simultaneous transmission. We discuss this in more detail in the next section.

2.4 Why Link-Based View is Not Suitable for WANETs?

For simplicity, consider a single source multicast is used to reach a subset of nodes in the network from a given source s. One way to achieve this is that node s simply increases its transmission range to such an extent that it can reach all the group members. In general this "single hop" approach is not universally applicable since nodes usually have a limited maximum range and multiple hops are needed to reach all the nodes in the multicast group. Further, as we will see later, use of multiple

[1]We are assuming a single wireless communication channel is used by all the nodes. Further, we assume that medium access techniques is employed by a node to get access to the channel.

hops may lead to savings in overall energy consumption for the multicast message distribution.

Since a multi-hop solution is needed, the problem becomes that of constructing a minimum cost multicast tree. In wired networks, multicast trees are built based on link costs. However, no predefined "links" exist in a wireless network. Specially, the ones which employ power control. The number of links incident on a node depends upon the power level at which the node transmits. Simply, assuming that a node always transmits at its maximum transmission power may result in too many links and a non-optimal solution. Further, due to broadcast nature of wireless transmission a single transmission may reach multiple nodes. This makes it difficult to assign a "link cost" to each link since now one has to figure out how many nodes were reachable in a single transmission. This in turn depends upon the power level employed by each node and the node distribution.

Based on the above reasons, it's more appropriate to take a node-based view of wireless networks, where in costs are only assigned to the nodes. Consequently, the problem of constructing an energy-efficient multicast tree becomes equivalent to constructing a multicast tree with *minimum/maximum summation of node cost.* We will present the existing algorithms for constructing energy-efficient multicast trees in Section 4. In the next section, we present the node costs which are suitable for constructing energy-efficient multicast trees.

3 Energy Metrics and Cost Models

In general, there are two different criteria used for energy optimization:

1. **Total Energy Consumption (TEC):** is the total energy consumption of the system to complete a given task.
2. **System Lifetime (SL):** is the volume of task that can be completed with a given energy level in the system.

For example, consider the WANET shown in Figure 2 which consists of three nodes labeled 1, 2, and 3. Assume that all the three nodes belong to a multicast group with node 1 as the source node. Depending upon whether the energy optimization goal is TEC or SL, as illustrated in Figures 3 and 4, the source based multicast tree optimizing these goals is different. Figure 3 shows the multicast tree which minimizes TEC and Figure 4 shows the multicast tree that maximizes SL. Minimum overall energy consumption is achieved by node 1 simply forwarding multicast messages to both nodes 2 and 3 using one local forwarding at the cost of 10 **energy units (EU)**[2] per packet. However, assuming that each node initially has 200

[2]EU can be replaced by appropriate energy units such as J (Joules) or mJ (milli Joules).

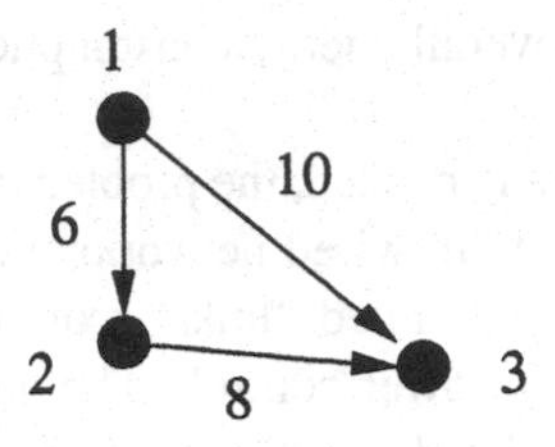

Figure 2: A topology of wireless network with 3 nodes where node 1 is the source node. The figure shows all the existing wireless links. Labels associated with the edges are the average energy cost of a node for transmitting packets over the link, such as energy cost for node 1 transmitting packet to node 2 is 6 EU/packet. Assume the residual battery energy at all three nodes is 200 EU.

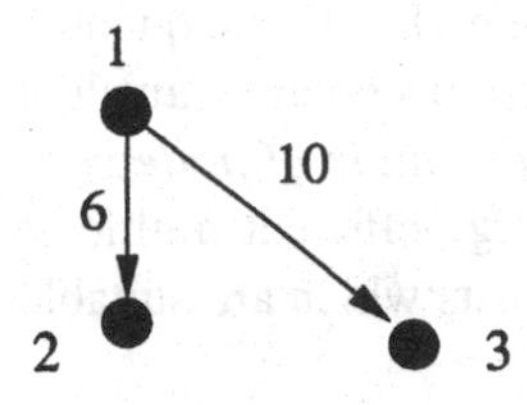

Figure 3: Minimum Energy multicast tree generated from Figure 2. Energy cost of the tree = $\max\{6, 10\} = 10$ EU/packet. Lifetime of the tree = $\frac{200}{10} = 20$ packets.

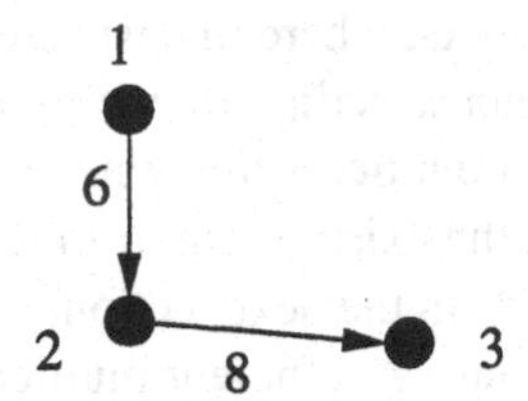

Figure 4: Maximum Lifetime multicast tree generated from Figure 2. Energy cost of the tree = $6 + 8 = 14$ EU/packet. Lifetime of the tree = $\min\{\frac{200}{6}, \frac{200}{8}\} = 25$ packets.

EU of residual battery energy, this strategy is not optimal for optimizing SL, since node 1 can forward only 20 multicast packets (200 EU ÷ 10 EU/packet) before it dies. Figure 4 shows the multicast tree which has a better lifetime since it can be used to forward 25 packets instead of just 20 packets.

3.1 Node Cost

The definition of node cost depends upon the optimization goals: minimizing energy consumption or maximizing multicast tree lifetime, as well as the type of multicast tree being used: source-based or group-shared. For minimizing energy consumption, node cost is the node energy cost, i.e. average energy cost (in terms of packets) of a node. Let E_i denote the average energy cost of node i in a multicast tree for one multicast packet transmission. For maximizing multicast tree lifetime, the expected remaining "life" of the node itself should be used as node cost. The **lifetime of a node** in a multicast tree, is the maximum number of multicast packets that may be processed (transmitted and/or received) by the node. If we use R_i to denote the residual battery energy of node i, then the lifetime of node i in a multicast tree $LT_i = \frac{R_i}{E_i}$.

3.1.1 Node Cost in Source-based Multicast Tree

In this section we restrict the discussion to calculation of a single node's energy cost for source-based multicast trees. As we have seen earlier, a node's lifetime can be derived from the node's energy cost and its residual battery energy. Let E_i^{RF} be the maximum energy cost (per bit) of the link between nodes i and its children for a multicast packet transmission. In a source-based tree a source node does not incur any reception cost and leaf nodes don't incur any transmission cost, however, a forwarding node incurs both the cost. Hence, node i's energy cost in a source-based multicast tree, E_i, can be modeled as follows:

$$E_i = \begin{cases} E_i^{RF} + E^T + E^R & \text{if node } i \text{ is not a leaf node;} \\ E_i^{RF} + E^T & \text{if node } i \text{ is the source node;} \\ E^R & \text{otherwise.} \end{cases} \tag{3}$$

In Section 4, we will discuss how BIP/MIP [3], EWMA [14], S-REMiT [15] algorithms use Equation 3 as node energy cost to construct minimum-energy source-based multicast trees.

Considering reliable multicast, link error rate needs to be incorporated into the node energy cost. Let e_i be the link error rate for node i to forward the multicast packet to all of node i's children reliably, and e_{i_parent} be the link error rate of node i's parent to forward the packet to all of its own children. The average number of retransmissions to achieve a successful transmission for a link with error probability e can be modeled as $\frac{1}{1-e}$ under the assumption that independent retransmission failures are independent of each other. Hence, Equation 3 can rewritten as follows to incorporate the energy overhead of retransmissions:

$$E_i = \begin{cases} \frac{E_i^{RF} + E^T}{1-e_i} + \frac{E^R}{1-e_{i\_parent}} & \text{if node } i \text{ is not a leaf node;} \\ \frac{E_i^{RF} + E^T}{1-e_i} & \text{if node } i \text{ is the source node;} \\ \frac{E^R}{1-e_{i,parent}} & \text{otherwise.} \end{cases} \tag{4}$$

RBIP algorithm [16] uses Equation 4 as node energy cost to construct minimum-energy source-based multicast trees.

3.1.2 Node Cost in Group-shared Multicast Tree

In contrast to source-based multicast tree, node's energy cost in a group-shared multicast tree is not only decided by the tree links attached to node but also decided by where message is coming from. For example, in Figure 2, if multicast packet coming from node 1, energy cost for node 2 to forward the packet is 8 EU/packet. If the multicast packet coming from node 3, energy cost for node 2 to forward the packet is 6 EU/packet. Assume the message generation rates at nodes 1 and 3 are 7 packets/second and 13 packets/second, respectively. Then the average energy cost of node 2 in the group-shared tree is $\frac{8*7+6*13}{7+13} = 6.7$ EU/packet. So node's cost in group-shared multicast tree needs to incorporate message generation rates of all the multicast source nodes in the group.

We use terminal node to denote a node whose degree on the tree is equal to 1. Assume that the tree links that are incident on a node are ordered in descending order of their energy cost. We use $L_i[s]$ to denote the s-th highest energy cost tree links incident on node i and $E_i[s]$ to denote the energy cost (per bit) of $L_i[s]$. If $L_i[s]$ does not exist, then $E_i[s] \equiv 0$. Suppose L_i is the branch on which the multicast message arrives to be forwarded by node i. Based on the energy cost model in Section 3.1, the energy cost (per bit) of node i, EC_i, for forwarding message arriving on its incident branch L_i in multicast tree T is:

$$E_i = \begin{cases} E^T + E_i[2] + E^R & \text{if } L_i = L_i[1] \text{ and node } i \text{ is not a terminal node} \\ E^R & \text{if } L_i = L_i[1] \text{ and node } i \text{ is a terminal node} \\ E^T + E_i[1] & \text{if } L_i \neq L_i[1] \text{ and } i \text{ is the source node} \\ E^T + E_i[1] + E^R & \text{otherwise} \end{cases} \tag{5}$$

If a multicast message arrives on branch $L_i[1]$ and node i is not a terminal node, the energy cost of i to forward this message is $E^T + E_i[2] + E^R$; however, if i is a terminal node, the energy cost of i is only the reception energy cost E^R. Otherwise, if a multicast message arrives on a branch other than branch $L_i[1]$ and i is not the multicast source, the energy cost of node i to forward this message is $E^T + E_i[1] + E^R$. In the cases that node i is the multicast source, the energy cost of i is just the cost for sending the multicast message to all of its connected tree neighbors, that is $E^T + E_i[1]$.

3.2 Multicast Tree Cost

Similar to node cost, multicast tree cost is also dependent on the above two optimization goals. Note that these two different optimization goals may conflict with each other. Minimum energy consumption multicast tree may result in rapid depletion of energy at intermediate nodes possibly leading to network partition and interruption of the multicast service. On the other hand, a maximum lifetime multicast tree may not include all of the minimum energy routes. For example, Figure 2 is the network topology of three nodes, and Figure 3 and Figure 4 are the minimum energy and maximum lifetime multicast tree, respectively. The two different multicast tree costs are defined as follows:

3.2.1 Minimizing Energy Consumption:

The Total Energy Consumption (TEC) by all the transceivers in the multicast tree T for multicasting one packet is,

$$TEC(T) = \sum_{i \in T} E_i.$$

Using this definition, we can describe the problem of minimizing energy consumption of multicast tree as follows: Let T_G denote the set of all possible multicast trees for a fixed multicast group G. The minimum energy consumption multicast tree T^* is:

$$T^* = \arg \min_{\forall T \in T_G} \{TEC(T)\} = \arg \min_{\forall T \in T_G} \{\sum_{\forall i \in T} E_i\}.$$

This cost requires to carefully choose intermediate transceivers such that the overall energy consumed for the multicasting packet is minimized. Optimizing this cost has been proven to be a NP-complete problem [1, 2]. It is difficult to achieve this cost as it is not easy to select the appropriate intermediate nodes.

3.2.2 Multicast Tree Lifetime

Multicast tree lifetime can be defined as the time duration starting from beginning of multicast service until the first node in the network fails due to battery energy exhaustion. This cost is very important for some critical applications, such as battlefield ad hoc networks. The lifetime of multicast tree is :

$$LT(T) = \min_{i \in T} LT_i.$$

Using this definition, the problem of maximizing lifetime of a multicast tree can be stated as follows: Let T_G denote the set of all possible multicast trees for a fixed multicast group G. The maximum lifetime multicast tree $T^\diamond$ is:

$$T^\diamond = \arg \max_{\forall T \in T_G} \{LT(T)\} = \arg \max_{\forall T \in T_G} \{\min_{\forall i \in T} \frac{R_i}{E_i}\}$$

However, optimizing for this cost is also very difficult, because we need to select the nodes involved in multicasting in such a way that the energy of all the nodes are depleted uniformly. In other words, given a network, we need to determine some nodes that are bottleneck nodes meaning that their energy gets depleted faster than other nodes. The lifetime of the bottleneck nodes determine the lifetime of the entire multicast tree. So optimizing this cost is equivalent to maximizing the lifetime of bottleneck nodes. This problem is similar to the "load balancing" problem where tasks need to be sent to one of the many servers available so that the response time is minimized – this is know to be a NP-complete problem [8].

4 Constructing Energy-Efficient Multicast Trees

In this section, we discuss algorithms for constructing energy-efficient multicast trees. As we discuss in Section 2.4, the problem of constructing energy-efficient multicast tree in a WANET is NP-complete [1, 2]. Several heuristic algorithms have been developed. In the following, we discuss several protocols for constructing energy-efficient source-based multicast trees as well as group-shared multicast trees using power control technique.

4.1 Protocols

The protocols for constructing energy-efficient multicast distribution trees can be categorized as: centralized and distributed protocols. A centralized protocol has two limitations: first, it needs global knowledge which may introduce high communication overhead especially in large scale networks; second, it would be very "expensive" if it runs repeatedly to adapt to the dynamic changes in the network, such as the remaining battery power at nodes. Since this would involve periodic sending of state information to a centralized node – a costly operation for a energy-constrained network. Given these drawbacks of a centralized approach, distributed protocols have been designed which are more suitable for the energy-efficient multicast tree problem in WANETs [14, 3, 17].

4.1.1 Centralized Protocols

Broadcast Incremental Power (BIP) algorithm is a centralized algorithm to construct a minimum-energy source-based broadcast tree [3]. BIP algorithm uses the node energy cost in Equation 3, but it neglects E^T and E^R. So BIP algorithm only considers the energy cost on the link for multicast packet transmission. BIP is similar to Prim's algorithm for constructing minimum cost spanning tree, except that it considers only "incremental energy cost" in deciding which link to add to the tree.

It constructs the tree starting from the source node, and then incrementally absorbs other nodes in the network sequentially as follows.

1. **Incremental Energy Cost Calculation:** For each node i which is already in the tree, and each node j which is not yet in the tree, the incremental energy cost associated with adding node j as node i's children is: $\Delta_{i,j} = E_{i,j} - E_i$.

2. **Absorb Node with Minimum Incremental Cost:** Find nodes i and j with the minimum value of $\Delta_{i,j}$. Absorb node j to the tree as node i's children.

As an extension of BIP, Multicast Incremental Power (MIP) algorithm is proposed for building a source-based multicast tree by eliminating all redundant transmissions (the transmissions that are not used to reach any member of the multicast group from the BIP tree) by pruning the tree [3]. To extend the network lifetime, BIP/MIP incorporates the initial battery level and residual battery level in the node cost computation [18]. To achieve reliable multicasting, Reliable BIP (RBIP) algorithm is proposed by incorporating the link error rate into the node cost for building minimum energy source-based broadcast/multicast tree [16]. It has been have proven that the approximation ratio of Minimum Spanning Tree (MST) [3] is between 6 and 12, and the approximation ratio of BIP is between $\frac{13}{3}$ and 12 [17].

Embedded Wireless Multicast Advantage (EWMA) algorithm also constructs a minimum-energy source-based broadcast tree [14]. Similar to BIP/MIP, EWMA algorithm also uses the node energy cost in Equation 3, but it neglects E^T and E^R. EWMA uses MST as the initial tree. Then, EWMA refines the initial tree to minimize TEC of the source-based broadcast tree. For example, the refinement at node i as follows.

1. **New Transmission Energy Selection:** Node i selects a downstream node, say node j. Then node i increases its transmission energy to cover all of node j's children. The incremental energy of node i is

$$\Delta E_i^j = max_{k \in j's\ children}\{E_{i,k}\} - E_i.$$

Calculate the energy $Gain$ as

$$Gain_i^j = \sum_{k \in Eliminated_i} E_k - \Delta E_i^j,$$

where $Eliminated_i$ is the set of nodes whose transmissions were eliminated when node i increased its power level.

[3]MST is built based on the link cost. So MST may not be minimum energy multicast tree in WANET.

Table 1: Characteristics of current approaches for constructing energy-efficient multicast tree in WANET.

Algorithms	Source-based	Group-shared	Min. Energy	Max. Lifetime	Reliablity	Dist./Centralized
EWMA [14]	X		X			C,D
BIP/M-IP [3, 18]	X		X	X		C
Dist-BIP-A, Dist-BIP-G [19]	X		X			D
RBIP [16]	X		X		X	C
G-REMiT [20], S-REMiT [15]	X	X	X			D
L-REMiT [21]	X			X		D

2. **Exclude Transmissions:** Selects the node j with highest positive $Gain$. Then increase transmission energy of node i to cover j, eliminate all the transmissions at node i's downstream, which are already covered by node i.

The refinement phase is from source node leaf node in EWMA.

4.1.2 Distributed Protocols

Now let us discuss some distributed algorithms for constructing energy-efficient multicast trees.

BIP algorithm has two distributed versions: Distributed-BIP-All (Dist-BIP-A) and Distributed-BIP-Gateways (Dist-BIP-G) [19]. In Dist-BIP-A algorithm, each node first constructs a BIP tree locally. Then the source node starts to broadcast its local tree structure to all of its neighbors. When a node hears from a node that already in the tree it broadcasts its own locally generated BIP tree to all of its neighbors. After several iterations, the Dist-BIP-A tree covers all of the nodes in the network. Dist-BIP-G assumes two-hop neighbors' information are available at every node. In Dist-BIP-G algorithm, source node starts to build BIP tree locally to cover its two-hop neighbors. And *gateway* node is used to denote a neighbor of the source node that connects one or more of the source node's two-hop neighbors. Only the gateway nodes of node i broadcasts the locally generated BIP tree. This reduces the number of local broadcasts leading to conservation of bandwidth and reduction in contention. Both Dist-BIP-A and Dist-BIP-G algorithms have slightly worse performance than the centralized version of the BIP algorithm.

EWMA algorithm also has a distributed version, called EWMA-Dist [14]. EWMA-Dist is organized by rounds in the refinement phase. Every node in EWMA-Dist has two-hop neighbor information. In each round, a node tries to decrease the total energy consumption (from node's point of view) by increasing its transmission power to eliminate some transmissions of its children nodes. In EWMA-Dist, the refinement starts from the source node and ends at the leaf nodes.

REMiT (Refining Energy efficiency of Multicast Trees) is a suite of distributed algorithms for enhancing energy-efficiency of multicast trees [15]. REMiT algorithms refine energy-efficiency of a pre-existing multicast tree by switching some tree nodes from one branch (forwarding) node to another. REMiT algorithms are categorized along energy-metric dimension (minimizing energy-consumption or maximizing lifetime) and multicast-tree type dimension (source-based or group-shared tree). For example, S-REMiT [15] and G-REMiT [20] are used for minimizing energy-consumption of a source-based and group-shared multicasting tree, respectively. And L-REMiT [21] can extend the tree-lifetime of a source-based multicast tree.

Now let us use S-REMiT as an example to discuss how to construct an energy-efficient source-based multicast tree in REMiT. It is divided into two phases. In the **first phase**, S-REMiT builds an initial multicast tree using available tree building algorithms, such as [22]. In the **second phase**, S-REMiT selects a node in the tree, say node i. Let x to denote the parent node of node i. Following is the S-REMiT algorithm at node i used to refine the multicast tree.

1. **New parent selection:** Node i selects a new parent node j with the highest positive energy $Gain$: $Gain_i^{x,j} := (E_i + E_x + E_j) - (E_i' + E_x' + E_j')$, where E_i and E_i' is the energy cost of node i before and after node i changes its parent node from x to j, respectively. With the help of one-hop neighbor information, node i calculates E_i and E_i' using Equation 3, respectively.

2. **Switch to New Parent:** Node i deletes the tree link between nodes i and x and adds a tree link between nodes i and j.

The above steps is called as *a refinement* at node i. Every node in the tree does its one refinement sequentially. Every node finishes one refinement, is called *one refinement round* of S-REMiT, If there is no energy $Gain$ in a refinement round of S-REMiT, S-REMiT stops the refinement process. Then S-REMiT prunes the source-based multicast tree by deleting the redundant transmissions that are not needed to reach the members of the multicast group from the tree. After the pruning, S-REMiT will be terminated.

L-REMiT algorithm also uses refinement on the initial tree to extend lifetime of source-based multicast tree. The lifetime of a multicast tree is decided by the bottleneck nodes, which are the nodes with minimum lifetime in the tree [21]. Then L-REMiT only needs to refine the lifetime of bottleneck nodes as much as possible so that the multicast tree's lifetime is extended as follows: L-REMiT selects a bottleneck node in the tree, say node x. Let node i be the farthest (costliest) child node of node x. Following is the L-REMiT algorithm at node i used to refine the multicast tree.

1. **New parent selection:** Node i selects a new parent node j with the highest positive lifetime $LGain$: $LGain_i^{x,j} := min\{LT'_i, LT'_x, LT'_j)\} - LT_x$, where LT_x and LT'_x is the lifetime of node x before and after node i changes its parent node from x to j, respectively.

2. **Switch to New Parent:** Node i deletes the tree link between nodes i and x and adds a tree link between nodes i and j.

After each refinement step, a new bottleneck node needs to be recomputed. The refinement is then applied to the new bottleneck node. Refinement continues till no positive lifetime gain is achieved.

Similar to S-REMiT, G-REMiT algorithm uses refinement on the initial tree to construct energy-efficient group-shared multicast tree. In a group-shared tree, when a node switches its connected tree branch to another node, energy cost of some other nodes which are even not involved in the tree branch switching may also be affected [20]. So G-REMiT is more complex in node cost computation and tree branch switching operation than S-REMiT. Table 1 shows the characteristics of the various protocols we discussed in this section.

4.2 Qualitative Analysis

In this section, we will analyze the current energy-efficient multicast tree building algorithms. For the comparison, we use propagation model $E_{i,j}^{RF} = K(r_{i,j})^{\alpha}$, where $r_{i,j}$ is the Euclidean distance between i and j, K is a constant dependent upon the properties of the antenna and α is a constant which is dependent on the propagation losses in the medium.

Let us first compare some of the current algorithms in building minimum energy source-based multicast tree. BIP/MIP algorithm constructs minimum energy multicast tree by adding nodes in the tree sequentially. So the sequential order will

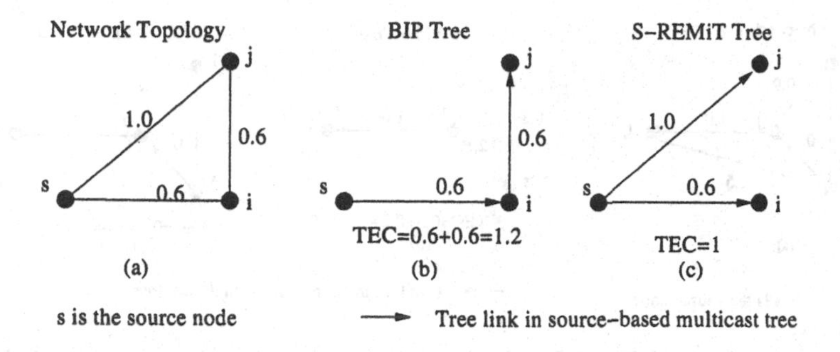

Figure 5: BIP vs S-REMiT in building source-based multicast tree. Label associated with the link is the energy cost of the link assuming that $E^T = 0$ and $E^R = 0$.

affect the performance of BIP/MIP algorithm. Also when adding a node in the tree, BIP/MIP tries to minimize the additional cost to absorb the node in the tree by minimizing the increment cost of the new link. In other words, BIP/MIP uses a link-based approach to deal with a node-based problem. Let us use an example to illustrate the pitfalls of BIP/MIP algorithm. A network topology is shown in Figure 5(a). The multicast tree generated by BIP/MIP and S-REMiT is shown in Figures 5(b) and (c), respectively. For presentation simplicity, we assume that $E^T = 0$ and $E^R = 0$ (this assumption is also used for comparison between EWMA and S-REMiT). As shown in Figure 5(b), if source node s adds nodes j first or adds nodes i and j together, then TEC of the multicast trees will be lower than if it adds node i first. So the scheme of adding nodes sequentially is not very efficient for constructing source-based energy efficient multicasting tree. EWMA algorithm is a greedy algorithm, it tries to reduce the number of downstream transmitting nodes as much as possible when there is a chance to reduce the TEC of the multicast tree. However, it neglects the contribution of nodes, whose depth are same or larger, for minimizing TEC of the tree. For example, EWMA generates a multicast tree shown in Figure 6(b). Since the depth of node j is same as node i, node j has no chance to be i's parent in EWMA algorithm. But in S-REMiT, node j will become node i's parent as shown in Figure 6(c). This is because every node has the same chance to switch its parent in S-REMiT. So S-REMiT may perform better than BIP/MIP and EWMA algorithm for building energy efficient source-based multicast tree.

Second, we discuss the current algorithms in building minimum energy group-shared multicast tree. Most of the current algorithms only try to minimize energy consumption of source-based multicast tree. A node in a source-based tree only forwards the multicast message from its parent node to its children nodes. But a

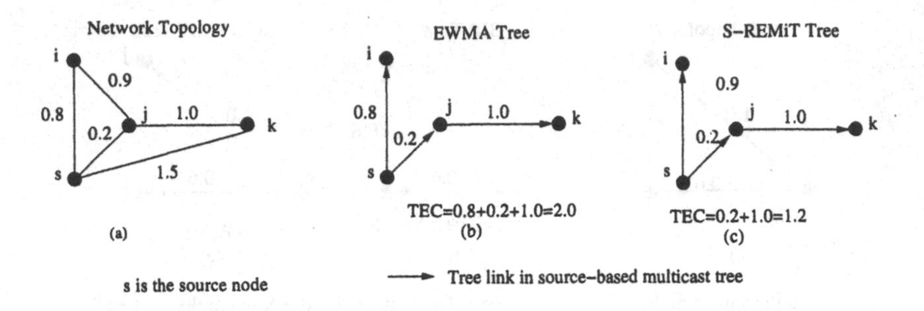

Figure 6: EWMA vs S-REMiT in building source-based multicast tree. Label associated with the link is the energy cost of the link assuming that $E^T = 0$ and $E^R = 0$.

node in a group-shared tree needs forward the multicast message to all of its connected tree neighbors except the node from which it received the multicast message. Therefore, the energy cost of the node in source-based tree is only decided by the energy cost of the links to its farthest children [14, 15]. Because the energy cost of the node in group-shared tree is not only decided by the energy cost of the links which are between the node and its connected tree neighbors but also decided by where the message is coming from. In other words, BIP/MIP and EWMA algorithms should not use energy cost of source-based multicast tree to build minimum energy group-shared multicast tree. And if a minimum energy source-based tree is used as a group-shared tree, it may not be energy efficient any more. So when the group-shared multicast tree is more energy-efficient to a specific source node, the tree will be less energy-efficient to other source nodes.

For example, in Figure 7, nine nodes are distributed in a region with dimensions 15x8. Nodes 1, 2,..., and 8 are group nodes, node 9 is not a group node, and all the group nodes have different multicast message generation rate. Node 1 is the source node when we applying MIP and EWMA algorithms. From Figure 7, we find that TEC of EWMA is highest. It means that EWMA algorithm is not energy-efficient compared with other algorithms. This is because EWMA algorithm tries to minimize the energy consumption of node 1's source-based multicast tree by minimizing the depth of the tree as much as possible. But once node 1's source-based tree is used as group-shared tree, the group-shared tree is not energy-efficient when other nodes also serve as multicast sources. However, since distributed version of EWMA (EWMA-Dist) cannot reduce the depth of the tree as much as EWMA centralized counterpart, EWMA-Dist performs better than EWMA centralized version when the tree is used as a group-shared tree. As shown in Figure 7, TEC of MIP tree is a little smaller than MST but higher than G-REMiT algorithm. It means

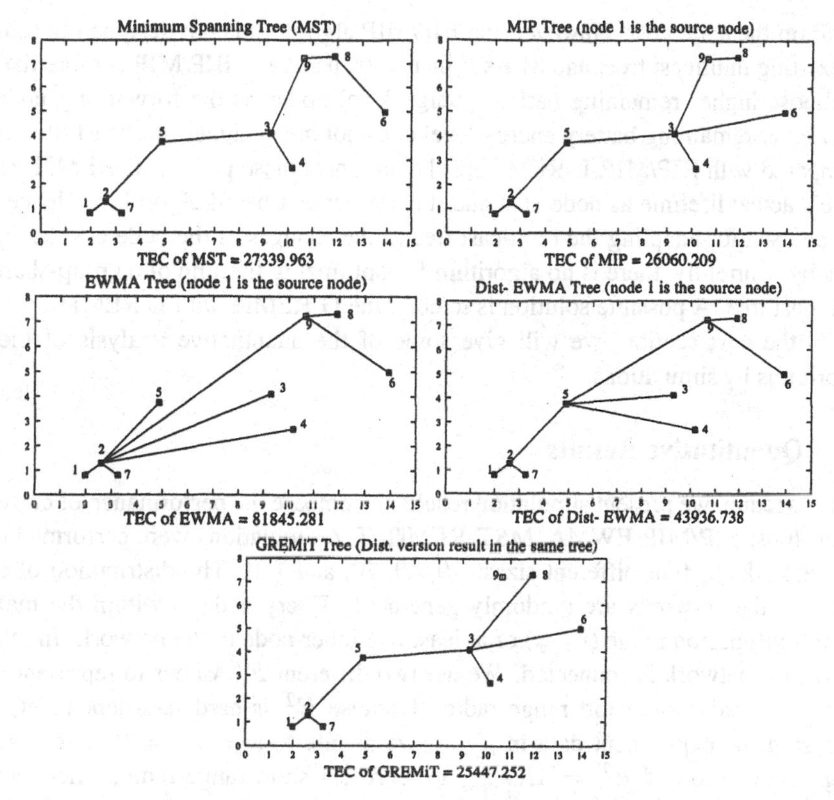

Figure 7: Different Multicast Trees generated by the algorithms with propagation loss exponent $\alpha = 2$, and $\lambda_1 = 86kb/s, \lambda_2 = 84kb/s, \lambda_3 = 80kb/s, \lambda_4 = 98kb/s, \lambda_5 = 44kb/s, \lambda_6 = 97kb/s, \lambda_7 = 13kb/s, \lambda_8 = 46kb/s$ (Node 1,2,..,8 are group nodes, but node 9 is not a group node).

that MIP algorithm performs better than MST algorithm but worse than G-REMiT algorithm.

Third, we present some results for current algorithms for optimizing lifetime of source-based multicast tree: BIP/MIP and L-REMiT. BIP/MIP uses a node cost C_i ($C_i = E_i(\frac{R_i(0)}{R_i(t)})^\beta$, where E_i is average energy cost of node i in the multicast tree for one multicast packet transmission, $R_i(0)$ and $R_i(t)$ are the residual battery energy of node i at time 0 and t, respectively, and $0 \leq \beta \leq 2$ is the weighting factor). However, this node cost C_i for BIP/MIP algorithms has following three limitations: 1) C_i is not node i's actual lifetime cost, even when $\beta = 1$; 2) Node cost is a function of time, so BIP/MIP algorithms should periodically refine the multicast tree based on the node cost. However, the BIP/MIP algorithm is not designed to refine an existing multicast tree; and 3) As β increases, using C_i, BIP/MIP is more likely to choose higher remaining battery energy level nodes as the forwarding nodes. But higher remaining battery energy level does not mean higher multicast lifetime. Compared with BIP/MIP, L-REMiT [21] overcomes these pitfalls. L-REMiT uses node's actual lifetime as node cost and it is refinement-based algorithm. Hence, it can be used for adapting the multicast tree to the changes of the node cost as time goes by. Currently, there is no algorithm for optimizing lifetime of a group-shared multicast tree. A possible solution is to combine G-REMiT with L-REMiT.

In the next section, we will give some of the quantitative analysis of these algorithms by simulations.

4.3 Quantitative Results

In this section, we present simulation results to compare the performance of current algorithms: BIP/MIP, EWMA, MST, REMiT. The simulations were performed using networks of four different sizes: 10, 40, 70, and 100. The distribution of the nodes in the networks are randomly generated. Every node is within the maximum transmission range (r_{max}) of at least one other node in the network. In other words, the network is connected. We use two different E^T values to represent the long range radio and short range radio. Because E^T is hardware dependent, by analyzing the experiment data in [5, 6], we decide to use $E^T = 0$ to represent long range radio and $E^T = 4Kr_{max}^{\alpha}$ to represent short range radio, where K is constant dependent upon the properties of the antenna and α is a constant which is dependent on the the propagation losses in the medium. We ran 100 simulations for each simulation setup consisting of a network of a specified size to obtain average TEC, the propagation loss exponent α is varied from 2 to 4.

Let the five types of algorithms be denoted by T_{alg}, where $alg \in A = \{$REMiT, EWMA, EWMA-Dist, MST, BIP/MIP$\}$. We use $TEC(T)$ to denote the TEC value of a tree T. We use TEC of multicast tree to define the performance metric: $Normalized\,TEC$ with algorithm alg is:

$$\frac{TEC_{alg}}{TEC_{best}},$$

where $TEC_{best} = \min(TEC(T_{alg})), alg \in A$. Using this metric, we can compare the minimum energy multicast trees obtained from REMiT, EWMA, EWMA-Dist, MST, and BIP/MIP are in terms of their TEC value.

We also use *Normalized Lifetime* as the performance metric:

$$\frac{LT(T_{alg})}{LT(T_{best})},$$

where $LT(T_{best}) = \max\{LT(T_{alg})\}, alg \in A$.

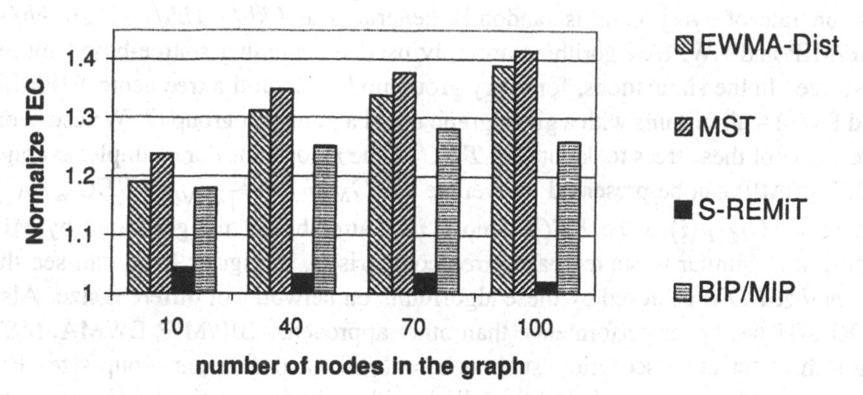

Figure 8: Average *Normalized TEC* for long range radios when 50% nodes are in multicast group for source-based multicast tree ($\alpha = 2, r_{max} = 10, K = 1, E^T = 0, E^R = 0.1 * K(r_{max})^\alpha$).

Figure 8 shows the average *Normalized TEC* (shown on the vertical axis) achieved by the algorithm on networks of different size (horizon axis). The figure show that the solutions for minimum energy source-based multicast tree obtained by S-REMiT have, on the average, lower *Normalized TEC* than the other solutions (This is also true for other scenarios, such as $\alpha = 3, 4$, different group size, short range radios.).

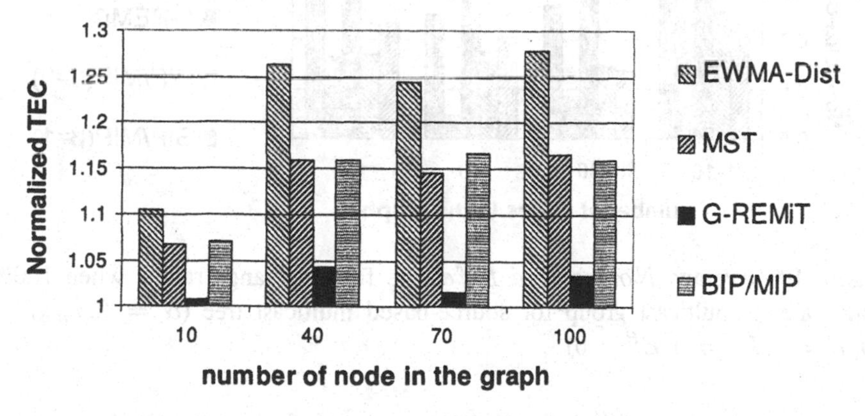

Figure 9: Average *Normalized TEC* for long range radios when 100% nodes are in multicast group for group-shared multicast tree ($\alpha = 2, r_{max} = 10, K = 1, E^T = 0, E^R = 0$).

To evaluate the performance of algorithms in building group-shared multicast tree, we assume that the link bandwidth is $100kb/s$, and multicast message generation rate of every node is randomly generated as $0kb/s, 1kb/s, ..., 100kb/s$. BIP/MIP and EWMA algorithms are only used for building source-based multicast tree. In the simulations, for every group node we build a tree using BIP/MIP and EWMA algorithms with a given graph o and a multicast group G. We use average TEC of these trees to denote the TEC of the algorithm. For example, average TEC of MIP can be presented as average $TEC_{MIP} = \frac{1}{|G|}\sum_{\forall i \in G} TEC_{MIP}(i)$, where $TEC_{MIP}(i)$ is the TEC of node i's source-based tree generated by MIP algorithm. Similar to source-based tree comparison, in Figure 9, we can see the *Normalized TEC* achieved by these algorithms on networks of different size. Also G-REMiT has better performance than other approaches BIP/MIP, EWMA, MST algorithms for other scenarios, such as $\alpha = 3$ or 4 and different group size. But for short range radios, we found that all algorithms have very similar performance for constructing group-shared multicast tree. The main reason for such behavior is that the node's E^T is substantially greater than the maximum value of Kr_{max}^{α}. So the energy saving by transmission power control is very little. Consequently, the simplest and reasonable approach for short range radio is that every node tries to cover as many nodes as possible.

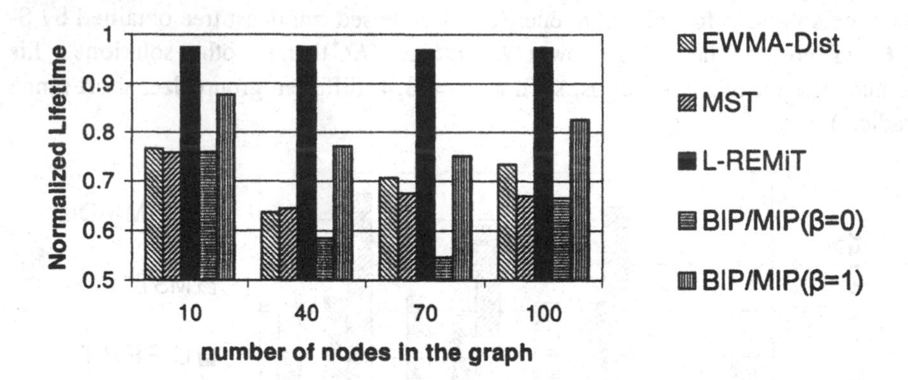

Figure 10: Average $Normalized\ Lifetime$ for long range radios when 100% nodes are in multicast group for source-based multicast tree ($\alpha = 2, r_{max} = 10, K = 1, E^T = 0, E^R = 0$).

In Figure 10, we can see L-REMiT has, on the average, higher *Normalized Lifetime* than those multicast tree obtained by other algorithms. This is also true for other scenarios, such as $\alpha = 3, 4$, different group size, and short range radios.

According to the simulations results, we find that REMiT algorithms have better performance than BIP/MIP, EWMA, and MST for various scenarios.

5 A Framework for Energy-Efficient Multicast

In this section, we present a general framework for developing energy-efficient multicast schemes, which takes into account application-specified QoS (such as reliability and delay constraints), while adapting to various network parameters (such as interference, mobility and link error rate). The framework is shown in Figure 11. Following are the components in the framework for energy efficient multicasting:

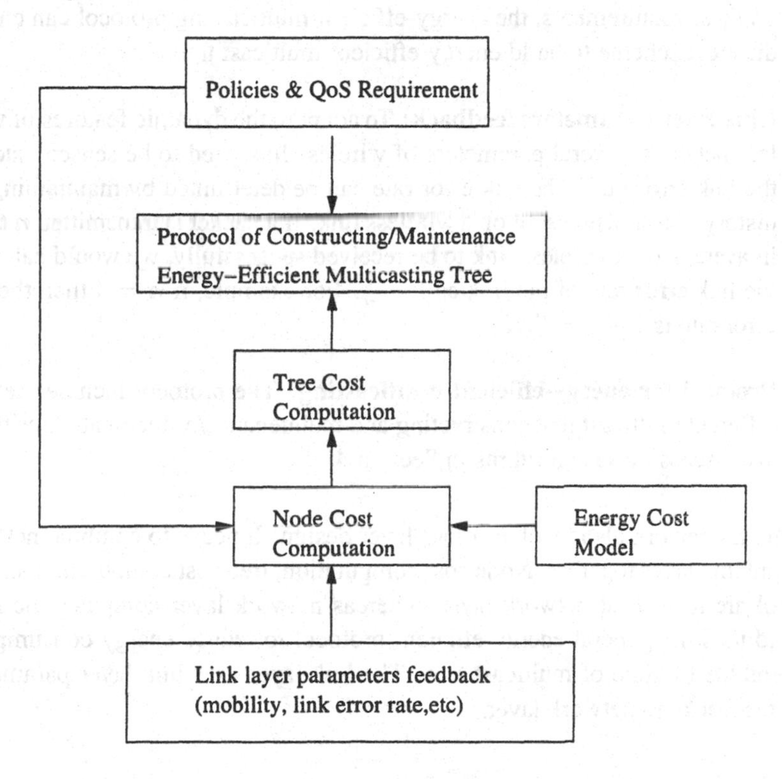

Figure 11: A Framework for Energy-Efficient Multicast.

1. **Energy Cost Model:** We have discussed energy cost model in Section 3. According to the nature of wireless transceivers, different energy cost metric can be applied.
2. **Node Cost Computation:** We have discussed this issue in Section 3.1. As shown in Figure 11, node cost computation uses application specified QoS (such as delay constraints), application specified policies (such as minimize

energy consumption and maximize lifetime), energy cost model, and wireless network parameters (such as link error rate) as inputs.

3. **Tree Cost Computation:** Using results of node cost computation, we can compute cost of the multicast tree as shown in Section 3.2.

4. **Policies and QoS Requirements:** Applications can specify optimization goals and QoS constraint (such delay and reliability) in the framework. Based on these requirements, the energy-efficient multicasting protocol can choose different scheme to build energy-efficient multicast tree.

5. **Link layer parameters feedback:** To adapt to the dynamic features of wireless network, several parameters of wireless link need to be sensed, such as the link error rate. The link error rate can be determined by maintaining the history of retransmission on a wireless link. If a packet is transmitted n times in average on a wireless link to be received successfully, we would calculate the link error rate of the link as $1 - \frac{1}{n}$. For example, if $n = 4$ then the link error rate is $1 - \frac{1}{4} = 0.75$.

6. **Protocol for energy-efficient multicasting:** The protocol includes several different multicast tree constructing and maintenance/refinement algorithms. We discuss these algorithms in Section 4.

This framework also involves cross layer design. It needs to combine network layer and link layer together. Node cost computation, tree cost computation and the protocol are located at network layer, whereas network layer computes the node cost and uses the general energy-efficient protocol to reduce energy consumption or extend the lifetime of multicast tree. The link layer uses link layer parameters as the feedbacks to network layer.

6 Conclusions

In this chapter, we presented general techniques for conserving energy in wireless communication. Then we described the metrics for modeling cost of energy consumption and different optimization goals for energy-efficient multicasting. We also discussed the existing solutions for constructing energy-efficient multicast distribution trees for WANETs and gave the qualitative analysis and simulations results of these solutions. Finally, we presented a framework for energy-efficient multicasting.

7 Acknowledgments

We thank the anonymous reviewers for their insightful comments which helped to improve the quality of the chapter. This work is supported in part by NSF grants ANI-0123980 and ANI-0196156.

References

[1] A. E. F. Clementi, P. Crescenzi, P. Penna, G. Rossi, and P. Vocca, On the complexity of computing minimum energy consumption broadcast subgraphs, *Proceedings of 18th Annual Theoretical Aspects of Comp. Sc. (STACS)* Vol. 2010 (Springer-Verlag, 2001) pp. 121-131.

[2] F. Li and I. Nikolaidis, On minimum-energy broadcasting in all-wireless networks, *Proceedings of the 26th Annual IEEE Conference on Local Computer Networks (LCN 2001)* (Tampa, Florida, Nov. 2001) pp. 193-202.

[3] J. E. Wieselthier, G. D. Nguyen, and A. Ephremides, On the construction of energy-efficient broadcast and multicast tree in wireless networks, *Proceedings of IEEE INFOCOM 2000* (Tel Aviv, ISRAEL, Mar. 2000) pp. 585-594.

[4] J. E. Wieselthier, G. D. Nguyen, and A. Ephremides, Resource-limited energy-efficient wireless multicast of session traffic, *34th Annual Hawaii Int'l Conf. System Sciences* (Maui, Hawaii, Jan. 2001) pp. 3460-3469.

[5] R. Min and A. Chandrakasan, Top five myths about the energy consumption of wireless communication, *ACM Sigmobile Mobile Communication and Communications Review (MC2R)* Vol.6 No.4 (Oct. 2002).

[6] L. M. Feeney and M. Nilsson, Investigating the energy consumption of a wireless network interface in an ad hoc networking environment, *Proceedings of IEEE INFOCOM* (Anchorage, AK, Apr. 2001) pp. 1548-1557.

[7] C. E. Jones, K. M. Sivalingam, P. Agrawal, and J. C. Chen, A survey of energy efficient network protocols for wireless networks, *Wireless Networks* Vol.7 No.4 (2001) pp. 343-358.

[8] S. Singh, C. S. Raghavendra, and J. Stepanek, Power-aware broadcasting in mobile ad hoc networks, *Proceedings of PIMRC'99 Conference* (Osaka, Japan, Sep. 1999) pp. 22-31.

[9] K. M. Sivalingam, J. C. Chen, P. Agrawal, and M. B. Srivastava, Design and analysis of low-power access protocols for wireless and mobile ATM networks, *Wireless Networks* Vol.6 No.1 (2000) pp. 73-87.

[10] M. Zorzi and R. R. Rao, Energy constrained error control for wireless channels, *IEEE Personal Communications* Vol.4 (Dec. 1997) pp. 27-33.

[11] C. C. Chiang, M. Gerla, and L. Zhang, Adaptive shared tree multicast in mobile wireless networks, *Proceedings of IEEE Globecom'98* (Sydney, Australia, Nov. 1998) pp. 1817-1822.

[12] S. J. Lee, W. Su, J. Hsu, M. Gerla, and R. Bagrodia, A performance comparison study of ad hoc wireless multicast protocols, *Proceedings of the IEEE INFOCOM 2000* (Tel Aviv, Israel, Mar. 2000) pp. 565-574.

[13] P. Karn, MACA - a new channel access method for packet radio, *Proceedings of the 9th ARRL Computer Networking Conference* (Ontario, Canada, 1990).

[14] M. Cagalj, J. P. Hubaux, and C. Enz, Minimum-energy broadcast in all-wireless networks: NP-Completeness and distribution issues, *Proceedings of ACM MobiCom 2002* (Atlanta, Georgia, Sep. 2002) pp. 172-182.

[15] B. Wang and S. K. S. Gupta, S-REMiT: An algorithm for enhancing energy-efficiency of multicast trees in wireless ad hoc networks, *Proceedings of IEEE 2003 Global Communication Conference (GLOBECOM 2003)* Vol.6 (San Francisco, CA, Dec. 2003) pp. 3519-3524.

[16] S. Banerjee, A. Misra, J. Yeo, and A. Agrawala, Energy-efficient broadcast and multicast trees for reliable wireless communication, *IEEE Wireless Communications and Networking Conf. (WCNC)* (New Orleans, Louisiana, Mar. 2003).

[17] P. J. Wan, G. Calinescu, X. Y. Li, and O. Frieder, Minimum-energy broadcast routing in static ad hoc wireless networks, *Proceedings of the IEEE INFOCOM 2001* (Anchorage, Alaska, Apr. 2001) pp. 1161-1171.

[18] J. E. Wieselthier, G. D. Nguyen, and A. Ephremides, Resource management in energy-limited, bandwidth-limited, transceiver-limited wireless networks for session-based multicasting, *Computer Networks* Vol.39 No.2 (2002) pp. 113-131.

[19] J. E. Wieselthier, G. D. Nguyen, and A. Ephremides, Distributed algorithms for energy-efficient broadcasting in ad hoc networks, *IEEE Military Communications Conf.* (Anaheim, CA, Oct. 2002).

[20] B. Wang and S. K. S. Gupta, G-REMiT: An algorithm for building energy efficient multicast trees in wireless ad hoc networks, *Proceedings of 2nd IEEE International Symposium on Network Computing and Applications (NCA-03)* (Cambridge, MA, Apr. 2003) pp. 265-272.

[21] B. Wang and S. K. S. Gupta, On maximizing lifetime of multicast trees in wireless ad hoc networks, *Proceedings of International Conference On Parallel Processing (ICPP-03)* (Kaohsiung, Taiwan, China, Oct. 2003) pp. 333-340.

[22] R. G. Gallager, P. A. Humblet, and P. M. Spira, A distributed algorithm for minimum weight spanning trees, *ACM Trans. Programming Lang. & Systems* Vol.5 No.1 (Jan. 1983) pp. 66-77.

IP3S: a Framework for Power-Conserving Multicast and Broadcast Algorithms in Static Wireless Ad Hoc Networks

Hung Q. Ngo, Dazhen Pan, Shiva-Shankar Ramanna, and Suchita Kaundin
Computer Science and Engineering Department,
State University of New York at Buffalo, Amherst, NY 14260, USA.
E-mail: {hungngo, dpan, sramanna, skaundin}@cse.buffalo.edu

Contents

1 Introduction

Energy conservation in ad hoc networks is of paramount importance. In some applications, in fact, energy is entirely non-renewable [10, 23]. Moreover, in many typical ad hoc network applications such as mobile conferencing, emergency services, and battle field communications, multicasting and broadcasting are the most natural communication primitives. They have received special attention from various researchers in recent years [2–5, 18–22, 26–28, 30–34].

The power-aware broadcast/multicast problems we are facing are assumed to be *source initiated* (as opposed to *topology control oriented* [6, 16]), namely some distinguished node among a set of given nodes initiates the process of finding a power assignment vector which indicates the transmission power level at each node in the network. Also, we assume *omnidirectional antennas* are used so that given a power assignment at a node, say u, all nodes within a certain radius of u are reachable from u in a single hop. The power assignment needs to satisfy the condition that all nodes in the network (in the broadcast case) or all nodes in a multicast group (in the multicast case) are reachable from the source node, possibly in several hops. The problem is to find such a power assignment vector with minimal total power.

The problem of finding an energy-optimal power assignment, for either the broadcast or multicast version, is **NP**-hard, as repeatedly shown by many authors [2, 3, 18]. Hence, this is naturally a difficult problem.

All the works cited above deal with static ad hoc wireless networks. Mobility adds a whole new dimension to the problem. In this chapter, we briefly survey existing researches on the problem, then discuss in more details ideas and results of some of our recent works on this problem [19,21,22,26]. We hope to eventually extend our ideas to cope with mobility in a near future.

When the network topology is known globally or at some central control node (e.g. a base station), we have the centralized (or globalized) version of the problem. In the distributed (or localized) version, only local information exchanges between neighboring nodes are allowed in order to compute the power assignment. Good centralized algorithms are interesting theoretically, and often challenging. They could also give insights into designing distributed ones. Last but not least, they serve as benchmarks to compare different (distributed) algorithms.

However, in the context of ad hoc networks, good distributed algorithms are ideal. Limited resources of ad hoc nodes make it imperative that nodes should use information from one- or two-hop neighbors to make routing decisions.

The literature on the broadcast version of this problem is quite large, as shall be subsequently reviewed. On the other hand, relatively very little is known on the multicast version except for the obvious idea of pruning some broadcast tree. Pruning does not give good performance once the multicast group sizes start to drop below a certain threshold, as shall be demonstrated in this chapter.

We shall present a group of algorithms for both broadcast and multicast problems. It is quite natural to consider both problems at once, as insights from designing an algorithm for one problem are often very helpful to solve the other. Broadcasting is a special case of multicasting and most known multicast algorithms simply prune the broadcast tree constructed from an energy-efficient broadcast algorithm. Most importantly, all our algorithms share the same core idea, and the distributization is also similar.

Our algorithms are based on a critical and interesting idea called *incremental power with potential power saving* (**IP3S**), which can be applied to many algorithms, not just our own.

The rest of this chapter is organized as follows. Section 2 formally describe the problem and related notions. Section 3 reviews related works and introduces the IP3S framework. Section 4 describes our main ideas in more details. Section 5 discusses broadcast algorithms, while section 6 are for multicast algorithms. Section 7 gives simulation results and some analyses. Lastly, Section 8 concludes the chapter and discusses several open research problems arising from this work.

2 Preliminaries

2.1 Communication Model

We assume that all nodes are equipped with omnidirectional antennas with adjustable power levels. There are two different basic assumptions: (1) the power levels can continuously be adjusted from 0 to some level p_{max}; (2) the power levels can only be chosen from a given discrete set $\{0, p_1, \ldots, p_m\}$ of power levels. In fact, when the nodes are heterogeneous it is possible that each node has its own power level set in case (2), or different p_{max} in case (1).

For presentation clarity, we shall restrict ourselves to case (1). Algorithms and protocols with assumption (1) can easily be extended to handle case (2) in a variety of ways. A simple strategy is to assume (1) and then "round" an assigned power level up or down to the closest available level from the given set. Another strategy is to, instead of vary power levels continuously, vary the levels in a discrete manner. Whatever the strategy we choose, algorithms' performances are not changed by much if the granularity of the power level set is fine enough [33].

The most common, admittedly simplistic, attenuation model assumes that signal power falls proportional to d^α, where d is the signal traveling distance, and α is an environmentally dependent real constant between 2 and 4 [10,24,31]. Suppose a node u is transmitting with power $p[u]$. A node v of distance d_{uv} from u can properly receive the signal from u if $p[u] \geq \gamma d_{uv}^\alpha$. Here, γ represents the receiver's power threshold for signal detection, often normalized to be 1. Thus, from here on we assume that v can properly receive u's signal iff $p[u] \geq d_{uv}^\alpha$.

We also assume, for simplicity of presentation, that nodes are homogeneous with respect to their power level sets or their p_{max}. Our algorithms work in the same way when network nodes are heterogeneous.

Let us continue with the u, v example above. If $p_{max} < d_{uv}^\alpha$, then it is impossible to get from u to v via a single hop. In this case, multi-hop transmission is necessary. Multi-hop transmission often also saves the total power usage.

We are now ready to define the broadcast/multicast problems formally. We first describe the problems in a highly general setting, then discuss the specialization leading back to the original problem. Graph theoretic terminologies we use here are fairly standard (see, e.g., [29]). Terminologies and concepts for NP-complete theory and approximation algorithms can be found in [12,14].

2.2 The Minimum Energy Consumption Broadcast Subgraph (MECBS) Problem

In the problem, we are given a directed graph $G = (V, E)$ with a symmetric cost function $c : E \to \mathbb{R}^+$ on its edges, namely $c(u, v) = c(v, u), \forall(u, v) \in E$. A distinguished vertex $r \in V$, called the source node, is also given. A *power assignment vector* is a function $p : V \to \mathbb{R}^+$, which assigns to each node of G some "power level". The *reachability graph* $G_p = (V, E_p)$ given a power assignment vector p is defined as follows. The directed graph G_p has the same set of vertices as G. There is an edge from u to v in G_p if and only if $p[u] \geq c(u, v)$, that is, the power assigned to u is at least the cost to reach v.

We are to find a power assignment vector p such that there is a directed path from r to every node in G_p, so as to minimize the sum $\sum_{v \in V} p[v]$. A slightly more realistic variation of the problem also has a given $p_{max} > 0$ and requires that $p[v] \leq p_{max}, \forall v \in V$.

When the graph G is a complete graph whose nodes are points on a d-dimensional Euclidean space, $d \geq 1$, and $c(u, v) = d_{uv}^\alpha$, $\alpha \in [2, 6]$, we denote the problem as

$\mathbf{MECBS}[N_d^\alpha]$. The case when $d = 2$ is of most interest, and is where most known results come from.

In this paper we focus our description on the case $d = 2$, although our algorithms and protocols work in the same way for the general d case. One reason for this restriction is that most known algorithms were designed for ad hoc networks on a 2-dimensional plane, and thus it is natural to compare our algorithms with others' in this particular case.

2.3 The Minimum Energy Consumption Multicast Subgraph (MECMS) Problem

MECMS is similar to MECBS except for the fact that, in this case, we have a subset $R \subseteq V - \{r\}$, called the *multicast group*. The problem is to find a power assignment vector p such that there is a directed path, in G_p, from the source r to every node in the multicast group R, so as to minimize the sum $\sum_{v \in V} p[v]$.

3 Overview of Existing Works

3.1 The Broadcast Case

Wieselthier et al. [31] studied several heuristics and studied their performances by simulations for the MECBS$[N_2^\alpha]$ problem. The algorithms they studied include: MST (minimum spanning tree), BIP (Broadcast Incremental Power), and SPT (Shortest Path Tree). Although MST does not work well as compared to most other algorithms, the advantage is that it does have a constant approximation ratio [28]. Moreover, there are standard distributed algorithms to compute MSTs [11] of a graph. It can be shown that BIP works better than MST analytically [28]. It is also quite easy to turn BIP into a distributed algorithm, in much the same way distributed MST works.

Wan et al. [28] and Clementi et al. [8] gave upper and lower bounds on the performance ratios of several of these heuristics. In particular, the MST heuristic for $\mathbf{MECBS}[N_2^2]$ was shown to have approximation ratio between 6 and 12; the BIP heuristic for the same problem has approximation ratio between $\frac{13}{3}$ and 12; the SPT approximation ratio is at least $\frac{n}{2}$. One can envision applying the same greedy heuristic as Chvatal's approximation algorithm for the weighted SET-COVER problem [7] to solve $\mathbf{MECBS}[N_d^\alpha]$. We refer to this heuristic as the ABIP (Average Broadcast Incremental Power) algorithm. Basically, instead of adding a new node each time as BIP does, ABIP could add a few nodes at a time, as long as the average incremental cost is the least among all choices. Unfortunately, ABIP has worst-case performance ratio at least $\frac{4n}{\ln n} - o(1)$ [28] for the $\mathbf{MECBS}[N_2^2]$

problem. We have implemented ABIP and simulation results show that it is not as good as MST on average. Cagalj et al. [2] gave the EWMA (Embedded Wireless Multicast Advantage) algorithm which tries to modify an MST to form a better power assignment. The basic idea is to start from the root node, check to see if expanding the root's power to cover the children of one of its children would save some power. The saving comes from the fact that once the children of a node are cover, the power assignment at that node could be reduced to 0. This process propagates itself until all nodes are covered. One advantage of EWMA is that it is a better modification of MST, hence its performance ratio is at least as good as MST, which is upper bounded by 12 for the $\mathbf{MECBS}[N_2^2]$ problem. The authors also showed by simulations that it works better than BIP on average. Additionally, the authors also gave a distributed protocol to implement this algorithm. Distributed MST is run first, then distributed EWMA is run by exchanging information between 2-hop neighbors. The disadvantage of this distributed protocol is that it is still highly complicated, requiring two rounds of computations.

All the algorithms mentioned above are centralized. Recently, Julien Cartigny et al. [5] presented the related neighborhood graph (RNG) broadcast oriented protocol (RBOP), which is distributed. In [22, 26], we presented a distributed algorithm called DIP3S which works better than RBOP. We also gave the best (shown by simulation) centralized algorithms for the broadcast problem.

The reader is also referred to [3,4,18,28] for some analytical results on the general graph version of the problem. For the general version of $\mathbf{MECBS}$, Caragiannis et al. [3] found an approximation algorithm with performance ratio of $10.8 \ln n$, by a reduction of $\mathbf{MECBS}$ to the *node-weighted connected dominating set* problem, which has a $1.35 \ln n$-approximation [13]. In the same paper, the authors also devised a polynomial time solution to the $\mathbf{MECBS}[N_1^\alpha]$ problem. Liang [18] also addressed the same problem by a reduction to the Steiner Tree problem [15], however the approximation ratio is about $O(\log^3 n)$, which is not as good.

As far as negative results are concerned, unless $\mathbf{NP} \subseteq \mathbf{DTIME}(n^{O(\log \log n)})$, $\mathbf{MECBS}$ can be shown to be inapproximable within a ratio of $(1-\epsilon) \ln n$ by a reduction to the SET COVER problem [14] or the CONNECTED DOMINATING SET problem [28].

3.2 The Multicast Case

The easiest approach to build a multicast tree is to *prune* a broadcast tree, in much the same way the Internet multicast protocols work (DVMRP, PIM, etc.).

Henceforth, we append a prefix "P-" before a broadcast algorithm's name to denote the multicast algorithm based on pruning. (For example, P-MST is the pruning version of MST.)

Wieselthier et al. [31] have experimented the P-BIP, P-SPT, P-MST heuristics, and found out that for very small group sizes, P-SPT outperforms the other two algorithms, while for moderate to large group sizes, P-BIP performs the best.

Wan et al. [27] analytically showed that in the worst case P-MST, P-SPT, and P-BIP have linear performance ratios in terms of the number of nodes in the network. Hence, they do not perform well theoretically. In the same paper, the authors also proposed an analog of MST for the multicast case, called the *shortest path first* (SPF) algorithm. The algorithm starts from the root node r, grows out a set of covered nodes S with $S = \{r\}$ initially, and each time it finds a shortest path from any node in S to any uncovered node in R (the multicast group). This way, after each iteration at least a new node in M is covered. The analog for BIP was called minimum incremental path first (MIPF), which works in much the same way as SPF, but we pick a new path which yields the least incremental power. It could be shown that both SPF and MIPF have constant approximation ratios.

Another algorithm based on Steiner minimum trees (SMT) [15] was also given in the same paper. The SMT-based algorithm, on the other hand, was shown to have approximation ratio at most $12\left(1 + \frac{\ln 3}{2}\right)$, based on a relatively new result on SMT by Robins and Zelikovsky [25]. One drawback of the SMT algorithm is that it is quite complicated and impractical for our purpose.

In two recent papers [19], we proposed the centralized MIP3S algorithm which, by simulations, is shown to outperform all known algorithms, and the distributed DMIP3S algorithm which is the best distributed algorithm so far, which also compares favorably with other centralized ones.

3.3 The IP3S Framework

Our works were based on a simple yet important observation: building a (broadcast or multicast) tree starting from the source often involves increasing the power level at a certain node, say v, (or at a set of nodes), to cover some new nodes until all nodes (in the multicast group) are covered. The power expansion at v, however, could make the current power assignment at some other nodes redundant, in the sense that v's new power level covers nodes which have been covered before. Hence, by increasing v's power, we might be able to reduce some other nodes' powers! This idea is referred to as *incremental power with potential power saving* (IP3S).

To implement this idea, a certain kind of data structure and invariance have to be maintained in order to not disconnect the current graph. As long as the invariance holds, the idea can be applied to any algorithm on this problem.

A group of algorithms based on the IP3S idea are proposed as summarized in Table 1. We have also devised, implemented, and experimentally validated the per-

	Centralized	Distributed
Broadcasting	IP3S, IP3S-b [22]	DIP3S [22, 26]
Multicasting	P-IP3S, MIP3S [19, 22], MIP3S-b	DMIP3S, P-DIP3S [21, 22]

Table 1: Algorithms we devised based on the IP3S idea. "P-":Pruning, "D":Distributed, "M":Multicasting, "-*b*":Variation

formances of all these algorithms. Very extensive simulations are made to compare these algorithms with known ones. The simulations show that our centralized algorithm works the best and our distributed algorithms work better than most known centralized algorithms.

Especially for multicasting, we show that, for large multicast groups of size about 65% of the total number of nodes, it is better to run DIP3S, which was designed for the broadcast case, then prune back (P-DIP3S). On the other hand, smaller group sizes require an entirely different strategy to be energy-efficient.

4 The IP3S Idea

In many known algorithms, a central theme has been to iteratively "grow" a set S of covered nodes starting from the source, i.e. initially $S = \{r\}$. At each step, a node, say v, in S is chosen according to some objective function and $p[v]$ is expanded to cover one or more new nodes. The new nodes are then added into S, until all nodes in the broadcast (multicast) group are in S. The invariance is to keep S as the set of nodes reachable from the source r with the current power assignment vector.

The IP3S idea is a nice observation applied to this crucial "expansion step" of an algorithm. Consider a potential expansion at v, illustrated in Figure 1(a). Many known algorithms pick a node v with the least incremental power, disregarding the fact that *a larger expansion might actually be better*! The reason for this is simple, as shown in Figure 1(a). Nodes a and b, for example, were covered by node u before the expansion of v. After the potential expansion, however, $p[u]$ could be reduced significantly if nodes a and b are the farthest of the nodes which u cover. Hence, we can write a procedure which reduces all such u's powers right after each expansion.

There is a problem with the idea as it is, however. Consider the situation in Figure 1(b). After reducing $p[u]$ so that u does not cover a and b, node v may not be reachable from the source anymore. What happens was that we broke the links (u, a) and (u, b). When all directed paths from r to v contain either (u, a) or (u, b), then breaking up (u, a) and (u, b) shall disconnect v from r. (Note also that, in general the paths from r to v may take a few hops until they get to a and/or

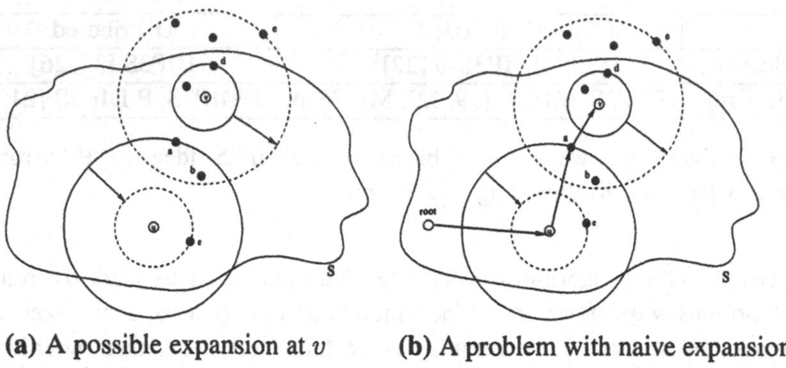

(a) A possible expansion at v and reduction at u

(b) A problem with naive expansion: potential disconnectivity.

Figure 1: Illustration of the IP3S idea.

b, and then a few more hops before reaching v. The situation in the figure is only illustrative.)

To cope with this *potential disconnectivity* problem, we adopt variations of the following solution: each $v \in S$ maintains locally an (r, v)-path corresponding to the current power assignment. The check for disconnectivity could then be done locally at the common neighbors of u and v like a and b, which are involved in the potential power reduction. This way, only one-hop local information is needed if the idea is to be made distributed.

It is intuitively clear that, if v was to maintain all possible paths from r to it, then more such u's can reduce their power. This is because the current (r, v)-path that v is maintaining might contain (u, a) or (u, b), while there exists another (r, v)-path which does not. In this paper, however, we choose the former solution to reduce the complexities of our protocols, and keep the total amount of information kept at each node to a minimal degree.

Two sensible applications of the IP3S idea are:

(a) take into account the potential power saving when deciding which v to expand.

As an example, one strategy for the broadcast problem is to *pick a $v \in S$ so that the increase in power at v* **minus** *the total power saving at all the u's is as small as possible*. Note that the value of this objective function can even be negative! The centralized version of this strategy, called IP3S [1], proves to work very well, beating all known centralized algorithms. Our

[1]We will use IP3S to denote both the essential idea of all our algorithms and the centralized

implementation of IP3S takes time at most $O(n^3 \lg n)$ (Section 5.1).

(b) as strategy (a) often involves extra computations not suitable for distributed algorithms on limited-resource ad hoc network, we can do the power reduction alone. This strategy is influential in our distributed algorithm designs, to be discussed in the following 2 section.

5 Broadcasting

5.1 Centralized Algorithm

The following is a straight implementation of IP3S for the broadcast problem.

There are a few natural assumptions and data structures:

1. The input consists of a set V of n nodes and a root node $r \in V$
2. There is a power level p_{max} which is the upper bound on nodes' power levels. (This assumption can be relaxed easily for heterogeneous networks, where nodes have their own p_{max}.)
3. Each node $v \in V$ has an array N_v of neighbors reachable from v at p_{max}. Let n_v denote the number of reachable neighbors of v. We also maintain a corresponding array L_v of power levels needed to reach the nodes in N_v. Thus, $L_v[i]$ is the least power level for v to get to node $N_v[i]$. We also assume the L_v are sorted in increasing order. Thus, if $p[v] = L_v[i]$, then v would "cover" all neighbors in $N_v[1..i]$, and no more. Obviously, $L_v[n_v] \leq p_{max}$.

   There are a variety of ways a node can construct its L_v list in practice, such as utilizing GPS, time delay, or signal strength measurements.
4. For each node $v \in V$, there is an index $f[v]$ for which $N_v[f[v]]$ is the farthest neighbor currently covered by $p[v]$. Implicitly, $f[v] = 0$ if $p[v] = 0$. We want $p[v] = L_v[f[v]]$, so we also assume $L_v[0] = 0$.
5. Each node $v \in V$ has a pointer $\pi[v]$ which points to the parent node on a path from the root to v.
6. Each node $v \in V$ has a set E_v of edges along a path from the root to v. For instance, $(\pi[v], v) \in E_v$, $(\pi[\pi[v]], \pi[v]) \in E_v$, and so on.

   We choose a set data structure since searching on a set is more efficient than on a list, while the traversing time is the same.

---

algorithm for broadcasting. This should be able to be distinguishable from the context.

The data structures described above are not necessarily the best for a centralized algorithm. However, we describe the centralized version of IP3S in a particular way which shall facilitate the discussion on its distributive version.

We maintain a set S of nodes which are reachable from r so far, given the current power assignment p. Initially $S = \{r\}$ and $p[v] = 0, \pi[v] = \text{Nil}, E_v = \emptyset, \forall v \in V$. When $S = V$, the algorithm terminates.

Let us consider one particular step of the algorithm. For each $v \in S$, recall that $p[v]$ is the current power assigned to v. We need to pick a node $v \in S$ to extend its power. Suppose we want to increase v's power from $p[v]$ to p to cover a few more neighbors. Let W be the set of new nodes covered, and U be the set of parents of nodes in $W \cap S$.

For each $u \in U$, if a is the farthest node in N_u currently covered by $p[u]$, and if $(u, a) \notin E_v$, then $p[u]$ can be reduced to one level closer to u, and $\pi[a]$ now can be pointed to v. Of course, this reduction is only done hypothetically to see how potential saving could be obtained by this increase in power at v. This process is repeated at u until no more reduction is possible. Also, the process is done at each $u \in U$.

The objective value at v is then $(p - p[v])$ **minus** the total potential saving at the u's. All these are done by passing a FALSE flag to the **Reduce-Power** subroutine, which will pretend this increase and return the resulting power increment.

We vary p from $p[v]$ to p_{max} to pick the best objective value δ_v. Then, a node $v \in S$ with the least δ_v is picked to expand its power. We use v_s to denote this node. After v is picked, the actual reduction is done using the **Reduce-Power** with a TRUE flag, as shown in Algorithm 5.1.

Algorithm 5.1. IP3S

1: $S \leftarrow \{r\}$;
2: ***for each*** $v \in V$ ***do***
3: $\quad p[v] \leftarrow 0$; $\pi[v] \leftarrow Nil$; $f[v] \leftarrow 0$; $E_v \leftarrow \emptyset$
4: ***end for***
5: ***while*** $|S| \neq n$ ***do***
6: $\quad v_s \leftarrow Nil$; $i_s \leftarrow Nil$; $\delta_s \leftarrow +\infty$
7: $\quad$ ***for each*** $v \in S$ ***do***
8: $\quad\quad i_v \leftarrow Nil$; $\delta_v \leftarrow +\infty$ *// temporary variables*
9: $\quad\quad$ ***for*** $i \leftarrow f[v] + 1$ **to** n_v ***do***
10: $\quad\quad\quad$ *// try to cover an uncovered node*
11: $\quad\quad\quad$ ***if*** $N_v[i] \notin S$ ***then***
12: $\quad\quad\quad\quad$ *// let's pretend we increase* $p[v]$ *to* $L_v[i]$
13: $\quad\quad\quad\quad$ $\Delta \leftarrow$ **Reduce-Power** $(v, i, FALSE)$
14: $\quad\quad\quad\quad$ *// now we've got the objective value* Δ *if we increase* $p[v]$ *up to* $L_v[i]$

```
15:          if δ_v > Δ then
16:             δ_v ← Δ;  i_v ← i
17:          end if
18:       end if
19:    end for
20:    if δ_s > δ_v then
21:       v_s ← v;  δ_s ← δ_v;  i_s ← i_v
22:    end if
23: end for
24: if v_s = Nil then
25:    return DISCONNECTED
26: else
27:    for i ← f[v_s] + 1 to i_s do
28:       if π[N_v[i]] = Nil then
29:          π[N_v[i]] ← v_s
30:          E_{N_v[i]} ← E_{v_s} ∪ {(v_s, N_v[i]}
31:       end if
32:    end for
33:    f_{v_s} ← i_s;  p[v_s] ← L_v[f[v_s]] // new power at v_s
34:    Reduce-Power(v_s, i_s, TRUE)
35: end if
36: end while
```

Algorithm 5.2. Reduce-Power$(v, i, FLAG)$

```
1: Δ ← L_v[i] − p[v]
2: W ← N_v[f[v] + 1, ..., i]
3: U ← {π[w] | w ∈ W ∩ S}
4: for each u ∈ U do
5:    k ← f[u]    // index to the farthest node that u covers
6:    while k ≠ 0 and N_u[k] ∈ W do
7:       a ← N_u[k] // for clarity
8:       if (u, a) ∉ E_v then
9:          k ← k − 1
10:         if FLAG = TRUE then
11:            // the flag tells us to do the actual reduction
12:            f[u] ← k; p[u] ← L_u[k]; π[a] ← v; E_a ← E_v ∪ {(v, a)}
13:         else
14:            Δ ← Δ − (L_u[k + 1] − L_u[k])
15:         end if
16:      else
```

17: $\quad\quad k \leftarrow 0;$ *// Getting here means* $(u, a) \in E_v$
18: ***end if***
19: ***end while***
20: ***end for***
21: ***if** FLAG = FALSE **then***
22: **return** Δ
23: ***end if***

There are standard set data structures [9] which take $O(\lg n)$-time for searching and updating. Hence, the roughest analysis of this algorithm yields a running time of $O(n^4 \lg n)$.

The above implementation uses the first strategy of applying IP3S idea which takes into account the potential power saving when calculating the cost function. If we use the second strategy, we will get IP3S-b which is much less complicated and only sacrifices the performance a little.

5.2 Distributed Algorithm

DIP3S is the distributed version of IP3S-b. As we discussed in Section 4, the "-b" version of IP3S is more suitable to be distributed. Since all distributed algorithms proposed in this paper use this version, we will just ignore the "-b" for them.

Special techniques are used to make IP3S distributed, these techniques are also useful for designing the distributed multicast algorithm (see section 6.2).

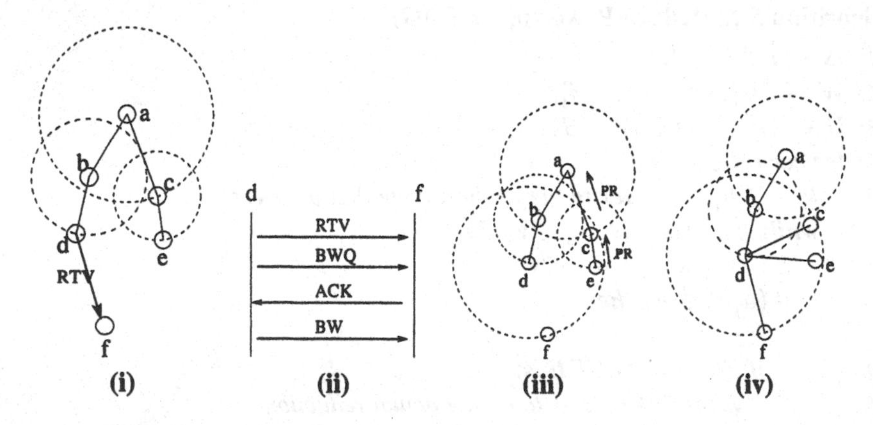

Figure 2: Message communication in a typical node expansion

We grow a broadcast tree starting from the root. Each node maintains a list of power values (or distances) required to cover any of its neighbor. This information

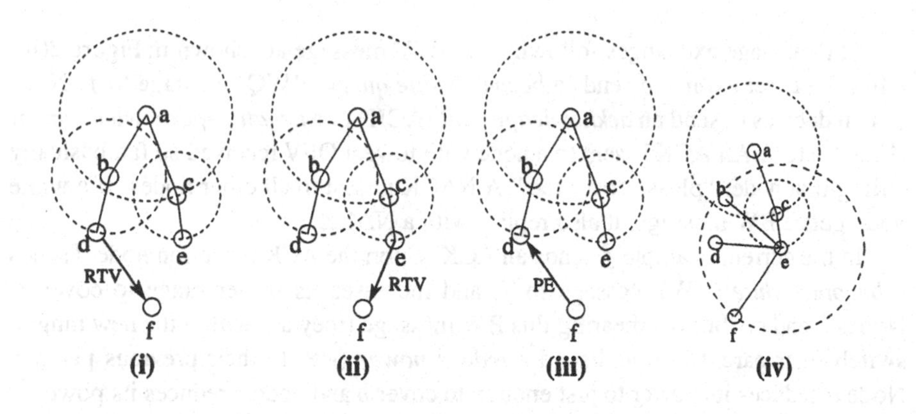

Figure 3: Preemption process in DIP3S

can be obtained locally in a variety of ways, including signal strengths, time delay, GPS, or microwave distance [1].

Let S be the set of nodes in the current tree. Initially $S = \{r\}$ (source-initiated). Nodes in S are "white," or active nodes, the rest are "black." Unless all its neighbors are white, every white node independently computes an object function value (OFV), which is the amount of incremental power required to cover its nearest uncovered neighbor. (This step of the algorithm is similar to that of BIP.) The white nodes then set a timer proportional to OFV^2. (The values OFV^2 is mapped linearly to an interval from 0 to a maximum timer value Δ.)

At the end of the timer, a node "expands" its power to cover the nearest uncovered neighbor. (Since nodes operate independently, the neighbor might have been covered already during the timer. We shall see that a simple NACK and the preemption mechanism prevent this redundant expansion.) The newly covered node becomes a child of the covering node, and also becomes white. Several mechanisms are implemented so that nodes get updated information on their newly covered neighbors, and re-compute their timers, as illustrated below.

A typical message sequence required to complete a covering process are illustrated in Figure 2. Figure 2(i) shows part of some broadcast tree at a particular time. Think of node a as handling a branch of the current tree. Nodes b and c are children of a, according to the current power assignment. Similarly, d and e are b's and c's child, respectively. As soon as d gets covered by b it calculates its OFV to cover node f, which is d's nearest uncovered neighbor. Node d sends a *report timer value* (RTV) message to f, and sets a timer proportional to OFV^2. The RTV packet contains the value of the OFV. If any of a, b, c, e has f as their closest uncovered neighbor, they would do the same.

The message exchanges following the RTV message are shown in Figure 2(ii). After the timer expires d sends a *become white query* (BWQ) message to f. Node f then decides to send an *acknowledgement* (ACK) or a *negative-acknowledgement* (NACK) to d. An ACK is sent to a node with lowest OFV received so far, basically telling that node: "please cover me." A NACK is sent to all other nodes. If a white node gets a BW message, it also replies with a NACK.

In the current example f sends an ACK. Upon the ACK reception node d sends a *become white* (BW) message to f, and increases its power range to cover f. Nodes c and e, after overhearing this BW message (they are within the new range), switch their parent to d and send a *reduce power* (RP) to their previous parents. Node a reduces its power to just enough to cover b and node c reduces its power to zero. The final broadcast tree is shown in Figure 2(iv).

To speed up the updating of information, we use the concept of *preemption*, as illustrated in Figure 3. Node f receives an RTV message from e following the RTV message from d. It then compares the OFV values in both the RTV messages and selects node e to cover itself. It then sends a *preemptive* (PE) message to node d following which node d recalculates its OFV. Figure 3(iv) shows the final broadcast tree.

6 Multicasting

6.1 Centralized Algorithm

After IP3S has constructed a broadcast tree rooted at r, one can easily prune the tree back to cover up to the multicast members only, in much the same way that IP-multicast algorithms work. The resulting algorithm is referred to as P-IP3S.

The other two algorithms in this category are MIP3S and MIP3S-b. Recall that we have a subset $R \subseteq V - \{r\}$ of nodes where multicast data from r are to be delivered to. When $|R|$ is large, building the entire broadcast tree, then prune a few branches makes sense, and proves to work very well. However, when $|R|$ is small, building the broadcast tree is too greedy. A good broadcast tree looks more at the global picture of overall saving, hence pruning it back does not necessarily gives a good local solution.

MIP3S and MIP3S-b are based on the shortest path first (SPF) idea [27] and the potential power saving idea.

The idea of SPF is to build the multicast tree more locally. We also maintain a set S of nodes reachable from r so far, which is initialized to be $S = \{r\}$. We "grow" S until $R \subseteq S$. In each iteration of SPF, we find a "shortest" path P between **any** node in S and **any** node in $R \setminus S$ (see Figure 4). Mathematically, this shortest path is a shortest path between S and $R \setminus S$. In the figure, the black

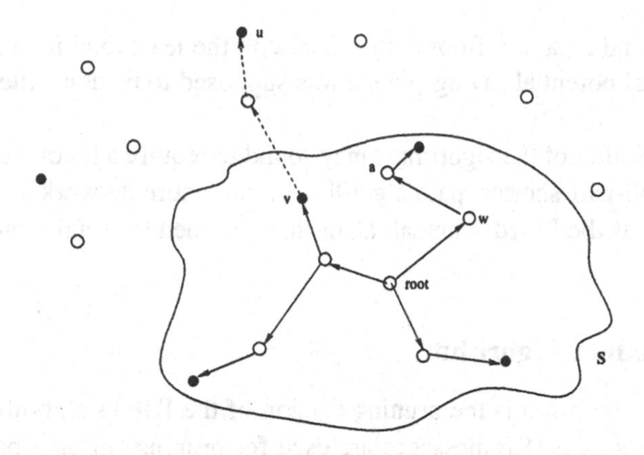

Figure 4: Find a shortest path between S and $R \setminus S$

nodes represent nodes in R. For each pair (x, y) of nodes on the plane (or any d-dimensional plane), the distance between x and y is the minimum power to get from x to y. The term "shortest path" is taken in this sense. In our case, we can just take the "length" of an (x, y)-edge to be d_{xy}^{α}, where d_{xy} is the Euclidean distance between x and y, and α the power attenuation factor.

To this end, let us see how the potential power saving idea can be applied here. In Figure 4, after the shortest path from v to u is found, v's power level is supposed to be expanded to cover the next node on the path. However, this expansion allows v to also cover a, which was previously covered by w. Hence, we can reduce w's power significantly. This power reduction can be applied systematically on each node whose power level is increased, as described in the previous section.

Another twist we implemented was that we find a shortest *incremental* path between S and $R \setminus S$. In Figure 4, for example, the "length" of the new arc from v to the next node in the path might be large, yet the incremental power imposed on v might be very small.

After the new path is found, all nodes on the path are assigned with the power level needed to reach the next node on the path. The set S is now expanded to include all nodes on the path along with all nodes covered by the new power assignment.

What we have just described is the algorithm we name MIP3S-b, whose actions in each iteration can be summarized as follows.

1. Find a shortest incremental path P from S to $R \setminus S$.
2. Assign each node on P the new power level to reach the next node.
3. For each node v along P, run the power reduction method as in IP3S.

The other variation – MIP3S – requires a little bit more work. In step 1 of each iteration, we find a path P from S to $R \setminus S$ with the least total incremental power **minus** the total potential saving (which was supposed to be done afterward in step 3).

The description of the algorithms may sound to require a lot of work. However, if we run an all-pair shortest path algorithm on the entire network as a preprocessing step, such as the Floyd-Warshall algorithm [9], then the total running time is at most $O(n^3)$.

6.2 Distributed Algorithm

The P-DIP3S algorithm is the pruning version of the DIP3S algorithm. Members send prune messages (RP messages are used for pruning) to their parents, propagating the power reductions back as far as possible. Since the pruning process is distributed in nature, P-DIP3S is quite easy to implement.

Another algorithm is the distributed MIP3S, called DMIP3S. Distributizing MIP3S can be done in the same manner as distributizing IP3S. A difference is that each node maintains a routing table indicating the next node along a shortest path to any node in the network. We have many options to build the routing table, including distance-vector like algorithms. As the network is static, this routing table only need to be constructed once, and can be used again and again for any multicast source and groups. Also, notice that the metric is not the hop-count, but the total power needed.

Every white node keeps a list of uncovered multicast members, just as in DIP3S where every node knows which of its neighbors are not yet covered. New white nodes will inherit this information from their parents when they become white, i.e. become active for the first time. This list is updated when the node overhears the ACK from some multicast member, receives a NACK to itself, or a list update from the root. The list update from the root shall be explained later. A NACK (in either DIP3S or DMIP3S) means that the node to be covered has been covered or has chosen someone else to cover itself. Note that, this mechanism allows the list of uncovered multicast members not to have to be up-to-date at all time.

In DIP3S, every white node tries to cover its nearest uncovered neighbor. In DMIP3S, each white node tries to cover its nearest uncovered multicast member, possibly via several hops. This can be done easily by checking its own routing table and uncovered member list. In fact, it is useless to construct a new path from a node v to a multicast member w, if the node after v in the path, say u, is already white. The reason is that u would have done it by itself. Hence, each node only needs to find paths through one of its non-white neighbors.

The timer setup and message communication mechanisms are almost the same. One difference is that, in DIP3S all communications are one hop away, while in

DMIP3S they are normally multiple hops away. If we imagine the shortest path as one "virtual" hop or "tunnel," then the idea will be clearer. Also, the destination and source should always be embedded in the message body to ensure the proper forwarding and identification.

The power reduction step is done slightly differently, however. The BW message is sent by every node on the path to the newly covered member. A white node will switch its parent and send a *reduce power* to its previous parent after overhearing the BW message. A black node would become white and starts its timer. The BW message also contains the sender's list of uncovered multicast members.

In this multicasting case, if a node reduces its power to zero and it is not a multicast member then it can be removed from the set S. The node shall be inactive, but keep its color white until it receives a STOP message from the root. (This step is analogous to the pruning step in P-DIP3S.) Its parent may be able to further reduce its power via the RP messages.

A white node will remove the member's entry from its uncovered multicast member list if it receives PE, NACK targeting itself, or overhears ACK from that member. It will stop the algorithm when the list is empty or when a STOP message is received. For fast termination of the whole algorithm, we let ACKs always be forwarded back along the current tree to the root r. Lastly, r will send a list update message forward along the tree when it receives the ACKs from all multicast member. The last list update message is also a STOP message, for nodes to release their resources.

Another problem in multicasting is that the member group is dynamic. Members are joining and leaving the multicast session all the time.

For members who are leaving, they can simply send a RP message to their parent to remove themselves from the tree. And their parents can propagate back this reduce power message. It is the same as pruning.

For new members who are joining, they will need to broadcast a "Join Request" message. Nodes who receive this message, but not in the multicasting tree, will just relay this message to their neighbors. If a node in the multicasting tree receives this message, it will mark the new member as uncovered and try to cover it as talked above.

7 Simulation Results

7.1 Broadcasting

Five algorithms DIP3S, IP3S, BIP, EWMA and RBOP are to be compared.

The number of nodes varies from 10 to 100 with an increment of 10 at each step. All these nodes are uniformly distributed on a $10 * 10$ square. For each network instance, three different values of the propagation loss exponent α were tested: 2, 3 and 4.

The distributed algorithms were implemented in C with YACSIM [17]. YACSIM is a C based library of routines that provides discrete event and random variate facilities.

In the simulations done in previous works [2, 31], a node's maximum power level is always assumed to be sufficient to cover all the nodes. In our simulations, the p_{max} is set to be enough to cover 25%, 50%, 75%, and 100% of the region. Here, to cover $x\%$ of the region means p_{max} can only be used to reach the nodes within a distance of $x\%$ of the maximum possible distance of any two nodes. This is the length of the diagonal of the 10×10 square, which is $10\sqrt{2}$.

Varying p_{max} has several values. The obvious one is that in reality nodes may not have enough power to reach the furthest nodes. Secondly, even when a node can reach really far, restricting its power to a certain percentage of the maximum possible power could save energy.

The source node is selected randomly among all the nodes.

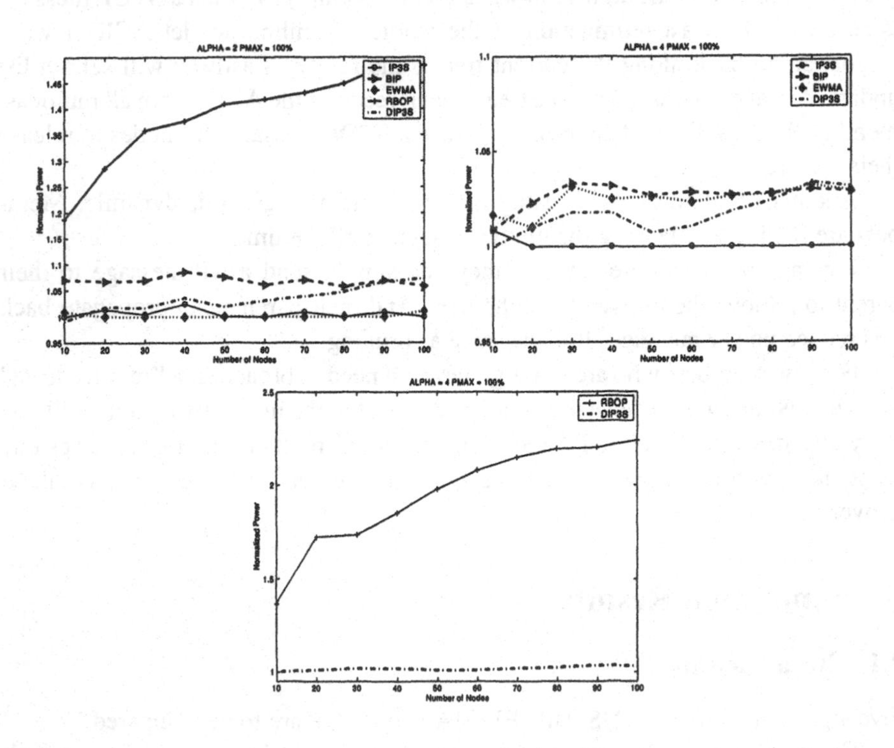

Figure 5: p_{max} is enough to cover 100% of the region

For each simulation configuration, which is a combination of network size n, propagation loss exponent α, and node's power level p_{max}, we generate 100 random instances and run the above algorithms on them.

The performance metric we adopt is the total power of the multicast tree. To compare among different algorithms, we use the idea of *normalized power* [31]. Let

$$\mathcal{A} = \{\text{DIP3S, IP3S, BIP, EWMA and RBOP}\}$$

be the set of all five algorithms to be evaluated. Let $T_A(I)$ denote the total power of the multicast graph of a network instance I, computed by algorithm $A \in \mathcal{A}$. Then, the normalized power of algorithm A on instance I is

$$T'_A(I) = \frac{T_A(I)}{\min\{T_A(I), A \in \mathcal{A}\}}.$$

As indicated in [31], this metric has the advantage that it does not depend on the size of the region being tested. If nodes are distributed in a larger region, then the overall power consumption is scaled by a certain factor, which does not effect the normalized powers.

We present the results for the cases when $\alpha = 2, 4$, and maximum power levels 100%, 50%, as shown in Figures 5, 6, 7, 8. **The rest of the graphs follow the exact same trend.** Different figures may have different scales in order to make the figures easy to identify.

From the simulations, we can make the following conclusions:

- IP3S works better than all known centralized algorithms.
- For all values of α, for varying p_{max} and network sizes, DIP3S works better than centralized BIP, RBOP, and almost as good as centralized EWMA.
- Since RBOP trades the power efficiency for less information exchange, the resulting tree is not very power efficient. It depends on the lasting time of the multi/broadcasting session. A power efficient tree is preferred when the session is long.

7.2 Multicasting

For the case of multicasting, eight algorithms P-IP3S, P-DIP3S (distributed), DMIP3S (distributed), MIP3S, SPF, P-BIP (also called MIP in [31]), P-EWMA, and P-RBOP are to be compared.

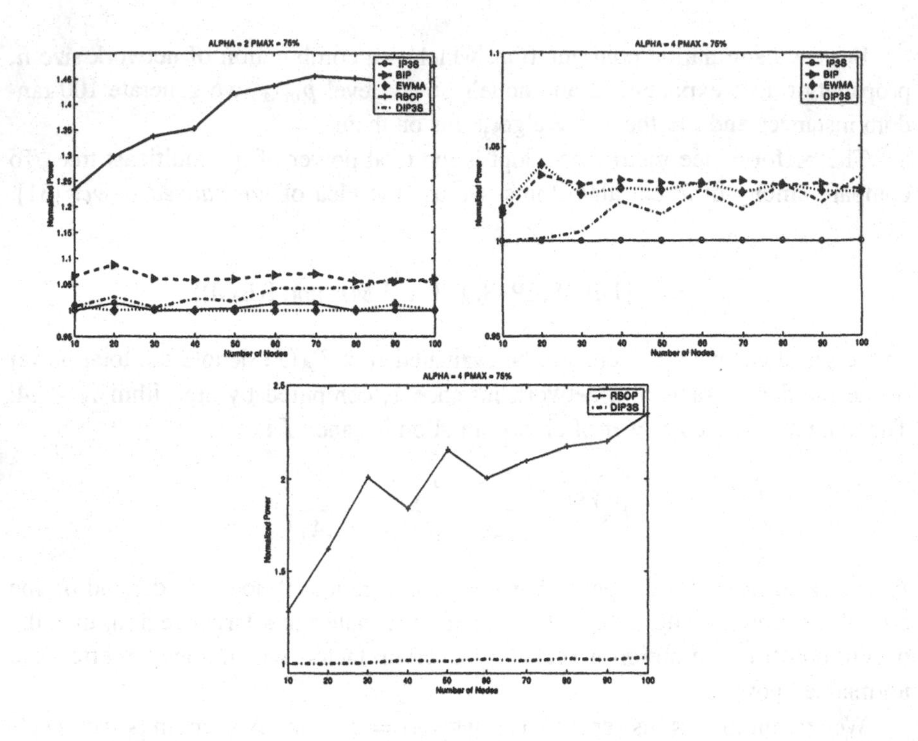

Figure 6: p_{max} is enough to cover 75% of the region

Although some of these algorithms were designed for the broadcast problem, we implement their pruning versions to see how they compare to ours when multicast group sizes are large. For large group sizes, the multicast problem becomes the broadcast problem.

All the settings are still the same with the broadcasting, except now we have multicast group. The multicast group members R are also selected randomly. We vary the size of R from 20% of all nodes to 100% of all nodes, at a step of 20% each.

We present the results for the cases when $\alpha = 2, 4$, for multicast group sizes 20%, 40%, 60%, 80%, 100% and maximum power levels 100%, 50%, as shown in Figures 9, 10, 11.

Also, Table 2 shows the variances of the power ratios T'.

From the simulations, we can make the following conclusions:

- Within the range of small multicast groups ($\leq$ 65% the total number of nodes) or of large groups ($\geq$ 65%) the relative algorithms' performances

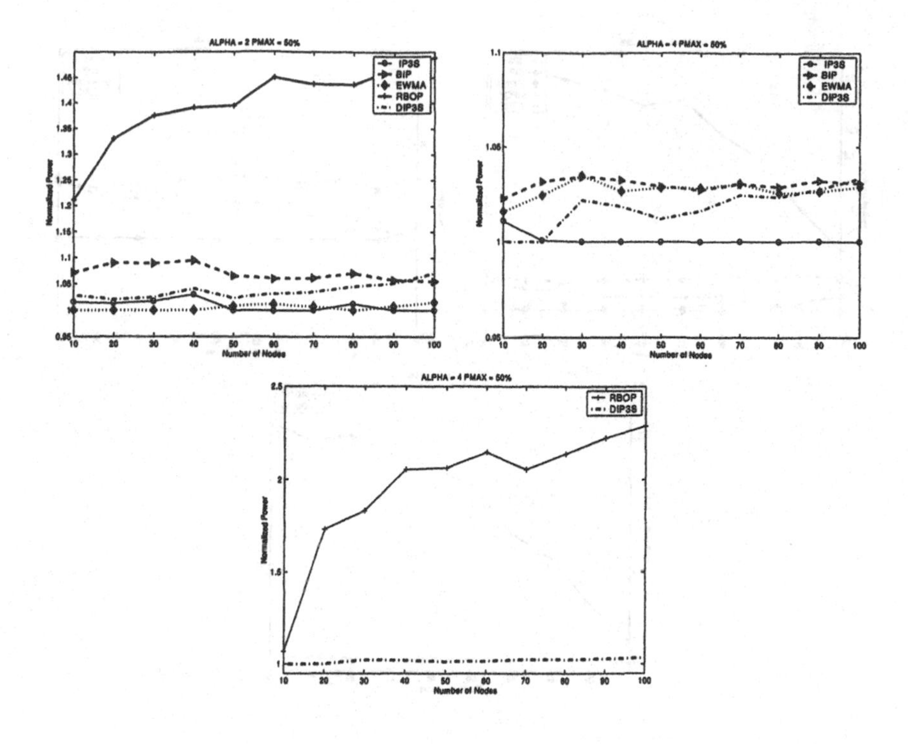

Figure 7: p_{max} is enough to cover 50% of the region

are roughly the same no matter what the value of α is. However, when α increases the performance difference between algorithms becomes less. This is possibly due to the fact that large values of α forces nodes to use large power levels, leaving not much room for algorithms to leverage.

- The relative performances of the algorithms are also quite similar independent of the number of network nodes.

- The relative performances of the algorithms are the same no matter what the value of p_{max} is.

- When group sizes are very large (80 to 100% of the total number of nodes), i.e. when the problem becomes the broadcast problem, our centralized P-IP3S algorithm works the best amongst all known algorithms. The distributed version P-DIP3S is worse than centralized EWMA and P-IP3S, but it is much better than the other distributed algorithm P-RBOP. P-DIP3S is in fact better than the centralized P-BIP (same as MIP).

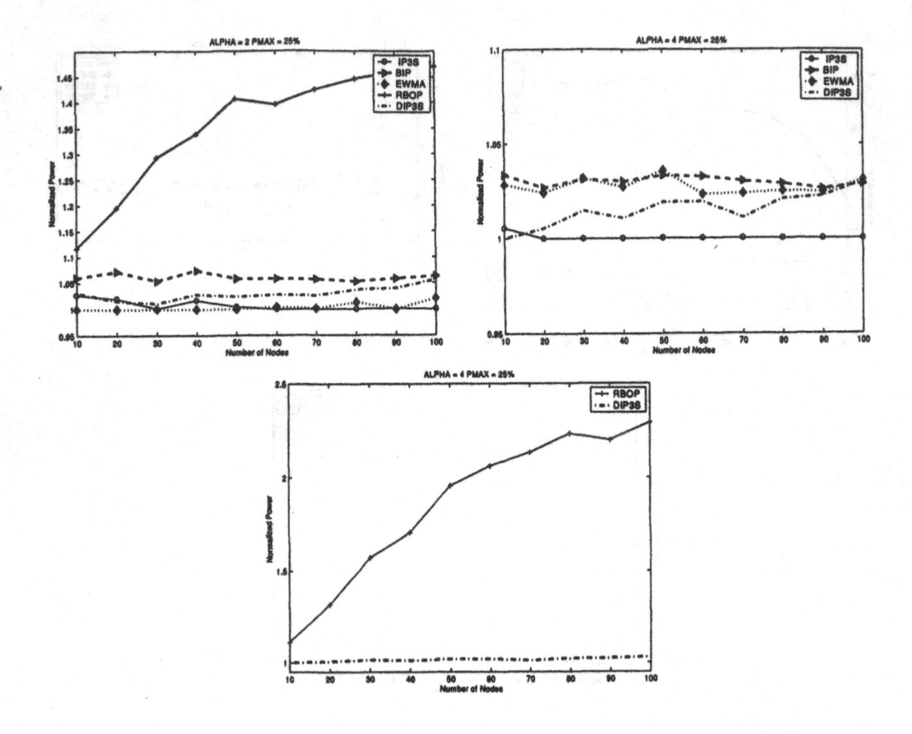

Figure 8: p_{max} is enough to cover 25% of the region

- There is a group size cut point of around 60 to 65%, where DMIP3S starts to work very well. In fact, the distributed DMIP3S works better than all known multicast algorithms, except for our own centralized MIP3S.

- The centralized MIP3S algorithm is the best multicast algorithm in the group.

- P-RBOP's performance is highly unstable for large values of α (4).

- MIP3S and DMIP3S perform very consistently (stable) with respect to the best algorithm for each network instance. This fact can be seen from the variance table.

Why all algorithms based on IP3S work so well? There is always tradeoff between performance and resource usage. The second strategy of applying IP3S removes the redundancy in the power assignment when a node increases its power level. The first strategy not only does that, but also tries to find a better power assignment by enlarging the search space.

30-Nodes Networks								
α	MIP	P-EWMA	P-IP3S	SPF	MIP3S	D-MIP3S	P-DIP3S	P-RBOP
2	0.0403	0.0816	0.0659	0.0194	0.0036	0.0063	0.0418	0.0764
4	0.0142	0.0222	0.0442	0.0043	0.0012	0.0011	0.0352	1.4378
60-Nodes Networks								
α	MIP	P-EWMA	P-IP3S	SPF	MIP3S	D-MIP3S	P-DIP3S	P-RBOP
2	0.0219	0.0366	0.0250	0.0053	0.0006	0.0024	0.0296	0.0756
4	0.0093	0.0389	0.1073	0.0023	0.0005	0.0008	0.1524	0.7802
100-Nodes Networks								
α	MIP	P-EWMA	P-IP3S	SPF	MIP3S	D-MIP3S	P-DIP3S	P-RBOP
2	0.0115	0.0297	0.0245	0.0025	0.0001	0.0018	0.0135	0.0647
4	0.0045	0.0177	0.0355	0.0015	0.0004	0.0021	0.0669	0.9306

Table 2: variances of normalized tree power for 100 instances of 30,60,100-nodes networks with maximum power 100% and group size 20%

Since it's a static network, the saving on the cost of long-term operation will be much higher than the cost spent on the initial searching and calculation.

8 Conclusions and Future Works

In this chapter we have discussed the problem of devising algorithms to solve the minimum energy consumption broadcast/multicast problems in static wireless ad hoc networks. A brief overview of existing works, along with a thorough explanation of the IP3S framework were presented.

By extensive simulations, for broadcasting, we have shown that IP3S outperform all known algorithms, and DIP3S – the distributed version of IP3S-b is even better than most know centralized broadcasting algorithms.

For multicasting, we show that DMIP3S outperform most known centralized algorithms when multicast groups are small, about 60% to 65% or less of the total number of nodes. In fact, MIP3S - the centralized version of DMIP3S - is the best centralized multicast algorithm.

On the other hand, for large group sizes, i.e. when the multicast problem tends to the broadcast problem, P-DIP3S performs the best among the distributed algorithms and better than most of the centralized ones.

There are several problems we are working on, arising from the ideas of this chapter. Analyzing the approximation ratios of the centralized versions is a major challenge. Good lower bound on the approximation ratios should also be devised. Another major challenge is to devise good power-conserving distributed multicast and broadcast algorithms under mobility.

9 Acknowledgements

Hung Q. Ngo is partially supported by NSF CAREER Award CCF-0347565. We would like to thank Ms. Purnima Mavinkurve, Mr. Amit Chandak, and Mr. Vikas P. Verma for various discussions and implementation help.

References

[1] A. Benlarbi, J.-C. Cousin, R. Ringot, A. Mamouni, and Y. Leroy, Interferometri c positioning systems by microwaves, *Proceedings of Microwaves Symposium (MS)* (2000).

[2] M. Cagalj, J.-P. Hubaux, and C. Enz, Minimum-energy broadcast in all-wireless networks: Np-completeness and distribution issues *Proceedings of the 8th annual international conference on Mobile computing and networking (MOBICOM)* (Atlanta, Georgia, USA, ACM Press, 2002) pp. 172-182.

[3] I. Caragiannis, C. Kaklamanis, and P. Kanellopoulos, New results for energy-efficient broadcasting in wireless networks, *Proceedings of the 13th International Symposium on Algorithms and Computation (ISAAC '02)* (Springer Verlag, Lecture Notes in Computer Science, Nov. 2002) pp. 332-343.

[4] I. Caragiannis, C. Kaklamanis, and P. Kanellopoulos, Energy-efficient wireless network design, *Proceedings of the 14th International Symposium 2002 on Algorithms and Computation (ISAAC '03)* (Springer Verlag, Lecture Notes in Computer Science, Nov. 2003) pp. 585-594.

[5] J. Cartigny, D. Simplot, and I. Stojmenovic, Localized minimum-energy broadcasting in ad-hoc networks, *Proceedings of the Twenty-Second Annual Joint Conference of the IEEE Computer and Communications Societies (INFOCOM)* (2003).

[6] T. Chu and I. Nikolaidis, Energy efficient broadcast in mobile ad hoc networks, *Proceedings of the 1st International Conference on Ad-Hoc Networks and Wireless (ADHOC-NOW)* (2002) pp. 363-374.

[7] V. Chvátal, A greedy heuristic for the set-covering problem, *Math. Oper. Res.* Vol.4 (1979) pp. 233-235.

[8] A. E. F. Clementi, P. Crescenzi, P. Penna, G. Rossi, and P. Vocca, On the complexity of computing minimum energy consumption broadcast subgraphs, *STACS 2001, Lecture Notes in Comput. Sci.* Vol.2101 (Springer, Berlin, 2001) pp. 121-131.

[9] T. H. Cormen, C. E. Leiserson, R. L. Rivest, and C. Stein, *Introduction to Algorithms*, (MIT Press, Cambridge, MA, 2nd edition, 2001).

[10] A. Ephremides, Energy concerns in wireless networks, *IEEE Wireless Communications* Vol.9 (2002) pp. 48-59.

[11] R. Gallager, P. Humblet, and P. Spira, A distributed algorithm for minimum-weight snapping trees *ACM Trans. on Prog. Lang. and Systems* Vol.5 (1983) pp. 66-77.

[12] M. R. Garey and D. S. Johnson, Computers and intractability (W. H. Freeman and Co., San Francisco, Calif., 1979), A guide to the theory of NP-completeness, A Series of Books in the Mathematical Sciences.

[13] S. Guha and S. Khuller, Improved methods for approximating node weighted Steiner trees and connected dominating sets, *Inform. and Comput.* Vol. 150 (1999) pp. 57-74.

[14] D. S. Hochbaum, ed., *Approximation Algorithms for NP Hard Problems*, (PWS Publishing Company, Boston, MA, 1997).

[15] F. K. Hwang, D. S. Richards, and P. Winter, The Steiner tree problem, Vol.53 of *Annals of Discrete Mathematics*, (North-Holland Publishing Co., Amsterdam, 1992).

[16] L. M. Kirousis, E. Kranakis, D. Krizanc, and A. Pelc, Power consumption in packet radio networks (extended abstract), *Proceedings of STACS* (1997) pp. 363-374.

[17] A. M. Law and W. D. Kelton, *Simulation Modeling and Analysis*, (McGraw Hill, USA, 1991).

[18] W. Liang, Constructing minimum-energy broadcast trees in wireless ad hoc networks, *Proceedings of the third ACM international symposium on Mobile ad hoc networking & computing (MOBIHOC)* (Lausanne, Switzerland, ACM Press, 2002) pp. 112-122.

[19] P. M. Marvinkurve, H. Q. Ngo, and H. Mehra, MIP3S: Algorithms for power-conserving multicasting in wireless ad hoc networks, *Proceedings of the*

11th IEEE International Conference on Networks (ICON) (Sydney, Australia, 2003).

[20] H. Q. Ngo, D. Pan, and V. Verma, Power-conserving algorithms and protocols in ad hoc networks, in X. Cheng and D.-Z. Du (eds.) *Ad Hoc Networking: Recent Advances* (Kluwer Academic Publishers, 2003).

[21] D. Pan, P. M. Marvinkurve, H. Q. Ngo, V. Verma, and A. Chandak, DMIP3S: Distributive routing algorithms for power-conserving multicasting in static wireless ad hoc networks, to appear in *Proceedings of the 2004 Workshop on High Performance Switching and Routing (HPSR)* (Phoenix, Arizona, 2004).

[22] D. Pan, H. Q. Ngo, P. M. Marvinkurve, V. Verma, and A. Chandak, IP3S: Centralized and distributive routing algorithms for power-conserving multicasting in wireless ad hoc networks, (submitted 2004).

[23] C. E. PERKINS, *Ad Hoc Networking*, (New Jersey, USA, Pearson Education, Dec. 2000).

[24] T. Rappaport, *Wireles Communications: Principles and Practices*, (Prentice Hall PTR, New Jersey, USA, 2nd ed., Dec. 2001).

[25] G. Robins and A. Zelikovsky, Improved Steiner tree approximation in graphs, *Proceedings of the Eleventh Annual ACM-SIAM Symposium on Discrete Algorithms* (San Francisco, CA, 2000) pp. 770-779.

[26] V. P. Verma, A. Chandak, and H. Q. Ngo, DIP3S: A distributive routing algorithm for power-conserving broadcasting in wireless ad hoc networks, *Proceedings of the Fifth IFIP-TC6 International Conference on Mobile and Wireless Communications Networks (MWCN)* (Singapore, World Scientific, 2003) pp. 159-162.

[27] P.-J. WAN AND G. CALINESCU, Minimum-energy multicast in routing in static ad hoc wireless networks, (manuscript, 2001).

[28] P.-J. Wan, G. Calinescu, X.-Y. Li, and O. Frieder, Minimum-energy broadcast routing in static ad hoc wireless networks, *Proceedings of the Twentieth Annual Joint Conference of the IEEE Computer and Communications Societies (INFOCOM)* Vol.2 (2001) pp. 1162-1171.

[29] D. B. West, *Introduction to graph theory*, (Prentice Hall Inc., Upper Saddle River, NJ, 1996).

[30] J. E. Wieselthier, G. D. Nguyen, and A. Ephremides, Multicasting in energy-limited ad-hoc wireless networks, *Proceedings - IEEE Military Communications Conference MILCOM* Vol.3 (1998) pp. 723-729.

[31] J. E. Wieselthier, G. D. Nguyen, and A. Ephremides, On the construction of energy-efficient broadcast and multicast trees in wireless networks, *Proceedings of the Nineteenth Annual Joint Conference of the IEEE Computer and Communications Societies (INFOCOM)* Vol.2 (2000) pp. 585-594.

[32] J. E. Wieselthier, G. D. Nguyen, and A. Ephremides, Algorithms for energy-efficient multicasting in static ad hoc wireless networks, *Mobile Networks and Applications* Vol.6 (2001) pp. 251-263.

[33] J. E. Wieselthier, G. D. Nguyen, and A. Ephremides, The effect of discrete power levels on energy-efficient wireless broadcast in ad hoc networks, *Proceedings of the 13th IEEE International Symposium on Personal, Indoor and Mobile Radio Communications* Vol.4 (2002) pp. 1655-1659.

[34] J. E. Wieselthier, G. D. Nguyen, and A. Ephremides, Energy-efficient multicasting of session traffic in bandwidth- and transceiver-limited wireless networks, *Cluster Computing* Vol.5 (2002) pp. 179-192.

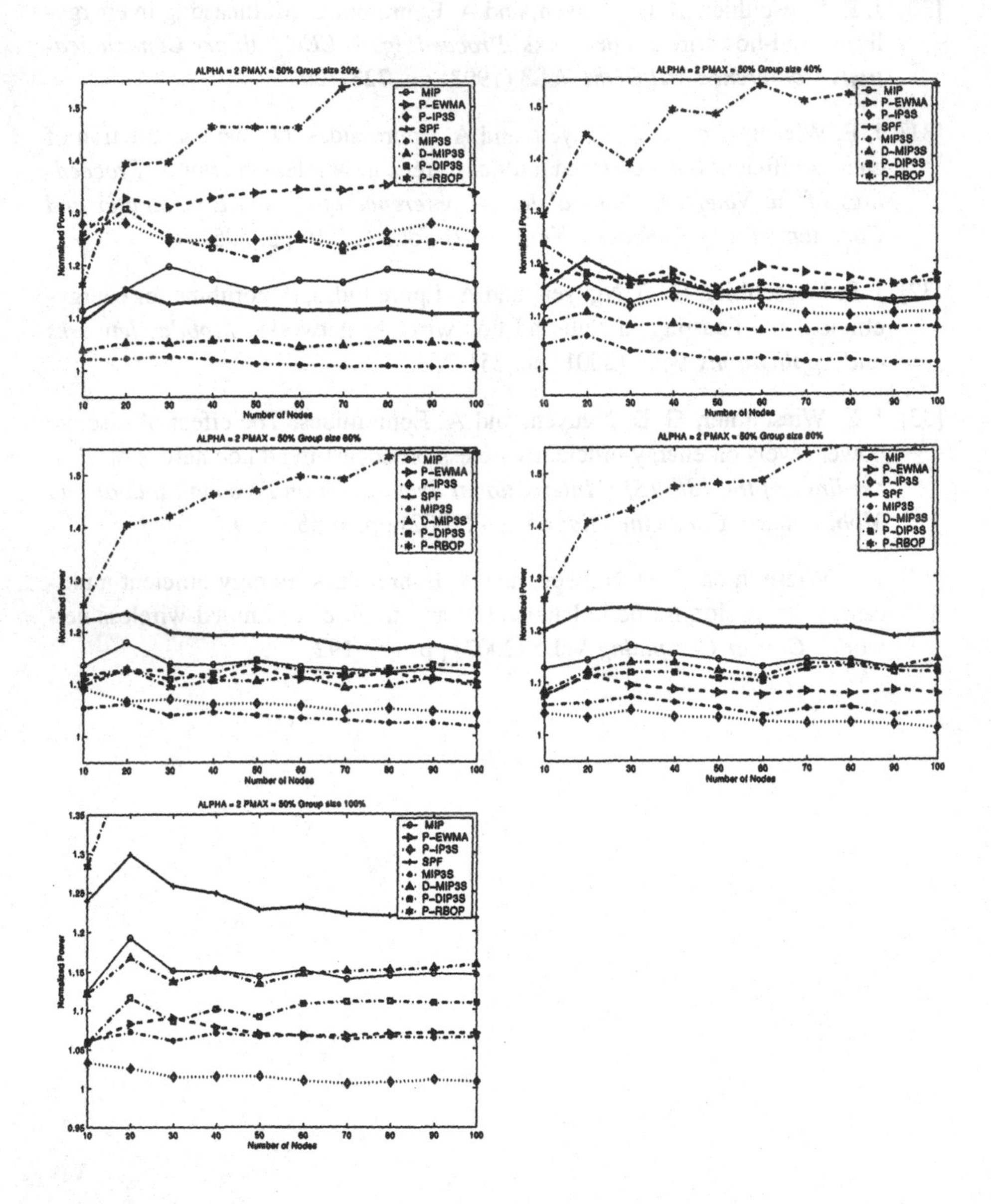

Figure 9: The effect of different groups sizes. p_{max} is 50% of the region, $\alpha = 2$, group size is 20%, 40%, 60%, 80%, 100%.

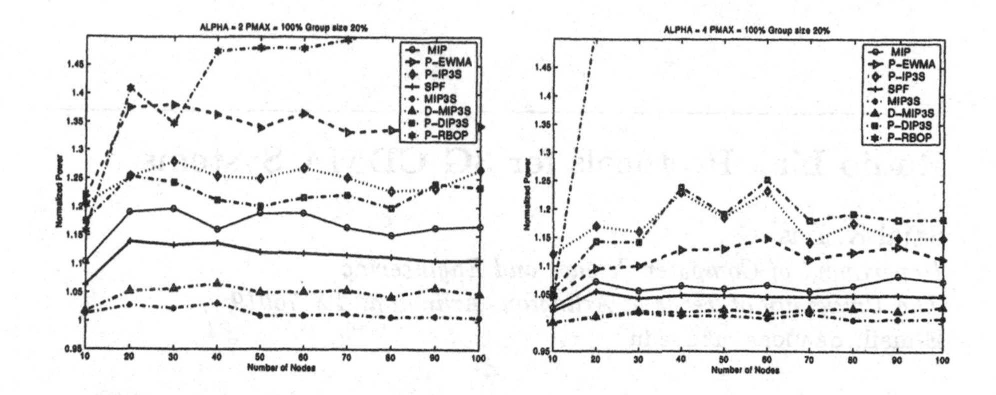

Figure 10: The effect of different values of α. Both scenarios have p_{max} is 100% of the region, group size is 20%, $\alpha = 2, 4$.

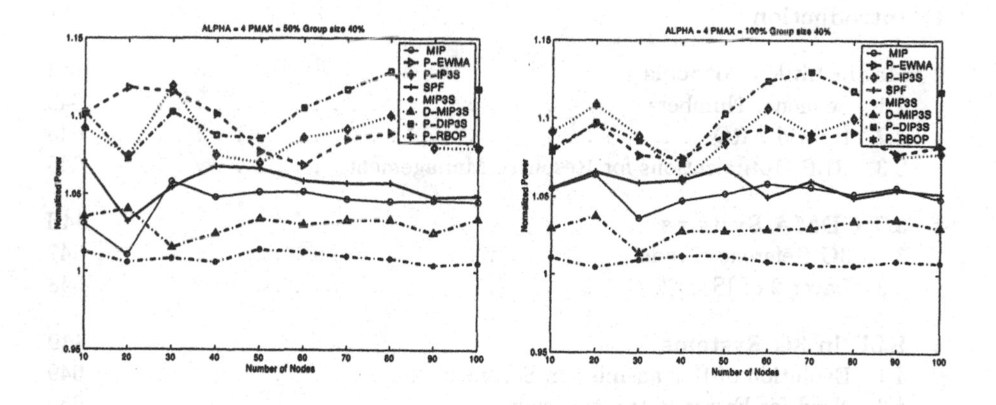

Figure 11: The effect of different values of p_{max}. Both scenarios have group size is 60%, $\alpha = 4$, the p_{max} is 50%, 100% of the region respectively.

Radio Link Protocols for 3G CDMA Systems

Sajal K. Das
Department of Computer Science and Engineering
The University of Texas at Arlington, Arlington, TX 76019
E-mail: `das@cse.uta.edu`

Mainak Chatterjee
Department of Electrical and Computer Engineering
University of Central Florida, Orlando, FL 32816
E-mail: `mainak@cpe.ucf.edu`

Contents

References

1 Introduction

With the proliferation of the World Wide Web (WWW) in our daily life, a number of *wireless data* services such as voice, audio, video streaming, file and web downloading also need to be supported in the wireless access networks. Though the usage of voice services has been increasing and the cost of operation per user is decreasing, the rate of wireless voice penetration is slowing down as the market is reaching saturation. The revenue growth from exclusively voice services is expected to fall in the coming years while wireless data services promise to offer significant growth opportunity for carriers all over the world. In other words, the provisioning of data services will become increasingly important for the revenue growth in wireless networks.

All the emerging wireless data networking technologies today rely on the Internet Protocol (IP) because IP is still the most dominant internetworking protocol. Moreover, the already existing Internet infrastructure should be exploited as much as possible to defray the cost of overlaying wireless technology. The advances in IP coupled with the provisioning for quality of service (QoS) for multimedia applications makes IP a good choice for cellular providers to deliver the service to the already existing huge customer base.

No matter which wireless data technology emerges as the dominant radio interface, the fact remains that it will rely on the IP-based network because of its omni-presence. Transport control protocol (TCP), the commonly used transport layer protocol for IP, is still the major suite for IP and provides reliable end-to-end transmission [24] in the wireline domain. The design of TCP has been such that it performs well in wireline networks where the channel error rates are extremely low and any occurrence of congestion is due to loss of packets only. However, when TCP is used in the wireless domain, which is characterized by high bit error rate, the performance of TCP severely degrades. Any packet loss at the wireless link is also interpreted as congestion by TCP, which responds to it by reducing the transmission window size, initiating the congestion control mechanism and resetting the retransmission time [15]. The congestion control mechanism designed for wireline networks causes an unnecessary reduction in the TCP throughput. To deal with this behavior of TCP, several schemes have been proposed to alleviate the effects of non-congestion related losses over wireless links. One such scheme is the *radio link protocol* (RLP) which is primarily meant for cellular networks. Radio link protocols are employed at the last hop the communication (i.e., the wireless link between the base station and the mobile terminals) to increase the reliability of the wireless link.

The rest of the chapter is organized as follows. In Section 2, we demonstrate the working of radio link protocols and discuss hybrid ARQs. In Section 3, we present the evolution of the CDMA systems and the 3G network architecture. Since RLP is employed at Layer 2, we discuss the features and functions of Layer 2 of IS-2000. Section 4 shows the need for better retransmission schemes for the 3G CDMA systems and shows how retransmissions at the MAC layer can help improve the performance of the RLP.

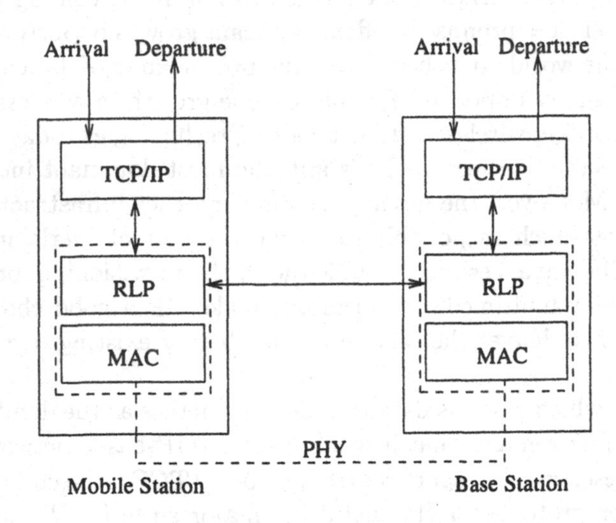

Figure 1: Simplified stack

2 Radio Link Protocols

As discussed earlier, radio link protocols are used in cellular data networks to shield the effect of the loss over wireless links from the TCP layer [9, 14]. The RLP is generally employed at the data link layer between the TCP/IP layer and the physical channel as shown in Figure 1. It segments down an upper layer packet (a TCP segment in this case) into several RLP frames before transmitting over the wireless channel as shown in Figure 2. A physical layer header is added to the RLP frame before it is mapped on to a physical layer frame for transmission. The fragmentation is done to reduce the granularity of the transmission, i.e, in case of any error, an RLP frame which is of a smaller size is affected rather than the whole TCP segment. In case of an RLP frame loss during transmission, the RLP uses an Automatic Repeat reQuest (ARQ) error recovery mechanism to retrieve the lost RLP frame. The process for recovery of erroneous frames is initiated by the receiver by requesting retransmission of the missing or damaged frames. The recovery of the erroneous frames should be done before the TCP timer expires for the TCP throughput to remain unaffected.

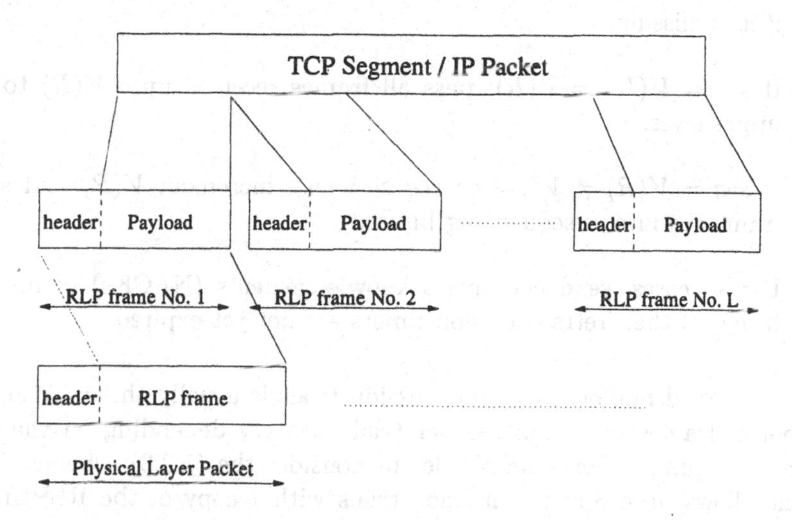

Figure 2: Fragmentation of TCP segments into RLP frames

2.1 Sequence Numbers

During the data transfer phase, the RLP maintains a sending sequence number count, $V(S)$ and two other sequence numbers for receiving, $V(R)$ and $V(N)$. All operations on these sequence numbers are carried out in modulo arithmetic. $V(S)$ is incremented whenever a new RLP frame is sent out. In other words, $V(S)$ is the sequence number of the frame to be sent. $V(R)$ is the sequence number of the next new frame expected to be received. $V(N)$

is the sequence number of the frame needed for sequential delivery, i.e,. $V(N)$ is the oldest sequence number of the missing frames. If *seq* denotes the sequence number of a newly received frame, then the RLP transmission procedure can be described by the following rules.

- If $seq < V(N)$, or if the frame is already stored in the re-sequencing buffer, discard the frame.
- If $seq = V(N)$, update $V(N)$ to the next oldest missing frame sequence number. Pass received frames up to $V(N) - 1$ to the upper layer.
- If $V(N) < seq < V(R)$, store the frame *seq* in the re-sequencing buffer if it is missing.
- If $seq = V(N) = V(R)$, pass all frames received upto $V(R)$ to the upper layer.
- If $seq = V(R) \neq V(N)$ or $seq > V(R)$, increment $V(R)$ and store frame *seq* into re-sequencing buffer.
- For all cases, send negative acknowledgements (NACKs) of missing frames if their retransmission timers are not yet expired.

The allowed number of retransmission trials is usually three. Also, the number of frames retransmitted per trial can vary depending on the performance required. For example, let us consider the (1,2,3) scheme. This scheme allows upto 3 retransmission trials with 1 copy of the RLP frame being sent on the first trial. If that frame is not received correctly, then 2 copies of the same frame are sent on the second trial. This increases the probability of correct reception at the cost of higher redundancy. If neither frame is received correctly, then 3 copies of the same frame would be sent on the third and final trial. The RLP aborts the frame recovery process after the allowed number of retransmissions are exhausted and forwards the frames which were waiting for the missed frame to the upper layer. The upper layer (TCP, for example) would reassemble the frames to form a packet and would detect an error. The upper layer then would invoke its own retransmission scheme to recover the erroneous packet. Another example of a retransmission scheme would be (1,1,1,1,1), where the maximum number of retransmissions is 5 with one copy of the frame being sent each time.

2.2 Hybrid ARQ

The RLP uses a number of retransmission schemes like (1,1,1,1,1), (1,2,3), (1,1,2,3), etc [14], depending on the channel conditions and the performance required for the session it is supporting. Oftentimes, hybrid ARQs are also used to enhance the performance of RLPs. Hybrid ARQs incorporate certain *forward error correction* (FEC) schemes through which it ensures that there is a higher probability of the packets reaching the receiver end.

The transmitter on the receipt of a NACK or a time-out will trigger a retransmission. It might so happen that a packet which has been successfully received, had the ACK damaged. In that case, the transmitter will time-out and re-send the packet resulting in duplication of the packet at the receiver buffer. Now the question arises about the usefulness of retransmission. That is, will the retransmitted packet be on-time at the re-sequencing buffer for it to be passed on to the higher layers? This can only be possible if the round trip time (RTT) is sufficiently low and the packet can be accommodated in the re-sequencing buffer. If the RTT is high and retransmission is not feasible, then the RLP frames can be made more robust by adopting FEC schemes. The combination of FEC and ARQ is known as the hybrid ARQ. If the two schemes are matched to the channel conditions, then the hybrid ARQ can significantly change the system performance. The FEC reduces the number of retransmissions by correcting the detectable and correctable errors. However, all errors cannot be corrected and the receiver sends for a retransmission request of the RLP frame rather than passing on the uncorrected frame to the upper layers to be corrected by TCP. Thus, by properly combining FEC and ARQ, the overall system throughput and reliability can be increased. Differential RLP [11] allows the frames from the same TCP segment to be treated differently with respect to the ARQ and FEC schemes, i.e., RLP frames obtained from the same TCP segment are encoded with different FEC schemes and undergo different ARQ mechanisms.

2.3 RLP Optimizations for Resource Management

To better manage and utilize the resources available to the system, the RLP must optimize certain parameters. The selection of the RLP frame size is a critical issue for the performance of the RLP. Since a fixed size header is added to every RLP frame, it is desirable to have large frame size so that the data goodput is high. But under lossy conditions, it might not be a good idea to have large frame size because any loss would result in a retransmission of equivalent amount of data. Though a smaller frame size seems reasonable, the overhead due to headers can be appreciably high. The same problem arises for selecting the FEC scheme. For better detection and correction of errors of the received RLP frames at the receiver, a robust FEC scheme must be employed. This implies more overhead due to the redundant bits and as a result the effective data rate will be penalized. On the other hand, if the FEC scheme is not that strong the receiver might not be able to recover from the errors that frequently and more retransmissions will be required. From the application's perspective, the different optimizations performed at the RLP will be manifested in the form delay and throughput. In other words, if an application needs a certain level of QoS, the RLP can tune its parameters in such a manner that the expectation of the application is met. The RLP can also be made to dynamically adapt to the fluctuations in the wireless channel conditions and provide the same level of performance.

3 3G CDMA Systems

There has been a great deal of debate in recent times about what have been referred to as the third generation (3G) wireless systems and services. This debate has focused a spotlight on wireless data capabilities, since a distinguishing feature between "2G" and "3G" systems is the ability of 3G to support a wide variety of information-intensive services. The plans for 3G systems are expected to deliver significantly richer wireless data services.

To bring the WWW traffic to the wireless mobile devices, it is important that a suitable protocol or standard be chosen to cater to the growing demands of data services over wireless channels which could handle a wide variety of multimedia traffic with different QoS requirements. CDMA based technologies have shown to provide many benefits to carriers and consumers, including better voice quality, broader coverage and stronger security. Today, many leading wireless service providers use CDMA to provide high-quality voice and data services to over 100 million users worldwide. As new generations of technologies and standards are being developed, CDMA clearly stands out as the most prominent 3G technology.

The first commercial launch of the code division multiple access (CDMA) systems by QUALCOMM in 1995 catered to voice services only; since then CDMA has become one of the world's fastest-growing wireless technologies. Second generation (2G) technologies that are still being used for delivering wireless data services include include GSM, TDMA, and IS-95-A CDMA networks. To this extent, many operators are planning to use CDMA as the 3G choice. The specifications of 3G systems are defined in the IMT-2000 standard of the International Telecommunications Union (ITU) [2]. The two most important CDMA technologies referred are cdma2000 and wideband CDMA (WCDMA). The cdma2000 [7] system is just being deployed in commercial networks and supports voice and data on the same 1.25 MHz carrier. A higher data rate evolution, called cdma2000 1X-DV, has been standardized by the Telecommunications Industry Association (TIA) and will be added to the IMT-2000. The WCDMA technology which supports both high-rate packet data and high-rate circuit-switched data on a single 5 MHz carrier is specified by the 3rd Generation Partnership Project (3GPP) [1] and is rapidly emerging as the global 3G radio access technology. Current WCDMA specifications support data rates up to 2 Mbps in indoor/small-cell-outdoor and up to 384 Kbps with wide-area coverage, which is in full agreement with the IMT-2000 requirements.

In 3G systems, extensive efforts have been made to mitigate the losses in the wireless link by local retransmissions. For example, link layer transmissions protocols such as RLP and RLC (radio link controller) are in proposition for 3G-1X [6] and UMTS (Universal Mobile Telecommunications System) [4], respectively. As of today, link layer transmissions have been incorporated in cdma2000 [5] and WCDMA [3]. These mechanisms have helped in improving the packet loss probability in the wireless link, thereby managing to mitigate the adverse impact of channel losses on TCP. Though these mechanisms succeed in concealing losses from TCP, they increase the delay variability which is a hindrance in maintaining the assured QoS [10].

3.1 3G Reference Model

A simplified architecture of a 3G wireless network is shown in Figure 3. The base stations are connected to the Radio Network Controller (RNC). The RNC performs CDMA specific functions like handoffs, encryption, power control, etc. It also performs link layer retransmission using RLP in 3G1X system or using RLC (radio link control) in UMTS. In the 3G1X system, the RNC is connected to a PDSN (Packet Data Service Node) using a GRE (Generic Routing Encapsulation) tunnel, which is a form of tunnel in IP. In the UMTS system, the RNC is connected to a SGSN (Serving GPRS Support Node) using a GTP (GPRS tunneling protocol) tunnel (another form of tunnel in IP); the SGSN is connected to the GGSN (Gateway GPRS Support Node), again through a GTP tunnel.

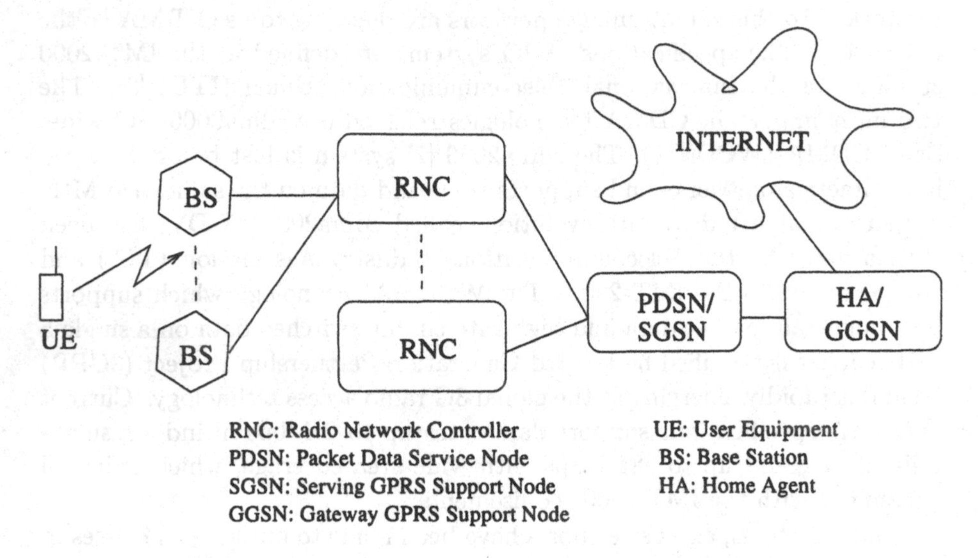

Figure 3: 3G Network Architecture

In this architecture, the RNC receives an IP packet through the GTP/GRE tunnel from the SGSN/PDSN. The RNC fragments this packet into a number of radio frames and then performs transmission and local retransmissions of these frames using the RLP (RLC) protocol. The base station receives the radio frames from the RNC and then schedules the transmission of the radio frames on the wireless link. There are various scheduling disciplines which might or might not take into consideration the channel state information. The user equipment (UE) receives the radio frames and if it discovers that a frame is lost or corrupted, then it requests for a local retransmission of that frame using the RLP (RLC). Since, the RNC maintains per-user queue of the radio frames, the RNC can do a local retransmission and the lost frames could be recovered quickly.

3.2 Layer 2 of IS-2000

For the sake of completeness, let us show how the RLP is implemented at the Layer 2 (link layer) of the IS-2000 protocol stack, which is outlined in TIA/EIA/IS-2000.4-A (March 2000). The link layer provides protocols to support and control data transport services. It is divided into two sublayers, the Link Access Control (LAC) and the Media Access Control (MAC). The layering as described in IS-2000 is shown in Figure 4. The LAC sublayer provides an interface for transporting data over the air between peer upper layer entities. The LAC employs a number of different protocols to match the quality of service requirements of each upper layer entity to the characteristics of the MAC sublayer in order to provide scalable transmission reliability capabilities. It utilizes various end to end reliable ARQ protocols that use sequence numbering, ACKs/NACKs and retransmission of lost or damaged frames to provide reliable services. The MAC sublayer provides a control function that manages resources supplied by the physical layer and coordinates their usage by various LAC service entities. This function is needed to resolve contention between LAC service entities within a single mobile station, and between competing mobile stations. The MAC sublayer also provides multiplexing and QoS control. This can be done by prioritizing requests fairly, and resolving conflict messages. This QoS control mechanism can help to balance the varying QoS requirements of multiple concurrent services.

It can be noted that the services which do not use TCP, such as voice calls, video streaming and VoIP, cannot use the services offered by RLP. This is because these services are mostly real-time in nature and cannot afford any retransmission and would thereby bypass RLP.

4 RLP in 3G Systems

4.1 Evolution of Retransmission Schemes

Numerous retransmission schemes have been proposed to enhance TCP throughput over wireless links. We will outline only those schemes that have been proposed for CDMA systems, particularly for the standardized protocols. Bao [9] investigated the performance issues related to TCP and RLP interaction in the CDMA protocol stack. In [22], the impact of TCP source activity on the call admission control for IS-95 was studied. The support of data services over the IS-95 physical channels using RLP was proposed in [17]. For IS-99, the performance evaluation of TCP over RLP was shown in [16] and performance for circuit mode data services was shown in [14]. Several studies have been made for the cdma2000 system. The performance of TCP over the cdma2000 RLP was shown in [20]. A NACK based hybrid ARQ scheme was proposed in [25].

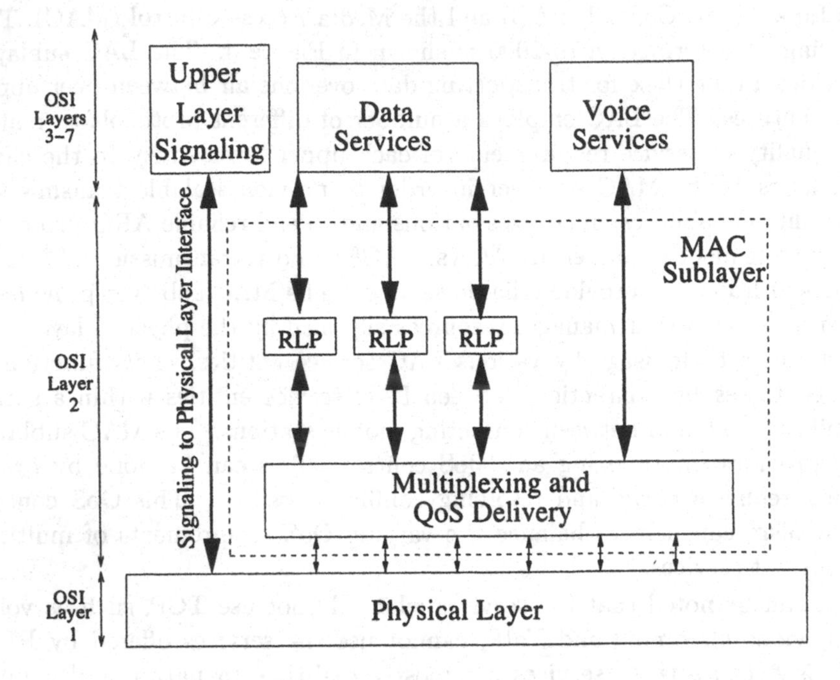

Figure 4: Layering of IS-2000

4.2 Need for Faster Retransmissions

Radio link protocols are usually sufficient to shield the physical layer impairment from the TCP, but might fail to do so if the application has very strict delay requirements [22]. It is generally considered that *real-time* traffic can tolerate some amount of errors because of the nature of such applications. For example, if packets are dropped in voice communication, human intelligence is able to find the consistency in speech. Similarly, in video transmission if packets or frames are lost, the decoder at the receiver does an interpolation between frames to conceal the loss. These kinds of loss concealment can be done as long as the losses are within a non-observable limit. But sometimes it might so happen that the losses due to wireless channel conditions are so high that error concealment becomes almost impossible. In such cases, interpolation or interpretation of a signal may not only be difficult but impossible. Therefore, it is recommended that the damaged packets are salvaged as much as possible and also as quickly as possible.

4.3 MAC Retransmissions for cdma2000

To avoid the delay associated with retransmissions at the RLP, a faster lower layer MAC-ARQ can be used. The faster retransmissions can provide a better round trip time for real-time applications. Since the number of transmissions allowed at the MAC layer is finite, it does not completely eliminate the possibility of having missing or damaged frames. If the MAC-ARQ fails to deliver a frame correctly even after retransmitting the maximum allowed number of times, the responsibility is passed on to the RLP layer to retrieve the frame. Thus we get two layers of retransmission reliability as shown in Figure 5. We will briefly discuss the retransmission at the MAC layer of cdma2000 [12] and WCDMA [13] protocol standards.

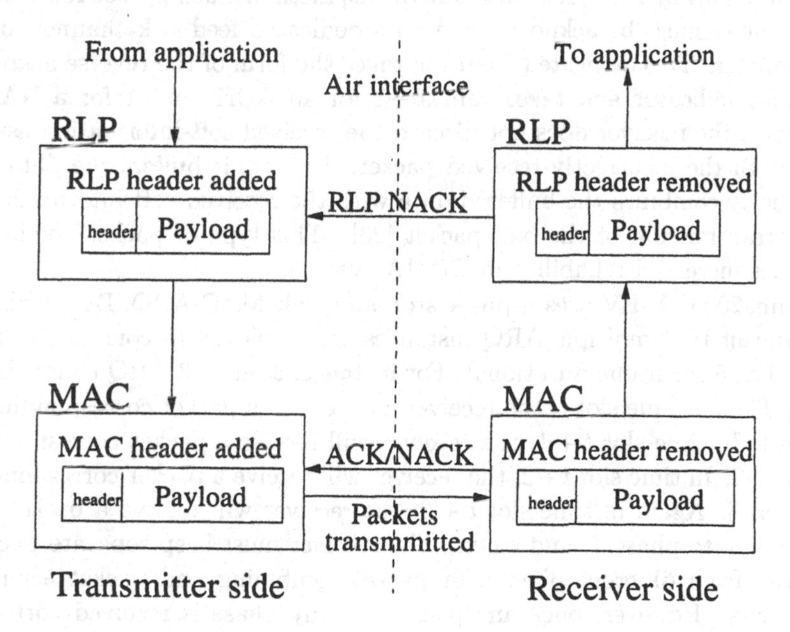

Figure 5: Two layer reliability

The fast ARQ mechanism (also called MAC-ARQ) in cdma2000 1X-EV is important for ensuring that some performance loss can be recovered. There are two reasons why RLP cannot provide the functionality needed for MAC-ARQ in cdma2000. First, in the process of selecting the base station with the strongest signal for the cell selection, the RLP terminates at the last network element (for example, base station) resulting in network delays in servicing retransmission requests at the RLP layer. Fast cell site selection (FCSS), a feature in cdma2000, provides mobility support for best effort data services. This is achieved by the mobile terminal by "echoing" information about packets it has recently received over the shared channel during a transition form one base station to another so as to provide fast synchronization of the packet transmission queues between the new and the old base stations. Second, the forward shared channel (FSHCH), which carries payload for best effort data users (in both time and code multiplexed mode) might contain several protocol data units (PDUs), not all of which come from the RLP. In fact, some PDUs might come directly from the layer above RLP. As a result, MAC-ARQ provides retransmissions quickly in cdma2000 which employs a stop-and-wait hybrid ARQ method. In this method, each packet received by the receiver must be acknowledged on a dedicated feedback channel to the transmitter. This dedicated feedback takes the form of the reverse acknowledgment indicator and takes values $+1$ for an ACK, or -1 for a NACK. However, the receiver does not discard the received soft-information associated with the incorrectly-received packet. Rather, it buffers the data and coherently combines the buffered data with the received soft information of the retransmission of the bad packet [23]. This type of packet combining provides increased reliability in CDMA systems.

cdma2000 1X-EV uses n-phase stop-and-wait MAC-ARQ. By "n-phase" it is meant that multiple ARQ instances are employed in consecutive time slots (i.e., 5-ms frame durations). For instance, assume 3 ARQ channels are used. Then in time slot t, the receiver will receive a packet corresponding to phase 1. In time slot $t+1$, the receiver will receive a packet corresponding to phase 2. In time slot $t+2$, the receiver will receive a packet corresponding to phase 3. Again in time slot $t+3$, the receiver will receive a packet corresponding to phase 1, and so on. The receiver must keep separate packets received from different phases for packet combining and packet acknowledgements. However, once any packet for any phase is received correctly, the receiver may deliver the packets to the higher layers (e.g., RLP). The timing diagram for this scheme is explained in Section 4.4.

4.4 Fast ARQ

The transmitter transmits one RLP packet in each 5 ms physical layer frame and waits for the ACK. If the ACK does not arrive in 20 ms (equal to 4 ARQ phases, which is specific to cdma2000), then the frame is retransmitted im-

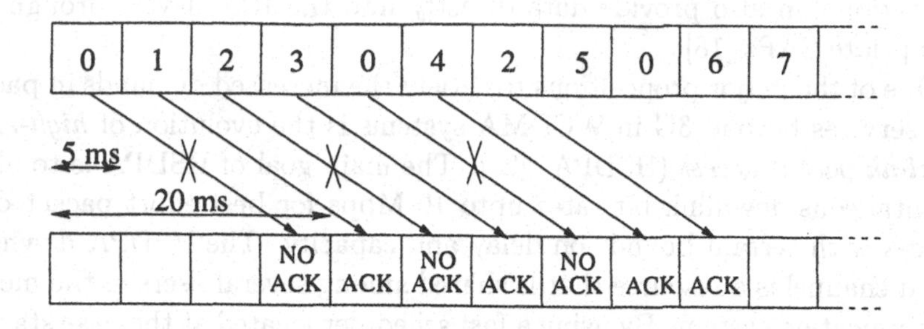

Figure 6: NARQP = 4 Timing

mediately. It can be seen from Figure 6 that frames with sequence numbers $0, 1, 2, 3, \cdots$ are being transmitted. Frames 0 and 2 are undergoing retransmission because of non-receipt of ACKs. The ACK timers of these packets are again reinitialized. If we consider the down link, then it is not necessary that the mobile station will deliver the ACK/NACK precisely at the slot boundaries, each of which is 5 ms. The actual physical layer boundary is more precise. This is due to the fact that normally when using coherent receivers in the reverse link (for example in IS-2000), the base-station suffers from some processing delay. The number of retransmission trials allowed is varied between 1 and 3. If a packet is not successfully received or combined at the receiver even after the maximum number of MAC retransmissions, then the RLP retransmission is triggered.

4.5 MAC Retransmissions for WCDMA

The way wideband CDMA (WCDMA) handles the MAC layer retransmissions is somewhat different from cdma2000. The radio interface protocol stack for WCDMA has a dedicated radio link control (RLC) layer to take care of the RLP functionality [19]. The stack primarily consists of three layers. Layer 1 is the physical layer (PHY). The main components of layer 2 are the MAC and the RLC. Layer 3 contains the radio resource control (RRC) which is mainly responsible for the radio resource allocation to the user equipments (UEs). It also does all the control plane and user plane signalling between the UTRAN (Universal Terrestrial Radio Access Network) and the UEs, and tries to deliver the negotiated quality of service (QoS) to the UEs. In order for the RRC to do so, it sends control signals to all the sublayers in Layer 2 through the service access points (SAPs). Any application can also provide data directly into the RLC layer through the appropriate SAPs [18].

One of the major propositions to satisfy the increased demands in packet data services beyond 3G in WCDMA systems is the evolution of *high-speed downlink packet access* (HSDPA) [21]. The main goal of HSDPA is to allow instantaneous downlink bit rates upto 10 Mbps for best-effort packet data services with certain bounds on delay and capacity. The HSDPA downlink shared channel is a resource that is shared among several users in the mobile communication system. By using a fast scheduler located at the base station, the HSDPA channel can be assigned to the user with the currently available best channel, i.e., the user that can transmit with the highest data rate. The idea behind this is that all users shall only use the HSDPA channel when their own downlink is good, and let other users utilize the channel when it is anyway bad. Some of the basic principles used in HSDPA are fast link adaptation, fast scheduling and fast retransmissions of erroneously received packets. A similar "n-phase" MAC-ARQ with packet combing is also used in WCDMA.

5 Summary

As the future generations cellular networks become more data-centric due to the pre-domination of TCP/IP oriented services in the Internet, sophisticated and reliable link layer protocols will be required to support end-to-end services over the wireless links and to extend transport layer protocols such as TCP to the mobile hosts. The use of radio link protocols at the link layer to provide an acceptable error performance is now a standard practice. The main function of the RLP is to conceal the channel related losses from TCP by quickly recovering the dropped packets by means of local retransmissions. With the delay requirements of the data services over 3G CDMA networks becoming more stringent, faster and hybrid radio link protocols are being developed. Also, retransmissions at the MAC layer offer added reliability to the radio link protocols and make them more robust to mitigate losses over the wireless link which in turn makes data service provisioning in 3G networks more efficient.

References

[1] http://www.3gpp2.org

[2] http://www.itu.org

[3] 3G Partnership Project, Release 99.

[4] Third Generation Partnership Project, RLC Protocol Specification (3G TS 25.322, 1999).

[5] TIA/EIA/cdma2000, Mobile Station- Base Station Compatibility Standard for Dual-Mode Wideband Spread Spectrum Cellular Systems, Washington: TIA, (1999).

[6] TIA/EIA/IS-707-A-2.10, Data Service Options for Spread Spectrum Systems: Radio Link Protocol, Type 3, (Jan. 2000).

[7] TIA/EIA/IS-2000-2, Physical Layer Standard for cdma2000 Spread Spectrum Systems, (Mar. 2000).

[8] Proposed 1XTREME Physical Layer Delta Specification, Source: Motorola, Nokia, LSI Logic, Texas Instruments and Dot Wireless, Contribution to to 3GPP2, (Aug. 2000).

[9] G. Bao, Performance evaluation of TCP/RLP protocol stack over CDMA wireless links, *ACM Wireless Networks Journal* Vol. 2 (1996) pp. 229-237.

[10] M.C. Chan and R. Ramjee, TCP/IP performance over 3G wireless links with rate and delay variation, *Proceedings of ACM Mobile Computing and Networking (MobiCom)* (2002) pp. 71-82.

[11] M. Chatterjee, S. Ganguly and J. Sarkar, Improving RLP performance by differential treatment of frames, *Proceedings of IEEE International Conference on Communications (ICC) 2004*.

[12] M. Chatterjee, S.K. Das and G.D. Mandyam, Two-layer retransmission reliability for CDMA2000 1X-EV, *Proceedings of IEEE Vehicular Technology Conference (VTC)* Vol. 1 (2002-Spring) pp. 75-79.

[13] M. Chatterjee, G.D. Mandyam and S.K. Das, Fast ARQ in high speed downlink packet access for WCDMA systems, *Proceedings of European Wireless* (Feb. 2002), pp. 451-457.

[14] A. Chockalingam and G. Bao, Performance of TCP/RLP protocol stack on correlated fading DS-CDMA wireless links, *IEEE Transactions on Vehicular Technology* Vol. 49 No. 1 (Jan. 2000) pp. 28-33.

[15] D.E. Comer, Internetworking with TCP/IP, Vol. 1 (Prentice-Hall, 1991).

[16] A.S. Joshi, Umesh M.N., A. Kumar, T. Mukhopadhyay, K. Natesh, S. Sen, and A. Arunachalam, Performance evaluation of TCP over radio link protocol in TIA/EIA/IS-99 environment, *IEEE International Conference on Personal Wireless Communication* (1999) pp. 216-220.

[17] J.M. Harris and M. Airy, Analytical model for radio link protocol for IS-95 CDMA systems, *Proceedings IEEE 51st Vehicular Technology Conference (VTC)* Vol. 3 (2000-Spring) pp. 2434-2438.

[18] C. Geßner, R. Kohn, J. Schniedenharn and A. Sitte, Layer 2 and Layer 3 of UTRA-TDD, *Proceedings IEEE 51st Vehicular Technology Conference (VTC)* Vol. 2 (2000-Spring) pp. 1181-1185.

[19] H. Kaaranen, A. Ahtiainen, L. Laitinen, S.K. Naghian and V. Niemi, *UMTS Networks: Architecture, Mobility and Services*, (Wiley & Sons Publishing, 2001).

[20] F. Khan, S. Kumar, K. Medepalli, and S. Nanda, TCP performance over CDMA2000 RLP, *Proceedings IEEE 51st Vehicular Technology Conference (VTC)* Vol. 1 (2000-Spring) pp. 41-45.

[21] S. Parkvall, E. Dahlman, P. Frenger, P. Beming and M. Persson, The evolution of WCDMA towards higher speed downlink packet data access, *Proceedings IEEE 51st Vehicular Technology Conference (VTC)* Vol. 3 (2001-Spring) pp. 2287-2291.

[22] S. K. Sen, J. Jawanda, K. Basu, N.K. Kakani, and S.K. Das, TCP source activity and its impact on call admission control in CDMA voice/data Network, *Proceedings of Fourth ACM/IEEE International Conference on Mobile Computing and Networking (MobiCom)* (1998) pp. 276-283.

[23] S. Souissi and S.B. Wicker, A diversity combining DS/CDMA system with convolutional encoding and Viterbi decoding, *IEEE Transactions on Vehicular Technology* Vol. 44 No. 2 (May 1995) pp. 304-312.

[24] W.R. Stevens, *TCP/IP Illustrated*, (Addison-Wesley, 1994).

[25] Umesh M.N., A.S. Joshi, A. Kumar, T. Mukhopadhyay, A NAK based hybrid type II ARQ scheme for cdmaOne/cdma2000 systems, *IEEE Vehicular Technology Conference* Vol. 5 (Fall 1999) pp. 2596-2600.

A Survey on Improving TCP Performance over Wireless Networks

Xiang Chen, Hongqiang Zhai, Jianfeng Wang, and Yuguang Fang
Department of Electrical and Computer Engineering
University of Florida, Gainesville, FL 32611
E-mail: {xchen@ecel, zhai@ecel, jfwang@, fang@.ece}.ufl.edu

Contents

1 Introduction

As a result of the advancement of wireless technology and the proliferation of handheld wireless terminals, recent years have witnessed an ever-increasing popularity of wireless networks, ranging from wireless Local Area Networks (WLANs) and wireless wide-area networks (WWANs) to mobile ad hoc networks (MANETs). In WLANs (e.g., the Wi-Fi technology) or in WWANs (e.g., 2.5G/3G/4G cellular networks), mobile hosts communicate with an access point or a base station that is connected to the wired networks. Obviously, only one hop wireless link is needed for communications between a mobile host and a stationary host in wired networks. In contrast, there is no fixed infrastructure such as base stations or access points in a MANET. Each node in a MANET is capable of moving independently and functioning as a router that discovers and maintains routes and forwards packets to other nodes. Thus, MANETs are multi-hop wireless networks by nature. Note that MANETs may be connected at the edges to the wired Internet.

Transmission control protocol (TCP) is a transport layer protocol which provides reliable end-to-end data delivery between end hosts in traditional wired network environment. In TCP, reliability is achieved by retransmitting lost packets. Thus, each TCP sender maintains a running average of the estimated round trip delay and the average deviation derived from it. Packets will be retransmitted if the sender receives no acknowledgment(ACK) within a certain timeout interval (e.g., the sum of smoothed round trip delay and four times the average deviation) or receives duplicate acknowledgments. Due to the inherent reliability of wired networks, there is an implicit assumption made by TCP that any packet loss is due to congestion. To reduce congestion, TCP will invoke its congestion control mechanisms whenever any packet loss is detected. Since TCP is well tuned, it has become the de facto transport protocol in the Internet that supports many applications such as web access, file transfer and email. Due to its wide use in the Internet, it is desirable that TCP remains in use to provide reliable data transfer services for communications within wireless networks and for those across wireless networks and the wired Internet. It is thus crucial that TCP performs well over all kinds of wireless networks in order for the wired Internet to extend to the wireless world.

Unfortunately, wired networks and wireless networks are significantly different in terms of bandwidth, propagation delay, and link reliability. The implication of the difference is that packet losses are no longer mainly due to network congestion; they may well be due to some wireless specific reasons. As a matter of fact, in wireless LANs or cellular networks, most packet losses are due to high bit error rate in wireless channels and handoffs between two cells, while in mobile ad hoc networks, most packet losses are due to medium contention and route breakages, as well as radio channel errors. Therefore, although TCP performs well in wired networks, it will suffer from serious performance degradation in wireless networks if it misinterprets such non-congestion-related losses as a sign of congestion and consequently invokes congestion control and avoidance procedures, as confirmed through analysis and extensive simulations carried out in [4, 5, 7, 18-21]. As TCP performance deteriorates more seriously in ad hoc networks compared to WLANs or cellular networks, we divide wireless networks into two large groups: one is called one-hop wireless networks that include WLANs and cellular networks and the other is called multi-hop wireless networks that include MANETs.

To understand TCP behavior and improve TCP performance over wireless networks, given these wireless specific challenges, considerable research has been carried out and many schemes have been proposed. As the research in this area is still active and many problems are still wide open, this chapter serves to pinpoint the primary causes for TCP performance degradation over wireless networks, and cover the state of the art in the solution spectrum, in hopes that readers can better understand the problems and hence propose better solutions based on the current ones.

This chapter is organized as follows. We present in Section 2 a brief overview of TCP congestion control mechanisms and some current performance enhancement techniques. As the challenges TCP is facing differ in one-hop and multi-hop wireless networks and so do the solutions, it is suitable to separate them into two sections. Section 3 starts by identifying the challenges imposed on the standard TCP in one-hop wireless networks, followed by the classification of some existing solutions according to their design philosophy. Among the solutions, there are four large categories. The first class of schemes attempts to improve TCP performance by splitting a TCP connection into two at the base station or access point. Relying on an intelligent proxy located at the base station enforcing tasks such as local retransmission or ACK suppression/regulation, the second class eliminates the negative effects of wireless links on TCP. The approaches in the third class aim at hiding the characteristics of wireless links from TCP by providing a reliable link layer. The last category resolves the problems by slightly modifying TCP at the end systems, e.g., selective acknowledgment enabling or fast retransmission. In each class, the solutions are discussed in certain details. The structure of Section 4 is similar to that of Section 3, except that TCP performance over MANETs is the focus. Similarly, current solutions can also be grouped into three camps, according to their design philosophy. The first camp incorporates network feedback information into their designs to modify TCP's response to non-congestion-related packet losses while the second camp attempts to do so without explicit feedback. Unlike the previous two, the third one starts by tuning the lower layers in order for TCP to operate normally, while leaving TCP intact. With the understanding that current solutions fail to improve on some critical issues such as fairness, Section 5 gives some suggestions on future research issues. Finally, concluding remarks are given in Section 6.

2 Overview of TCP

Before we dive into the detailed discussion of questions such as *why TCP performs poorly in wireless networks*, *how TCP performance can be improved*, it is necessary to prepare the reader by presenting an overview of not only the basic functionality of TCP but also the state-of-the-art in TCP. The basic functions of TCP as a transport layer protocol include flow control, error recovery and congestion control, while the state-of-the-art techniques include fast retransmission and recovery, selective acknowledgment, etc., mainly focusing on how to promptly and effectively respond to network congestion.

2.1 Basic Functionality of TCP

It is well known that TCP is a connection-oriented transport protocol that is aimed at guaranteeing end-to-end reliable ordered delivery of data packets over wired networks. For this purpose, basic functionalities such as flow control, error control, and congestion control are indispensable. While these functions have a clean-cut definition of their own, in practice they are closely coupled with one another in TCP implementation.

In TCP, a sliding window protocol is used to implement flow control, in which three windows are used, namely, *Congestion Window*, *Advertised window*, and *Transmission Window*. Congestion window indicates the maximum number of segments (Without causing confusion, the term segment and packet are used interchangeably henceforth) that the sender can transmit without congesting the network. As shown next in details on congestion control, this number is determined by the sender based on the feedback from the network. Advertised window, however, is specified by the receiver in the acknowledgements it. Advertised window indicates to the sender the amount of data the receiver is ready to receive in the future. Normally, it equals to the available buffer size at the receiver in order to prevent buffer overflow. Transmission window means the maximum number of segments that the sender can transmit at one time without receiving any ACKs from the receiver. Its lower edge indicates the highest numbered segment acknowledged by the receiver. Obviously, to avoid network congestion and receiver buffer overflow, the size of transmission window is determined as the minimum of the congestion window and the receiver's advertised window.

To notify the sender that data is correctly received, TCP employs a cumulative acknowledgement mechanism. In other words, upon the receipt of an ACK, the sender knows that all previously transmitted data segments with a sequence number less than the one indicated in the ACK are correctly received at the receiver. In the case that an out-of-order segment (identified on the basis of sequence numbers) arrives at the receiver, a duplicate ACK is generated and sent back to the sender. It is important to note that in wired networks, an out-of-order delivery usually implies a packet loss. If three duplicate cumulative ACKs are received, the sender will assume the packet is lost. A packet loss is also assumed if the sender does not receive an ACK for the packet within a timeout interval called retransmission timeout (RTO),

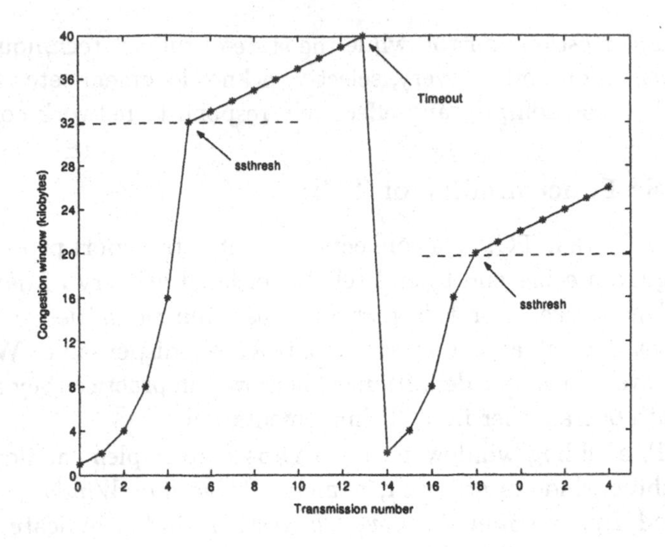

Figure 1: TCP congestion window dynamics ([44])

which is dynamically computed as the estimated round-trip time (RTT) plus four times the mean deviation. By retransmitting the lost packet, TCP achieves reliable data delivery.

It turns out that in wired networks, almost all the packet losses are due to network congestion rather than transmission errors. Thus, in addition to retransmission, TCP responds to packet losses by invoking its congestion control mechanism. TCP congestion control is also based on the sliding window mechanism described above and consists of two major phases: *slow start* and *congestion avoidance*. In the slow start phase, the initial congestion window size (*cwnd*) is set to one maximum segment size (MSS) and is incremented by one MSS on each new acknowledgement. After *cwnd* reaches a preset threshold (*ssthresh*), the congestion avoidance starts and it is increased linearly, i.e., it is increased by one segment for each RTT. Upon a timeout, *ssthresh* is set to the half of the current transmission window size (but at least two segments) and the congestion window is reduced to 1 MSS. Then slow start mechanism starts again. This procedure is also called the additive increase and multiplicative decrease algorithm (AIMD, [25]). The entire congestion control algorithm is illustrated in Fig. 1. Note that the sender reacts to three duplicate ACKs in a different way, which is described in *fast retransmission and fast recovery* in the next subsection.

2.2 State-of-the-Art in Standard TCP

Most of the progress made in TCP is centered on error recovery and congestion control. Representative innovations include fast transmissions and fast recovery [42], selective acknowledgements [31], random early detection (RED, [17]) in routers, and explicit congestion notification (ECN, [39]). Notice that depending on what features are included, there are several TCP flavors, including TCP Tahoe, TCP Reno, TCP New Reno, etc. Among them, TCP Reno is by far most widely deployed. Next, we briefly describe these innovations in the following.

2.2.1 Fast Retransmission and Fast Recovery

As noted earlier, a packet can be assumed lost if three duplicate ACKs are received. In this case, TCP performs a fast retransmission of the packet. This mechanism allows TCP to avoid a lengthy timeout during which no data is transferred. At the same time, *ssthresh* is set to one half of the current congestion window, i.e., *cwnd*, and *cwnd* is set to *ssthresh* plus three segments. If the ACK is received approximately one round trip after the missing segment is retransmitted, fast recovery is entered. That is, instead of setting *cwnd* to one segment and starting with slow start, TCP sets *cwnd* to *ssthresh*, and then steps into congestion avoidance phase. However, only one packet loss can be recovered during fast retransmission and fast recovery. Additional packet losses in the same window may require that the RTO expire before retransmission.

2.2.2 Selective Acknowledgment

Owing to the fact that fast retransmission and fast recovery can only handle one packet loss from one window of data, TCP may experience poor performance when multiple packets are lost in one window. To overcome this limitation, recently the selective acknowledgement option (SACK) is suggested as an addition to the standard TCP implementation.

The SACK extension adopts two TCP options. One is an enabling option, which may be sent to indicate that the SACK option can be used upon connection establishment. The other is the SACK option itself, which may be sent by TCP receiver over an established connection if SACK option is enabled through sending the first option.

The SACK option contains up to four (or three, if SACK is used in conjunction with the Timestamp option used for RTTM [24]) SACK blocks, which specifies contiguous blocks of the received data. Each SACK block consists of two sequence numbers which delimit the range of data the receiver has received and queued. A receiver can add the SACK option to ACKs it sends back to a SACK-enabled sender. In the event of multiple losses within a window, the sender can infer which packets have been lost and should be retransmitted using the information provided in the SACK blocks. A SACK-enabled sender can retransmit multiple lost packets in one RTT instead of detecting only one lost packet in each RTT.

2.2.3 Random Early Detection

Random Early Detection (RED) is a router-based congestion control mechanism that seeks to detect incipient congestion and notify some TCP senders of congestion by controlling the average queue size at the router. To notify the TCP senders of congestion, the router may mark or drop packets, depending on whether the senders are cooperative. As a response, the senders should reduce their transmission rate. This is done in two algorithms. The first algorithm is to compute the average queue size by using exponential weighted moving average. If we denote by avg and q the average queue size and the current queue size, respectively, then $avg = (1 - wq) \times avg + wq \times q$, where wq is the queue weight. The other algorithm is to compute the packet-marking or packet-dropping probability p_a. If avg falls in between min_{th} and max_{th}, the packet marking probability $p_b = max_p(avg - min_{th})/(max_{th} - min_{th})$ and the final marking probability $p_a = p_b/(1 - count * p_b)$, where max_p and $count$ are design parameters, respectively, denoting the maximum value for p_b and the number of packets having arrived since last packet marking or dropping. If avg exceeds max_{th}, $p_a = 1$, which means that the router marks or drops each packet that arrives. Through control over the average queue size prior to queue overflow, RED succeeds in preventing heavy network congestion and global synchronization as well as improving fairness. Notice that numerous variants of RED have been proposed to improve various performance of the original RED [16, 29, 33, and 34].

2.2.4 Explicit Congestion Notification

Most of current Internet routers employ traditional "drop-tail" queue management. In other words, the routers drop packets only when the queue overflows, which could lead to the undesirable global synchronization problem as well as heavy network congestion. Recently, active queue management (AQM) mechanisms have been proposed since they can detect congestion before the queue overflows at the routers and inform TCP senders of the congestion, thereby avoiding some of these problems caused by the "drop-tail" policy. In the absence of Explicit Congestion Notification (ECN), however, the only choice that is available to AQM for indicating congestion to end systems is to drop packets at the routers. With ECN, AQM mechanisms have an alternative to allow routers to notify end systems of congestion in the network.

ECN requires some changes to the header of both IP and TCP. In the IP header, an ECN field with two bits is used. By setting this field to specific bits, the router can send an indication of congestion to end systems. For TCP, two new flags in the Reserve field of the TCP header are specified. By manipulating these two flags, the TCP sender and the TCP receiver can enable ECN via negotiation during connection setup; the receiver can inform the sender if it receives congestion indications from intermediate routers; and the sender can inform the receiver that it has invoked congestion control mechanisms [39].

3 TCP in One-Hop Wireless Networks

In this section, we focus on TCP performance in one-hop wireless networks, which typically include wireless LAN and wireless cellular networks. We first summarize some challenges adversely affecting TCP performance. Then, some representative schemes proposed to improve TCP performance are described. Notice that in this chapter we focus on how to improve TCP performance, so some schemes such as WTCP [41], which attempts to propose a totally different transport layer protocol, are not presented here since it is not an improvement scheme based on TCP.

3.1 Challenges

Compared with wired networks, one-hop wireless networks have some inherent adverse characteristics that will significantly deteriorate TCP performance if no action is taken. In essence, these characteristics include bursty channels errors, mobility and communication asymmetry.

3.1.1 Channel Errors

In wireless channels, relatively high bit error rate because of multipath fading and shadowing may corrupt packets in transmission, leading to the losses of TCP data segments or ACKs. If it cannot receive the ACK within the retransmission timeout, the TCP sender immediately reduces its congestion window to one segment, exponentially backs off its RTO and retransmits the lost packets. Intermittent channel errors may thus cause the congestion window size at the sender to remain small, thereby resulting in low TCP throughput.

3.1.2 Mobility

Cellular networks are characterized by handoffs due to user mobility. Normally, handoffs may cause temporary disconnections, resulting in packet losses and delay. TCP will suffer a lot if it treats such losses as congestion and invokes unnecessary congestion control mechanisms. The handoffs are expected to be more frequent in next generation cellular networks as the micro-cellular structure is adopted to accommodate an increasing number of users. Thing could be worse if TCP cannot handle handoffs gracefully. Similar problems may occur in wireless LAN, as mobile users will also encounter communication interruptions if they move to the edge of the transmission range of the access point.

3.1.3 Asymmetry

In one-hop wireless networks, the wireless link between a base station and a mobile terminal in nature is asymmetric. Compared with the base station, the mobile terminal has limited power, processing capability, and buffer space. Another asymmetry stems from the vastly different characteristics of wired links and wireless links. The former is reliable and has large bandwidth while the latter is error-prone and has limited and highly variable bandwidth. For example, the bandwidth of a typical Ethernet is 10Mbps (100Mbps or even higher for fast Ethernet) while the highest bandwidth for 3G networks is only about 2Mbps. Therefore, the wireless link is very likely to become the bottleneck of TCP connections.

3.2 Current Solutions

The quest to overcome the deficiency of TCP over wireless links has been courting extensive efforts. Among the various solutions proposed to improve TCP performance, there are four major categories: *split-connection solutions*, *proxy-based solutions*, *link-layer solutions*, and *end-to-end solutions*. The split-connection solutions attempt to improve TCP performance by splitting a TCP connection into two at the base station so that the TCP connection between the base station and the mobile host can be specially

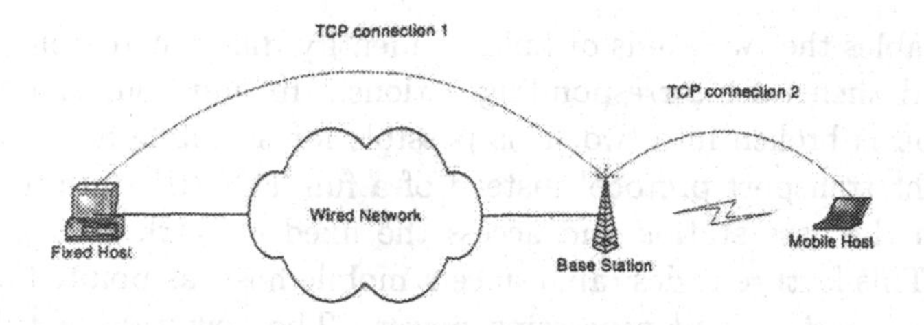

Figure 2: I-TCP, splitting a TCP connection into two connections

tuned for the wireless links. Realizing the base station is a critical point, approaches based on proxy put an implicit or explicit intelligent agent at the base station, detecting packet losses over wireless links and taking corresponding actions (such as duplicate ACK suppression and/or local retransmission) to ensure the TCP sender responds correctly. For the third category, a reliable link layer is built by adopting some link error recovery mechanisms, seeking to hide link errors from the TCP sender. Unlike the previous three classes, the end-to-end approaches enhance TCP by using SACK to quickly recover from multiple packet losses or by predicting incoming handoffs to avoid unnecessary congestion control invocation. Next, some representative schemes in each category are presented.

3.2.1 Split-Connection Solutions

Indirect TCP: Indirect-TCP (I-TCP) [7] protocol proposed by Bakre and Badrinath suggests that any TCP connection from a mobile host (MH) to a machine on the fixed network (FH) should be split into two separate connections: one between the MH and its base station (BS) over the wireless medium and the other between the BS and the FH over the fixed network, as shown in the Fig. 2. A packet sent to MH is first received by BS, it then sends an acknowledgment to FH and then the packet is forwarded to MH. If MH moves to a different cell while communicating with an FH, the whole connection information maintained at the current BS is transferred to the new BS and the new BS takes over thereafter. The FH is unaware of this indirection and is not affected when this switch occurs. Also, since the end-to-end connection is split, the TCP connection over the wireless link can use some wireless-link-aware TCP variation, which may be tailored to handle wireless channel errors and handoff disruption.

From the above description, we see that I-TCP separates the congestion control functionality on the wireless link from that on the fixed network, which enables the two kinds of links to identify different reasons for packet losses and then take corresponding actions. In addition, since the TCP connection is broken into two, it is possible for a mobile host to use some lightweight transport protocol instead of a full TCP/IP suite to communicate with the base station and access the fixed network through the base station. This feature is desirable since a mobile host, as pointed out earlier, has limited battery and processing power. The downside of this scheme, however, is the following. First, I-TCP violates the end-to-end semantics of TCP acknowledgments, as both the wired part and the wireless part of a connection have their own acknowledgments. Second, control overhead is considerable as the base station needs to maintain a significant amount of state for each TCP connection and all the state information needs to be transferred to the new base station in the event of a handoff, which could result in a long delay.

M-TCP: M-TCP [9] is another split-connection approach that breaks up a TCP connection between a FH and a MH into two parts: one between the FH and the BS, and the other between the BS and the MH. What makes it different from I-TCP is that it manages to preserve TCP end-to-end semantics.

M-TCP is assumed to operate upon the underlying three-level architecture shown in Fig. 3. A mobile host (MH) communicates with the BS in the cell. Several BSs are controlled by a supervisor host (SH), which, serving a gateway, is connected to the wired network. The authors opt for this architecture for two reasons. The first is that the functionalities at a BS can be transferred to SH, which may reduce the cost of the network as one SH is in charge of several BSs; the other is that the number of handoffs is greatly reduced since a MH roaming from one cell to another need not perform handoffs as long as the two cells are controlled by the same SH.

Another important assumption made by M-TCP is that a relatively reliable link layer is operating underneath M-TCP to recover losses such that the bit error rate over wireless links is low. The implication of this assumption is that TCP performance degradation is mainly due to frequent disconnections caused by handoffs.

M-TCP operates as follows. Assume that the MH has acknowledged bytes up to sequence number x, the SH sends an ACK for bytes up to $x-1$ to the TCP sender. Note that this is different from I-TCP in that the SH only sends ACKs to the sender when it receives ACKs from the MH. If the SH does not receive ACKs beyond x for some time, the SH will assume this is due to temporary wireless link outage. Therefore, it sends an ACK for the last byte x with a zero window size. Upon receiving this ACK,

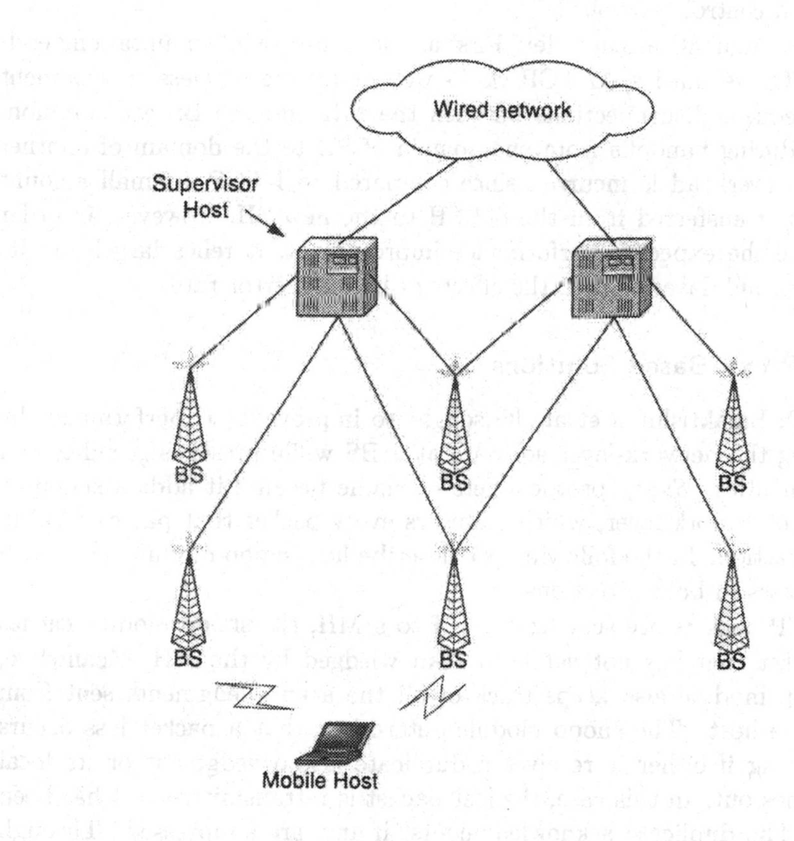

Figure 3: Three-level architecture underlying M-TCP

the sender will enter into *persist* mode, freezing all its transmission states such as RTO and congestion window. When the wireless link is regained, the MH will notify the SH by sending a greeting packet. The SH, in turn, informs the sender of this reconnection, allowing the sender to resume its transmission from the frozen state. Through this way, the adverse efforts of disconnections on TCP performance are gracefully eliminated since no congestion control is invoked.

Some comments are in order. First and foremost, while maintaining end-to-end TCP semantics, M-TCP works well under the wireless environment where frequent disconnections between the MH and the BS are common. Second, during handoffs from one domain of SH to the domain of another SH, little overhead is incurred since compared to I-TCP, a small amount of state is transferred from the old SH to the new SH. However, in order to achieve the expected performance improvement, it relies largely on its underlying link layer to hide the effects of high bit error rate.

3.2.2 Proxy-Based Solutions

SNOOP: Balakrishnan et al. [6] sought to improve TCP performance by modifying the network-layer software at a BS while preserving end-to-end TCP semantics. *Snoop* protocol gets its name because it adds a snooping module to network layer, which monitors every packet that passes a BS in either direction. In the following, we describe how snoop module deals with packet losses in both directions.

If TCP packets are sent from a FH to a MH, the snoop module caches each packet that has not yet been acknowledged by the MH. Meanwhile, the snoop module also keeps track of all the acknowledgments sent from the mobile host. The snoop module determines that a packet loss occurs by detecting if either it receives a duplicate acknowledgment or its local timer times out. In this case, the lost packet is retransmitted if it has been cached. The duplicate acknowledgments, if any, are suppressed. Through this way, unnecessary congestion control mechanism invocations are avoided since packet losses due to wireless channel errors are hidden from the FH.

In the case that packets are transmitted from an MH to an FH, since the MH cannot tell whether a packet loss is due to errors on the wireless link or due to congestion elsewhere in the network, TCP SACK option is used. At the BS, when the snoop module notices a gap in the inbound sequence numbers of the packets sent from the MH, selective acknowledgements are sent to the MH. Upon receiving such SACKs, SACK-enabled MH will retransmit the lost packets for local loss recovery.

SNOOP is also designed to handle handoffs. Several BSs near a MH will form a multicast group and buffer some latest packets sent from the FH. Prior to a handoff, the MH will send control messages to determine that a BS with strongest signal should be the primary one, i.e., the one forwarding packets to the MH, and that all other BSs just buffer packets. Therefore, both handoff latency and packet losses are reduced.

The major merit of this approach is that, it improves TCP performance through performing local retransmissions across the wireless link without affecting end-to-end TCP semantics. On the other hand, although TCP performance during handoffs may be improved, considerable overhead is incurred for maintaining the multicast BS group and state transfer from one BS to another. Finally, it is worth noting that special care needs to be taken to handle the interaction of the snoop module retransmission and TCP end-to-end retransmission because SNOOP is similar to link-level retransmission approaches over the wireless links.

Ack Regulator: Since link layer enhancement schemes are shown to successfully improve TCP performance over wireless link, they have been adopted in 3G wireless networks. For example, reliable link-layer protocols such as RLP [49] and RLC [47] are respectively used in 3G1X [48] and UMTS [46]. However, as pointed out in [11], these link layer protocols also introduce increased delay and rate variability, which may cause bursty ACK arrivals (called ACK compression [56]) and consequently degrade TCP throughput. This effect becomes more pronounced as some channel state based scheduling schemes [8] is used in 3G wireless networks as well.

To reduce such negative effects, Chan and Ramjee ([11]) proposed a network-based solution called *Ack Regulator*, which is implemented at the radio network controller (RNC) to regulate the flow of ACKs sent from the mobile host to the TCP sender. Notice that since most applications like web browsing mainly use the downlink, this solution is designed for TCP connections toward the mobile hosts. The key idea is that, the RNC should control the number of ACKs sent back to the sender each time a data packet is transmitted to the mobile host or an ACK arrives from the mobile host, such that there is at most one packet loss due to buffer overflow in one window of transmitted packets. In this way, the TCP sender operates mainly in the congestion avoidance phase. In this scheme, the RNC maintains a data queue for each TCP flow from the sender to the mobile host and an ACK queue from the mobile host to the sender. By monitoring the current available buffer space and estimating the number of future incoming data packets, the RNC decides how many ACKs it should send to the sender each time. Significant performance improvement has been reported, which, nevertheless, is achieved at the expense of increased complexity and buffer space at the RNC.

Advertised Window Control: While a great deal of effort has been made on the IEEE 802.11 MAC, little research work has been focused on the interaction of TCP with WLANs. However, to fully understand TCP behaviors over WLAN is very important, as TCP is the de facto transport layer protocol for most applications over WLANs.

In [38], the work by Pilosof et al. has shed some light in better understanding TCP fairness over WLANs. Through analysis and simulation, it is discovered that the upstream (from the mobile host to the base station) and downstream (from the base station to the mobile host) TCP flows do not fairly share the wireless medium, with a throughput ratio between them as high as ten times, in favor of upstream flows. They discovered that this ratio is sensitive to the buffer size at the base station. In particular, TCP unfairness may fall into four different regions as the buffer size is varying. Part of the reason is that given the TCP receiver window size, the downstream TCP window size fails to reach the receiver window size if some data packets are lost due to insufficient buffer, while upstream TCP window size can reach the receive window size because it can tolerate some ACK losses.

Thus, they proposed to modify the receiver's advertised window field in the ACKs when they pass through the base station. More precisely, given that there are n TCP flows in the WLANs and the buffer size at the base station is B, the advertised window size will be set to the minimum of the original advertised window size and $\lfloor B/n \rfloor$. Simulations and experiments show that the throughput ratio between upstream and downstream TCP flows is almost 1 after adopting this change.

3.2.3 Link-layer Solutions

AIRMAIL: Since TCP performance degradation is partly due to the high bit error rate of the wireless link, it is intuitive to shield TCP from such errors. With a reliable link-layer protocol in place, unnecessary TCP congestion control invocation due to channel errors can be avoided, and hence TCP performance is improved. Based on this idea, a reliable link-layer protocol named AIRMAIL (AsymmetrIc Reliable Mobile Access In Link-layer) was proposed in [4]. In AIRMAIL, two well-known link error recovery techniques, i.e., forward error correction (FEC) and automatic repeat request (ARQ) are employed. Moreover, in order to accommodate the asymmetry lying between the two ends of a wireless link, i.e., the BS and the MH, AIRMAIL purposely devises some asymmetric ARQ error control and window-based flow control techniques as shown in the following:

- Timers are always at the BS regardless of whether it is transmitting or receiving. Thus all timer-related operations are conducted in the BS.

- The base station receiver sends its status to a mobile transmitter periodically. However, this is not the case for the mobile receiver to send its status to the base station transmitter. Rather, the mobile receiver sends status messages based on an event-driven mechanism. The difference in the mechanisms of sending status messages is justified by the power constraint at the mobile host.

In addition to ARQ, three levels of FEC, namely bit-level FEC, byte-level FEC, and packet-level FEC, have been employed to provide increased error correction capability under different mobile environment.

Several comments on AIRMAIL are in order. First, since AIRMAIL only involves changes at the link layer, no modifications need to be made to TCP. Obviously, it fits in well with the layered structure of network protocols. Second, when designing ARQ techniques, AIRMAIL takes into account the asymmetry between the BS and the MH, a desirable feature which may relieve the requirement of computing power on the mobile host and prolong the battery life of the mobile host as well. Third, the drawback of AIRMAIL is that it cannot account for temporary disconnections due to handoff. Thus, even though it succeeds in reducing bit error rate over wireless links, it can do little to prevent TCP from timing out when an acknowledgment is not received on time because of long disconnections. In fact, this observation might apply to various link layer approaches. Finally, the interaction between link-layer retransmissions and end-to-end retransmissions can be complicated, as shown in [14]. It showed that link-layer retransmission protocols only improve TCP performance when the packet error rate exceeds a certain threshold. Further study on the interaction is needed in order to improve TCP performance with the aid of the link-layer enhancement.

TULIP: TULIP (Transport Unaware Link Improvement Protocol) is a TCP-unaware link layer protocol that works upon the MAC layer [36]. Because TULIP is targeted for half-duplex wireless links, to avoid collision between two opposite data streams, it only passes one packet at a time to the MAC layer. The procedure is described as follows. After receiving a TCP packet from the upper layer, TULIP passes it to the MAC layer. When starting to transmit the packet, the MAC layer notifies TULIP by sending a signal *TRANS*. Upon reception of *TRANS*, TULIP starts a timer Δt_1, which is estimated as the time duration between the beginning of data packet transmission and the end of the reception of a link-layer ACK (or a link-layer ACK piggybacked with a data packet). In the case that Δt_1 is underestimated because of packet length variations, the MAC layer will inform TULIP by sending another signal *WAIT*, specifying the additional time Δt_2. Readers are referred to [36] for details on how to set Δt_2 as it involves the specific MAC layer mechanism. To locally recover lost or corrupted packets due to channel errors, a link-layer selective acknowledgment mechanism is used to retransmit packets, which is assigned high priority compared to normal data packets in order for fast recovery. Moreover, to save bandwidth over the wireless link, a mechanism called MAC acceleration is introduced to piggyback a TCP ACK with the TULIP ACK. Through simulation it is shown that TULIP achieves a bit better performance compared to SNOOP in the environment where errors are exponentially distributed over the wireless channel.

3.2.4 End-to-end Solutions

Fast Retransmission: Fast retransmission is perhaps the simplest end-to-end scheme to improve TCP performance. Based on the observation that TCP encounters unacceptably long pauses in communication during handoffs which cause increased delays and packet losses, fast retransmission was proposed by Caceres and Iftode to overcome this problem [10]. The MH will send duplicate ACKs to the TCP sender immediately after the handoff process is completed. In this way, the TCP sender can begin retransmission without waiting for the timeout, hence preventing serious throughput drop.

Selective Acknowledgement: As described earlier, the TCP selective acknowledgment mechanism can allow a SACK-enabled sender to retransmit in one RTT multiple lost packets in one transmission window and hence avoid continuous timeouts. However, this mechanism does not distinguish the reasons for packet losses and still assumes all losses are caused by congestion. Consequently, TCP congestion control procedures are inappropriately called for, which throttles the sender's transmission rate. As shown in [5], SACK is useful over the error-prone wireless link where losses occur in bursts.

Freeze-TCP: It is observed that most current TCP schemes require base stations to monitor the TCP traffic and actively participate in flow control in order to enhance performance. However, such schemes might be undesirable or even useless for several reasons. First, to be compatible with currently existing infrastructure, it is ideal that no modification should be made to intermediate nodes, since such nodes may belong to other organizations and hence are unavailable for modification. Second, as network security is becoming increasingly important, end-to-end traffic is likely to be encrypted and hence inaccessible to intermediate nodes. As a result, some schemes such as SNOOP, I-TCP or M-TCP can no longer work in such scenarios since they all require the base station to access the traffic before taking actions. Finally, overly relying on mediation at the intermediate nodes may cause a significant amount of control overhead, creating network bottlenecks under heavy traffic load.

To overcome these deficiencies, the author in [22] proposed Freeze-TCP, a true end-to-end TCP enhancement scheme. The key idea of this scheme is to exonerate the base station from intervening in the end-to-end TCP connections. By constantly observing its received signal strength, a mobile host can predict a temporary disconnection due to handoffs or fading. Once such an event is predicted, the mobile host sends an ACK to the TCP sender with a zero advertised window size. Upon reception of such an ACK, the sender enters a persist mode. That is, the sender freezes all retransmission timers and sends *zero window probes* (ZWP) until the mobile host advertises a non-zero receiving window size. Since ZWPs are sent out with exponentially backoff, it is possible that the sender remains idle even the mobile host has recovered from the disconnection. To tackle this problem, the same technique as in [10] is employed. Namely, as soon as the mobile host knows that it has reconnected, it will send three duplicate ACKs to the sender, forcing the sender to start fast retransmission. The main advantage of this scheme is that it improves TCP performance without any modification to the intermediate nodes. However, it could be easily seen that the actual performance depends largely on the accuracy with which the mobile host predicts an impending disconnection.

4 TCP in Mobile Ad Hoc Networks

TCP performance in mobile ad hoc networks is the focus of this section. It is expected that compared to one-hop wireless networks, TCP will encounter more serious difficulty in providing end-to-end communications in mobile ad hoc networks, as MANETs are, in essence, infrastructureless, self-organizing multi-hop networks, and lacking centralized network management. Next, we present the main problems in ad hoc networks, followed by recent solutions.

4.1 Challenges

Some salient characteristics of mobile ad hoc networks, which seriously deteriorate TCP performance, include the unpredictable wireless channels due to fading and interference, the vulnerable shared media access due to random access collision, the hidden terminal problem and the exposed terminal problem, and the frequent route breakages due to node mobility. From the point of view of network layered architecture, these challenges can be broken down into five categories, i.e., a) the channel error, b) the medium contention and collision, c) the mobility, d) the multi-path routing, and e) congestion, whose adverse impacts on TCP is elaborated next.

4.1.1 Channel Errors

The effects of channel errors in ad hoc networks are similar to those in one-hop wireless networks except that they are more serious, since a TCP connection now may consist of multi-hop wireless links, unlike the situation in cellular networks or wireless LAN where only the last hop is wireless. Accordingly, the congestion window size at the sender may shrink more dramatically due to channel errors in several wireless hops, resulting in even lower throughput in ad hoc networks.

4.1.2 Medium Contention and Collision

Contention-based medium access control (MAC) schemes, such as the IEEE 802.11 MAC protocol [9], have been widely studied and incorporated into many wireless testbeds and simulation packages for wireless multi-hop ad hoc networks, where the neighboring nodes contend for the shared wireless channel before transmitting. There are three key problems, i.e., the hidden terminal problem, the exposed terminal problem, and unfairness. A hidden node is the one that is within the interfering range of the intended receiver but out of the sensing range of the transmitter. The receiver may not correctly receive the intended packet due to collision from the hidden node. An exposed node is the one that is within the sensing range of the transmitter but out of the interfering range of the receiver. Though its transmission does not interfere with the receiver, it could not start transmission because it senses a busy medium, which introduces spatial reuse inefficiency. The binary exponential backoff scheme always favors the latest successful transmitter and results in unfairness. These problems could be more harmful in multi-hop ad hoc networks than in Wireless LAN as ad hoc networks are characterized by multi-hop connectivity.

MAC protocols have been shown to significantly affect TCP performance [20, 21, 40, 45, 51, and 54]. When TCP runs over 802.11 MAC, as [54] pointed out, the instability problem becomes very serious. It is shown that collisions and the exposed terminal problem are two major reasons for preventing one node from reaching the other when the two nodes are in each other's transmission range. If a node cannot reach its adjacent node for several times, it will trigger a route failure, which in turn will cause the source node to start route discovery. Before a new route is found, no data packet can be sent out. During this process, TCP sender has to wait and will invoke congestion control algorithms if it observes a timeout. Serious oscillation in TCP throughput will thus be observed. Moreover, the random backoff scheme used in the MAC layer exacerbates this [20]. Since large data packet sizes and back-to-back packet transmissions both decrease the chance of the intermediate node to obtain the channel, the node has to back off a random period of time and try again. After several failed tries, a route failure is reported.

TCP may also encounter serious unfairness problems [20, 40, 45, and 54] for the reasons stated below:

- Topology causes unfairness because of unequal channel access opportunity for different nodes. As shown in Fig. 1, where the small circle denotes a node's valid transmission range and the large circle denotes a node's interference range, all nodes in a 7-node chain topology experience different degree of competitions. There are two TCP flows, namely flow 1 from node 0 to 1 and flow 2 from node 6 to 2. The transmission from node 0 to node 1 experiences interference from three nodes, i.e., nodes 1, 2, and 3, while the transmission from node 3 to node 2 experiences interference from five nodes, i.e., nodes 0, 1, 2, 4, and 5. Flow 1 will obtain much higher throughput than flow 2 due to the unequal channel access opportunity.

- The backoff mechanism in the MAC may lead to unfairness as it always favors the last successfully transmitting node.

- TCP flow length influences unfairness. Longer flows implies longer round trip time and higher packet dropping probability, leading to lower and more fluctuating TCP end-to-end throughput. Through this chain reaction, unfairness is amplified, as high throughput will become higher and low one lower.

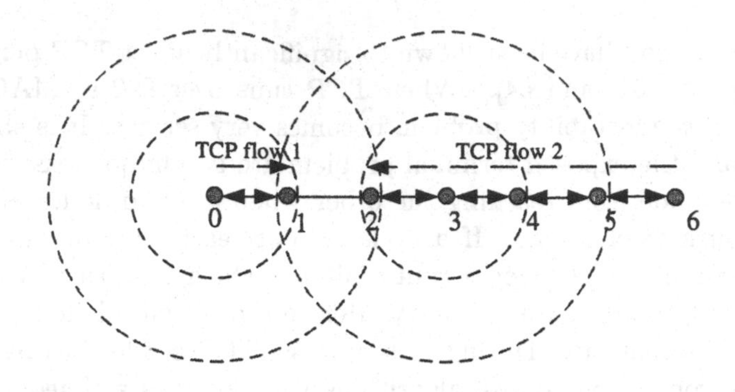

Figure 4: Node interference in a chain topology

4.1.3 Mobility

Mobility may induce link breakage and route failure between two neighboring nodes, as one mobile node moves out of the other's transmission range. Link breakage in turn causes packet losses. As we said earlier, TCP cannot distinguish between packet losses due to route failures and packet losses due to congestion. Therefore, TCP congestion control mechanisms react adversely to such losses caused by route breakages [1, 15, and 27]. Meanwhile, discovering a new route may take significantly longer time than TCP sender's RTO. If route discovery time is longer than RTO, TCP sender will invoke congestion control after timeout. The already reduced throughput due to losses will further shrink. It could be even worse when the sender and the receiver of a TCP connection fall into different network partitions. In such a case, multiple consecutive RTO timeouts lead to inactivity lasting for one or two minutes even if the sender and receiver finally get reconnected.

Fu et al. conducted simulations considering mobility, channel error, and shared media-channel contention [19]. They indicated that mobility-induced network disconnections and reconnections have the most significant impact on TCP performance comparing to channel error and shared media-channel contention. TCP NewReno merely achieves about 10% of a reference TCP's throughput in such cases. As mobility increases, the relative throughput drop ranges from almost 0% in a static case to 100% in a highly mobile case (when moving speed is $20m/s$). In contrast, congestion and mild channel error (say 1%) have less visible effect on TCP (with less than 10% performance drop compared with the reference TCP).

4.1.4 Multi-path Routing

Routes are short-lived due to frequent link breakages. To reduce delay due to route re-computation, some routing protocols such as TORA [35] maintain multiple routes between a sender-receiver pair and use multi-path routing to transmit packets. In such a case, packets coming from different paths may not arrive at the receiver in order. Being unaware of multi-path routing, TCP receiver would misinterpret such out-of-order packet arrivals as congestion. The receiver will thus generate duplicate ACKs that cause the sender to invoke congestion control algorithms like fast retransmission (upon reception of 3 duplicate ACKs).

4.1.5 Congestion

It is known that TCP is an aggressive transport layer protocol. Its attempt to fully utilize the network bandwidth makes ad hoc networks easily go into congestion. In addition, due to many factors such as route change and unpredictable variable MAC delay, the relationship between congestion window size and the tolerable data rate for a route is no longer maintained in ad hoc networks. The congestion window size computed for the old route may be too large for the newly found route, resulting in network congestion if the sender still transmits at the full rate allowed by the old congestion window size.

Congestion/overload may give rise to buffer overflow and increased link contention, which degrades TCP performance. As a matter of fact, [28] showed the capacity of wireless ad hoc networks decreases as traffic and/or competing nodes arise.

4.2 Current Solutions

As is shown in the previous section, there is a magnitude of research work on improving TCP performance over one-hop wireless networks. However, many of these mechanisms are designed for infrastructure-based networks and depend on the base stations in distinguishing the error losses from congestion losses. Since mobile ad-hoc networks do not have such an infrastructure, they are hard to be applied in mobile ad-hoc networks directly.

More recently, several schemes have been proposed to improve TCP performance over mobile ad hoc networks. We classify the schemes into three groups, based on their fundamental philosophy: TCP with feedback schemes, TCP without feedback schemes, and TCP with lower layer enhancement schemes. Through the use of feedback information to signal non-congestion-related causes of packet losses, the feedback approaches help TCP distinguish between true network congestion and other problems such as channel errors, link contention, and route failures. On the other end of the solution spectrum, TCP without feedback schemes makes TCP adapt to route changes without relying on feedback from the network, in light of the concern that feedback mechanisms may bring about additional complexity and cost in ad hoc networks. The third group, lower layer enhancement schemes, starts with the idea that TCP sender should be hidden from any problems specific in ad hoc networks while lower layers such as routing layer and MAC layer need to be tailored with TCP's congestion control algorithms in mind. As expected, this idea guarantees that TCP end-to-end semantics is maintained for ad hoc networks to seamlessly internetwork with the wired Internet. In the following, we present some representative schemes according to the aforementioned taxonomy.

4.2.1 TCP with Feedback Solutions

TCP-F: In the mobile ad hoc networks, topology may change rapidly due to the movement of mobile hosts. The frequent topology changes result in sudden packet losses and delays. TCP misinterprets such losses as congestion and invokes congestion, leading to unnecessary retransmission and loss of throughput. To overcome this problem, TCP-F (TCP-Feedback) [12] was proposed so that the sender can distinguish between route failure and network congestion. Similar to Freeze-TCP and M-TCP discussed above, the sender is forced to stop transmission without reducing window size upon route failure. As soon as the connection is reestablished, fast retransmission is enabled.

TCP-F relies on the network layer at an intermediate node to detect the route failure due to the mobility of its downstream neighbor along the route. A sender can be in an *active* state or a *snooze* state. In the active state, transport layer is controlled by the normal TCP. As soon as an intermediate node detects a broken route, it explicitly sends a route failure notification (RFN) packet to the sender and records this event. Upon reception of the RFN, the sender goes into the snooze state, in which the sender completely stops sending further packets, and freezes all of its timers and the values of state variables such as RTO and congestion window size. Meanwhile, all upstream intermediate nodes that receive the RFN invalidate the particular route in order to avoid further packet losses. The sender remains in the snooze state until it is notified of the restoration of the route through a route reestablishment notification (RRN) packet from an intermediate node.

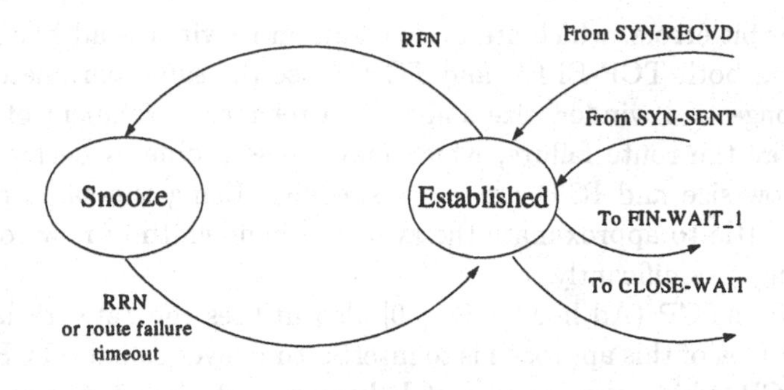

Figure 5: The TCP-F state machine [12]

Then it resumes the transmission from the frozen state. The state machine of TCP-F is shown in Fig. 5.

TCP-ELFN: Holland and Vaidya proposed another feedback-based technique, the *Explicit Link Failure Notification* (ELFN) [23, 32]. The goal is to inform the TCP sender of link and route failures so that it can avoid responding to the failures as if congestion occurs. ELFN is based on DSR [26] routing protocol. To implement ELFN message, the route failure message of DSR is modified to carry a payload similar to the "host unreachable" ICMP message. Upon receiving an ELFN, the TCP sender disables its congestion control mechanisms and enters into a "stand-by" mode, which is similar to the snooze state of TCP-F mentioned above. Unlike TCP-F using an explicit notice to signal that a new route has been found, the sender, while on stand-by, periodically sends a small packet to probe the network to see if a route has been established. If there is a new route, the sender leaves the stand-by mode, restores its RTO and continues as normal. Recognizing most of popular routing protocols in ad hoc networks are on demand and route discovery/rediscovery is event driven, periodically sending a small packet at the sender is appropriate to restore routes with mild overhead and without modification to the routing layer.

Through explicit route failure notification, TCP-EFLN and TCP-F allow the sender to instantly enter snooze state and avoid unnecessary retransmissions and congestion control which wastes precious MH battery power and scarce bandwidth. With explicit route reestablishment notification from intermediate nodes or active route probing initiated at the sender, these two schemes enable the sender to resume fast transmission as soon as possible. But neither of these two considers the effects of congestion, out-of-order packets, or bit errors, which are quite common in wireless ad hoc networks. In addition, both TCP-ELFN and TCP-F use the same parameter sets including congestion window size and RTO after reestablishment of routes as those before the route failure, which may cause problems because congestion window size and RTO are route specific. Using the same parameter sets helps little to approximate the available bandwidth of new route if the route changes significantly.

ATCP: ATCP (Ad hoc TCP) [30] also utilizes the network layer feedback. The idea of this approach is to insert a thin layer called ATCP between IP and TCP, which ensures correct behavior in the event of route failures as well as high bit error rate. The TCP sender can be put into a *persist* state, *congestion control* state or *retransmit* state, respectively, corresponding to the packet losses due to route breakage, true network congestion or high bit error rate. Note that unlike the previous two feedback-based approaches, packet corruption caused by channel errors has also been tackled. The sender can choose an appropriate state by learning the network state information through explicit congestion notification (ECN) messages and ICMP "Destination Unreachable" messages.

The state transition diagram for ATCP at the sender is shown in Fig. 6. Upon receiving a "Destination Unreachable" message, the sender enters into the persist state. The TCP at the sender is frozen and no packets are sent until a new route is found, so the sender does not invoke congestion control. Upon receipt of an ECN, congestion control is invoked without waiting for a timeout event. If a packet loss happens and the ECN flag is not set, ATCP assumes the loss is due to bit errors and simply retransmits the lost packet. In case of multi-path routing, upon receipt of duplicate ACKs, TCP sender does not invoke congestion control, realizing multi-path routing shuffles the order in which segments are received. So ATCP works well when the multi-path routing is applied.

ATCP is considered to be a more comprehensive approach in comparison with TCP-F and TCP- ELFN in that it accounts for more possible sources of deficiency including bit errors and out of order delivery due to multi-path routing. Through re-computation of congestion window size each time after route reestablishment, ATCP may adapt to change of routes. Another benefit of ATCP is that it is transparent to TCP, and hence nodes with and without ATCP can interoperate.

In summary, as shown by the simulations, these feedback-based approaches improve TCP performance significantly while maintaining TCP's congestion control behavior and end-to-end TCP semantics. However, all these schemes require that the intermediate nodes have the capability of de-

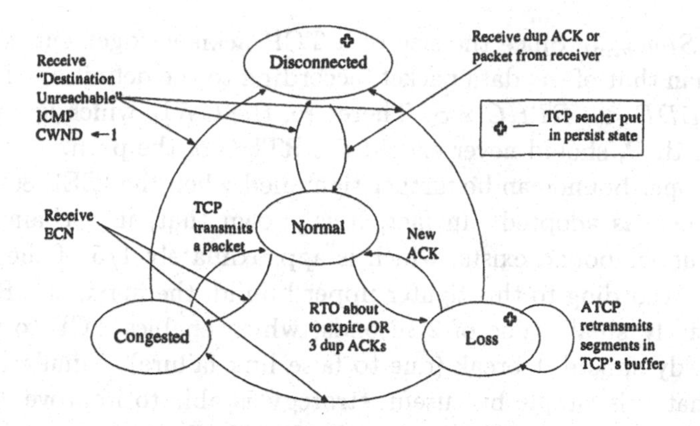

Figure 6: State transition diagram for ATCP at the sender [30]

tecting and reporting network states such as link breakages and congestion. Enhancement at the transport layer, network layer, and link layer are all required. It deserves further research on the ways to detect and distinguish network states in the intermediate nodes.

4.2.2 TCP without Feedback Solutions

Adaptive Congestion Window Limit Setting: Based on the observation that TCP's congestion control algorithm often over-shoots, leading to network overload and heavy contention at the MAC layer, Chen et al. [13] proposed an adaptive congestion window limit (CWL, measured in the number of packets) setting strategy to dynamically adjust TCP's CWL according to the current round-trip hop-count (RTHC) of the path, which can be obtained from routing protocols such as DSR. More precisely, the CWL should never exceed the RTHC of the path.

The rationale behind this scheme is very simple, as shown in the following. It is known that to fully utilize the capacity of a network, a TCP flow should set its CWL to the bandwidth-delay product (BDP) of the current path, where a path's BDP is defined as the product of the bottleneck bandwidth of the forward path and the packet transmission delay in a round trip. On the other hand, the CWL should never exceed the path's BDP in order to avoid network congestion. In ad hoc networks, if we assume the size of a data packet is S and the bottleneck bandwidth along the forward and return paths is the same and equal to bw_{min}, it can be easily seen that the delay at any hop along the path is less than the delay at the bottleneck link, i.e., S/bw_{min}. Since the size of a TCP acknowledgement is normally smaller than that of the data packet, according to the definition of the BDP, we know $BDP <= RTHC \times S$. Therefore, the CWL, which is bounded by the path's BDP, should never exceed the RTHC of the path.

This upper bound can be further tightened when the IEEE 802.11 MAC layer protocol is adopted. In fact, it is shown that, in a chain topology, a tighter upper bound exists, which is approximately 1/5 of the RTHC of the path. According to this tighter upper bound, the maximum RTO is set to a relatively small value of 2 seconds, which enables TCP to probe the route quickly should it break (due to false link failure). Simulation results showed that this simple but useful strategy is able to improve TCP-Reno performance by 8% to 16% in a dynamic MANET.

TCP-DOOR: TCP-DOOR [50] attempts to improve TCP performance by detecting and responding to out-of-order (OOO) packet delivery events and thus avoiding invoking unnecessary congestion control. By definition, OOO occurs when a packet sent earlier arrives later than a subsequent packet. In ad hoc networks, OOO may happen multiple times in one TCP session because of route changes.

In order to detect OOO, ordering information is added to TCP ACKs and TCP data packets. OOO detection is carried out at both ends: the sender detects the Out-of-Order ACK packets and the receiver detects the Out-of-Order data packets. If the receiver detects OOO, it should notify the sender, considering the fact that it is the sender who takes congestion control actions. Once the TCP sender knows of an OOO condition, it may take one of the two responsive actions: temporarily disabling congestion control and instant recovery during congestion avoidance. The first action means that, whenever an OOO condition is detected, TCP sender will keep its state variables such as RTO and the congestion window size constant for a time period T_1. The second action means that, if during the past time period T_2, the TCP sender has already entered the state of congestion avoidance, and it should recover immediately to the state prior to such congestion avoidance. The main reason is the detection of OOO condition implies that a route change event has just occurred.

However, OOO can be detected only after a route has recovered from failures. As a result, TCP-DOOR is less accurate and responsive than a feedback-based approach that is able to determine whether congestion or route errors occur, and hence report to the sender at the very beginning. Furthermore, it may not work well with multi-path routing since multi-path routing may cause OOO as well. Therefore, it is concluded that TCP-DOOR may work as an alternative to the feedback-based approach to improve TCP performance over ad hoc network, if the latter is not available.

Fixed RTO: In TCP congestion control, TCP doubles the RTO and retransmits the oldest unacknowledged packet when the retransmission timer expires. Although this exponential backoff mechanism of the RTO could handle network congestion gracefully, it is no longer suitable in MANETs when the loss of packets or ACKs is caused by temporary route breakages, as discussed earlier. In such a case, the RTO should be recalculated, if possible, according to the new route instead of being doubled. Furthermore, when the new route is established, TCP sender should start the transmission immediately instead of waiting for the expiration of retransmit timer.

In the fixed RTO approach [15], no feedback from lower layers is needed. Rather, a heuristic is employed to distinguish route failures and congestion. When timeouts occur consecutively, i.e., an ACK is not received before the second RTO expires, the sender assumes a route failure rather than network congestion takes place. Therefore, the unacknowledged packet is retransmitted again without doubling the RTO. The RTO remains fixed until the route is re-established and the retransmitted packet is acknowledged. By adopting this strategy, the TCP sender avoids waiting for a long period of time before attempting to retransmit. This fast retransmission would force routing protocol especially like AODV [37] and DSR to repair routes fast, which in turn leads to a large congestion window on average and high TCP throughput. Actually, this technique complements TCP-DOOR.

4.2.3 Lower Layer Enhancement Solutions

Routing Layer Enhancement: A framework termed *Atra*, due to Anantharaman et al., aims to improve TCP performance over ad hoc networks by enhancing routing layers [3]. Three mechanisms, called *Symmetric Route Pinning* (SRP), *Route Failure Prediction* (RFP), and *Proactive Route Errors* (PRE), are introduced to minimize the probability of route failures, to predict route failures in advance, and to minimize the latency in conveying route failure information to source, respectively. Since asymmetric path would increase the probability of route failure for a connection, in the first mechanism, the ACK path of a TCP connection is always kept the same as the data path. Based on the progression of signal strengths of packet receptions from the concerned neighbor, the second mechanism enables the node to predict the occurrence of link failure more accurately. Finally, with PRE, when a link failure is detected, all sources that have used the link in the past certain period are informed of the link failure. This mechanism reduces the latency involved in the route failure information delivery and consequently reduces the number of packet losses and also triggers early alternate route computations.

Link Layer Enhancement: Fu et al. [18] have discussed the interaction between TCP and 802.11 MAC. Their studies reveal two interesting results. First, given a specific network topology and flow pattern, there exists a TCP window size, say W^*, at which TCP throughput is maximized since the best spatial reuse can be achieved; further increasing the window size will reduce throughput. However, the standard TCP protocol does not operate around W^*, typically with an average window much larger than W^*. As a result, TCP experiences throughput reduction due to reduced spatial reuse and increased packet loss. In the simulated scenarios, 4% to 21% throughput reduction from maximum throughput is observed. Second, most packet drops experienced by TCP are not due to buffer overflow, but due to link-layer contention that are incurred by hidden terminals. They showed that contention drops exhibit a load-sensitive loss feature: as the injected TCP packets exceed W^* and further increase, the link dropping probability becomes non-negligible and increases accordingly; after the injected TCP packets exceed another threshold W, the link dropping probability saturates and flattens out. It turns out that the link-layer dropping probability is not significant enough to make the average TCP window oscillate around W^*, which subsequently leads to suboptimal TCP throughput.

Therefore, two link layer techniques were proposed in [18] to improve TCP efficiency: a *Link-RED* (Random Early Detection) algorithm to tune the wireless link's packet dropping probability and an adaptive link-layer pacing scheme to reduce the medium contention. The Link-RED algorithm attempts to maintain the optimum congestion window size at the TCP sender. At the link layer each node measures the average number of the retries for recent packet transmissions. Normally, when the TCP sender increases the congestion window size and injects more packets into the network, this average number will increase, as more packets will aggravate medium contention. The head-of-line packet is dropped from the buffer or marked as congested with a probability calculated based on this average number. Once it detects packet losses or the congestion flag in the ACKs, the TCP sender invokes the congestion control algorithm that could help maintain the congestion window size around the optimum value and hence improve TCP's throughput.

The goal of adaptive link-layer pacing is to alleviate the medium contention especially when the congestion window size exceeds the optimum value. It is enabled from within the Link-RED algorithm. When a node (which just sends a packet) notices its average number of retries is less than a predefined threshold, it calculates its backoff time as usual. Otherwise, it increases the backoff period by an interval equal to the transmission time of the previous data packet, and backs off accordingly.

Neighborhood RED: As described in the previous subsection on challenges, TCP exhibits serious unfairness in ad hoc networks as a result of the combination of MAC-inherent problems such as medium contention, the hidden terminal problem, and the exposed terminal problem. As these problems are likely to exist in nodes which are located in a neighborhood, Xu et al. [53] proposed a scheme named *neighborhood RED* (NRED) that seeks to improve TCP fairness from the point of view of a neighborhood. By definition, a node's neighborhood consists of the node itself and the nodes which can interfere with this node's signal. To make things simpler, a node's neighborhood considered in the scheme comprises the node itself and its one-hop and two-hop neighbors.

The key idea of NRED is that each node forms a distributed queue of a neighborhood based on the individual queues maintained at every node located in the node's neighborhood, and the RED scheme can be applied to the distributed queue to address the fairness issue, as it has proven to be effective, in wired networks, in improving fairness among TCP flows by controlling average queue size at routers.

The NRED scheme boils down to three algorithms, namely, *Neighborhood Congestion Detection* (NCD), *Neighborhood Congestion Notification* (NCN), and *Distributed Neighborhood Packet Drop* (DNCP). Instead of counting on each node actively advertising its own queue size information and then measuring the neighborhood queue size, which may cause a large amount of overhead or even aggravate congestion, NCD intelligently gets around the difficult task by monitoring channel utilization. Normally, channel utilization can serve as an indicator of the queue size, based on the observation that channel utilization around a node is likely to increase when the queues at its neighboring nodes build up. An early congestion is assumed to take place as the channel utilization exceeds a certain threshold. If congestion is detected, the node will calculate the packet dropping probability and send it in a NCN packet to its neighbors, provided certain conditions are met in order to avoid "overreaction". The neighbors, upon the reception of such notification, will drop some packets according to DNCP.

Simulation studies show that the NRED can improve TCP fairness to some extent in ad hoc networks. However, the price paid is that the aggregate throughput in the network is actually reduced, which shows there is still room for further improvement.

5 Future Research Directions

At this point, after we discussed the challenges and visited some representative solutions, it is well recognized that in order for TCP to deliver a comparable performance in wireless networks to that in wired networks, quite a few critical issues need to be addressed. Note that compared with its one-hop counterparts, ad hoc networks require more efforts to handle as things are much more complicated. In this section we discuss some of these open issues for which searching for a better solution demands special efforts. It is worthy noting that we do not mean to list all. Rather, we concentrate on those that we believe are most important.

5.1 TCP Fairness

TCP unfairness becomes pronounced in wireless LANs [38]. In mobile ad hoc networks, the unfairness problem is more severe. It is shown that in a mobile ad hoc network with multiple flows, the throughput can be significantly different among competing flows. This phenomenon is particularly evident when comparing flows of short paths to those of long paths [20]. Compared with the considerable effort paid to improve TCP end-to-end throughput, fairness is a critical issue that deserves more attention. In fact, this insufficiency can be seen from the number of proposed scheme targeted for fairness: among all the schemes we present in this chapter, only *Advertised Window Control* and *Neighborhood RED* address this issue, although a few schemes have touched upon fairness. Since bandwidth over wireless links is very limited bandwidth compared with that over wired links, it is crucial for every flow to fairly share the bandwidth in wireless networks. Therefore, more mature approaches are highly expected.

5.2 Interactions among Different Layers

Layered network architecture brings a myriad of advantages. At the same time, it requires a close look at the interactions among different layers when designing a good scheme. Currently, many solutions are focused on one specific layer, attempting to isolate the problem and solve it. It is true that TCP might perform better with a highly effective and efficient link layer or routing layer, e.g., an MAC protocol which can quickly resolve medium contentions, or a mobility-aware routing protocol which can gracefully handling route changes. However, this approach may be problematic or even counterproductive, as suggested in [14]. Furthermore, as many factors such as bursty channel errors, medium access contention, and route breakage are all contributing to TCP throughput deterioration in mobile ad hoc networks, a unified solution is justified which takes into account the interaction among different layer. We thus argue that a cross-layer approach seems more desirable and promising.

5.3 Compatibility with the Wired Internet

For the purpose of internetworking with the wired Internet as required in future pervasive mobile computing, whatever TCP is designed for ad hoc networks should be fully compatible with the Internet. This quest for compatibility translates into two requirements for future research. First, TCP's end-to-end semantics must be maintained. Second, TCP performance should be considered when TCP connections span both the wired networks and mobile ad hoc networks.

6 Conclusions

As the assumption made by TCP that any packet loss is due to network congestion is no longer valid in wireless networks, TCP performs poorly in such networks. In this chapter, we point out the major reasons for this performance degradation. In particular, factors such as error-prone wireless channels and handoffs result in the poor TCP performance over one-hop wireless networks, while, aside from these factors, other factors such as medium access contention, frequent route changes, and breakages are considered to lead to the poor TCP performance over multi-hop wireless networks. Compared with one-hop wireless networks, we can see it is more difficult to make TCP perform well in multi-hop wireless networks.

This chapter presents the state-of-the-art in recent efforts to improve TCP performance. Given the reasons, almost all the proposed schemes attempt to achieve better TCP performance with either of the two ideas: TCP should be capable of distinguishing non-congestion-related packet losses from congestion caused packet losses such that corresponding actions can be taken to deal with the losses; or non-congestion-related losses should be reduced such that TCP can work normally without any modifications. Interestingly enough, there seems little study attempting to combine these two ideas.

Again, we choose to present the proposed schemes after separating those for one-hop wireless networks from those for multi-hop wireless networks for the purpose of clarity. In the realm of one-hop wireless networks, there are four groups of schemes, i.e., split-connection approaches, proxy-based approaches, link-layer enhancement approaches, and end-to-end approaches. According to [5], a TCP-aware reliable link-layer protocol such as SNOOP performs best. However, TULIP is claimed to deliver better performance than SNOOP under some circumstances. In case of frequent and long disconnections, M-TCP appears to perform well. In the realm of multi-hop wireless networks, there are also three groups of schemes, namely, TCP with feedback approaches, TCP without feedback approaches, and lower layer enhancement approaches. In conclusion, feedback-based schemes seem to be able to react more quickly to non-congestion-related packet losses, thus to be more effective in enhancing TCP performance [50]. However, the price to be paid is that they are more difficult to implement, for they require end nodes and intermediate nodes to cooperate with each other. On the other hand, approaches without feedback information are relatively simple to implement, although the performance gain may not be high enough. Meanwhile, some solutions by enhancing the link layer and routing layer shed insights into how to reduce non-congestion-related losses in order to improve TCP performance.

Finally, although some encouraging improvements have been reported by employing the proposed schemes, none of them can work well in all scenarios and meet all the challenges mentioned. Therefore, there is still much work to be done in the near future. To serve as guidance for future research, some critical issues regarding improving TCP performance and fairness are identified.

References

[1] A. Ahuja, S. Agarwal, J. P. Singh and R. Shorey, Performance of TCP over different routing protocols in mobile ad-hoc networks, *IEEE Vehicular Technology Conference 2000* Vol.3 pp. 2315-2319.

[2] I. Ali, R. Gupta, S. Bansal, A. Misra, A. Razdan and R. Shorey, Energy efficiency and throughput for TCP traffic in multi-hop wireless networks, *IEEE INFOCOM'02* (New York 2002).

[3] V. Anantharaman, S.-J. Park, K. Sundaresan and R. Sivakumar, TCP performance over mobile ad-hoc networks: a quantitative study, *To appear in Wireless Communications and Mobile Computing Journal (WCMC), Special Issue on Performance Evaluation of Wireless Networks* (2003).

[4] E. Ayanoglu, S. Paul, T. F. LaPorta, K. K. Sabnani and R. D. Gitlin, AIRMAIL: a link-layer protocol for wireless networks, *ACM Wireless Networks* (Feb. 1995).

[5] H. Balakrishnan, V. Padmanabhan, S. Seshan and R. Katz, A comparison of mechanisms for improving TCP performance over wireless links, *Proceedings of ACM SIGCOMM'96* (Aug. 1996).

[6] H. Balakrishnan, S. Seshan and R. H. Katz, Improving reliable transport and handoff performance in cellular wireless networks, *ACM Wireless Networks* (Dec. 1995).

[7] A. Bakre and B. R. Badrinath, I-TCP: indirect TCP for mobile hosts, *Proc. 15th International Conf. On Distributed Computing systems (ICDCS)* (May 1995).

[8] P. Bhagwat, P. Bhattacharya, A. Krishna and S. K. Tripathi, Enhancing throughput over wireless LANs using channel state dependent packet scheduling, *IEEE INFOCOM'96* (San Francisco, Mar. 1996).

[9] K. Brown and S. Singh, M-TCP: TCP for mobile cellular networks, *ACM computer communication review* Vol.27 No.5 (Oct. 1997).

[10] R. Caceres and L. Iftode, Improving the performance of reliable transport protocols in mobile computing environments, *IEEE JSAC* Vol.19 No.7 (Jul. 2001).

[11] M. C. Chan and R. Ramjee, TCP/IP performance over 3G wireless links with rate and delay variation, *MobiCom'02* (Sep. 2002).

[12] K. Chandran, S. Raghunathan, S. Venkatesan and R. Prakash, A feedback-based scheme for improving TCP performance in ad hoc wireless networks, *IEEE Personal communications* Vol.8 No.1 (Feb. 2001) pp. 34-39.

[13] K. Chen, Y. Xue and K. Nahrstedt, On setting TCP's congestion window limit in mobile ad hoc networks, *IEEE ICC'03* (Anchorage, Alaska, May 2003).

[14] A. DeSimone, M. C. Chuah and O. C. Yue, Throughput performance of transport-layer protocols over wireless LANs, *Proc. Globecom '93* (Dec. 1993).

[15] T. D. Dyer and R. V. Boppana, A comparison of TCP performance over three routing protocols for mobile ad hoc networks, *ACM Mobihoc* (Oct. 2001).

[16] S. Floyd and K. Fall, Router mechanisms to support end-to-end congestion control, *LBL Technical report* (Feb. 1997).

[17] S. Floyd and V. Jacobson, Random early detection gateways for congestion avoidance, *IEEE/ACM Transaction on Networking* Vol.1 No.4 (Aug. 1993).

[18] Z. Fu, P. Zerfos, H. Luo, S. Lu, L. Zhang and M. Gerla, The impact of multihop wireless channel on TCP throughput and loss, *IEEE INFOCOM'03* (San Francisco, Mar. 2003).

[19] Z. Fu, X. Meng and S. Lu, How bad TCP can perform in mobile ad-hoc networks, *IEEE Symposium on Computers and Communications* (Italy, Jul. 2002).

[20] M. Gerla, R. Bagrodia, L. Zhang, K. Tang and L. Wang, TCP over wireless multihop protocols: simulation and experiments, *Proceedings of IEEE ICC'99* (Vancouver, Canada, Jun. 1999).

[21] M. Gerla, K. Tang and R. Bagrodia, TCP performance in wireless multihop networks, *Proceedings of IEEE WMCSA'99* (New Orleans, LA, Feb. 1999).

[22] T. Goff, J. Moronski, D. S. Phatak and V. Gupta, Freeze-TCP: a true end-to-end TCP enhancement mechanism for mobile environment, *IEEE INFOCOM'00* (Tel-Aviv, Mar. 2000).

[23] G. Holland and N. H. Vaidya, Analysis of TCP performance over mobile ad hoc networks, *MOBICOM'99* (Seattle, Aug. 1999).

[24] V. Jacobson, R. Braden and D. Borman, TCP extensions for high performance, *RFC 1323*, (May 1992).

[25] V. Jacobson and M. Karels, Congestion avoidance and control, *Proceedings of ACM SIGCOMM'88* (Aug. 1988).

[26] D. B. Johnson, D A. Maltz and Y. Hu, The dynamic souce routing protocol for mobile ad hoc networks, *IETF Internet Draft.* http://www.ietf.org/internet-drafts/draft-ietf-manet-dsr-08.txt, (2003).

[27] D-K. Kim, C.-K. Toh and Y. Choi, TCP-BuS: improving TCP performance over wireless ad hoc networks, *IEEE Comsoc Journal On Communications And Networks (JCN)* Vol.3 No.2 (2001).

[28] J. Li, C. Blake, D. S. J. De Couto, H. Lee and R. Morris, Capacity of ad hoc wireless networks, *MobiCom'01* (Rome, Italy, Jul. 2001).

[29] D. Lin and R. Morris, Dynamics of random early detection, *ACM Computer Communication Review* Vol.27 No.4 (Oct. 1997).

[30] J. Liu and S. Singh, ATCP: TCP for mobile ad hoc networks, *IEEE JSAC* Vol.19 No.7 (Jul. 2001).

[31] M. Mathis, J. Mahdavi, S. Floyd and A. Romanow, TCP selective acknowledgement options, *RFC 2018* (Oct. 1996).

[32] J. P. Monks, P. Sinha and V. Bharghavan, Limitations of TCP-ELFN for ad hoc networks, *MOMUC* (2000).

[33] T. J. Ott, T. V. Lakshman and L. H. Wong, SRED: stabilized RED, *Proceedings IEEE INFOCOM '99* (New York, Mar. 1999).

[34] R. Pan, B. Prabhakar and K. Psounis, CHOKe: a stateless active queue management scheme for approximating fair bandwidth allocation, *Proceeding of INFOCOM'00* (Tel-Aviv, Mar. 2000).

[35] V. D. Park and M. S. Corson, A highly adaptive distributed routing algorithm for mobile wireless networks, *Proceedings of IEEE INFOCOM'97* (Kobe, Japan, Apr. 1997).

[36] C. Parsa and J. J. Garcia-Luna-Aceves, Improving TCP performance over wireless networks at the link layer, *ACM Mobile Networks and Applications* Vol. 5 (2000).

[37] C. E. Perkins, E. M. Belding-Royer and S. Das, Ad hoc on demand distance vector (AODV) routing, *IETF RFC 3561*

[38] S. Pilosof, R. Ramjee, D. Raz, Y. Shavitt and P. Sinha, Understanding TCP fairness over wireless LAN, *IEEE INFOCOM'03* (San Francisco, Mar. 2003).

[39] K. Ramakrishnan, S. Floyd and D. Black, The addition of explicit congestion notification (ECN) to IP, *RFC 3168*, Sep. 2001.

[40] E. Royer, S. J. Lee and C. Perkins, The effects of MAC protocols on ad hoc network communication, *IEEE WCNC* (Chicago, IL, Sep. 2000).

[41] P. Sinha, N. Venkitaraman, R. Sivakumar and V. Bharghavan, WTCP: a reliable transport protocol for wireless wide-area networks, *ACM MobiCom'99* (Aug. 1999).

[42] W. Stevens, TCP slow start, congestion avoidance, fast retransmit, and fast recovery algorithms, *RFC 2001* (Jan. 1997).

[43] W. Stevens, *TCP/IP Illustrated* Vol.1 (Addison-Wesley, 1996).

[44] A. S. Tanenbaum, *Computer Networks*, (4th Edition, Prentice-Hall International, Inc. 2002).

[45] K. Tang and M. Gerla, Fair sharing of MAC under TCP in wireless ad hoc networks, *Proceedings of IEEE MMT'99* (Venice, Italy, Oct. 1999).

[46] *Third Generation Partnership Project*, (Release 1999).

[47] *Third Generation Partnership Project*, RLC protocol specification (3G TS 25.322:), (1999).

[48] *TIA/EIA/cdma2000*, Mobile station - base station compatibility standard for dual-mode wideband spread spectrum cellular systems, Washington: Telecommunication Industry Association, (1999).

[49] *TIA/EIA/IS-707-A-2.10*, Data service options for spread spectrum systems: radio link protocol type 3, (Jan. 2000).

[50] F. Wang and Y. Zhang, Improving TCP performance over mobile adhoc networks with out-of-order detection and response, *ACM MobiHoc'02* (Lausanne, Switzerland, Jun. 2002).

[51] H. Wu, Y. Peng, K. Long, S. Cheng and J. Ma, Performance of reliable transport protocol over IEEE 802.11 wireless LAN: analysis and enhancement, *IEEE INFOCOM'02* (New York, Jun. 2002).

[52] K. Xu, S. Bae, S. Lee and M. Gerla, TCP behavior across multihop wireless networks and the wired internet, *ACM WoWmoM'02* (Sep. 2002).

[53] K. Xu, M. Gerla, L. Qi and Y. Shu, Enhancing TCP fairness in ad hoc wireless networks using neighborhood RED, *ACM MobiCom'03* (Sep. 2003).

[54] S. Xu and T. Saadawi, Does the IEEE 802.11 MAC protocol work well in multihop wireless ad hoc networks? *IEEE Communications Magazine* (Jun. 2001).

[55] S. Xu and T. Saadawi, Revealing TCP unfairness behavior in 802.11 based wireless multi-hop networks, *IEEE PIMRC'01* (Oct. 2001).

[56] L. Zhang, S. Shenker and D. Clark, Observations on the dynamics of a congestion control algorithm: the effects of two-way traffic, *Proceedings of ACM SIGCOMM'91* (Sep. 1991).

[54] S. Xu and T. Saadawi, Does the IEEE 802.11 MAC protocol work well in multihop wireless ad hoc networks? *IEEE Communications Magazine* (Jun. 2001).

[55] S. Xu and T. Saadawi, Revealing TCP unfairness behavior in 802.11 based wireless multi-hop networks, *IEEE PIMRC'01* (Oct. 2001).

[56] L. Zhang, S. Shenker and D. Clark, Observations on the dynamics of a congestion control algorithm: the effects of two-way traffic, *Proceedings of ACM SIGCOMM'91* (Sep. 1991).